AF553785

PLANT PRODUCT PHARMACEUTICALS

ENCYCLOPAEDIA OF BIOPHARMACEUTICAL

Vol. 4

PLANT PRODUCT PHARMACEUTICALS

By

Dr. S.K. Prasad

School of Studies of Zoology & Biotechnology

Vikram University

Ujjain

DISCOVERY PUBLISHING HOUSE PVT. LTD.

NEW DELHI-110 002

Published by:
Tilak Wasan
DISCOVERY PUBLISHING HOUSE PVT. LTD.
4383/4B, Ansari Road, Darya Ganj
New Delhi-110 002 (India)
Phone : +91-11-23279245, 43596064-65
Fax : +91-11-23253475
E-mail : discoverypublishinghouse@gmail.com
namitwasan9@gmail.com
sales@discoverypublishinggroup.com
web : www.discoverypublishinggroup.com

First Edition: 2011

Reprinted: 2016

ISBN: 978-81-8356-736-7

Plant Product Pharmaceuticals

© Author

All rights reserved. No part of this publication should be reproduced, stored in a retrieval system, or transmitted in any form or by any means: electronic, mechanical, photocopying, recording or otherwise, without the prior written permission of the author and the publisher.

This book has been published in good faith that the material provided by authors is original. Every effort is made to ensure accuracy of material, but the publisher and printer will not be held responsible for any inadvertent error(s). In case of any dispute, all legal matters are to be settled under Delhi jurisdiction only.

Printed at:
Infinity Imaging Systems
Delhi

Preface

The present title "Encyclopaedia of Biopharmaceutical" has been written for those in the pharmaceutical research and those responsible for the education and training in pharmaceutical science and technology of graduate and undergraduate students. Medicine is an ever changing science. As new research and clinical experience broaden our knowledge, changes in treatment and drug therapy are required. This branch of life science has progressed enormously in recent years and the significant advances in therapeutics and an understanding of the need to optimize during delivery in the body have brought about an increased awareness of the valuable role played by the dosage forms. This statement is as true as it was back in ninteenth century and perhaps more so, given the increasing emphasis being placed on discovery, development, and use of large molecular entities as therapeutic and diagnostic agents. Development of these abilities requires an integration of knowledge, skills, attitudes, and values that can be acquired only through structured learning process including independent study, hands on practice and the availability of advanced literature. This tittle has designed to meet such needs of learners in the health professions.

In the last two decades, the pharmaceutical industry has experimented and successfully adopted several integrated and multidisciplinary approaches in the research areas of dring compound screening, toxicological evaluation, and pharmaceutical product development. The book is written in a concise style that facilitates an in-depth level of understanding of the essential concepts. The objectives of the present title are three folds: (i) to serve as a useful tool to help guide scientists in research and development by out-lining the theory and successful practice of in vitro - in vivo correlation, (ii) to help formulators apply the tool in designing and developing prototypes that enable selection of clinical formulations, and (iii) to help formulate strategy(ies) for product life-cycle management.

To make the work more comprehensive and informative, the author has consulted many authoritative books, research journals, abstracts, monographs etc., so there can be no claim to originality except in the manner of treatment.

The author expresses his thanks to his friends and colleagues whose continue inspirations have initiated him to bring out this book.

The author expresses his gratitude to Mr. Wasan and staff of M/s Discovery Publishing House Pvt. Ltd. for their whole hearted co-operation in the publication of this book.

Author

CONTENTS

1

Introduction

Botanical or herbal products have been used extensively as drug treatments in the complementary and alternative medical (CAM) system in many regions of the world, but have not been subjected to the same rigorous evaluation by regulatory agencies as that for modern nonbotanical pharmaceutical agents. To facilitate further development of new drugs from botanical sources, the Center for Drug Evaluation and Research (CDER) of U.S. Food and Drug Administration (FDA) has published a draft Guidance for Industry: Botanical Drug Product in August, 2000. The Guidance has since been published in its final form in June of 2004. The new regulatory approaches in the Guidance take into consideration the unique features of the botanical drugs and the substantial past human experiences.

Regulatory Objectives

To support approval as a nonbotanical drug, adequate and well-controlled clinical studies are required. Not only must the treatment be shown to be effective, with real patient benefits in morbidity and/or mortality to justify the safety risk of adverse reactions as observed, but the clinical data must also provide practical instructions for use, which can be reasonably followed by health care professionals. There is no reason why these requirements should be different for botanical drugs, because patient suffering and treatment benefits are independent of medical theory or practice. Regardless of medical systems and terminology used, courses of diseases/conditions should be established and treatment effects must be clinically meaningful. Incorporation of alternative medical practice into the clinical studies will be acceptable if the new set of instructions derived from such studies will be practical.

The objective of the new FDA regulatory approaches to new botanical drug development is not to create an additional category of products different from dietary supplements or nonbotanical drugs. Instead, the goal is to confer the botanical new drugs the same degree of confidence in quality and clinical usefulness, as that of nonbotanical drugs, and ultimately to bring the botanical drugs into the mainstream medical use.

Distinctive Features of Plants

From a regulatory perspective, botanical drugs have some unique features that demand special considerations. Clinically, there is a large quantity of anecdotal experience about the efficacy, which is not supported by modern scientific data, nor can it easily be dismissed. Likewise, extensive human usages also suggest that most of the botanical preparations are possibly safe, but confidence in such presumed safety can only be based on mostly poorly documented data. The pre-existing availability of most botanical products, although not marketed as drugs, also creates difficulties in their regulation. Concern about the quality of botanical products poses another regulatory challenge. Most botanical products are complex and variable mixtures of constituents too numerous to characterize individually. For many preparations, active ingredients have not been identified and it is thus difficult to quantify strength or potency of the botanical product. Because of the biological nature of botanical products, the assessment of their impurity and stability is often more problematic than that of nonbotanical pure drugs.

Beyond the natural mixtures in one part of a single plant, many botanical drugs are combinations of multiple botanical products. While the rationale of combining many plants is not easy to understand and contributions from ingredients remain to be elucidated, it is also not clear whether the ratio and composition have been optimized in many widely used formulations.

Plant Guideline

In the Guidance, the botanical drug products are defined as those that contain as ingredients vegetable materials, which may include plant materials, algae, macroscopic fungi, or combinations thereof, that are used as drugs. It may be available as (but not limited to) a solution (e.g., tea), powder, tablet, capsule, elixir, topical, or injectable. In the current version, fermentation products, highly purified (or chemically modified) botanical substances, allergenic extracts and vaccines that contain botanical ingredients are excluded. In essence, the Botanical Guidance provides that

1. Further purification of the botanical preparations is not required,
2. Identification of active ingredient(s) is not essential,
3. Chemistry, manufacturing and control (CMC) is extended to raw materials, not just regulations of drug substance and product, and
4. Nonclinical testing may be reduced or delayed for products with extensive history of human use.

It should be emphasized that, in the above new approaches, only different types of information are used in part of the safety assessment. This does not imply that overall standards for quality consistency and clinical efficacy/ safety are more or less stringent than that of nonbotanical drugs.

In general, requirements in CMC and nonclinical studies for initiating a clinical study of botanical drugs depend to a large extent on the marketing history, known safety concerns (if any), the degree of modification from past use, and the scale of proposed clinical trials. For a small early phase study, animal toxicity may not be needed if the preparation and usage are the same as in prior human experiences. On the other hand, for large scale, more definitive trials, greater assurance in product quality, consistency, and reproducibility is necessary, as well as the safe use in the clinical setting of the protocol.

As evidence of prior human use, documentation of marketing history (with volume of sales) and review of past and current references, compendia, and literatures should be provided. It is understandable that many of these publications are in variable format and quality, and often do not consist of modern scientific data. In this respect, the Agency will accept all types of documentation for consideration and determine the validity of support in individual cases.

As noted above, all botanical drugs contain many potentially active molecular entities and many preparations are combinations of multiple parts/plants. Because the current policy on fixed-dose combination products requires demonstration of contribution to overall efficacy and safety from each active ingredient, it could be impractical or impossible for botanicals to comply with this regulation in the development as a drug product. This issue was not addressed in the current version of Botanical Guidance. While the Agency is considering revision of the fixed-dose combination regulation to accommodate the difficulties encountered by the botanical drugs, the sponsor is encouraged to consult the CDER for assistance.

Plants Review Team in Center for Drug Evaluation and Research

To acquire and consolidate regulatory experiences on botanical drugs, a dedicated BRT has been established in CDER. The BRT will provide scientific expertise on botanical issues to other reviewing staff to ensure consistent interpretation and implementation of the Botanical Guidance and related policies. In addition, the BRT has the following functions:

1. Participates in all phases of reviews, meetings, and decision- making processes for all botanical drug applications and submissions as a collaborative scientific discipline, and serves as an expert resource for CDER on all botanical issues.
2. Collects information, maintains a database of botanical applications, and performs periodic analysis on the status of botanical new drug development.
3. Responds to external constituents who have general botanical drug development questions.
4. Responds as the expert resource for CDER to issues and meeting requests from the Office of the Commissioner and interfaces on common botanical issues and fosters communication with the FDA Center for Food Safety and Applied Nutrition and the National Center for Complementary and Alternative Medicine and the Office of Dietary Supplements at the National Institutes of Health (NIH).
5. Interfaces with external professional regulatory and scientific groups, makes presentations, and participates in workshops to promote and enhance botanical drug product development and knowledge.

The botanical reviews performed by the BRT cover the following area:

1. *Medicinal plant biology*: methods and problems in species identification; potential misuse of related but incorrect species.
2. *Pharmacology and toxicology of medicinal plants used in the proposed studies*: activities based on old, alternative theories and/or modern testing.
3. *History of prior human uses*: therapeutic effects in the CAM system; potential toxicities from past experience.

The BRT experts serve as members of the review teams, and provide scientific opinion on the botanical drug product in a role similar to that of other disciplines such as chemistry. Currently, the BRT is a team of experts consisting of a medical officer as team leader, a pharmacognosy reviewer, and a project manager.

Review Processes for Plant Applications

To implement the Botanical Guidance, a new set of review processes for botanical applications have been delineated in a new CDER MAPP. As described in the MAPP, the BRT will respond only to general inquiries on botanical-related issues and interpretation of the Botanical Guidance and related policies. For questions about individual botanical drug product with specific clinical indication, the sponsor should submit the application to the new drug review divisions in charge of the therapeutic area in the CDER's Office of New Drugs, and the applications will remain under the divisions' administration. For specific botanical drug applications, all regulatory decisions will be the responsibilities of the new drug division and all regulatory actions will be issued by the division directors. Communication between sponsors and all review team members, including BRT staff, will be conducted through the project manager of the new drug division. As a member of the review team, the BRT experts provide scientific opinion on the botanical drug product in a role similar to that of experts in other disciplines.

In principle, all botanical submissions will be managed in the same manner as nonbotanical drug products by all review disciplines. That is, primary and secondary reviews in CMC, pharmacology/

toxicology, biopharmaceutics, and clinical and statistical issues will be conducted by the respective primary reviewers and team leaders. To ensure consistency across different new drug review divisions, the supporting disciplines in CMC, pharmacology/toxicology, and biopharmaceutics have also designated one to three senior staff to serve as expert consultant(s) in botanical issues for the review divisions in each area. These review processes for botanical applications have been tested in CDER with approximately 100 submissions. Collaborations between BRT and the new drug divisions have been smooth and productive.

Plant Drug Applications in Center for Drug Evaluation and Research

As of April 30, 2004, there are a total of 203 botanical drug applications in CDER, including 167 investigational new drug (IND) applications and 36 pre-IND consultations. At least 75% of the total botanical applications were submitted after 1999, and about two per month were received by the Agency recently. Of these, 43% are commercial development programs and the remaining are academic research projects. These botanical submissions are distributed in all 14 therapeutic divisions, with most activities aggregated in the oncology, antiviral, and dermatology–dental drug areas. A great majority of botanical sponsors have taken advantage of the pre-IND consultation service provided by FDA. As a result, most IND applications were successful with initial submission and few (less than 20) were placed on clinical hold for safety concerns. However, despite the early success, many development programs and research projects have subsequently been suspended for various reasons. As of the above-mentioned cutoff date, nearly two-thirds (66%) of INDs still remain active (have not been placed on clinical hold, inactivated by FDA, or withdrawn by sponsor for lack of activities). To date, there have been no submissions of NDAs to FDA for marketing approval of botanical prescription drugs.

Challenges in the Review of Plant Drug Applications

Not surprisingly, quality of the botanical products is a frequent review issue in our regulatory experiences. Some sponsors had not presented accurate name/identification of the botanical plants and/or description of manufacturing processes. Because of the recent incidence of diethylstilbestrol containing PC-SPES, adulteration of botanicals with active chemical drugs has become a serious concern for both the study supporter (e.g., NIH) and the regulatory agency. For many botanical preparations, there are often uncertainties in the identity of the plant species and/or consistency of botanical raw materials. Complicated manufacturing processes add further possible variation to the drug substance and final products. While most contamination problems unique to botanicals are resolvable, purity/potency and stability are more difficult technical issues, without knowing the identity of active ingredients.

Both the industry and the regulatory agency have realized that, as complex mixtures, it is usually difficult to define or characterize botanical preparations and differentiate among similar products. The tough task for the reviewing staff is thus finding out how to apply the set of regulations designed for highly pure, small molecular entities to a less well-defined botanical system. Apparently, some allowance of imprecision will be needed for CMC of botanical drugs without sacrificing the therapeutic consistency of different batches. As provided in the Botanical Guidance, clinical studies have been permitted for many botanical preparations prior to a complete set of conventional animal toxicity testing. The decisions were not difficult for submissions with substantial and well-documented history of past human use. But some other applicants had not presented an adequate summary of the past human experiences and had failed even to document well-known toxicity of the herbal ingredients. Between these two extremes, how to adjust the requirements of animal toxicity data and substitute that with large quantity but poor quality of human experiences is another big challenge to the regulatory agency in the review of botanical applications. As noted above, all available information on the historical use of botanical preparations will be accepted for safety consideration. But for FDA clinical reviewers, such experiences are often poorly documented and difficult to interpret or correlate with the paradigm of conventional medicine. In the alternative medical system, almost all the diagnoses to be treated with herbal medicine are defined in imprecise and foreign terms. Typically, one herbal medicine is indicated for numerous seemingly unrelated conditions, most of which are symptomatic relief without clear mechanisms. For these reasons, integrating all the background information into the overall safety assessment for botanical applications has been difficult and required active participation of the BRT.

Prospects of Further Development

As noted above, there has been no botanical product approved as prescription new drug by the FDA. The slow pace of progress in botanical new drug development has been increasingly disappointing, possibly for the following reasons:

1. The industry is still struggling with technical difficulties in bringing a complex and ill-defined system to comply with regulatory requirements set for precision of pure chemical drugs.
2. Some of the diseases and conditions selected as indications for the botanical drugs are difficult to study. There are few exciting products to satisfy serious and unmet medical needs.
3. Many sponsors were inexperienced in new drug development and unrealistic about the resources required for the complicated processes.
4. There is no effective protection of intellectual property right and little incentive for further development of pre-existing preparations available on the market (albeit not yet as drugs).

Thus, while the Agency will in general use previous uncontrolled human experiences to expedite limited early stage testing to assess the therapeutic potential of herbal medicines, the overall progress has been slow. However, the technical difficulties in quality controls should be resolvable, and more clinical trials should be initiated. The sponsors of botanical applications should be prepared to go through a complicated scrutiny, the same as that for nonbotanical drugs, and plan ahead with an assessment of difficulties in clinical testing. Lastly, although market exclusivity may not be strictly enforceable for well-known botanical preparations, benefit of the first FDA-approved botanical drugs may still be significant but underestimated for the sponsor.

2

Plant Drugs

Nature, as a biochemist, is unparalleled. Mankind has harvested the benefits of her combinatorial talents for thousands of years as a source of novel medicinals. For more than half a century, however, natural products have been relegated to source materials for "*chemical libraries*" from which new "leads" are mined. In a grants announcement, the National Institutes of Health (NIH) stated: "Chemical libraries are a mainstay of drug discovery. Well-crafted libraries, consisting of collections of anywhere from a few compounds to millions of them, can help scientists sort quickly through a haystack of possibilities to find the shining needle that may be developed into a lifesaving drug". Once an "*active*" or "*new chemical entity*" (NCE), is found, it is isolated, "*optimized*," and then screened in receptor assays for further clinical development. To enhance the odds of finding clinically useful NCEs, the pharmaceutical industry has developed modern techniques, which include proteomics, bioinformatics, "high- throughput" screening techniques, combinatorial chemistry, and computer-aided design and prediction of drug toxicity and metabolism. These new approaches are based upon increasing knowledge of receptor sites in normal and disease states.

But "botanicals" do not fit easily into this development paradigm. As a drug class, botanicals have no rival for their structural complexicity or diversity of effects. Defined by their heterogeneity, botanicals have multiple actives, which in some cases are unknown and in others are too numerous to evaluate. Screening these products against individual receptor targets cannot begin to describe their rich activity profiles. Similar to biologic products, such as vaccines, the potential activity of botanicals as pharmaceutical products may be best illuminated in bioassays and, more importantly, in living systems.

Mainstream U.S. pharmaceutical development has thrived on regulatory policies that evolved over decades of experience with single NCEs. Therefore, the heterogeneous nature of botanicals engenders discomfort and uncertainty in those who are used to single-chemical entity drugs and the NCE regulatory structure. But the interest in botanicals and other heterogeneous products as pharmaceuticals has inspired a change in the U.S. perspective, resulting in a new regulatory paradigm that draws upon historical precedent and novel interpretations of regulatory policies. Only recently has the U.S. Food and Drug Administration (FDA) developed a regulatory definition of a "botanical," as any product that "contains ingredients of vegetable matter or its constituents as a finished product". For the purposes of US regulation, botanicals include drug products derived from one or more plants, algae, or macroscopic fungi, but does not include a highly purified or chemically modified substance derived from such a source. In the United States, botanicals can be regulated as foods, drugs, biologics, cosmetics, and medical devices. However, no botanical "*pharmaceuticals*," or more precisely, botanical "new" drugs are being marketed at this time. A "new" drug is defined by the Federal Food, Drug and Cosmetic

(FD&C) Act as any drug marketed after 1938 that is "...not generally recognized as safe (GRAS) and effective under the conditions prescribed, recommended or suggested in the labeling" or one that has become GRAS, but "which has not... been used to a material extent or for a material time". A "new" drug must be proven to be safe and effective for its intended use prior to marketing in the United States. To study a "new" drug, the product sponsor must file an "Investigational New Drug" (IND) application with the FDA, exempting the product from the requirements of safety and efficacy while it is being tested in an investigational setting. After sufficient evidence is obtained, the sponsor can submit the data to the FDA as part of a "*New Drug Application*" (NDA). If the information is deemed adequate, the FDA can approve the product for marketing.

Therefore, to be marketed as a pharmaceutical, a botanical must traverse the modern regulatory process resulting in an approved NDA. The fact that no botanical is currently "*NDA-approved*" has caused many to speculate whether a botanical could ever make it through the rigorous U.S. drug development process, and still others to ask "why" one would choose to pursue this avenue, given the panoply of regulatory options already available to botanicals in the United States.

Regulatory Options

When asked to describe a "botanical" product, food products come quickly to the mind. Fruits, vegetables, grains, herbs and spices, condiments, and teas are easily recognized as products from plant sources. Dietary supplements, defined by the Dietary Supplement Health and Education Act (DSHEA) of 1994, belong to the regulatory category of food products permitted to contain ingredients that are "... herbs or other botanicals, ... dietary substances or concentrates, metabolites, constituents, extracts, or combination of these ingredients ...." Products such as ginkgo, St. John's wort (SJW), and ginseng are examples of botanicals currently marketed in the United States as dietary supplements. Less commonly recognized as botanicals are the allergenic vaccines derived from grasses and pollens, which are regulated as biologics, and the dental alginates, poultices, and adhesives that are medical devices. Finally, botanicals are often ingredients of cosmetics, such as aloe-containing hand lotions and herbal shampoos.

Historical Perspective

To understand the current regulatory milieu of botanicals as drugs in the United States, one must revisit their historical use. Botanical medicine is an integral part of U.S. history. The democratic processes that governed this new nation extended to the practice of the healing arts. At the beginning of the 19th century, no single medical profession existed. Samuel Thomson, who had no formal education, received a patent for his system of "*botanic medicine*," which he described in the "New Guide to Health," published in 1822. His followers, known as "Thomsonians," practiced a form of naturopathy using plant-based medicines such as *Lobelia inflata* or "*Indian tobacco*"—a violent emetic to purge the system of obstructions—and red pepper, to induce perspiration. After Thomson's death in 1843, his disciples formed another botanic sect, known as the "*Eclectics*," deriving their name from their assimilation of the "best" from the various schools of medicine that were developing at the time.

By 1900, most drugs in the United States were derived from natural sources. Ingredients were pulverized and extracted with hot water and alcohol, and administered in the form of teas, suspensions, emulsions, and syrups. Fine powders were prepared for pills or ointments. Ingredients were combined in proprietary mixtures called "patent" medicines. Hawked by their inventors as "cure-all" medicinals, these secret formulas were not only concocted by uneducated consumers, but were also sold directly to the unsuspecting public. Preparations often contained dangerous and addictive substances, such as morphine, cocaine, and opium, which resulted in severe and sometimes fatal reactions. In 1906, the first comprehensive federal legislation in the United States was passed to address the quality and safety of food and drug products. This new law, known as the Pure Foods and Drugs Act, prevented the importation of "adulterated and spurious drugs and medicines," and required that all ingredients be identified on the drug label, thus ending the era of "secret" nostrums.

Passage of the 1906 act heralded the beginning of a new regulatory environment that would transform the U.S. drug industry. Many so-called drug "manufacturers" disappeared overnight, while others succeeded in complying with the new regulations. Those who did change would grow into a new industry, gaining further momentum with the passage of the comprehensive 1938 FD&C Act. Fueled by the 1937 tragedy in which a hundred or so individuals died following ingestion of a tainted formulation of an "*Elixir of Sulfanilamide*," the new act now required that drugs be demonstrated as safe in animals and humans prior to being marketed. This new Act also considerably expanded the federal government's jurisdiction over the food and drug industries and has become the cornerstone of modern food and drug regulation in the United States.

By this time, however, botanicals as a product class had the reputation of being mostly palliative. Many had a slow onset of action and were used to treat signs and symptoms, without improving the underlying disease process. A notable exception was quinine, an extract of the bark of the South American cinchona tree, traditionally used to ward off the symptoms of malaria. One natural product, however, would irrevocably change the U.S. drug industry. From its discovery as a product of fermentation, to its isolation, purification, and synthesis, penicillin set the stage for an entirely new industry based on single chemical entities. Penicillin belonged to a class of antibiotics that became known as the "miracle" drugs. Unlike the traditional botanicals, these drugs displayed a rapid onset of action and demonstrated remarkable therapeutic efficacy that would elevate the public's expectations for all future pharmaceuticals.

Following the passage of the FD&C Act, FDA was authorized to permit NDAs for "new" drugs, but the agency could not approve them affirmatively. One such NDA was for the botanical rauwolfia (Rauwolfia serpentina), first marketed in the United States in 1953. Used in India for centuries, root of rauwolfia was sold in the United States as a treatment for hypertension. More than a dozen NDAs were subsequently recorded for the drug, all of which were discontinued by 1982, as better antihypertensives came to market.

Passage of the Drug ("Kefauver–Harris") amendments in 1962 further tightened the federal government's control over the regulation of pharmaceuticals, increasing FDA's role in the testing of investigational drugs and providing the agency greater powers of enforcement. The legislation was precipitated by the thalidomide disaster that left hundreds of European infants malformed after their mothers took the drug as a sleeping aid during pregnancy. Drugs were now required to demonstrate proof of safety and efficacy for their labeled indications, prior to being marketed. Data collected from investigational studies would now be reviewed by FDA to determine whether the legal test of "*substantial evidence*" was met for approval. FDA was also required to review retrospectively all drugs that had entered the domestic market between 1938 and 1962. Those marketed under an NDA were reviewed by the Drug Efficacy Study Implementation or "DESI" program. Under DESI, more than 3400 drug products and 16,000 claims were reviewed. Thirty percent of the drugs lacked sufficient supporting evidence and were considered ineffective and removed from the U.S. market.

DESI also included 420 nonprescription ("over the counter" or "OTC") drugs, which had entered the US market through the "new" drug procedures during 1938 and 1962. OTC drugs not reviewed by the DESI program numbered in the hundreds of thousands and were considered to be "GRAS". The agency began its efficacy review of the OTC drugs in 1972. To conserve resources, FDA focused on active ingredients, which it grouped by therapeutic category. For each category, FDA promulgated regulations in the form of monographs, establishing conditions by which the ingredients were determined to be "generally recognized as effective" or "GRAE." Faced with the enormity of its task, FDA chose to limit its review to those ingredients that had U.S. marketing experience to support "material time" and "material extent". As a result of the agency's narrow interpretation of the statute, many botanical ingredients were ineligible for inclusion in the OTC review, because at that time they were only being marketed outside of the United States. However, products that had entered the U.S. market prior to 1938 were exempt from review, and included botanicals such as senna, cascara, and witch hazel, which continue to be marketed as OTC drugs today.

Complementary and Alternative Medicine

During the 1980s, U.S. interest in botanicals resurfaced under the guise of "complementary and alternative medicine" (CAM). By this time, botanicals had lost considerable credibility. Proponents of "Laetrile"—a concoction of cyanogenic glycosides extracted from peach pits—drew media attention when the FDA began seizing the product as an "unapproved" drug and as "ineffective cancer treatment". However, by 1991, in response to growing consumer interest, the U.S. Congress appropriated funds to the National Institutes of Health for the establishment of a federal research program for the scientific evaluation of CAM. Now identified as one of the "*biologically-based therapies*," botanical medicine became a research priority for the new NIH Office of Alternative Medicine (OAM).

This dramatic turnabout in the U.S. consumer attitude toward botanical medicine was not lost on foreign manufacturers. Many companies in Europe and Asia that had never stopped marketing botanicals were now eager to satisfy the growing U.S. demand. A rate-limiting step, however, was FDA's exclusion of foreign marketing experience as a threshhold criteria for inclusion in the OTC drug monograph process. Most botanicals had been discontinued from the US market in the earlier part of the century. Due to the 1962 amendments these same products would now have to undergo FDA approval as "new" drugs through the IND/NDA process—a process established based on single chemical entities. In July 1992, the European American Phytomedicine Coalition (EAPC), an alliance between the U.S. and European phytomedicine companies, submitted a petition to FDA requesting that foreign marketing histories be eligible for inclusion in the OTC review process. The agency took several years to respond to the petition, and when it did, its response was to request more information from the petitioners. This inaction led many in the industry to conclude that FDA either did not consider botanicals on equal footing with the single chemical entities, or would not seriously entertain their review under the IND/NDA process. According to Loren Israelsen, an attorney involved with the herb industry and EAPC cocounsel, "The issues raised by the EAPC are important policy considerations which deserve a thoughtful and affirmative response from FDA. We have tried to frame the problem and the solution squarely and FDA's silence is not only disappointing but gives support to the industry's belief that the Agency remains inflexible and unresponsive".

FDA's inaction may have been responsible in part for efforts leading to the passage of DSHEA in October 1994. The new law addressed "herbs and other botanicals" in the context of nutritional supplement ingredients. Over the next six years, botanical supplement sales doubled, peaking at over US $4 billion in 2000. Even so, the intent of Congress was clear from the language of DSHEA: unlike drugs, dietary supplements were not intended to diagnose, mitigate, treat, cure, or prevent disease, although similar to drugs they could make claims to "affect the structure or function of the body".

New Regulatory Paradigm

The legal limitations placed on dietary supplements with respect to disease claims did nothing to stem the increase in the consumer use of botanicals as an alternative means to prevent or treat disease conditions. A survey conducted in 1990 and repeated in 1997 found "*herbal medicine*" to be one of the leading alternative therapies responsible for a significant increase in CAM usage in the United States. In response to growing concerns over the safety and efficacy of botanicals, the OAM funded several grants that proposed to study botanicals for therapeutic indications. Although products could be purchased in local grocery and health-food stores, their evaluation as "new" drugs required that the trials be conducted under IND applications.

What criteria would the FDA use to determine whether the clinical trials could be allowed to proceed under an IND application? The test products were complex mixtures of botanical extracts with multiple or unknown actives. In the absence of a single known active, routine chemistry, pharmacology, and toxicology testing could not be easily conducted. The closest products resembling botanicals were biologics: vaccines and blood- derived products. Both product categories were defined by strict controls over the manufacturing process, rather than by chemical determination of the product in the final vial. Botanicals raised unique issues: not only were plant nomenclature and taxonomy not internationally harmonized, but also the common names for plants varied by country

FDA needed a new regulatory paradigm—one that would be consistent with current drug law, but could also address the unique characteristics and status of botanicals as a heterogeneous class of pharmaceuticals. Over the next six years, the FDA developed a "Draft Guidance for Industry on Botanical Drug Products," published on August 10, 2000, and finalized in June 2004. The FDA would also amend its regulations to allow foreign marketing experience as a basis for including foreign-marketed botanicals in the OTC drug monograph process.

Selecting a Route to Market

Although the new FDA guidance provides a broad outline for botanical drug development, it does not assume all botanicals to be drugs. Instead, manufacturers are presented with a "*decision-tree*" that describes the possible regulatory categories. How a product is regulated depends on several factors, which include product formulation and route of administration. Topically administered products can be sold as drugs, devices, or cosmetics, but not as foods or dietary supplements. Parenteral administration is reserved for drugs. Tablet or capsule formulations can be marketed as dietary supplements or drugs, but not as "conventional" foods.

Intrinsic safety can further define the regulatory possibilities. Foods, including dietary supplements, must be safe for the general public. Conventional foods are limited to ingredients that are "dietary," "GRAS," or approved food additives. In contrast, dietary supplements can contain both "dietary" ingredients and "new" dietary ingredients. "Dietary" ingredients must have been "present in the food supply in a form used for food, not chemically altered." Ingredients marketed in the United States after October 15, 1994, are considered to be "new dietary ingredients.""New" dietary ingredients must have "a 'history of use' or other evidence of safety ... that the new dietary ingredient would be reasonably expected to be safe under conditions of labeling." For products containing "*new dietary ingredients*," FDA must receive written notification at least 75 days prior to marketing, providing information to support the safety of the ingredients.

Unlike foods, including dietary supplements, the safety of a drug is based on a "benefit to risk ratio." "Benefit" is an assessment of the drug's efficacy, balanced against any negatives with respect to a particular indication in a target population. Thus, a drug that is deemed "safe" to treat leukemia may not be "safe" to treat osteoarthritis.

Intended Use

A defining principle of U.S. regulation is a product's "*intended use.*" "Intended use" is determined by labeling claims. "Labeling" encompasses not only the required elements that make up the printed label on the bottle, but also any direct or implied claims made by the manufacturer or distributor in the product packaging, advertising, and promotional materials. "Intended use" defines the product category. For example, the FD&C Act defines drugs as "articles intended for use in the diagnosis, mitigation, treatment, cure or prevention of disease or to affect the structure or function of the body." Drug products may bear "disease," "sign," or "symptom-related" claims ("prevents migraine"; "lowers blood pressure"; and "relieves cough and fever").

Similar to drugs, dietary supplements can make claims to "affect the structure or function of the body." However, supplements are specifically prohibited from making "disease" claims. A dietary supplement bearing "structure or function" claims is also required to carry the following disclaimer on the product label: "The FDA has not evaluated this claim. This product is not intended to diagnose, mitigate, treat, cure or prevent disease".

Advantages of the Drug Route

If a botanical is shown to reverse an abnormal test result, modify another drug's adverse event, or act synergistically with another modality, it will usually be best developed as a "drug." This is especially true if the botanical provides a distinct benefit for a patient population, but would pose safety concerns if used by the general public. In contrast to prevention, risk-reduction, or health-maintenance trials, therapeutic studies are often able to demonstrate clinical benefit with smaller numbers of subjects and with more tangible measures of outcome.

But sponsors, enticed by the low cost of market entry for dietary supplements, often dismiss the potential advantages of pharmaceutical development. Because the regulatory schema for a drug is significantly more complicated than that for foods, US law provides many protections for those who choose the pathway of drug development, not the least of which is the confidentiality of the process. From the filing of the IND through the NDA approval, exchanges between sponsors and FDA are kept confidential by the agency. This is in stark contrast to the very public process of food applications, petitions, and notifications.

Patents, trademarks, and copyrights provide additional proprietary protection, which may have a more profound impact on pharmaceuticals than on products sold in most other categories. Although food and cosmetic ingredients may be afforded some protection through composition and process patents, an underlying premise behind marketing of food products is their similarity to prior foods and ingredients with known histories of safe use. The further a "*new dietary ingredient*" strays from a traditional food ingredient, the more documentation will be necessary to ensure "safety," thus undermining the ease and minimal cost at which most food products are allowed to come to the market. In contrast, pharmaceuticals exploit differences, capitalizing on the nuances between molecular analogs and minor alterations in formulation, dosing, and usage. "New" drugs do not have "GRAS/E" status. Generic equivalents of "new" drugs usually enter the marketplace only after an innovator's patent protections have expired. Under the "Price Competition and Patent Term Restoration" Act (also known as the "Hatch-Waxman" Act), not only can the term of a drug's patent be extended or "restored" to account for market-time lost while the product is under regulatory review, but also it is in the FDA's purview to grant periods of marketing exclusivity during which the agency will not accept a competitive filing. For new clinical entities, the first product approved under an NDA can receive a five-year period of marketing exclusivity, and an additional five years is added to the expiration date of any patents. New indications, formulations, or routes of administration requiring additional clinical trials (beyond bioequivalency studies) are granted a three-year period of marketing exclusivity and patent term extension. Under the Orphan Drug Act of 1983, drugs for rare disorders for which the target indication occurs in less than 200,000 individuals annually in the United States, are granted the maximum term of seven years exclusivity and patent extension. Pediatric indications can provide an extra six months of protection. No similar provisions are available for dietary supplements or conventional foods.

Finally, "new" drug approval following the demonstration of safety and efficacy through the IND/NDA process brings with it an added benefit of medical and scientific acceptance that can significantly boost the marketing message. For a drug, clinical results may be conveyed in more direct language for advertising and promotion, rather than the restrictive wording delineated for supplement "structure or function" or "health" claims.

Cost

The ongoing debate on U.S. drug development costs is another reason sponsors look to alternative development strategies for their products. At issue are recent estimates arrived at through a survey of 10 large pharmaceutical companies. Average development costs for an NCE were estimated at US $403 million (in 2000), and when the time between investment and marketing is added, costs totaled a mind-boggling US $802 million. The figures were based on new synthetic chemicals not previously tested in humans, reflecting costs from all NCE candidates tested, including those that had failed. From the tens of thousands of chemicals generated in the discovery process, only a few hundred made it through the screening process, leaving only a dozen or so to undergo animal safety testing. In one review, attrition rates were reportedly due to "safety issues" (20.2%), "toxicology concerns" (19.4%), and "disappointing clinical efficacy" (22.5%). Another 39% of candidates were terminated for business reasons and "other factors". Cost increases often begin in the laboratory: "What big pharmaceutical companies have done is tested a lot of existing materials. But now they've run out of those materials. Today, you have to create new compounds and then test them".

But botanical products do not have to be "created," and many have already been shown to produce potentially useful biological effects in humans. More importantly, documentation of safe use in the target species—humans—is a monumental step that can shorten the development time considerably.

Whether a particular botanical should be developed as a pharmaceutical depends on the product and the business objectives of the sponsor. Sponsors should be cognizant of the costs incurred in "Good Manufacturing Practices upgrades"—modifying the product's manufacturing standards either to meet U.S. drug standards or to move from a food to a pharmaceutical-grade product. Regardless of the extent of prior human use, a botanical drug will likely be required to undergo additional safety and efficacy testing. Undoubtedly, the majority of costs of drug development is incurred in the conduct of clinical trials. However, unlike NCEs, FDA may permit an initial study to be conducted with a product purchased "off-the-shelf" as a dietary supplement.

A randomized, controlled "pilot" study may be the initial trial for a botanical drug under an IND, with an agreement between the agency and the product sponsor that at least one other large multicenter trial with a pharmaceutical-grade product will be conducted for the NDA, if initial results are promising. Animal toxicology testing may also be required to address safety questions that cannot be easily assessed in humans. Although not a requirement for foods, interactions of the botanical drug with other drugs and with foods must also be evaluated, depending on the indication and the potential for interactions. Under the Prescription Drug User Fee Act, the FDA can levy fees for the review of applications containing clinical trial data, although waivers can be sought for nonprofit sponsors and small businesses.

Whole is Greater than the Parts

Whether a botanical should be pursued as a drug or whether it should be "mined" for its actives, depends on the botanical and its constituents. Isolation, purification, and synthesis of single active chemical moieties from natural products have produced a substantial number of the pharmaceuticals in use today. By some accounts, 62% of the current NCE anticancer drugs are nonsynthetic, and more than half of the antihypertensive drugs can be traced to natural product structures or mimics. Determining what is "active," however, is not always straightforward. Most botanical extracts yield legions of constituents with diverse biochemical profiles and pharmacologic effects. A constituent identified as "active" for one particular effect may be "inactive" for others. So-called "inactive" ingredients also may contribute to the biological effects of a product indirectly, through modulation of the actives.

To ensure lot-to-lot consistency, standardization of extracts often relies on constituents as "*biomarkers*" for plant identity and potency. SJW-(*Hypericum perforatum*), a perennial shrub traditionally used as a mood enhancer and mild antidepressant, has been tested in dozens of clinical trials, with mixed results for efficacy. Some of its purported bioactive constituents include naphthodianthrones, including hypericin; flavonoids; phloroglucinols, including hyperforin; and essential oils. For many years, hypericin was presumed to be the active component. As a result most extracts were standardized based on hypericin concentration. Recent data, however, support other components such as hyperforin and the flavanoids, that may also contribute to the therapeutic efficacy of the SJW extracts. Because these secondary components were previously unaccounted for in the standardization of the former clinical test articles, and because these constituents are chemically unrelated to and their content within the plant varies independently of hypericin, it has been argued that the potency of these constituents in any particular batch was unlikely to be similar to that of other batches. This variability between batches could explain the observed differences in the clinical trial results.

For many botanicals, the "whole is greater than the parts." Indeed, individual constituents of an extract may actually produce contradictory effects. Oriental ginseng (*Panax ginseng*) root, widely used in traditional Chinese and Korean medicine, contains over 28 different ginsenosides. Although chemically similar in structure, various ginsenosides have been shown experimentally to produce opposite effects: hypothermia or hyperthermia, hypotension or hypertension, and hemolysis or inhibition of hemolysis—depending on the type of ginsenoside. One must, therefore, use caution in determining what is "active" and be aware that the properties— both positive and negative—of any particular constituent may not represent the biological activity of the extract as a whole. In summary, botanicals are as diverse as nature itself. They bring a wealth of possibilities for innovative new drugs. As a result of the options available to botanical producers, a number of development approaches exist in the United States, but no single paradigm. While many botanicals are best sold as "foods," others will find a more promising future as pharmaceuticals. Those that are systematically studied in scientifically designed trials and are able to demonstrate consistent, clinically relevant biological activity may traverse the drug development process to achieve the status of "new" drugs—the "*botanical pharmaceuticals*."

Chinese Plant Products

Many Asian communities throughout the world have used Chinese botanical products for centuries. Recently, the usage of these botanical products has also increased in Western societies. Although the use of Chinese botanical products is on the rise, the potential and significance of interaction with Western drugs is not widely recognized and well characterized. While there are very few adequate, well-controlled clinical studies designed to investigate the potential for interaction between Chinese botanical products and drugs, in the English literature, there are examples of documented interaction between commonly used Chinese botanical products and currently available Western drugs, and these case reports will be reviewed in this chapter. In addition, issues that are more pertinent to the evaluation of the importance and clinical relevance of Chinese botanical product–drug interactions will be discussed.

In contrast to theoretical, in vitro, or animal data, the reports reviewed in this chapter provide the clinicians more relevant information, including description of the time course, the magnitude of the suspected interaction, and clinical outcome of the patient. This not only allows an evaluation of the clinical significance, but also provides a basis for further evaluation with well-designed studies. Obviously, case reports have their own inherent limitations, including the existence of potential confounding variables and limited generalizability, which in view of the known product variability of active constituents or content, could be of particular importance in botanical product–drug interactions. Finally, it should be recognized that the occurrence of one or more case reports does not necessarily imply an absolute contraindication of concurrent use of the botanical product and prescription or over-the-counter drug.

Currently, most of the literature reports of Chinese botanical product– drug interaction in humans involve warfarin, likely a function of its narrow therapeutic index requiring close monitoring of therapy with international normalized ratio (INR) and the presence of coumarin derivatives in a number of Chinese botanical products rendering them with anticoagulant property. In addition, some Chinese botanical products also possess antiplatelet effects and have potential for adverse interactions with analgesic drugs such as aspirin or nonsteroidal anti-inflammatory drugs. Based on human, animal, and in vitro data, other Chinese botanical products such as hawthorn have also been reported to interact with a variety of drugs.

It should be noted that while the focus of most Chinese botanical product–drug interaction reports understandably is on the occurrence of adverse effects, not all interactions result in an undesirable effect. An example is *Salviae miltiorrhizae* (danshen), which has been reported by multiple clinicians to cause bleeding with the concurrent use of warfarin [section "*Salviae miltiorrhizae* (Danshen)"], whereas less is known about the potential benefits that might result from combining danshen with an aminoglycoside. Wang et al. had demonstrated in animals the potential usefulness of combining danshen with kanamycin to reduce aminoglycoside-induced free-radical generation in vitro and ototoxicity in vivo, without interfering with serum concentration or efficacy of kanamycin in mice. Other examples of potential beneficial botanical product–drug interaction that warrants further studies include the combined use of the Chinese medicinal plant *Tripterygium wilfordi* and cyclosporin, which is described elsewhere.

In addition to the more conventional nomenclature system of using the botanical name, e.g., *Angelica sinensis*, Chinese botanical products also can be identified by their pinyin name, e.g., dong quai for *A. sinensis*. Although most English literature refer to Chinese botanical products by their botanical names or pharmaceutical names, the corresponding pinyin names are often used instead in most Chinese herbal literature. Therefore, searching for literature information regarding Chinese botanical products should ideally include the pinyin names in the search strategies, especially if the source of information is primary Chinese herbal literature.

Chinese Plant Products and Warfarin

Angelica sinensis

Dong quai (dang gui, tang kuei) is the extract from the dried root of *Radix Angelicae sinensis*, which belongs to the family Umbelliferae. It has been used for many years as a Chinese botanical remedy for management of menstrual cramps, irregular menses, and menopausal symptoms, and the usual dosage range is 3 to 15 g per day of raw drug prepared as a hot water decoction or alcoholic infusion. Different preparations, including alcoholic extracts, tablets, and teas, are available to the consumer. Dong quai contains coumarins and also may inhibit platelet aggregation. Therefore, this Chinese botanical product could potentially enhance the pharmacologic effect of warfarin-like compounds.

Page and Lawrence reported a 46-year-old female patient with rheumatic heart disease and atrial fibrillation, who was referred to the anticoagulation clinic for warfarin therapy management. She was successfully managed with warfarin (Coumadin) 5 mg/day, which maintained her INR within the range of 2 to 3 for about two years. At a routine clinic visit, the patient was found to have an elevated INR of 4.05 compared to 1.89 a month earlier. The prothrombin time (PT) also increased over the same time period from 16.2 to 23.5 seconds. Because she did not show any evidence of clinical bleeding, and there were no readily identifiable sources for the increased laboratory values, including dosing error and abnormal liver function, she was instructed to withhold the warfarin dose for one day. The patient missed a follow-up clinic visit and only returned after another month had passed. At that time her PT and INR were further increased to 27 seconds and 4.9, respectively. The patient also disclosed that she had been taking dong quai at a dosage of 565 mg once to twice daily for four weeks to manage her perimenopausal symptoms. She was instructed to miss a day of warfarin dosing with no additional change of warfarin dosage, and also to discontinue consumption of dong quai. Two weeks after discontinuing the dong quai regimen, her PT and INR decreased to 21.6 seconds and 3.41, respectively, with further reduction to 18.5 seconds and 2.48 after an additional two weeks. Ellis and Stephens reported similar interaction in a brief case report describing a patient with a mitral valve replacement, who had been stabilized on warfarin (dosage regimen not reported) for 10 years. After taking an unknown quantity of dong quai for a month, she presented to the clinic with an INR of 10. Although details of the report were very few, significant bruising as a result of the interaction was shown in a photographic figure published with the case.

Although the exact mechanism is not known, the coumarin constituent of dong quai is likely responsible for the enhanced pharmacological effect. An animal study also suggested that the basis of the interaction is likely pharmacodynamic and not pharmacokinetic in nature. Nevertheless, it should be noted that dong quai belongs to the family Umbelliferae, and plants within this family contain furocoumarins, which have been reported to inhibit cytochrome P-450 (CYP) activity, especially CYP3A4. In this regard, it is of note that the extract of the dried root of *Radix Angelica dahurica*, another botanical product belonging to the Umbelliferae family, has been reported to increase the area under the plasma concentration–time curve (AUC) and to prolong the elimination half-life of tolbutamide in rats by 2.5- and 2.3-fold, respectively, which was likely a result of the inhibitory effect of its furocoumarin components on different CYP isoenzymes, including those belonging to the 2C subfamily. Based on in vitro studies, it has also been suggested that another dong quai component, sodium ferulate, might inhibit platelet aggregation and cyclooxygenase activity.

Salviae miltiorrhizae

Danshen, the dried root and rhizome of *S. miltiorrhizae*, is another Chinese botanical product used for its ability to alleviate menstrual irregularities, as well as for its vasodilative and hypotensive functions in a variety of cardiovascular conditions. The botanical product had also been shown to inhibit platelet aggregation in vitro. Danshen is widely available in different preparations for oral consumption, with usual dose range of 9 to 15 g per decoction. In addition, its increasing popularity is reflected by its availability even in Chinese cigarettes.

Clinicians from Hong Kong reported a case of potential danshen– warfarin interaction in a 48-year-old female with a history of rheumatic heart disease, atrial fibrillation, and mitral stenosis (11). The patient underwent successful transvenous mitral valvuloplasty for management of her medical conditions, and was discharged with 1 mg warfarin, as well as furosemide and digoxin. Since discharge the patient's warfarin dosage ranged from 2.5 to 3.5 mg daily with an INR of 1.5 to 3. Her last warfarin dose adjustment was an increase in dose to 4 mg daily in response to an INR of 1.35. Since then the patient also had intermittent influenza-like symptoms, for which she took botanical products with danshen as one of the main ingredients, every other day. When the patient presented to the emergency room several weeks later because of increased flu-like symptoms, her clotting profile was noted to be significantly abnormal with an INR exceeding 5.6 and PT above 60 seconds.

There was no other clinical source of clotting abnormality and the most likely cause of over-anticoagulation was believed to be an interaction between warfarin and danshen. Although the patient stopped taking both warfarin and all botanical products, and received fresh frozen plasma, her clotting abnormality persisted for more than five days. Nevertheless, the patient suffered no clinical evidence of bleeding and over the next four months, she was stable on a daily warfarin regimen of 3mg and an INR of 2.5. There were two additional reports of danshen–warfarin interaction in the literature. A 62-year-old man with rheumatic mitral regurgitation had been stabilized on warfarin 5mg with INR of about 3.0 over four weeks after discharge. His other medications included captopril, furosemide, and digoxin. The patient then started taking danshen daily to help his heart condition. Two weeks later, he was admitted to the hospital with an INR of 8.4. Both the warfarin and danshen regimens were stopped. Fresh frozen plasma and packed red blood cells were administered, eventually decreasing the INR to 2.0. Over the next two weeks, he was restarted on warfarin and the dose titrated back to the previous regimen of 5 mg/day, resulting in a stable INR of 3.

Another case involved a 66-year-old male patient who was stabilized on 2 to 2.5 mg of warfarin per day with INR of about 2. About nine days prior to admission to the hospital, the patient developed nonspecific chest wall pain, for which he self-treated with two to three topical applications of 15% methyl salicylate and two decoctions of danshen over the next few days. On the day of admission, his INR was found to be greater than 5.5, and his warfarin regimen was stopped, followed by administration of fresh frozen plasma and packed red blood cells. The INR was subsequently stabilized at 2.0 to 2.5.

These three cases suggested that danshen might potentiate the anticoagulant effect of warfarin, although information regarding consumption of other Chinese botanical products was not available for two of the three cases. In the report of Tam et al., the patient also self-medicated with topical application of methyl salicylate, which might have initially exaggerated the anticoagulant effect of danshen. In all three reports, the absence of identifiable precipitating factors and the temporal relationship between botanical product consumption and onset of exaggerated anticoagulation effect suggested an interaction between danshen and warfarin. In addition to inhibiting platelet aggregation and interfering with extrinsic blood coagulation, danshen was also shown to affect warfarin pharmacokinetics in an animal study. Chan et al. reported that single-dose administration of danshen in rats increased the AUC and maximal concentration of the R- and S-isomers of warfarin. In addition, concurrent administration of warfarin and danshen for three days also increased the steady-state *R*- and *S*-warfarin concentrations, with resultant increases in PT by 11 seconds.

Lycium barbarum

The dried fruit of *L. barbarum* L., a common Chinese botanical product belonging to the family of Solanacaea, is available in different tea formulations for its beneficial effects on the kidney and the liver. Lam et al. described a potential interaction between a concentrated Chinese herbal tea and warfarin in a 61-year-old Chinese woman. The patient had been stabilized on a weekly warfarin dosage regimen of 18 to 19 mg/week, with a therapeutic INR ranging between 2 and 3, for her recurring atrial fibrillation. There were no signs and symptoms of abnormal anticoagulation.

On a routine anticoagulation clinic evaluation, the patient's INR was elevated to 4.1 from 2.5 obtained at the prior monthly visit, albeit with no clinical evidence of bleeding. There was no reported change in any of her medication regimens, diet, or lifestyle. However, the patient indicated that she had been consuming one cup of a concentrated herbal tea made from dried fruits of *L. barbarum* L. several times a day, to manage blurred vision secondary to a sore eye. When she presented to the clinic, the vision problems had already resolved. The patient was advised to discontinue the herbal tea consumption, and the warfarin weekly dosage regimen was adjusted to 16mg with a resultant INR of 2.0, followed by 18 mg/week with a resultant INR of 2.2, before resumption of the original dose of 19 mg/ week with a resultant INR of 2.5.

This case suggested that the elevated INR might be related to the consumption of the herbal tea made from the dried fruits of *L. barbarum* L. The investigators further performed an experiment using human liver microsomes to investigate the effect of the tea on warfarin metabolism. Using method provided by the patient, the investigator produced the herbal tea by adding 5 g of the fruit to 100 mL of boiling water. The hot water decoction was eventually reduced in volume to about 30 mL. The extract was then filtered and the resulting filtrate used in microsomal incubation.

The investigators reported that the prepared tea inhibited the metabolism of the *S*-warfarin isomer by CYP2C9. Furthermore, based on the amount of tea ingested, the solids concentration of the prepared tea, and assumed values of bioavailability and volume of distribution, the plasma concentration of the inhibitory component was estimated to be much lower than the inhibitory concentrations calculated from the in vitro experiment. However, the investigators emphasized the lack of knowledge regarding actual measured inhibitory concentration at the active site of metabolism and the actual bioavailability and distribution volumes of the inhibitory component. In addition, whether *L. barbarum* also possesses an anticoagulant or antiplatelet effect is not known. Therefore, despite the time sequence of INR changes with the use of the herbal tea, it is not known whether the effect is related to altered CYP or non-CYP disposition variables or to an anticoagulant effect of the botanical product itself.

Chinese plant product quilinggao

It is not uncommon for Chinese botanical products to be combined in different preparations for a variety of uses. "*Quilinggao,*" also referred to as "Essence of Tortoise Shell," is a combination Chinese botanical product produced by different manufacturers and promoted for improving general health and reducing internal "*body heat.*" At times, consumers consider and take the product as a health food rather than an herbal medicine. Clinicians in Hong Kong recently reported a patient with clinical evidence of bleeding and over-anticoagulation after consuming different brands of quilinggao.

A 61-year-old man had been receiving warfarin for his atrial fibrillation and chronic rheumatic heart disease. With a warfarin regimen of 3 mg alternating with 3.5 mg every other day, his INR was mostly stabilized within the range of 1.6 to 2.8. On a routine clinic visit for INR monitoring, his INR was found to be greater than 6.0 and he had complained of gum bleeding and epistaxis over three days prior to the clinic visit. There were no reports of changes in medication adherence and dietary habit of vitamin K. Upon further questioning, the patient revealed that he had been taking the combination botanical product quilinggao daily for over three years with a change in the brand just one week prior to the clinic visit. Although he noticed bruising on his left leg five days after taking the new (second) brand of quilinggao product, he continued to consume one can per day until the clinic visit.

His warfarin therapy was withheld and his INR decreased to 2.9 and 1.9, three and five days later, respectively. His warfarin regimen was restarted at the same 3/3.5mg on alternate days as before. The patient was later discharged with an INR of 2.5. He was consulted regarding the possible adverse consequences of taking warfarin and quilinggao concurrently. However, immediately after discharge, he began drinking one can of another (third) brand of quilinggao product daily and three days later his INR was elevated to 5.2. The patient was readmitted to the hospital and warfarin therapy withheld, resulting in decreases of INR to 4.3 and 3.4, two and three days later, respectively. The warfarin therapy was eventually restarted after an INR of 1.9 was reached.

Among the different ingredients listed on the labels, from the first and the second brands of quilinggao products, Chuanbeimu (*Fritillaria cirrhosa*) in the first brand, as well as Beimu (*Fritillaria* spp.), Chishao (*Paeoniae rubra*, Chinese peony), Jinyinhua (*Lonicera japonica*), and Jishi (*Poncirus trifoliata*) in the second brand were constituents that had antiplatelet and/or antithrombotic effects. The potential interacting constituent(s) could not be identified with the third quilinggao product because the patient could not remember its brand name. There was no readily identifiable cause for over-anticoagulation during both hospitalizations, and the temporal relationship between consumption of the last two quilinggao products and the changes in the INR values suggested that an additive interaction between the botanical product and warfarin was the likely cause of the exaggerated pharmacologic effect. The difference in interaction outcome between the first and second quilinggao products could be related to the greater number of interacting botanical constituents present in the second brand product.

Camellia sinensis (Green tea)

Consumption of green tea is a common practice in Asian countries such as China and Japan, and its use as a dietary supplement in the United States has also increased significantly over the years, perhaps reflecting a belief that it may prevent carcinogenesis. Although it is not usually considered as a botanical product, dry green tea leaves contain as much as 1.4 mg of vitamin K per 100 g of dry leaves . Dietary intake of vitamin K facilitates clotting factor synthesis and is well known to antagonize the anticoagulant effect of warfarin. The following report described a probable case of interaction between warfarin and green tea.

A 44-year-old Caucasian male had been treated with warfarin for more than a year for prophylaxis of thromboembolic complications associated with his St. Jude mechanical valve replacement in the aortic position. The therapeutic goal was to maintain his INR within the range of 2.5 to 3.5, and a review of the patient's medication history indicated that a decrease in his INR was always associated with a reduction in warfarin dose. One month prior to clinic visit, his warfarin regimen was 7.5 mg/day and the INR was 3.2. At the clinic, the patient's INR was found to be 3.79 with the same warfarin dosage regimen. However, he was asymptomatic and there was no obvious reason for the increased INR. He was counseled on the importance of a consistent intake of vitamin K–containing foods and instructed to continue on the same dosage regimen.

About three weeks later, the patient returned to the clinic for INR monitoring, and the value reported the next day was 1.37. Multiple attempts to contact the patient failed and he was lost to follow-up for another month, at which time he returned to clinic for a recheck of INR, which was reported as 1.14. The patient reported no change in compliance, diet, medications, or disease states. Nevertheless, on further questioning about his diet, the patient indicated that a week prior to his previous INR of 1.37, he had begun drinking about one-half to one gallon of green tea each day. The patient did not show any signs and symptoms associated with suboptimal anticoagulation, and he was instructed to continue his warfarin regimen and stop the green tea consumption. One week later, the patient's INR was 2.55, and subsequent values were mostly within the target range.

The time course of green tea consumption and discontinuance suggests that the tea could partially account for the changes in the patient's INR values. Although brewed green tea was reported to only contain 0.03 μg of vitamin K per 100 g of brewed tea, this patient's copious consumption of the green tea would obviously provide an exogenous amount of vitamin K that exceeds the usual recommended daily dietary intake of 0.5 to 1.0 μg/kg of vitamin K. In addition, the final concentration of vitamin K in any brewed tea would be affected not only by the amount of dry tea leaves used for brewing, but also by the volume of water used to prepare the tea for consumption. In summary, this case highlights the importance of consistent dietary intake of vitamin K for patients receiving warfarin therapy, and the fact that less well-known sources of exogenous vitamin K could provide an amount that greatly exceeds the recommended range of daily dietary intake.

Panax ginseng

Decreased INR associated with the use of *P. ginseng* was reported in a 47-year-old patient who had been stabilized on warfarin. Another case of inadequate anticoagulation with ginseng product resulting in thrombosis on a mechanical aortic valve prosthesis was reported in a 58-year-old patient.

Chinese Plant Products and Phenprocoumon

Zingiber officinale (ginger) has been used for centuries by traditional medical practitioners in East Asian countries to manage symptoms of common cold and rheumatic and digestive disorders, as well as for prophylaxis in the management of nausea and vomiting. For these different purposes, ginger has been used either as fresh or dried root, or in different preparations including capsules, liquid extracts, powders, tablets, or teas. In addition, aqueous extract of ginger has been shown to inhibit thromboxane synthase in a dose-dependent manner, thereby resulting in the reduction of platelet aggregation. This hemostasis effect could be related to gingerols, the active ingredients of ginger. Therefore, the potential exists for ginger to interact pharmacologically with coumarin derivatives. While to date there has been no report of interaction between ginger and warfarin, ginger has been recently reported to interact with phenprocoumon, a coumarin derivative commonly used in most European countries.

Kruth et al. reported that a 76-year-old woman who had been stabilized on long-term phenprocoumon therapy with therapeutic INR values was admitted to the hospital, secondary to elevated INR of greater than 10, prolonged partial thromboplastin time (PTT) of 84.4 seconds (normal <35 seconds), and epistaxis. Although the patient took several concurrent medications for her medical problems, which included atrial fibrillation, hypertension, chronic heart failure, and osteoporcsis, none of the drugs is known to interact with coumarin derivatives. More importantly, there have not been any changes in any of her drug regimens. However, for several weeks before the bleeding incident, the patient had regularly taken several ginger preparations, including dried ginger and tea prepared from ginger powder. Ginger intake was discontinued and with administration of several doses of vitamin K, the patient's INR and PTT eventually returned to baseline values, enabling resumption of her normal doses of phenprocoumon, with no further recurrence of bleeding episodes.

The time course of ginger administration and the absence of other potential interacting drugs suggest that ginger might be the cause of over-anticoagulation in this patient. In vitro evidence of CYP2C9 involvement in phenprocoumon metabolism is not supported by human pharmacokinetic data. The effect of ginger on CYP activity is not known and a possible pharmacokinetic basis cannot be established at this time. Although the literature evidence of ginger's effect on hemostasis is conflicting, this case suggests that caution needs to be exercised with ginger use in patients receiving anticoagulants. In this regard, it is noteworthy that even though abnormal clotting function as well as mild clinical bleeding in a 25-year-old woman was attributed to several natural coumarin constituents in a herbal tea product, the herbal tea also contains one whole ginger root that might exaggerate the anticoagulant effect of the coumarin constituents.

Chinese Plant Products and Aspirin

As discussed above, the pharmacological action of ginger has prompted suggestion that concurrent use of ginger and aspirin or nonsteroidal anti-inflammatory drugs may exaggerate bleeding potential, especially if the amount of ginger used is larger than that found in usual food items. The following case report indicates such potential, and there are three human studies in the literature, investigating the effect of ginger on platelets.

An unspecified but potentially large amount of marmalade containing 15% raw ginger was consumed by a patient, resulting in significant inhibition of platelet aggregation, although the patient was asymptomatic. One week after discontinuation of the ginger supplement, platelet function was found to be normal. The investigator also performed an in vitro study and reported that ground raw ginger has the potential to inhibit platelet aggregation.

To determine the relevance of in vitro results to clinical setting, Srivastava extended his previous in vitro study to seven healthy female volunteers, who received 5 g of fresh ginger daily for one week. The serum thromboxane activity (thromboxane B_2 formation) at baseline was not significantly different compared to that obtained after one week of ginger administration. There were also no evidence of ecchymosis or reports of unusual bleeding episodes. Eight healthy male volunteers received a single 2 g dose of dried ginger (Schwartz spice) or placebo in a randomized, double-blind, crossover study.

Three blood samples were obtained before, and at 3 and 24 hours after dose administration. Ginger intake resulted in no significant effect on bleeding time, platelet count, and platelet aggregation compared to administration of placebo capsules.

In another study, 18 healthy subjects (nine men and nine women) received an extemporaneous formulation of vanilla custard containing 15 g of raw Brazilian ginger root, 40 g of cooked stem ginger, or placebo once daily for 14 days in a randomized crossover manner. Blood sampling was performed on days 12 and 14 of each treatment for determination of platelet thromboxane B_2 production ex vivo. There were no significant changes with either of the ginger preparations when compared to placebo. In addition, no treatment order effects were noted, although there were no washout periods between the three treatment phases.

The effect of ginger on platelet aggregation and the potential for increased bleeding have been cited as reasons to exercise caution in the use of ginger in patients receiving aspirin and nonsteroidal anti-inflammatory drugs. Although the three clinical investigations reviewed above are associated with the usual study limitations, including extrapolation of observed effects from healthy volunteers to patients, relevance of negative findings in a small number of subjects, variable range of ginger doses and preparations used in the subjects, as well as whether the doses studied represent equivalent doses found in dietary supplements or botanical preparations, there is insufficient evidence at this time to conclude an unequivocal significant antiplatelet effect associated with the use of ginger. Additional human studies similar to those conducted for St. John's wort would clarify the interaction potential of ginger.

Angelica sinensis

In addition to the presence of natural coumarin derivatives, phytochemical analysis found that dong quai also contains ferulic acid and osthole as ingredients. Ferulic acid was reported to have antithrombotic activity. Similarly, study using the closely related *Angelica pubescens* also found osthole to be antithrombotic. These two chemical constituents exert their antithrombotic effects by interfering with different pathways responsible for platelet activation. Ferulic acid inhibits the release of serotonin and adenosine diphosphate from platelets, as well as reduces the thromboxane A_2 production, resulting in impaired platelet aggregation. Osthole directly inhibits the conversion of arachidonic acid to thromboxane A_2. Therefore, even though currently there is no report of an interaction available in the English literature, dong quai may potentiate the risk of bleeding if used concurrently with aspirin or nonsteroidal anti-inflammatory drugs.

Chinese Plant Products and Digoxin

Crataegus pinnatifida

Hawthorn has long been used as a medicinal substance, and an extract such as WS 1442, a formulation of hawthorn leaves with flowers, has been evaluated in different studies for treatment of heart failure. Patients with New York Heart Association class II heart failure participated in a placebo-controlled, randomized, multicenter trial. They received 30 drops of the extract three times daily for eight weeks. At the end of the study, heart failure condition was improved. A meta-analysis of available clinical trials suggests that the extract is useful as an adjunct treatment for patients with mild to moderate heart failure. Therefore, it is likely that hawthorn products would be administered together with digoxin in clinical management of patients.

Although a synergistic interaction between hawthorn and digoxin has been reported in Chinese herbal literature, until recently, there has been no case report or pharmacokinetic/pharmacodynamic data available in the English literature. A recent study evaluated the interaction potential between hawthorn and digoxin. Eight healthy volunteers participating in the study received, in a randomized crossover manner, digoxin 0.25 mg daily for 10 days and digoxin 0.25mg daily concurrent with 450 mg twice daily of Crataegus extract WS 1442 for 21 days, with a three-week wash-out period between the two treatment phases. Based on pharmacokinetic analysis of digoxin concentration–time profiles from both the treatment periods, administration of the hawthorn preparation produced a slight and statistically insignificant change in any of the pharmacokinetic parameters. There were also no statistically significant differences in blood pressure, heart rate, and PR interval. The pharmacological results from this study appear to contradict the Chinese literature of synergism between hawthorn fruit and cardiac glycoside. However, hawthorn may increase digoxin's effect on myocardial contractility, although that was not measured in the study. Future studies confirming the lack of pharmacokinetic interaction and adverse additive effect would provide additional evidence that hawthorn and digoxin can be administered safely together.

Chinese Plant Products and Digoxin-like Immunoreactivity

As discussed earlier, many patients use danshen for a variety of cardiovascular uses. Currently, there is no literature report of interaction between danshen and digoxin. However, Chinese botanical products can interfere with clinical laboratory monitoring of digoxin serum concentrations via their digoxin-like immunoreactive components. For example, danshen contains more than 20 diterpene quinines with chemical structures similar to digoxin. Depending on the type of immunoassay used, both falsely elevated and falsely decreased concentrations have been reported. Similarly, the bufadienolide constituents of the Chinese botanical product lu-shen-wan also bear structural similarity with digoxin, resulting in serum digoxin concentration of about 0.9 ng/mL in patients who took lu-shen-wan pills. Even though it is not necessarily considered to be a real botanical product–drug interaction, it would be prudent to check for potential interference by serum digoxin concentration determination, when these Chinese botanical products are used together with digoxin.

Limitations of Current Literature

Although literature publications on the use of herbal medicine have increased over the years, there are relatively few retrievable literature reports regarding concurrent use of Chinese botanical products and prescription and/or over-the-counter medications, when compared to drug interaction reports associated with concurrent use of two or more Western prescription drugs. While this may simply reflect a lack of reporting system for the consumers, another likely reason could be that pertinent information is not readily available in the English literature. Pharmacodynamic and pharmacokinetic interaction cases have been published in Chinese herbal literature, including Herb-Drug Interaction and Combined Medication, Chinese Herbal Medicine, and Pharmacology and Application of Herbal Medicine, and attempts are currently undertaken to provide this information in the mainstream literature.

Most of the Chinese botanical product–drug interaction cases involving warfarin and/or salicylate described above did not result in clinical bleeding episodes. Nevertheless, these and other botanical product–drug interaction reports discussed in this chapter underscore the limitation of available evidence based on case reports and extrapolation of relevance to other botanical products, which could be confounded by patient-specific variables, details of individual report, as well as variability in the content of active constituents among different botanical products. These limitations of interpretation and extrapolation are further challenged by unique ways of prescribing and preparing Chinese botanical products by Traditional Chinese Medicine (TCM) practitioners and patients, respectively. These will be discussed in the following sections.

Specific Issues Regarding Evaluation of Chinese Plant Product—Drug Interactions

Prescribing vs. Over-the-Counter Use of Chinese Plant Product

Using traditional and acceptable ways of reviewing and analyzing the drug- drug interaction literature, it is tempting for clinicians to report a case and/ or review the literature of botanical product–

drug interaction based on the available information for an individual botanical product. However, many herbal remedies used by consumers contain multiple herbs, e.g., the Chinese botanical product quilinggao reviewed above, Ping Wei San, the Chinese medicine used for the management of gastrointestinal disorders, and the different Kampo medicine (traditional Chinese botanical prescriptions) available in Japan. Different constituents within a botanical formula or remedy could have multiple effects on an individual constituent that range from augmenting to antagonizing its intended effect, thereby posing limitation on the usefulness of research or report pertaining to an individual botanical constituent.

In contrast to over-the-counter use by consumers, Chinese botanical products prescribed by TCM practitioners or herbalists are usually in the form of a formula combination, designed to enhance or reduce the effects of different botanical products. Indeed, it is a common knowledge that TCM practitioners and herbalists prefer prescribing Chinese botanical products in the form of "*raw herbs*" or raw plant materials rather than fixed-formula products, so that modification to a formula can be used to achieve the desirable therapeutic effect with minimal adverse outcome for a specific patient. TCM practitioners and herbalists have contended that by using a balanced combination of Chinese botanical products and by taking a therapeutic approach that tailor to a patient's holistic needs, i.e., his or her physical and psychological loss of balance, the Chinese botanical products have minimal potential to interact with Western drugs. Obviously, this knowledge of appropriate combination of Chinese botanical products would not play a role in the consumer's choice of botanical remedies purchased over the counter. In addition, the TCM practitioner usually assesses the patient on a regular basis, and adjustment is then made to the ingredient within the formulation. In this regard, it is not much different from warfarin dose adjustment by a clinician based on the patient's INR and clinical status.

By the same token, not all Chinese botanical products are compatible with each other. Classic Chinese herbal texts have mentioned 18 Incompatibles and 19 Counteractions. The 18 Incompatibles refer to a classic list of 18 botanical product–botanical product interactions, whereas the 19 Counteractions list 19 botanical product combinations in which the effect of one botanical product counteracts that of the other. Given this complexity of modified effect, self-administration of Chinese botanical products without consultation with health care providers or TCM practitioners likely poses more risk than benefit. In addition, for conventional drugs, the magnitude of interaction is mostly dose related or concentration related. Undoubtedly, the dosage of the interacting constituent, whether known or yet to be identified, could vary from one manufacturer to another. Therefore patients switching between different Chinese botanical products might have different outcomes, according to the botanical product– drug interaction, as discussed above for the patient who experienced an apparent interaction between quilinggao products and warfarin.

Preparation of Chinese Herbal Medicine for Consumption

After the prescription is filled and taken home, the Chinese botanical products or formula are usually prepared prior to consumption. In contrast to an oral dosage form of a synthetic drug taken orally in its entirety, the raw botanical products can be ground and taken directly. More commonly, the botanical product or formula of multiple botanical products can be prepared either as a hot water decoction (extraction) or as a 35% to 45% alcoholic infusion. The decoction method of preparation involves boiling the botanical products in about 500 to 600 mL of water until the volume is reduced to one-half and drinking the supernatant of the resulting concentrate or "soup." The infusion method of preparation involves immersing the botanical products in liquor (usually ethanol) for a period of time and then drinking the supernatant. Because it is known that allicin, the major component of garlic, is destroyed when garlic is cooked in oil, how the boiling process would affect the metabolic activity of major constituents has not been studied systematically. Interestingly, Guo et al. had shown that in vitro CYP3A4 inhibition by seven botanical products was consistently greater with the infusion method of preparation. This was true regardless of the lots or geographic locations of the source of the botanical products. Therefore, this represents an additional complexity in evaluating the botanical product–drug or botanical product–botanical product interaction and the need for inquiring the method of preparation during interview of patients regarding their use of botanical products.

Chinese Fixed-Botanical Formulations

Commercial Chinese fixed-botanical formulations are manufactured and marketed in various dosage forms. It is usually not known to what extent individual botanical constituent(s) would be chemically altered during the manufacturing process, regardless of whether the final formulation contains a standardized amount of an active constituent. In addition, commercially available products might differ in their formulation of constituents. The case report by Page and Lawrence listed different over-the-counter dong quai–containing botanical supplements that are available in the United States. Some of these supplements contain not only dong quai, but also multiple botanical products such as ginger, licorice root, or Siberian ginseng. The patient described in their report took Nature's Way PMS formula, which contains dong quai, cramping bark, chaste tree berry, licorice root, and ginger, in addition to folic acid and several vitamins.

Although specific ingredient information may not be easily available or apparent in case report or literature review, it is important to take into consideration the multiple ingredients within a commercially available product or a Chinese botanical formula, when reporting cases of botanical product–drug interaction, so that clinicians can come to appropriate conclusions regarding the significance of the interaction and/or extrapolation of the result to other products. Another good example of this attempt to report constituents within a botanical product or formulation is the case of the quilinggao–warfarin interaction discussed above.

The fact that most Chinese herbal medicines are complex mixtures of multiple active constituents further complicates the interpretation of study data, as well as extrapolation to other botanical products. Japanese Kampo (traditional Chinese herbal mixtures) prescriptions have been used for many years to treat different chronic conditions and are presently manufactured in Japan as drugs with standardized quantities and qualities of constituents. Homma et al. evaluated the effect of three commonly used Japanese Kampo prescriptions, Sho-saiko-to, Saiboku-to, and Sairei-to, on prednisolone pharmacokinetics in humans. All three botanical prescriptions contain glycyrrhizin, a strong inhibitor of 11-β-hydroxysteroid dehydrogenase. Chen et al. had shown that glycyrrhizin decreased plasma clearance and increased AUC and concentration of prednisolone.

However, even though glycyrrhizin was present in all three Kampo prescriptions, Homma et al. reported differential effect with respect to changes in prednisolone pharmacokinetics. Concurrent administration of Sho-saiko-to resulted in a 17% decrease in prednisolone AUC. On the other hand, coadministration of Saiboku-to resulted in a 15% increase in prednisolone AUC. Sairei-to administration resulted in no appreciable change in prednisolone pharmacokinetics. Similarly, Sho-saiko-to increased the prednisone to prednisolone ratio, which reflects 11-β-hydroxysteroid dehydrogenase activity, whereas the ratio was decreased in the presence of Saiboku-to and not changed by administration of Sairei-to. Sho-saiko-to contains seven botanical products with glycyrrhizin being one of them, whereas a total of 12 botanical products including glycyrrhizin are present in Sairei-to, although the relative amount of glycyrrhizin differs between these two botanical products.

These results suggest that botanical prescriptions containing higher glycyrrhizin content or constituents other than glycyrrhizin might be responsible for this differential effect on prednisolone pharmacokinetics and 11-β-hydroxysteroid dehydrogenase activity. The report by Wong and Chan on warfarin–quilinggao interaction also illustrates these two limitations: the presence of multiple active or interacting constituents and the often present variation in the composition of the constituents between different manufacturers. In this regard, although Sho-saiko-to and Sairei-to were shown to not alter the pharmacokinetics of the quinolone ofloxacin in seven healthy volunteers, it remains to be determined whether other Kampo prescriptions such as Saiboku-to would produce the same negligible effect.

With an increasing number of consumers using traditional Chinese herbal medicines, mostly without the advice of health professionals or TCM practitioners, the likelihood of Chinese botanical product–drug interactions is potentially high. To date, the number of interaction reports remain relatively low and, fortunately, few cases reported adverse clinical outcome in patients. However, the low prevalence of interaction simply might reflect a lack of recognition of the interaction potential, scant information from the primary Chinese literature, insufficient number of patients taking Chinese botanical products and potent Western drugs at the same time, or a combination of these factors. As demonstrated by the numerous interaction reports involving warfarin, it is important for clinicians to inquire patients specifically about their use of botanical products, which most do not necessarily disclose during patient interview or medication review. Likewise, to understand further the magnitude and clinical significance of potential or reported interactions, it is important to have more pharmacokinetic and pharmacodynamic studies conducted with quality botanical products in healthy volunteers and/or patients.

3

Postmarketing AERS

St. John's wort is a member of the genus *Hypericum*, which has 400 species worldwide in Europe, West Asia, North Africa, North America, and Australia. In the Western United States, the use of St. John's wort is especially prevalent in Northern California and Southern Oregon. The commercially available product contains hypericum dry extracts or their by-products prepared from flowers gathered during the time of blooming or from dried parts above ground.

In modern European medicine, St. John's wort extracts are included in many over-the-counter and prescription drugs for management of mild depression, and have clinical implications for bed-wetting and nightmares in children. The extracts are included in diuretic preparations and the oil is taken orally using a teaspoon to help heal gastritis, gastric ulcers, and inflammatory conditions of the colon. The oil is also used extensively externally in burn and wound remedies.

Recent reports in animal studies by Rolli et al. and Muller et al. show that clinically used hypericum extract inhibited the synaptic reuptake of 5-hydroxytriptamine (5-HT), noradrenaline, and dopamine with an inhibition concentration at 50% (IC_{50}) around 2 μg/mL. The bioactive substance responsible for the inhibition is identified as hyperforin, from a study by Muller et al. The effect of hypericin as an inhibitor of MAO-A has not been confirmed; however, other ingredients such as flavonoid aglycone, quercetin, and quercitrin have been shown to inhibit MAO-A.

Overall significant benefits of St. John's wort for mild depression compared to a placebo, or equivalent efficacy compared to tricyclic antidepressants (maprotiline, imipramine, and amitriptyline), in mild-to-moderate depression have been reported. However, cautious interpretation of these studies is warranted due to methodological weaknesses. Most tested preparations varied in several major ingredients. An advantage hypericum has over other antidepressants is its favorable side-effect profile. Hypericum has been shown to be well tolerated in patients with the incidence of adverse reactions similar to that of a placebo. The most common adverse effects reported after short-term therapy are gastrointestinal symptoms, dizziness/confusion, and tiredness/sedation.

Drug Interactions with St. John's Wort

There were multiple official regulatory warnings regarding the risk of increased drug levels of CYP3A4 substrates as a result of interactions with St. John's wort. For example, the U.S. Food and Drug Administration (FDA) published a Public Health Advisory in 2000, alerting about the risk of drug interaction with indinavir, antiretroviral agents, and other drugs used to treat heart disease, depression, seizure, certain cancers, transplant rejection, and oral contraceptives. The European Agency for the Evaluation of Medicinal Products (EMEA) issued a Public Statement in 2000, on the risk of drug interactions between *H. perforatum* (St. John's wort) and cyclosporine, digoxin, oral contraceptives, theophylline, warfarin, and antiretroviral medicinal products such as protease inhibitors (PIs) and non-nucleoside reverse transcriptase inhibitors such as zidovudine, didanosine, and zalcitabine. The Australia Therapeutic Goods Administration published a Media Release in 2000 on interactions with indinavir, cyclosporine, warfarin, digoxin, theophylline, PIs, HIV non-nucleoside reverse transcriptase inhibitors, anticonvulsants, oral contraceptives, nefazodone, selective serotonin-reuptake inhibitors (SSRIs) such as citalopram, fluoxetine, fluvoxamine, paroxetine, sertraline, and antimigraine drugs. The Canadians marketed St. John's wort products as food, without health claims. The Irish Medicines Board subjected St. John's wort to prescription control. New Zealand's Medsafe issued a media release statement for St. John's wort interactions with antiepileptic drugs, PIs, immunosuppressive agents, antidepressants, antimigraine drugs, and oral contraceptives.

The FDA's Adverse Event Reporting System (AERS) at the Center for Drug Evaluation and Research (CDER) is an electronic database that currently contains over three million reports of suspected drug-related adverse events, both serious and nonserious outcomes, from all marketed drugs since 1969, which have been submitted to the agency. The reports are initiated on a voluntary basis from both United States and foreign sources that include both health care professionals and consumers and submitted to manufacturers or directly to the FDA. It is important to note that AERS reports are usually of variable quality and completeness and do not necessarily imply a direct causal relationship between drug exposure and the adverse event(s). In some cases, an analysis of such reports may suggest that they are the consequence of the treated underlying disease, other concurrent medical conditions, and/or concomitant medical product treatment. In addition, due to the voluntary nature of reporting, it is not possible to determine the actual incidence of drug-related events or determine the actual degree of risk associated with drug usage. Moreover, due to differences in reporting of adverse events for different type of drugs and other factors affecting reporting, a quantitative comparison of risk between products is highly problematic.

Up to 2001, AERS indicated up to 39 case reports of possible drug interactions between St. John's wort and a prescription drug. In these case reports, the potential drug interactions occurred mostly with oral contraceptives, antidepressants, cyclosporine, and sildenafil. All cases were reported between 1997 and 2000. Most of the reported cases were in females, with an age range between 17 and 73 (mean 42.5) years of age. Four reported hospitalization as a serious outcome. Examples of reported drug interactions with St. John's wort are summarized below.

Cyclosporine

Coadministration of St. John's wort with cyclosporine has resulted in a significant reduction in cyclosporine concentrations, which has led to graft rejection. Decreased cyclosporine drug levels were reported in five cases. Two cases from the United States, one from Australia, and two from Switzerland reported decreased cyclosporine levels or decreased therapeutic response while on St. John's wort concomitantly. The age ranges and doses of cyclosporine used in these five patients, where the data was reported, were 26 to 62 and 200 to 250 mg, respectively. Dose of St. John's wort was 300mg in one case and 600 mg in another, but unspecified in three cases. Time to onset ranged from two to seven weeks after St. John's wort administration, and was unspecified in one case. The Australian case reported that cyclosporine levels returned to within normal range two weeks after stopping St. John's wort. The two Swiss cases documented endomyocardial rejection with concurrent St. John's wort therapy.

Oral Contraceptive Hormones

The metabolism of the components of oral contraceptives, ethinyl estradiol and norethindrone, is thought to be mediated at least in part by intestinal and hepatic CYP3A. St. John's wort significantly increased the oral clearance of norethindrone and decreased the peak serum concentration of norethindrone. Likewise, the elimination half-life of ethinyl estradiol was significantly reduced, therefore potentially reducing oral contraceptive efficacy or even failure. The most frequently reported hormones possibly interacting with St. John's wort were levonorgestrel in combination with estradiol through increased 3A4 metabolism. There were ten case reports of breakthrough bleeding while Alesse-28 and St. John's wort product were used concomitantly. The bleeding occurred from nine days to four months after Alesse-28 was started. The bleeding continued up to seven days while taking Alesse. Age ranged from 33 to 53 ($n = 7$, mean 41). Two patients were instructed to double the doses of Alesse-28 for an unspecified number of days and the bleeding stopped. One discontinued the use of Alesse-28. One patient became pregnant and had a miscarriage. This subject resumed taking St. John's wort and Alesse-28 and conceived again.

There were three reports of breakthrough bleeding while patients were on norgestimate and ethinyl estradiol (Ortho-Cyclen). A 29-year-old female took Ortho-Cyclen 28 tablets for a few months and experienced moderate breakthrough bleeding and some abdominal pain. She began taking two capsules of St. John's wort 10 days prior to her breakthrough bleeding. A 25-year-old female had been taking Ortho-Cyclen for years. She started taking St. John's wort for 30 days and experienced a lot of breakthrough bleeding. She is led to believe that St. John's wort decreased the efficacy of the oral contraceptive. A female of unknown age taking Ortho-Cyclen experienced breakthrough bleeding 17 days after starting St. John's wort.

There was one report in a 22-year-old female of irregular menses and unintended pregnancy while on levonorgestrel and St. John's wort with unspecified dose or indication. A 32-year-old female developed PMS symptoms, breakthrough bleeding, and unintended pregnancy while taking St. John's wort about the same time that she was taking levonorgestrel and ethinyl estradiol.

Antidepressants

For venlafaxine, fluvoxamine, and fluoxetine, the most frequently reported adverse events while on St. John's wort were hypertension and potential serotonin syndrome. There were no reports of lack of effect for these SSRIs, although St. John's wort may decrease the drug levels by inducing 3A4. Because St. John's wort inhibits reuptake of serotonin, noradrenaline, dopamine, and MAO, these events were likely associated with increased serotonin and/or adrenaline levels. Possibly under a similar mechanism, addition of St. John's wort to the MAOI phenelzine was associated with hypertensive crisis in one case, and with mild serotonin syndrome in another patient taking the tricyclic, doxepine. It is unclear whether sertraline had any drug interaction with St. John's wort, although four cases reported depression or intermittent ineffectiveness. According to the labeling, sertraline goes through extensive first-pass *N*-demethylation and the extent of 3A4 inhibition by sertraline is not likely to be of clinical significance.

One case from the United States and one from the United Kingdom reported hypertension and manic reaction, respectively, while on venlaflaxine and St. John's wort concomitantly. The case of hypertension from the United Kingdom was a 56-year-old male who had taken venlafaxine 300 mg daily for management of depression. Prior to consuming St. John's wort, the patient's blood pressure readings were 120/82 mmHg and five weeks after St. John's wort blood pressures were elevated at 180/115 mmHg and 165/112 mmHg. A 29-year-old male received Effexor 150mg for the treatment of dysrhythmia. He decreased the dose and started taking St. John's wort, three tablets daily or every other day, without his prescribing physician's knowledge. He experienced a hypomanic episode that was described as sleeping poorly and feeling "wired" for two days with a lot of "energy" and inability to relax or calm down. A literature report from France described "*serotonin syndrome*" experienced by a 32-year-old male patient as malaise with anxiety, excessive sweating, chills, and tachycardia four days after St. John's wort therapy while on venlafaxine. St. John's wort dose was interrupted on day 4, and the symptoms regressed in three days without modifying the dosage of the antidepressant.

There were two cases of hypertension from the United States, or possible serotonin syndrome reported with fluvoxamine while on St. John's wort concomitantly. A 44-year-old male with obsessive-compulsive disorder received fluvoxamine and experienced severe hypertensive crisis (160–170/120 mmHg) after two tablets of St. John's wort. The physician stated that the reaction was probably due to the combination of fluvoxamine and St. John's wort, which has MAOI activity. A 38-year-old male was on fluvoxamine for approximately two months and hypericum 600 mg daily for approximately two weeks before reporting possible serotonin syndrome with severe bitemporal headache. He was hospitalized to rule out myocardial infarction. There were no electrocardiogram (EKG) changes or apparent causative pathology. Symptoms resolved on discontinuation of both drugs.

A 73-year-old female was treated with fluoxetine for depression for a long period of time and had a history of hypertension managed with multiple concomitant drugs: digoxin, enalapril, aspirin, isosorbide, amlodipine, carvedilol, metformin, and furosemide. The patient was treated with hypericum extract 425 mg for one time. Half an hour to one hour after the first dose, the patient experienced a hypertensive crisis (270/130 mmHg) during her stay in a rehabilitation center. The event was treated with nifedipine and abated (with blood pressures decreasing to 160/96 mmHg).

Five cases (ages 36, 48, 48 and 60 and one unknown; three males and two females) reported intermittent ineffectiveness with sertraline, including complaints of "does not seem to be working," anxiety attack, or worsening depression. In four cases, symptoms occurred after sertraline was added to the continuing St. John's wort therapy. In contrast to reports with other antidepressants, these cases did not report hypertension or possible serotonin syndrome. It is uncertain if the occasional events were possibly associated with the patients' unstable psychiatric status following sertraline therapy, or due to potential sertraline-related adverse events.

Potential drug interaction between sertraline and St. John's wort cannot be ruled out in one case that experienced manic depressive disorder symptoms one to two weeks after St. John's wort was started into sertraline therapy. The patient was treated with an antipsychotic and has had no problems after discontinuing St. John's wort and decreasing the sertraline dose.

Sildenafil

Four cases of lack of effect or impotence were reported in patients using sildenafil while on St. John's wort and other concomitant drugs (what are they? Are any of them significant from the standpoint of drug interaction?). The age range of the four male patients was between 55 to 73 years. Viagra doses were all 50 mg p.r.n. The St. John's wort dose was 600 mg daily in one case but unspecified in the other three cases. One case reported that Viagra did not work, but provided no additional details. Another two cases experienced facial flushing, headache, and ineffective Viagra treatment. One case indicated that Viagra 50 mg was used several times with only partial erection. The dose was increased to 100 mg with similar results. The fourth case summarized below had no other concomitant drug listed and indicated that the Viagra worked without St. John's wort, but did not work when St. John's wort was taken.

A physician reported that a 60-year-old male started sildenafil 50 mg while he was also taking 600 mg of St. John's wort daily for depression. When the patient increased the dose of St. John's wort to 1200 to 1800 mg daily for unknown reasons, the sildenafil was reported to be partially effective. Patient increased the dose of sildenafil to 100mg but it was completely ineffective. The physician suspected that a drug interaction caused the adverse events. No other significant medical history was noted.

Anticonvulsants—Carbamazepine

There were two reported cases with carbamazepine. One case was that of a 17-year-old female who reported increased levels of carbamazepine following three months of St. John's wort and carbamazepine 200 mg b.i.d. with a baseline level of 4.7 μg/mL. She became nauseated with flu-like symptoms on 12/5/98. After experiencing a seizure, dizziness, and disorientation the next day, she was hospitalized with a carbamazepine level of 36 μg/mL.

Another case was a female who reported with complaints of increased incidence of "*muscle twitching*" episodes during daytime hours. These included "slapping leg and turning head to right." Patient was on valproic acid and carbamazepine and St. John's wort was started 50 days prior to the occurrence of adverse events. No carbamazepine levels were reported. Because carbamazepine is also a CYP450 3A4 inducer, the role of St. John's wort is not clear from these two cases because only one case reported increased carbamazepine levels.

The available case reports in the FDA AERS support the published literature that there are pharmacokinetic interactions between St. John's wort and CYP3A4 and/or *p*-glycoprotein substrates, such as cyclosporine, levonorgestrel/estradiol and sildenafil, and pharmacodynamic interactions with the SSRIs or MAOI. Subsequent clinical studies including those conducted via a CDER clinical pharmacology research cooperative agreement provided mechanistic basis of many of these interactions.

Postmarketing Reporting Systems

In the United States, the use of products, including botanicals, thought to fall within the realm of complementary and alternative medicine is very common. It is difficult to obtain reliable estimates of use or to compare many of the current publications in this area because of diverse definitions for categorizing these products (e.g., dietary supplement, food supplement, herbal medicine, natural remedy, traditional medicine, etc.) in both the United States and elsewhere. A recent report on the use of complementary and alternative medicine by U.S. adults in 2002 indicated that approximately 19% of the population used "nonvitamin, nonmineral, and natural products," 19% used folk medicine, and 3% used megavitamin therapy in the past 12 months. All of these types of products are collectively referred to as "complementary and alternative health products" (CAHP). The regulatory classification of individual products, even those containing the same or similar botanical ingredients, may be different and can influence the safety of the product as is briefly discussed below.

Regulatory Classification of Plant Products

Botanical products, including those containing herbs, may be marketed as foods, dietary supplements, or drugs in the United States. Claims made by the manufacturer, particularly on the product label and labeling (information accompanying the product), determine how a product is regulated in the United States, and not necessarily what ingredients the product contains, how the doctor prescribes it, or how the consumer uses it. A product's regulatory classification is important because it determines what safety and effectiveness standards apply, the types of data needed to make this determination, and who makes the determination.

Products that make claims on their labels or labeling that state or suggest that it can treat, cure, prevent, mitigate symptoms, or diagnose a disease are drugs [prescription and over-the-counter (OTC) drugs] in the United States. Drugs must be shown to be safe and effective prior to marketing, and the Food and Drug Administration (FDA) makes these determinations. There are very specific requirements for drug manufacture and marketing. These include factors such as the purity, potency, and formulation of the ingredients in the finished drug, and the kinds of information that can or must appear on the product label (e.g., product claims, safety information, or warnings). Conventionally, foods are items ingested for flavor, taste, aroma, or nutrition. The standards for foods primarily involve the safety and suitability of the food to meet nutritional needs, rather than safety and effectiveness as a form of treatment. Furthermore, the current standards for manufacturing or holding foods are mainly sanitation standards.

In the United States, dietary supplements are a special category of foods as defined under the Dietary Supplement Health and Education Act of 1994. Unlike prescription and OTC medicines, dietary supplements are not reviewed by the FDA before marketing. Manufacturers also do not need to register before producing or selling their products. Manufacturers of dietary supplements are legally responsible for assuring the safety of their marketed products, and the FDA has the responsibility to take action against unsafe dietary supplement products after they reach the market. Except in the case of a new dietary ingredient, where the law requires premarket review for safety data and other information, a firm does not have to provide the FDA with the evidence that shows that its product is safe and effective. The FDA intends to publish minimum standards for manufacturing dietary supplements, which will focus on practices that ensure the identity, purity, quality, strength, and composition of dietary supplements. Dietary supplements can make claims about the effect of a product on the structure or function of the body, but may not make "disease" claims.

Worldwide, the majority of commercially available finished medical botanical products are regulated as drugs, although the raw botanicals themselves may be commercially available and fall outside regulatory schemes. In contrast, in the United States, the majority of marketed botanical products with any type of health information are sold as dietary supplements. Despite being labeled as a dietary supplement, botanical products with disease claims are unapproved drugs, because their safety and efficacy for a particular indication have not been proven prior to marketing. Currently there are a few botanicals in OTC drug products, but no prescription drug products in the United States. This situation may ultimately change, because a number of new drug applications on botanical products have been submitted to the FDA.

Safety Concerns Related to Plants

As noted by numerous recent publications, the use of CAHP has increased dramatically in recent years, with echinacea, ginseng, ginkgo biloba, garlic, glucosamine, St. John's wort, peppermint, fish oils/omega fatty acids, ginger, and soy being the most commonly used products for health reasons. Safety concerns related to botanicals fall into two general areas—those related to populations using them and those related to the actual product or its ingredients.

There are a number of reasons to be concerned about the potential safety of such widespread use of CAHP. These products are frequently used by vulnerable populations, including older adults, those with chronic disorders, children, and women during pregnancy and lactation. These products are also used by patients to treat a variety of chronic disorders that are difficult to medically manage (e.g., anxiety, depression, dementia and memory impairment, headache, weight loss, back disorders, chronic pain, prostatic hypertrophy, and cancer). Choice of a particular product for a particular condition is usually based on the claims made for the product and anecdotes of "historical" use, rather than conclusive scientific evidence that establishes the safety and efficacy of a particular product for a particular condition.

Concurrent use of CAHP with prescription medicines is common, with reported frequencies ranging from about 20% to 43%. Less is known about potential interactions with OTC medications, but this too is of concern, particularly with the increasing switch of prescription drugs to OTC status. In addition, as with other types of CAHP, there are issues related to the recognition and monitoring of adverse events related to OTC drug products. Concurrent use of CAHP with OTC and prescription drug products can result in therapeutic failures or adverse events. Although many supplements are commonly advertised as being "natural," this does not make them automatically safer or better than drugs or synthetic ingredients. In many cases, there is much less credible information about the effects of particular natural products or their ingredients, and there is more product variability. Product quality and variability are known safety concerns. Natural products can contain anything found in our environment—including pesticides, bacteria, molds, heavy metals, and other poisons—as has been documented in the literature.

Identification and Evaluation of Potential Product Interactions

Consumers frequently do not tell their health providers about their use of CAHP, health providers often fail to ask about the use of such products, and most of the purchases of these products occur outside a pharmacy—all of these factors enhance the likelihood of adverse product interactions and make detection of such interactions much more difficult.

Identification and evaluation of potential interactions is also difficult because there is a paucity of reliable scientific information about the effects of ingredients in CAHP, and this difficulty is compounded by product- related factors: many of the products are multi-ingredient, and as noted above there can be wide variability in the quality and consistency of ingredients and products. A number of recent reviews have tried to evaluate the credibility of data as it relates to drug–CAHP interactions, but these are limited to the published literature and do not consider information available in various adverse event reporting systems (e.g., FDA systems, Poison Control Centers). Because adverse events are underreported, and even fewer will ever be published in the scientific literature, it is not possible to estimate the true magnitude or significance of the drug–CAPH interactions, which is likely far greater than any current published estimates. Consequently, the results from these studies should be used with caution because the lack of documented cases in the scientific literature does not mean that a particular CAHP is safe or that interactions with drugs have not occurred or will not occur. It is critical, therefore, that health professional be cognizant of the use of CAHP by their patients and be on the alert for potential interactions. Health professionals would also increase the knowledge in this area if they diligently reported suspected adverse events.

Adverse Event Reporting at FDA: Medwatch Program

MedWatch is an umbrella program developed by the FDA to enhance the reporting of serious adverse events by health professionals, which are suspected to be related to the use of FDA-regulated products. Within the FDA, there are many different systems at various levels in the agency, which deal with adverse event reports (AERs), including mandatory (active) and voluntary (passive) surveillance, which may have different infrastructure and system requirements. The particular system utilized depends upon the regulatory authority for the particular product. There is no central system based on the type of ingredients (i.e., botanical); the AER goes ultimately to the center with regulatory responsibility for the particular product. Consequently, more than one center in the FDA may have information about particular product interactions; for instance, an adverse event associated with a dietary supplement might be voluntarily reported by a consumer or the adverse event might be a mandatory report from a drug manufacturer where a CAHP is listed as a concurrent exposure.

The FDA considers postmarketing surveillance, which includes adverse event reporting, as one of the most useful indicators or signals of potential safety problems associated with a product. Although premarketing clinical studies can reveal certain safety problems, a major portion of the information concerning product safety becomes known only after marketing, with widespread use in "real life" situations. Postmarketing surveillance can either be active, such as in mandatory reporting by manufacturers of adverse events, or can utilize more passive systems, including voluntary reporting of adverse events, evaluation of consumer use data, etc. For certain drugs (those subject to the new drug approval process), it is mandatory that the manufacturers report adverse events to the FDA. Additionally, health professionals may voluntarily report adverse events associated with medical products. For other drugs, including many OTC drugs, reporting is currently voluntary.

In general, most systems used to evaluate adverse events associated with foods are passive or voluntary and are in their infancy when compared to the more formal and elaborate pharmacoepidemiologic systems that exist for certain types of drugs. The FDA learns about problems with foods, including dietary supplements, through a wide variety of sources, including the FDA MedWatch program for health professionals' reporting of adverse events. Other reporters of adverse events include consumers, state and local health departments, professional societies, other federal agencies or groups, and industry representatives that contact the FDA via multiple mechanisms (written and electronic correspondence, telephone, etc.).

There have been increasing calls from health professionals and certain members of Congress to require dietary supplements manufacturers to report serious adverse events to the FDA. The recent Institute of Medicine Report on dietary supplement safety also recommended that Congress amend the Dietary Supplement Health and Education Act (DSHEA) to require manufacturers and distributors to report to the FDA in a timely manner any serious adverse events associated with the use of its marketed products.

Adverse Events and Risk Management

The FDA plays an important public health role in the identification and management of health risks associated with the use of products that it regulates. An important function of postmarketing surveillance systems is signal generation (i.e., the identification of new or emerging health risks). Signals from adverse event information become apparent in a variety of ways. These include the emergence of a specific pattern of signs and symptoms, which is occurring with the use of a particular product or ingredient, an increasing number of adverse events or a change in the pattern, seriousness, or severity of adverse events observed with the use of a particular product or ingredient, or the occurrence of an adverse event that is unexpected with the use of a particular product. These systems are important because they can provide data not found or available prior to marketing on adverse effects seen in special groups, adverse effects that occur with relative infrequency, and adverse effects that develop with chronic use or exhibit latency.

For all the recognized advantages of using adverse event reporting as a component of pharmacovigilance of drug interactions, including those occurring with botanical products, there are also well-recognized limitations to current reporting systems, which are generally "passive" in nature. These include substantial and unquantified underreporting, frequent incomplete or inaccurate information in submitted reports, lack of exposure data, inability to detect adverse events with a long latency period, and the absence of a control group for the specific exposure.

A number of variables influence the likelihood of an adverse event being reported. These include the length of time that a product has been marketed, the market share, experience and sophistication of the population using the product, and publicity about adverse events. Currently there is little incentive for health professional reporting of adverse events, which partially underlies the problem with underreporting. Lack of exposure data and the issue of underreporting preclude estimation of incidence rates. Causality assessment is difficult or impossible because of the quality of the data received and the lack of a comparator (control) group. Finally, comparisons of product safety cannot be directly obtained from adverse event data.

When signals become apparent from the routine review of data in an adverse event reporting, additional elements of risk assessment are implemented, generally on a case-by-case basis, to provide the agency with adequate information to appropriately manage any public health risks. These additional elements may include clinical or scientific evaluation of all adverse events reported as associated with a particular product or ingredient; market surveys to gather information on product use (directions for use, warnings, populations using product, etc.); sample analyses, where appropriate, to identify particular substances in a product, the amount of a substance, etc.; independent scientific and clinical reviews from scientists in other Centers, or outside the agency; and expert scientific advisory committees.

This information serves as the basis of any FDA actions to mitigate risk. These actions, depending on the nature and severity of the identified risks, may include changes in the product's warnings or directions for use, education of the public (consumer, health professional, industry), or withdrawal of the product from the market. Because of the very limited amount of information that is available to the FDA at premarketing or first marketing, the majority of the FDA's efforts related to dietary supplement safety are focused in the postmarketing period. Any efforts to improve the safe use of CAHP, therefore, will include mechanisms to improve the type, quality, and availability of data that are available on these products. Such efforts could include more centralized electronic databases for scientific data, including that obtained from postmarketing surveillance of adverse events. Because we are in an era of limited resources, such efforts will require the coordinated efforts of federal agencies, academia, other public health groups, and industry to be successful.

4

STANDARDIZATION IN PLANT PRODUCTS

The popularity of botanical products in the United States is reflected in a survey on complementary and alternative medicine that showed that American consumers had spent an estimated $5.1 billion on botanical products in 1997. In the same year, the global market for botanical medicinal products was estimated to be approximately $20 billion. It has been estimated that currently more than 1500 botanical products are available in the U.S. market alone. This popularity has been fueled, in part, by the perception that botanicals are naturally derived products, and hence are safe and devoid of adverse effects. This perception appeared to be justified by a paper summarizing the fatality of pharmaceutical drugs and botanical products in the 1981–1993 period, in which statistics compiled by the National Center for Health Statistics, the American Association of Poison Control Centers, Centers for Disease Control and Prevention, the Journal of the American Medical Association, and the U.S. Consumer Product Safety Commission showed an annual mortality rate of 100,000 deaths for pharmaceuticals and none for botanical products.

However, because the information covered was only to the end of 1993, this report did not take into consideration the subsequent fatalities attributed to *Ephedra* and ephedrine products. With the increase in the number of incidents of adverse reactions being reported, a database on adverse reactions of botanical products has been created as part of the World Health Organization (WHO) International Drug Monitoring System. In recent years, it has become increasingly apparent that even therapeutically safe botanical products can manifest toxic effects as a result of botanical product–drug interaction, when administered concomitantly with synthetic pharmaceutical agents. The best-documented examples have been cases involving grapefruit juice and St. John's wort with a variety of drugs. Grapefruit (*Citrus* × *paradisi* Macfad.) juice has been documented to interact with calcium channel blockers as well as to increase the level of cyclosporin in the blood of transplant patients.

St. John's wort (*Hypericum perforatum* L.), a botanical used in the management of mild to moderate depression, has been found to increase the effects of monoamine oxidase inhibitors or serotonin reuptake inhibitors; reduce the blood levels, and hence the pharmacological effects of anticonvulsants (carbamazepine and phenobarbitone), anticoagulants (warfarin and phenprocoumon), oral contraceptives, theophylline, digoxin, cyclosporin, HIV reverse transcriptase inhibitors (nevirapine and efavirenz), and protease inhibitors (indinavir); increase photosensitivity with other such drugs; prolong narcotic-induced sleeping time; and decrease the level of cyclosporin in organ transplant patients. The adverse effects recorded for these and some other botanical products are due to true pharmacological interactions. There are, however, botanical product–drug interactions reported for botanical products that may not be true pharmacological/physiological events, and that can be avoided if in-process quality control

(QC) is in place during the manufacturing process. Botanical products adulterated with synthetic drugs such as phenylbutazone, indomethacin, corticoid steroids, caffeine, acetaminophen, indomethacin, hydrochlorothiazide, ethoxybenzamide, theophylline, diazepam, chlorpheniramine maleate, ibuprofen, phenobarbital, mefenamic acid, prioxicam, salicylamide, diethylstilbestrol, and warfarin have led to botanical product–drug interactions. Multicomponent Chinese or Ayurvedic botanical remedies, known to contain heavy metals such as lead and mercury as active ingredients, can likewise lead to adverse events and/or botanical product–drug interactions.

On the other hand, a number of adverse event reports recorded in the literature are themselves erroneous in nature due to the quality of the assessment and reporting, with Siegel's report of the so-called "*ginseng-abuse-syndrome*" (GAS) being a prime example. In this report, the author simply recorded adverse reactions in patients who had ingested "ginseng" without reference to which of a number of plants having the same common name were actually ingested. Further, the author did not take into account the concomitant pharmaceutical drugs and drugs of abuse used by the patients being reported to have adverse drug reactions to "*ginseng.*"

In monitoring botanical product–drug interactions, the quality of the data obtained may be influenced by a number of factors. Among the important issues to be addressed include raw material source and sourcing practices; intrinsic and extrinsic factors affecting the occurrence and concentration of active or marker chemical constituents in both the starting and finished products; the meaning of the word "*standardization*"; the methods of chemical and biological analyses employed; the manufacturing practices employed; substitution; adulteration of botanical products with pharmaceutical drugs or contamination with foreign toxic substances; formulation of the dosage form; regulatory requirements; and clinical experimental design and data interpretation. In this chapter, the influence of these quality assurance (QA)/QC and standardization issues on botanical product safety will be examined.

Material Quality and Quality Control Issues

The quality of presently available botanical products varies from very high to very low. Our study on selected commercial ginseng products prepared from *Panax ginseng* C.A. Meyer, *P. quinquefolius* L., and *Eleutherococcus senticosus* Max. (Araliaceae) and marketed as botanical supplements in North America in the 1995–1998 period showed that 74% of these products met label claims, with the ginsenoside contents of the *P. ginseng* and *P. quinquefolius* products analyzed ranging from 0.00% to 13.54% and 0.009% to 8.00%, respectively. The eleutherosides B and E content of *E. senticosus* root powder and other formulated products also showed similarly large variations. Studies on the quality of St. John's wort products showed that hypericin content ranged from 22% to 165% and that silymarin content in milk thistle [*Silybum marianum* (L.) Gaertn.] products ranged from 58% to 116% of the labeled claims. These content variations not only will influence the efficacy, but also could affect the safety of botanical products, because chemically induced drug interactions may be active compound–concentration dependent. Why are there such wide variations in the content of active/marker compounds in these products? Are such variations intrinsic to botanical products or are they due to external factors, or both?

Intrinsic and Extrinsic Factors

It is well established that intrinsic and extrinsic factors including plant species differences, organ specificity, diurnal and seasonal variation, environment, field collection and cultivation methods, contamination, substitution, adulteration, processing, and manufacturing practices greatly affect botanical quality. Intrinsically, plants are dynamic living organisms, each of which is capable of being genetically influenced to be slightly different in its physical and chemical characters. For example, a study on the accumulation of hypericin in *H. perforatum* showed that narrow-leafed populations have greater concentrations than the broader-leafed variety; variations of phytochemicals are greater in wild than in domesticated populations of the same species, as exemplified by the results of studies on the content of artemisinin, an antimalarial agent, in *Artemisia annua* L.; on michellamine B, a compound with in vitro anti-HIV activity, in *Ancistrocladus korupensis* D.W. Thomas & R.E. Gereau; and on the essential oil composition of *Ocimum basilicum* L. Also, the secondary chemical constituents of medicinal plants differ qualitatively as well as quantitatively from species to species as demonstrated by the presence of structurally different alkylamides in the roots of *Echinacea angustifolia* D.C. and *E. purpurea* (L.) Moench, and by their total absence in *E. pallida* (Nutt.) Nutt.

Organ specificity is yet another intrinsic factor influencing chemical variation because the site of biosynthesis and the site of accumulation and storage are normally different. Chemical biosynthesis usually takes place in the leaves, and then the product synthesized is transported through the stems to the roots for storage, with the chemical profiles in these organs being different from each other. Accumulation and storage can also take place in the leaves, but to a much lower extent, and very infrequently in the stems. An example of site-specific accumulation, as well as species specificity, is that of the compounds considered responsible for the immunostimulant effect of *Echinacea* species. These compounds encompass five groups of chemicals: caffeic acid derivatives, alkylamides, polyacetylenes (ketodialkenes and ketodialkynes), glycoproteins, and polysaccharides. As indicated above, alkylamides are found in the roots of *E. angustifolia* and *E. purpurea*, but they are structurally different, and are totally absent in *E. pallida* roots. Polyacetylenes, on the other hand, are present abundantly in the roots of *E. pallida*, but are absent in *E. angustifolia* and *E. purpurea* roots. Whereas the glycoproteins and polysaccharides are present in the aerial parts of all three species, they occur only in minute quantities in the roots.

Diurnal variation and seasonal variation are other intrinsic factors affecting chemical accumulation in both wild and cultivated plants. Depending on the plant, the accumulation of chemical constituents can occur at any time during the various stages of growth. In a majority of cases, maximum chemical accumulation occurs at the time of flowering, followed by a decline beginning at the fruiting stage. The time of harvest or field collection can thus influence the quality, efficacy, and safety of the final botanical product.

With respect to extrinsic factors, there are many that can affect the quality of medicinal plants. It has been well established that environmental factors such as soil, light, water, temperature, and nutrients can affect phytochemical accumulation in plants. For example, alkaloid concentrations in *Atropa belladonna* L. have been found to vary from 0.3% to 1.3%, when grown in different areas of the world. Also, the silymarin content in milk thistle was found to be highest in the fruits of plants grown under 60% water/field capacity (1.39%) and nitrogen level of 100 (1.46%) per acre. The methods employed in field collection from the wild, as well as in commercial cultivation, harvest, postharvest processing, shipping, and storage can also influence the physical appearance and chemical quality of the botanical source materials. Contaminations by microbial and chemical agents (pesticides, herbicides and heavy metals) as well as by insects, animals, animal parts, and animal excreta during any of the stages of source plant material production and collection can lead to lower quality and/or unsafe source materials. Heavy-metal contamination can occur at the cultivation, postharvest treatment, or product-manufacturing stages. Lead and thallium contaminations have been reported in multicomponent botanical mixtures, and cases of lead, thallium, mercury, arsenic, gold, and cadmium poisoning from the consumption of such products have been documented.

Botanical source materials collected in the wild often include non- targeted species either by accidental substitution or by intentional adulteration. However, substitution and adulteration of cultivated botanicals can also occur. Substitution of *Periploca sepium* Bunge for *E. senticosus* (eleuthero) has been widely documented and is regarded as being responsible for the "*hairy baby*" case involving maternal/neonatal androgenization. Adverse reactions due to plantain (*Plantago ovata* Forskal) being contaminated by *Digitalis lanata* Ehr. during harvest is another example of accidental adulteration by human error.

Adulteration of botanical products with synthetic drugs represents another problem in product quality and botanical product–drug interactions. Foremost among the documented cases are multicomponent Chinese or Ayurvedic botanical remedies. Chemical analyses of some arthritis remedies have led to the finding that synthetic anti-inflammatory drugs such as phenylbutazone, indomethacin, and/or corticoid steroids have been added. In a classic study of chemical adulteration of traditional medicine in Taiwan, 23.7% (618 of 2609) of botanical remedy samples collected by eight major hospitals were found to contain one or more synthetic therapeutic agents, including caffeine, acetaminophen, indomethacin, hydrochlorothiazide, prednisolone, ethoxybenzamide, phenylbutazone, betamethasone, theophylline, dexamethasone, diazepam, bucetin, chlorpheniramine maleate, prednisone, oxyphenbutazone, diclofenac sodium, ibuprofen, cortisone, ketoprofen, phenobarbital, hydrocortisone acetate, niflumic acid, triamcinolone, diethylpropion, mefenamic acid, prioxicam, and salicylamide. The most frequent adulterants were caffeine, acetaminophen, indomethacin, hydrochlorothiazide, prednisone, and chloroxazone. Obviously, such adulterated botanical products are prime candidates for botanical product–drug interactions.

Besides the unintentional in-process adulteration of heavy metals, it is well established that Ayurvedic medicine and traditional Chinese medicine sometimes employ complex mixtures of plant, animal, and mineral substances, and it is not uncommon to find appreciable quantities of heavy metals such as lead, mercury, cadmium, arsenic, and gold in certain formulations.

With respect to the words/claims, "*active compound*," "*marker compounds*," "*standardization*," and "*standardized products*," the clinician should be vigilant about their meaning when monitoring botanical safety. In the case of prescription and over-the-counter (OTC) drugs, each product has a single, defined "active" chemical constituent, which is used to measure or standardize product quality and determine shelf life. The active principle(s) of botanical products/dietary supplements, on the other hand, are largely unknown. For example, there is no evidence that the marker compounds eleutherosides B and E are the active principles in eleuthero (*E. senticosus*). Presently, there is also considerable disagreement as to whether hypericin or hyperforin is the active antidepressant principle in St. John's wort, with the latter being the current leading candidate.

Further, even when the active principles of a medicinal plant have been identified, the compound may not be commercially available for use as a reference standard. Hence, major constituent(s) of the source plant, whether biologically active or not, are currently employed as marker compounds for the standardization of most of the botanical products marketed, so that one manufacturer may not use the same reference standard as another. Compounding the issue of standardizing is the meaning of a "*standardized extract*," which may refer to (i) an extract made to a consistent standard such as a ratio of the starting plant material to that of the dried extract, (ii) an extract manufactured to contain a specific concentration of a marker compound(s), or (iii) any one of a number of botanical, agricultural, and/or manufacturing process control measures in the production of a material of reasonable consistency. With these inconsistencies in the meaning of standardization and standardized products, variations in efficacy and adverse events/drug interactions can, and will, occur.

Regulatory Influence

Botanical product quality and safety can also be influenced by regulatory status, which varies from country to country. In some countries, botanical products are regulated as medicine and are subject to mandated standards of quality, whereas in the United States a majority of botanicals are marketed as dietary supplements. Good manufacturing practices (GMPs) are required in the production of prescription and OTC drugs, but the regulatory provisions under the Dietary Supplement Health and Education Act (DSHEA) of 1994 provide little assurance of identity, quality, or purity for botanical dietary supplements. Thus, botanical dietary supplement products have not been subjected to mandated QA/QC standards as in the case of prescription and OTC drugs. Although the Food and Drug Administration advanced a notice of proposed rulemaking on current good manufacturing practice in the labeling and manufacturing standards on dietary supplements in March 2003, such standards have not yet been implemented. Elsewhere in the world, e.g., in the European Union and in most of Asia and Southeast Asia, national policies exist, but in some countries, these products are totally unregulated. Consequently, product quality, efficacy, and safety differ internationally, nationally, and from product brand to product brand, and even from lot to lot within the same brand.

Quality Assurance and Quality Control

For effective monitoring of botanical product–drug interaction in clinical studies or application, the clinician must be aware that standards of quality for botanical products do not exist in many countries, including the United States. Therefore, the products being evaluated must be accurately defined as to the quality of botanicals employed, and information on the QC measures employed to ensure their quality must be taken into consideration. If such QC/QA information, including standardization and what is meant by the term, is lacking, it is not possible to attribute the drug interactions observed to the botanical product in the clinical study/use, and the data being published will be invalid and/or misleading.

Information on QC of the botanical product under investigation or in clinical use must be derived from measures taken, from the procurement of source material to the production of the final formulation. Whether by field collection from the wild or by cultivation, good agricultural and/or collection practices must be adhered to during the procurement process, because the quality of the finished botanical products is obviously directly related to the quality and safety of the raw materials. Hence, whether field-collected or produced by cultivation, the identification and authentication of plant species by a taxonomic botanist is critical to ensuring that the correct source material is acquired. It is essential that the plant materials are identified by their scientific names (Latin binomial), and a description of the macroscopic, microscopic, and organoleptic (sensory) characters be provided along with herbarium specimens, drawings, or photographs. In the field collection of medicinal plants, care must be exercised to avoid the acquisition of nontargeted species and to free the targeted source material of undesirable plant parts, soil, rock, insects, animals, animal excreta, and other contaminants. Postcollection treatments should mirror those accorded cultivated plant materials.

Due to their genetic and chemical content variations, the site and date should be recorded for each collection. The production of raw materials by cultivation should normally lead to more uniform botanical products due to greater genetic uniformity. The production of quality raw materials can only be assured by employing good agricultural practices such as those described in the recently published WHO Guidelines on Good Agriculture and Collection Practices. The harvested source materials must be processed to produce the finished products under GMPs. GMP procedures employed for the manufacture of botanical products involving, at the raw material production end, botanical taxonomic identification to assure species identification must be implemented. Otherwise the efficacy, safety, and botanical product–drug interaction reported for one medicinal plant may in fact be those caused by another botanical product. It should be noted that although common names are most frequently used by the source material producers/collectors, a common name may apply to more than one plant. For example, "ginseng" may refer to American ginseng [*P. quinquefolius* L. (Araliaceae)], Asian/Korean ginseng [*P. ginseng* C.A. Meyer (Araliaceae)], Russian/Siberian ginseng [*E. senticosus* (Rupr. & Maxim.)

Maxim. (Araliaceae)], Blue ginseng [*Caulophyllum thalictroides* L. Michx. (Berberidaceae)], Brazilian ginseng [*Pfaffia paniculata* Kuntz (Amaranthaceae)], Indian ginseng [*Withania somnifera* L. Dunal (Solanaceae)], or Wild Red American ginseng [*Rumex hymenosepalus* (Polygonaceae)]. Thus, the identification of the source material from which the botanical product is being monitored must be by its Latin binomial.

At the processing and manufacturing stage, macroscopic, microscopic, and organoleptic analyses and analytical procedures similar to those employed for the manufacture of conventional drugs to assure quality and purity by appropriate protocols must be used. Otherwise, the quality of the finished product under clinical investigation/use may be compromised, and this can lead to adverse events and/or botanical product–drug interactions. Microscopic and organoleptic examinations will help assure botanical identity and purity because each plant species possesses characteristic microscopic cellular features, and may have distinct sensory properties. Macroscopic examination will reveal the presence of deterioration and signs of contamination by molds, insects, rodents, and other animals, as well as by other plants.

As with pharmaceutical drugs, botanical products should be thoroughly evaluated biologically, employing not only in vitro methods, but also the more relevant in vivo animal studies, particularly with respect to acute and chronic toxicity.

Procedures for the QC analysis of active and/or marker chemical compounds in botanical products during the manufacturing and post- marketing surveillance processes can be accomplished by colorimetric, spectroscopic, and/or chromatographic methods. Colorimetric and direct spectroscopic methods are older analytical procedures that quantify the absorption of structurally related compounds at a specific wavelength of light, expressed as a concentration of a reference standard (marker), which is normally the active or major chemical constituent in that plant material. Because other plant constituents possessing the same absorbance are included in the measurement, a higher concentration is usually ascribed to the test material. The use of these procedures has recently been on the decline. Modern methods for the chemical analysis of secondary chemical constituent markers in botanical products involve some form of chromatography. Thin-layer chromatographic procedures have the advantages of being simple and rapid, and they can provide useful characteristic profile patterns and are inexpensive to use. However, their resolving power is limited and quantitative data for minor constituents is difficult to obtain. Gas chromatography can provide a high resolution of the more volatile complex mixtures, but is of limited value in the case of nonvolatile polar compounds, especially the polar polyhydroxylated and glycosidic compounds.

High-performance liquid chromatography (HPLC) is capable of resolving complex mixtures of polar and nonpolar compounds, and has become the chromatographic method of choice for the qualitative and quantitative analysis of botanical extracts and products. HPLC can be coupled with a range of analytical techniques including ultraviolet (UV) spectroscopy, mass spectrometry, nuclear magnetic resonance (NMR), and evaporative light- scattering detection (ELSD). Combined with HPLC, any of these techniques is capable of producing a "*fingerprint*" of the botanical product. However, some of these detection methods may be inappropriate for the quantitative determination of a specific active or marker compound. The literature is replete with HPLC methods for the analysis of specific compounds in more than 95% of the botanical extracts or products in the market. Detection by UV is readily available in most labs, and is carried out either with a single- or dual-wavelength, or a full spectrum (e.g., photodiode array) detector, and is the most appropriate technique for the routine analysis of compounds that contain a UV-active chromophore. Combined HPLC–mass spectrometry (LC–MS) and liquid chromatography–tandem mass spectrometry (LC–MS^n) is being used increasingly.

The advantage of these methods is that as each compound is being eluted, it is captured by the mass spectrometer and provides a molecular ion and/or major mass fragment, which can provide a specific identification of the eluting "peak." However, ionization techniques compatible with HPLC, such as electro-spray ionization, show a broad range of sensitivity to various compound classes. This technique is excellent for compounds such as alkaloids, phenols, and organic acids, but can be highly insensitive to others such as aliphatic hydrocarbons, sterols, and polysaccharides. All compounds containing protons, including virtually all medicinally significant phytochemicals, can be detected by NMR. This technique generally provides more structural information than any other single technique. However, LC–NMR is available in only a few labs worldwide, requires the use of deuterated solvents for chromatography, and will remain inaccessible and prohibitively expensive for routine use for some time. A vast majority of all plant secondary metabolites are detectable by ELSD, but this method provides no structural information. There can exist no standardization regime that would be universally applicable to all medicinal botanical products. Ideally, the formulated product should be chemically assayed for an active constituent, using an analytical method appropriate for the given compound class, and also biologically assayed for in vitro and/or in vivo activity, using assay(s) relevant to the intended use of the product.

Clinical Experimental Design and Data Interpretation

For effective botanical product–drug interaction monitoring, there is a most critical need for a well-designed clinical experiment that not only takes into account the aforementioned QC issues, but also includes a safety monitoring component that will enable the clinician to delineate between adverse reactions caused by botanical product–drug interactions and those by drug–drug interactions due to concomitant ingestions of multiple pharmaceutical drugs by the patient, or adverse events owing to idiosyncratic causes. A system designed for careful and rational interpretation of study data should be devised so as to avoid erroneous conclusions such as those reported on the so-called GAS. In monitoring botanical product–drug interactions, the quality of the data obtained may be influenced by a number of factors. Among the contributing factors are: raw material source and sourcing practices; intrinsic and extrinsic factors affecting the occurrence and concentration of active or marker chemical constituents in both the starting and the finished products; standardization and standardized products; the methods of chemical and biological analyses; manufacturing practices; substitution; adulteration of botanical products with pharmaceutical drugs or contamination with foreign toxic substances; regulatory requirements; and the quality of the clinical experimental design and data interpretation.

5

Pharmacokinetics of Plant Products

A general disillusionment with conventional medicines, coupled with the desire for a "natural" lifestyle has resulted in an increasing utilization of herbal medicinal products (HMPs) across the developed world. Sales of botanical products in the United States have increased sharply in recent years, according to industry reports. An estimated $4 billion was spent in health food stores in 2000 for botanical products in bulk, as well as capsules, tablets, extracts, and teas. A similar trend is noted for European countries. Many consumers use HMPs in a holistic manner and mainly on the basis of their empirical and traditional applications. The use of HMPs in an evidence-based approach is known as "*rational phytotherapy*," which is in contrast to traditional medical herbalism. To obtain "*rationality*," HMPs must meet acceptable standards of quality, safety, and efficacy. Besides quality and safety issues, establishing the pharmacological basis for efficacy of HMPs is a constant challenge for researchers worldwide. In general, pharmacology can be defined as the study of the interaction of biologically active agents with living systems.

The study of pharmacology can be further divided into two main areas: pharmacodynamics and pharmacokinetics. Whereas in recent years the number of studies investigating the pharmacodynamic effects of HMPs has increased rapidly, there is still limited information available regarding herbal pharmacokinetics. This might be due to the following reasons. The study of herbal pharmacokinetics is extraordinarily complex because HMPs are multicomponent mixtures, which contain several chemical constituents. Therefore, concentrations of single compounds in the final product are in the lower milligram range per dose. The resulting plasma concentrations are often in the microgram per liter to picogram per liter range. As a consequence, analytical methods determining bioavailability and pharmacokinetics of HMPs have to be sufficiently sensitive. Advanced techniques such as gas chromatography–mass spectrometry (GC–MS)/MS or high-performance liquid chromatography–MS/MS can be used nowadays to accomplish these goals. For the majority of these multicomponent mixtures, the active constituents are often unknown. In other words, a substance that is detectable in body fluids is not necessarily the active compound of an extract. Further, the different compounds will have a different bioavailability, thereby complicating the design of pharmacokinetic studies with HMPs. Natural compounds are often prodrugs that are metabolized in the digestive tract. Moreover, HMPs can contain large polar molecules that might be expected to have poor and unpredictable bioavailability.

Bioavailability is defined as the rate and extent of active substances in the blood stream after oral doses. The bioavailability of a substance depends on several factors: the pharmaceutical preparation, the size of the molecule, the fat/water solubility of the compound, factors within the gut, first-pass effects, interaction with food, and individual factors in the patient, such as the influence of pathological factors. Bioavailability of compounds in the plant extract might also be influenced by other components in the mixture, which are not active themselves but can act to improve the stability, solubility, or the half-life time of the active compounds. Some authors divide their components into active and accompanying substances (so-called coeffectors). Coeffectors have an influence on the physicochemical properties of active compounds of an extract and, as a consequence, on their biopharmaceutical parameters.

There are several examples in the literature showing that such coeffectors improve not only the solubility but also the bioavailability of single compounds. Saponines were shown to significantly increase the absorption of corticosteroids, some antibiotics, flavones, phytosterols, and silicic acid. The concentration of kavain and yangonin in mouse brain samples is higher after administration of a Piper methysticum extract than after administration of the purified single compounds in the same amount. Similarly, the oral bioavailability of kavain from an extract of *P. methysticum* is 10 times higher than that of pure kavain. The improved bioavailability of ascorbic acid from a Citrus extract compared to pure ascorbic acid is explained by an increased absorption and an improved stability of vitamin C in presence of several flavonoids contained in the *Citrus* extract. List et al. showed that the transport of L-hyoscyamin from the mucosal to the serosal side of the rat's isolated ileum is increased when a native extract prepared from the leaves of Hyoscyamus niger is used instead of pure hyoscyamin; they suggest unidentified flavonoid glycosides as the responsible compounds.

Unfortunately, most investigations on this topic give few or no information about the mechanism of interaction and, in particular, about the compounds involved. One approach to identify the chemical structure of a coeffector was recently performed by Butterweck et al. Taken together, although the study of herbal pharmacokinetics appears to be difficult, the information derived from such investigations will become an important issue to link data from pharmacological assays and clinical effects. In particular, a better understanding of the pharmacokinetics and bioavailability of natural compounds can help in designing rational dosage regimen; and it can help to predict potential botanical product–drug interactions. In addition, those studies would provide supporting evidence for the synergistic nature of herbal medicines and would further help in optimizing the bioavailability and, hence, the efficacy of HMPs. In the following chapter, pharmacokinetic studies that have been conducted for some of the top-selling HMPs worldwide are listed, including SJW, ginkgo, garlic, willow bark, milk thistle, and horse chestnut.

GINKGO BILOBA

G. biloba L. is a member of the Ginkgoaceae family, a gymnosperm that has survived unchanged from the Triassic period. In traditional Chinese medicine, the seeds (nuts) of *G. biloba* were used as an antitussive, expectorant, and antiasthmatic, and in bladder infection. In China, the leaves of *G. biloba* were also used for the treatment of asthma and cardiovascular disorders. Today, standardized concentrated extracts prepared from the leaves of *G. biloba* are used for the treatment of peripheral circulatory insufficiency, cerebrovascular disorders, geriatric complaints, and for Alzheimer dementia.

Interestingly, no preclinical or clinical work has been done investigating the pharmacology, therapeutic efficacy, or safety of crude ginkgo leaf preparations. Almost all of the existing data focus on dry extracts characterized by 22% to 27% flavonol glycosides, 5% to 7% terpene trilactones, and less than 5ppm ginkgolic acids (EGb 761). Other chemicals present in the extracts are hydroxykynurenic acid, shikimic acid, protocatechuic acid, vanillic acid, and p-hydroxybenzoic acid. The monograph published by the Commission E of the German Health Authorities states that acceptable extracts should have an herb-to-extract ratio in the average range of 50:1. Extracts should be prepared with an acetone–water mixture and then be purified further. This standardization process eliminates unwanted components that might have toxic effects. In particular, it has been shown that adverse effects such as allergies were related to ginkgolic acids. Therefore, extracts that are used in drug manufacture are free of

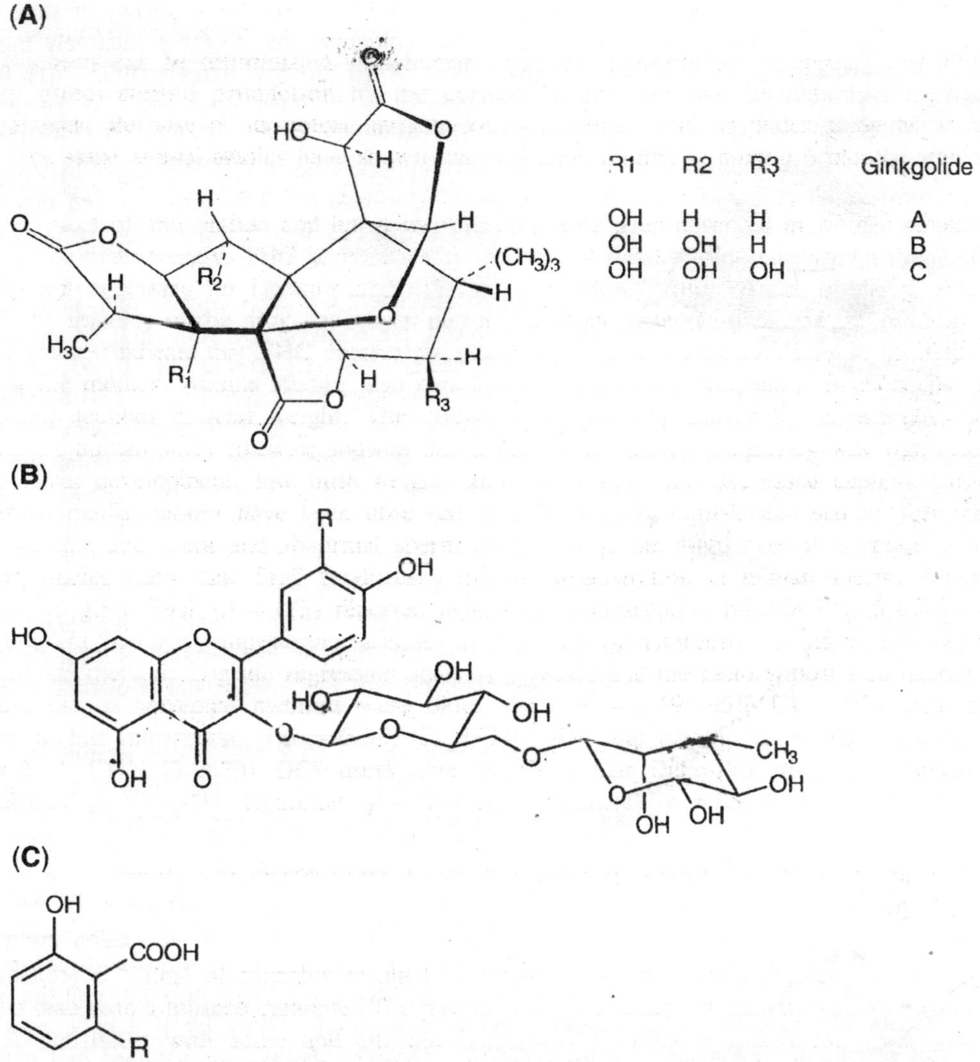

(A)

R1	R2	R3	Ginkgolide
OH	H	H	A
OH	OH	H	B
OH	OH	OH	C

(B)

(C)

Fig. 5.1. Structure of (A) ginkgolide A, ginkgolide B, bilobalide, (B) flavonolgycosides and (C) ginkgolic acids.

ginkgolic acids (less than 5 ppm). The question of whether ginkgolic acids possess allergenic potential or not is still discussed controversially, especially because it has been shown that leaf extracts, if taken orally, showed adverse effects even if they contained ginkgolic acids in a concentration of 1000 ppm, whereas a pure ginkgolic acid extract showed allergic effects.

Pharmacokinetics

Both human and animal pharmacokinetic studies have been done on ginkgo flavonol aglycones (quercetin, kaempferol, and isorhamnetin) and terpene rilactones (ginkgolide A and B and bilobalide).

Flavonol glycosides

Human clinical studies

In general, pharmacokinetic studies on flavonol glycosides are difficult to conduct because flavonoids are commonly present in the diet and their metabolites are numerous. An accurate pharmacokinetic assessment requires subjects to be maintained on a flavonoid-free diet for a period of time prior to dosing as has been done in the study by Pietta et al. Six volunteers received the relatively high single oral dose of 4 g (equivalent to 1 g flavonol glycosides) of ginkgo leaf extract (EGb 761) following seven days of a flavonoid-free diet. The following flavonol metabolites were found in the urine over three days: 4-hydroxybenzoic acid conjugate, 4-hydroxyhippuric acid, 3-methoxy-4-hydroxyhippuric acid, 3,4-dihydroxybenzoic acid, 4-hydroxybenzoic acid, hippuric acid, and 3-methoxy-4-hydroxybenzoic acid, which represented less than 30% of the flavonols administered. The authors noted that the very high dose of extract administered to the subjects made it difficult to extrapolate the results to normal clinical dosages (40–240mg extract daily).

The oral pharmacokinetics of the flavonol aglycones quercetin, kaempferol, and isorhamnetin in two healthy volunteers were studied by Nieder. Subjects were given 50, 100, and 300mg of ginkgo leaf extract–coated tablets (LI 1370). Peak plasma concentrations (C_{max}) of 25 to 30, 65, and 130 ng/mL, respectively, were achieved within two to three hours with half-lives ($t_{1/2}$) of two to four hours. Values returned to baseline 24 hours after intake. There was a linear relationship between the dose administered and the peak plasma level. In the study by Wocjcicki et al., the bioavailabilities of the same aglycones were determined using three different single oral dosage forms (capsule, liquid, and tablet). Results were similar to those of Nieder. However, the t_{max} was longer with the capsules. Values were back to baseline 24 hours after intake. The area under the curve (AUC) for the evaluated flavonoids did not differ significantly among formulations. The researchers concluded that the three formulations could be modeled by a one-compartment model with zero-order absorption without lag time, indicating that the aglycones were rapidly absorbed, and that the preparations had similar bioavailability. That the dosage form might have an influence of the bioavailability of ginkgolides and bilobalide was studied by Kressmann et al.

Increasing doses of a commercial special extract of *G. biloba* (LI 1370) were administered to two healthy volunteers (50, 100, and 300 mg). The amounts of kaempferol and quercetin were significantly higher compared to baseline. The flavonoids were metabolized and excreted primarily as glucuronic acid conjugates in urine.

Animal studies

In the study by Pietta et al., a single dose of ginkgo leaf extract (EGb 761) was administered orally to rats. Metabolites found in the urine represented less than 40% of the flavonoids administered. The presence of phenylalkyl acids in the rat urine but not in the human urine indicates that the flavonols were more extensively metabolized in humans than in rats. In the study by W tanabe et al., mice received a diet containing ginkgo leaf extract (EGb761; 36mg/kg daily) or a standard diet without the extract for four weeks. Afterwards, plasma levels of quercetin (12.0 ng/mL vs. 4.8 ng/mL), kaempferol (7.0 ng/mL vs. 3.2 ng/ mL), and isorhamnetin (49.6 ng/mL vs. 0 ng/mL) in both treatment groups were determined. The study indicates that these compounds can be absorbed intact into the blood stream.

Triterpene lactones

Human clinical studies

Mauri et al. investigated the pharmacokinetics of ginkgolides A, B, and bilobalide, after administration of ginkgo leaf extract (160 mg oral single dose) to 15 healthy subjects. The product given contained either a phospholipid complex (Ginkgoselect Phytosome) or not (Ginkgoselect) (both products contained 24% flavonol glycosides and 6% terpene trilactones; Indena SpA). Administration with the phospholipids enhanced maximum absorption (C_{max}) of total ginkgolides and bilobalide two- to threefold (from 85.0 to 181.8 μg/mL), but delayed t_{max} 1.5- to 2-fold. The AUC increased two- to threefold when the phospholipid complex was administered. In this single-compartment model, the mean elimination half-life ($t_{1/2}$) was approximately 120 to 180 minutes for all of the terpene trilactones, regardless of the product administered. Fourtillan et al. studied the pharmacokinetics of terpene trilactones in 12 healthy volunteers after single-dose intravenous (i.v.) or oral administration of ginkgo leaf extract (EGb 761), given with or without a meal. The authors could show that the consumption of a standard meal along with the oral dose of EGb 761 did not affect pharmacokinetic parameters. In a recent study, Drago et al. focused on the pharmacokinetics of two different dosage regimens for orally administered ginkgo leaf extract (Egb 761) in healthy volunteers. The subjects received either 40 mg twice daily or 80 mg once daily, with an interval of 21 days between cycles. It could be shown that a dosage of 40 mg twice daily resulted in a significantly longer $t_{1/2}$ (11.6 ± 5.2 vs. 4.3 ± 0.5) and mean residence time (MRT) (13.1 ±0.3 vs. 7.3 ± 0.6) than a single 80 mg dose. t_{max} was reached two to three hours after administration with both dosages. The authors conclude that the twice-daily dosage regimen with the lower dose is superior to that of a higher single daily dose.

The influence of the type of extract, the formulation, and dosage form on the bioavailability and pharmacokinetics of ginkolide A and B and bilobalide was recently investigated by Kressmann et al. Twelve healthy volunteers received either Ginkgol (containing Egb 761 = reference) or *G. biloba* capsules

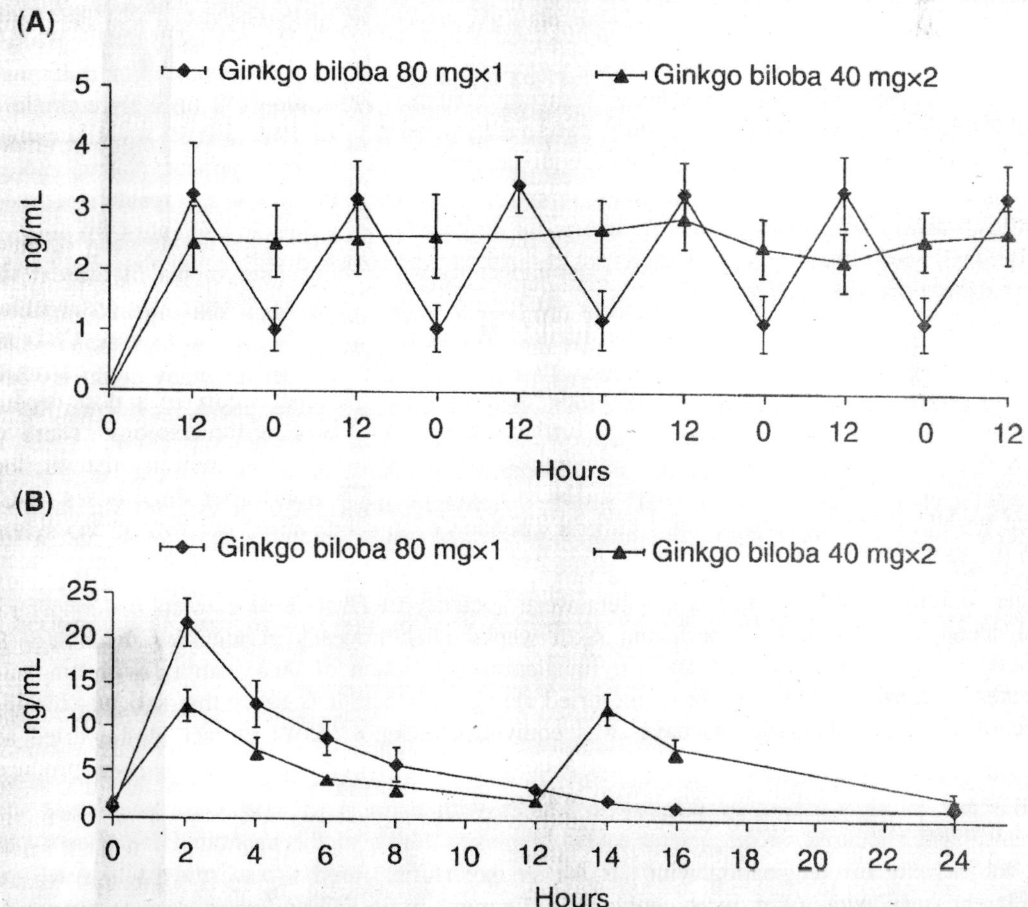

Fig. 5.2. Plasma concentrations of (ng/mL) of ginkgolide B (A) from day 1 to day 6 and (B) on day 7 after oral administration of 40 and 80 mg tablets of Ginkgo biloba extract.

(= test compound) containing another commercial dry extract in an open, single-dose crossover design study. All subjects received an oral dose of 120 mg extract under fasting conditions. Pronounced differences could be detected between the test and reference formulations regarding the bioavailability of the investigated constituents, ginkgolide A, ginkgolide B, and bilobalide. The authors clearly could show that the type of extract, the formulation, and dosage form influence the pharmacokinetics and bioavailability of potential active *Ginkgo* ingredients.

Animal studies

The bioavailability of ginkgolides A and B and bilobalide was studied in rats after a single oral administration of 30, 55, and 100 mg/kg Ginkgo leaf extract (EGb 761). The pharmacokinetics of these compounds was found to be dose-linear. Maximum plasma levels of ginkgolides A and B and bilobalide were reached in 30 to 60 minutes, with $t_{1/2}$ of ginkgolides A and B and bilobalide equaling 1.7, 2.0, and 2.2 hours, respectively, at the 30-mg dose and 1.8, 2.0, and 3.0 hours, respectively, at the 100-mg dose. Li and Wong examined the pharmacokinetics of two *Ginkgo* leaf extracts in rabbits: one standardized to 27% flavonoids and 6% terpenoids and specially prepared to yield at least 80% higher levels of ginkgolide B compared to other standardized extracts (BioGinkgo; Pharmanex), the other containing EGb 761 and standardized to 24% flavonoids and 6% terpenoids (Ginkoba, Pharmaton). Plasma concentrations of ginkgolides from the BioGinkgo extract exhibited peaks at two and five hours post-treatment with the 40 mg/kg dose and at one and five hours with the 60mg/kg dose. Mean C_{max} for the 40 and 60 mg/kg doses of the BioGinkgo extract were 18.8 ± 1.97 and 25.1 ± 3.39 μg/mL, respectively, demonstrating dose dependency. With the Ginkoba preparation (40 mg/kg), a single peak in plasma concentration was observed at three hours with mean C_{max} of 17.8 ± 0.59 μg/mL, similar to that of the former extract at the same dose. Twelve hours after the 40 mg/kg treatment, plasma ginkgolide levels were 2.6 times greater for BioGinkgo than for Ginkoba. The prolonged residence time and greater bioavailability was attributed to two factors: the slightly higher terpenoid content of the BioGinkgo preparation and, more importantly, the fact that the extract was enriched with ginkgolide B, which has a longer half-life than ginkgolide A.

St. John's Wort

Hypericum perforatum (Clusiaceae), commonly known as SJW, is used in many countries for the treatment of mild-to-moderate forms of depression. Several clinical studies provide evidence that SJW is as effective as conventional synthetic antidepressants. From a phytochemical point of view, *H. perforatum* belongs to one of the best-investigated medicinal plants. A series of bioactive compounds have been detected in the crude material, namely phenylpropanes, flavonol derivatives, biflavones, proanthocyanidins, xanthones, phloroglucinols, some amino acids, naphthodianthrones, and essential oil constituents.

Recent reports have shown that the antidepressant activity of *Hypericum* extracts can be attributed to the phloroglucinol derivative hyperforin, to the naphthodianthrones hypericin and pseudohypericin, and to several flavonoids. The role and the mechanisms of action of these different compounds are still a matter of debate. But, taking these previous findings together, it is likely that several constituents are responsible for the clinically observed antidepressant efficacy of SJW.

Pharmacokinetics

Single- and multiple-dose pharmacokinetic studies with extracts of SJW were performed in rats and humans, which focused on the determination of plasma levels of the naphthodianthrones hypericin and pseudohypericin and the phloroglucinol derivative hyperforin. Results from pharmacokinetic studies investigating plasma levels of different flavonoids after intake of SJW preparations are presently not available.

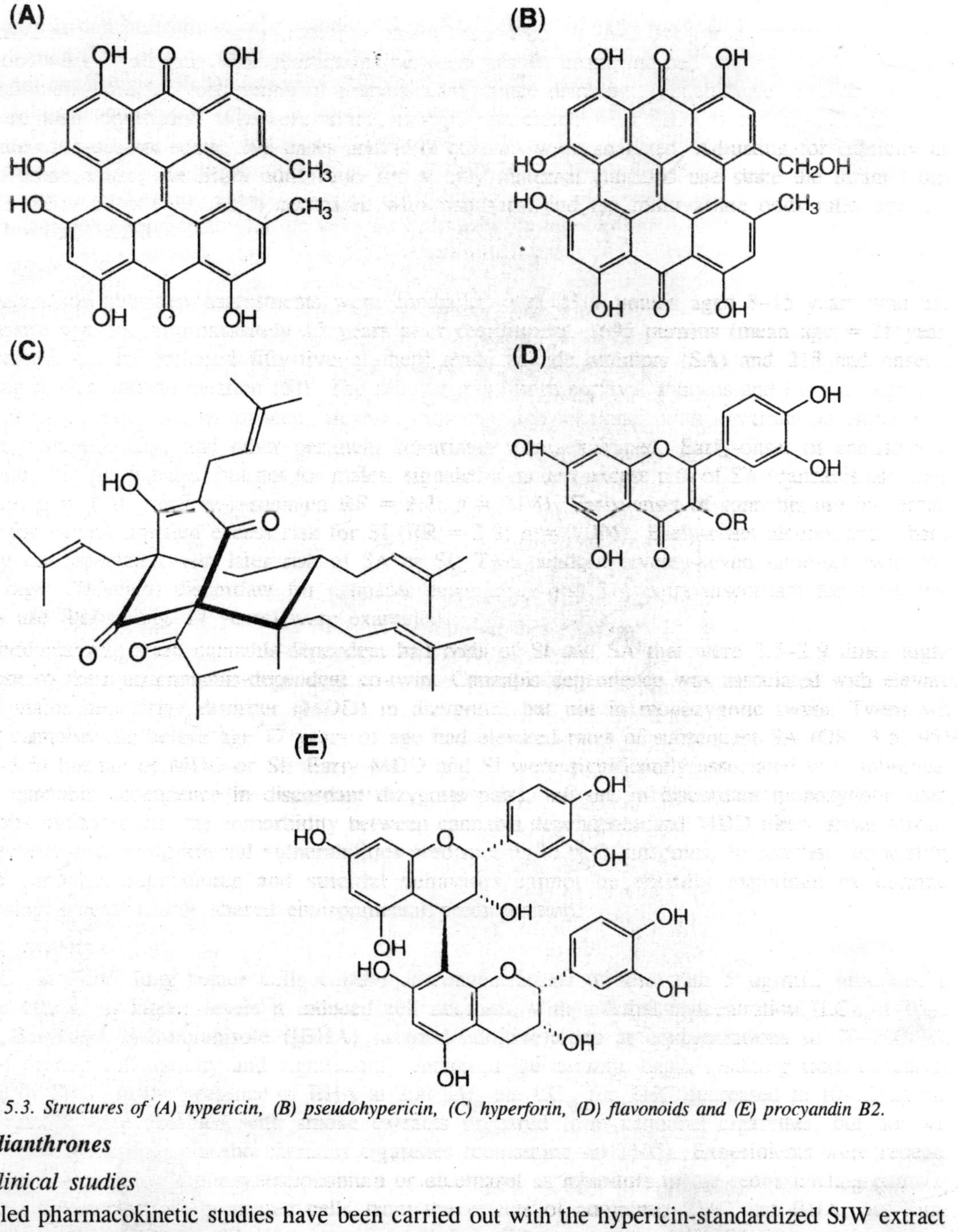

(A) (B) (C) (D) (E)

Fig. 5.3. Structures of (A) hypericin, (B) pseudohypericin, (C) hyperforin, (D) flavonoids and (E) procyandin B2.

Naphthodianthrones

Human clinical studies

Detailed pharmacokinetic studies have been carried out with the hypericin-standardized SJW extract LI 160. The preparation is reported to contain 300 mg of the dried extract of SJW, yielding 0.24% to 0.32% total hypericin. Administration of single oral doses of LI 160 (300, 900, and 1800 mg) to healthy male volunteers resulted in peak plasma hypericin concentrations of 1.5, 7.5, and 14.2 ng/mL for the three doses, respectively. Peak plasma concentrations were seen with hypericin between 2.0

and 2.6 hours and with pseudohypericin after 0.4 to 0.6 hours. The elimination half-life of hypericin was between 24.8 and 26.5 hours, and varied for pseudohypericin from 16.3 to 36.0 hours. The AUC showed a nonlinear increase on raising the dose—this effect was statistically significant for hypericin. Repeated doses of LI 160 (300 mg) three times daily resulted in steady-state concentrations after four days. Mean maximal plasma level during the steady-state treatment was 8.5 ng/mL for hypericin and 5.8 ng/mL for pseudohypericin. Kinetic parameters after i.v. administration of SJW extract (115 and 38 mg for hypericin and pseudohypericin, respectively) in two subjects correspond to those estimated after an oral dosage. Both hypericin and pseudohypericin were initially distributed into a central volume of 4.2 and 5.0 L, respectively. The mean distribution volumes at steady state were 19.7 L for hypericin and 39.3 L for pseudohypericin, and the mean total clearance rates were 9.2 mL/min for hypericin and 43.3 mL/min for pseudohypericin. The systemic availability of hypericin and pseudohypericin were roughly estimated to be 14% and 21%, respectively. In spite of their structural similarities, there were substantial pharmacokinetic differences between hypericin and pseudohypericin, which is not surprising considering the differences in the planarity of both molecules.

A placebo-controlled, randomized clinical trial with monitoring of hypericin and pseudohypericin plasma concentrations was performed to evaluate the increase in dermal photosensitivity in humans after application of high doses of SJW extract. The study was divided into a single-dose and a multiple-dose part. In the single dose crossover study, each of the 13 volunteers received either placebo or 900, 1800, or 3600 mg of the SJW extract LI 160. Maximum total hypericin plasma concentrations were observed about four hours after dosage and were 0, 28, 61, and 159 ng/mL, respectively. Pharmacokinetic parameters had a dose relationship that appeared to follow linear kinetics. In another study, the concentrations of hypericin and pseudohypericin in serum and skin blister fluid after oral intake (single and steady state) of relatively high doses of LI 160 were determined in 12 healthy volunteers. After a single oral administration of SJW extract (1800 mg), the mean serum level of total hypericin (hypericin þ pseudohypericin) was 43 ng/mL and the mean skin blister fluid level was 5.3 ng/mL. After steady-state administration (900 mg/day for seven days), the mean serum level of total hypericin was 12.5 ng/mL and the mean skin blister fluid level was 2.8 ng/mL. Serum levels of total hypericin were always higher than skin levels. However, the skin levels observed in this study are far below the hypericin skin levels that are estimated to be phototoxic (greater than 100 ng/mL).

Pharmacokinetics, safety, and antiviral effects of hypericin were studied in patients with chronic hepatitis C infection. The patients received an eight-weeks course of 0.05 and 0.10mg/kg hypericin orally once a day. The pharmacokinetic data revealed a long elimination half-life (mean values of 36.1 and 33.8 hours, respectively, for the doses of 0.05 and 0.10 mg/ kg) and mean AUC determinations of 1.5 and 3.1 μg/mL/hr, respectively. Because relatively high doses of 0.05 and 0.10 mg/kg/day were given, which will probably be not reached after oral intake of recommended doses of SJW extract preparations, it is not surprising that hypericin caused a considerable phototoxicity in this study.

Animal studies

Early pharmacokinetic studies in mice report that maximum plasma concentrations of hypericin and pseudohypericin were reached at six hours and were maintained for at least eight hours. The aqueous-ethanolic SJW extract used in this study contained 1.0 mg of hypericin. Pharmacokinetics and cerebrospinal fluid penetration of hypericin were studied after i.v. dose of 2mg/kg in monkeys. Mean peak plasma concentration of hypericin following this dose was 71.7 μg/mL (142 μM). Elimination of hypericin from plasma was biexponential, with an average terminal half-life of 26 $\pm$ 14 hours. The 2 mg/kg dose in nonhuman primates was sufficient to maintain plasma concentrations above 5.1 μg/mL (10 μM) for up to 12 hours (the in vitro concentration required for growth inhibition of human glioma cell lines is greater than 10 μM).

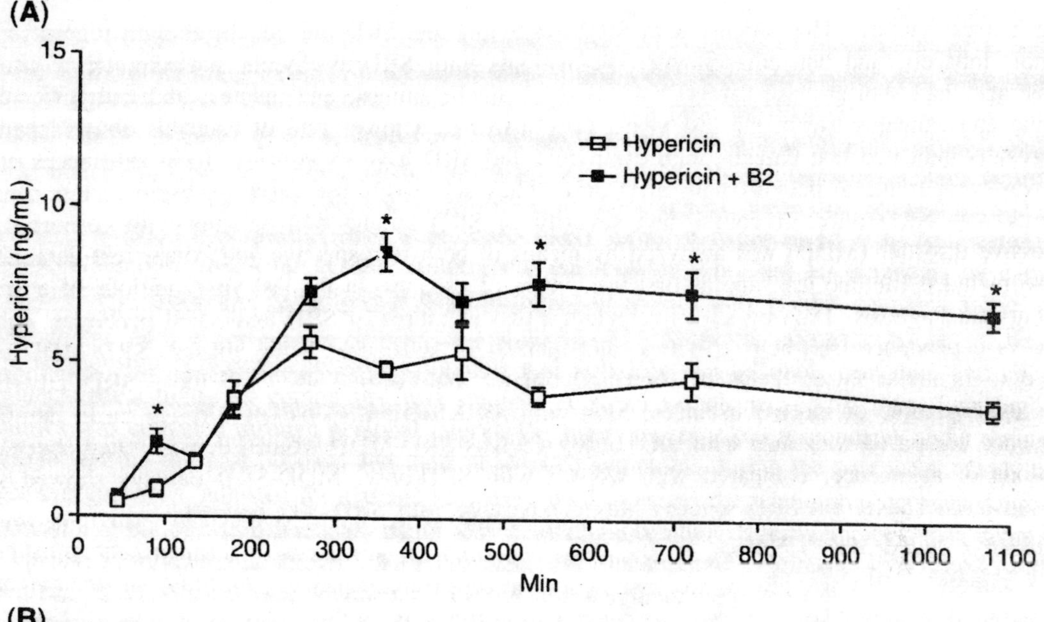

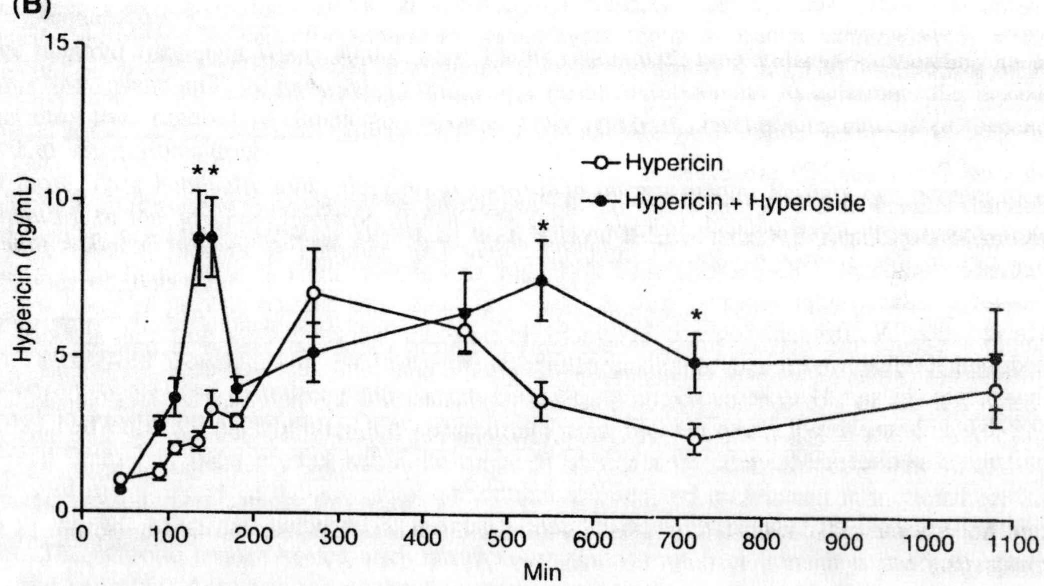

Fig. 5.4. (A) Plasma levels of hypericin in the presence and absence of procyanidin B2. (B) Plasma levels of hypericin in the presence and absence of hyperoside.

In general, the biological evaluation of hypericin in various test models is limited by its poor water solubility. It was shown in in vitro as well as in vivo studies that the water solubility of hypericin was remarkably enhanced in the presence of procyanidins or flavonol glycosides of SJW extract. In a recent pharmacokinetic study in rats, it was shown that procyanidin B2 as well as hyperoside increased the oral bioavailability of hypericin by approximately 58% (B2) and 34% (hyperoside). Procyanidin B2 and hyperoside had a different influence on the plasma kinetics of hypericin; median maximal plasma levels of hypericin were detected after 360 minutes (C_{max}: 8.6 ng/mL) for B2, and after 150

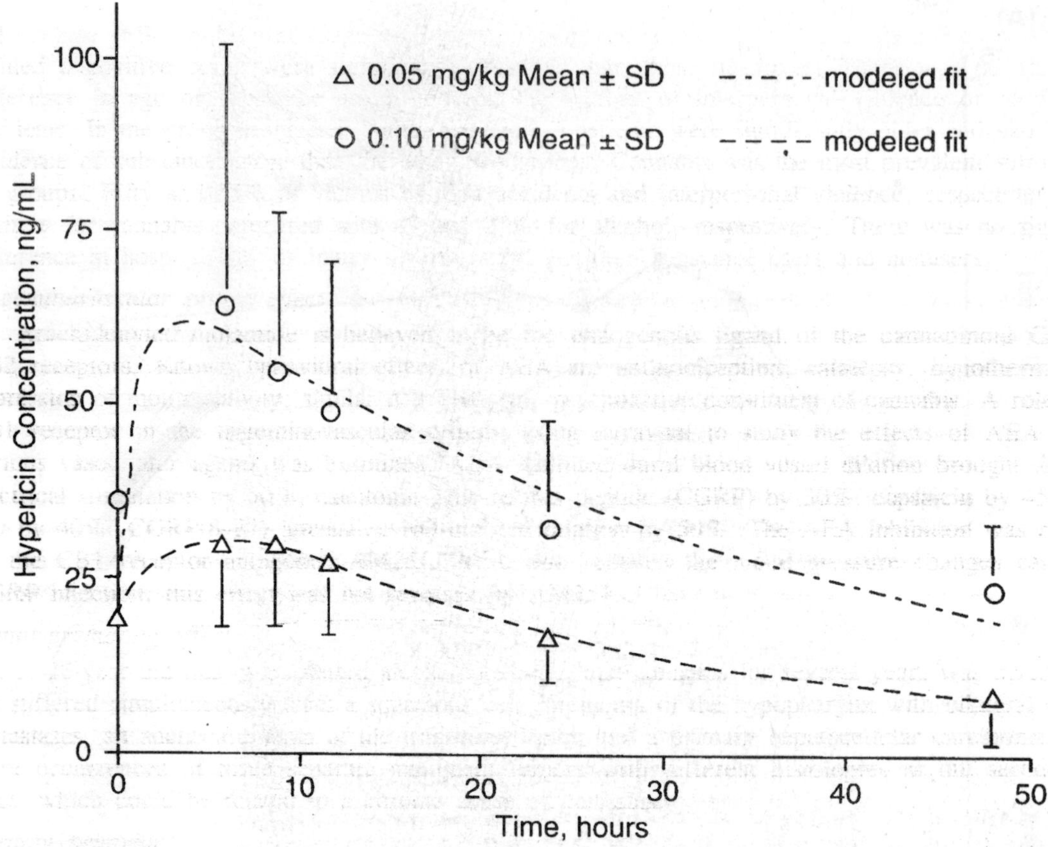

Fig. 5.5. Plasma concentrations of hypericin after oral administration of hypericin (0.05 and 1 mg/kg) to hepatitis C patients.

minutes (C_{max}: 8.8 ng/mL) for hyperoside. The authors suggest that treatment of patients with the entire SJW extract, depending on its composition, should be superior to the treatment with isolated compounds, because the extract provides not only different classes of active compounds, but also constituents that influence their bioavailability.

Hyperforin

Human clinical studies

Plasma levels of hyperforin were followed for 24 hours in two studies with healthy volunteers after administration of film-coated tablets containing 300 mg SJW extract representing 14.8 mg hyperforin. In the first crossover study, six male volunteers received 300, 600, or 1200 mg of a SJW extract preparation (WS 5572) after a 10-hour fasting time. Maximum plasma levels of 150 ng/mL (approximately 280 nM) were reached after 3.5 hours after intake of 300 mg SJW extract. Half-life and MRT were 9 and 12 hours, respectively. Hyperforin pharmacokinetics were linear up to 600 mg of the extract. Increasing the doses to 900 or 1200 mg resulted in lower C_{max} and AUC values than those expected from linear extrapolation of data from lower doses. In a repeated dose study with seven healthy volunteers, no accumulation of hyperforin in plasma was observed after intake of 900 mg/day SJW extract for seven days. The estimated steady-state plasma concentrations of hyperforin after intake of 3 × 300 mg/day was approximately 100 ng/mL (approximately 180 nM). The bioavailability of compounds of SJW was found to be influenced by the formulation characteristics.

An ethanolic SJW extract containing 5% hyperforin and 0.3% hypericin was administered as softgel capsules to 12 healthy volunteers. A second standard formulation in a two-piece hard gelatin capsule was also used for comparison purposes. C_{max} of hyperforin was 168.4ng/mL for the soft gelatin formulation and 84.3 ng/mL for the hard gelatin capsule. The t_{max} values for hyperforin were 2.5 hours for the soft gelatin capsule compared to 3.1 hours for the reference formulation, whereas the total AUCs were 1483 and 583.7 hr ng/mL, respectively. Taken together, the soft gelatin capsules exhibited a higher individual absorption when compared with the corresponding data for the hard gelatin capsule. This finding confirms former results, which show that the absorption from soft gelatin capsules is in general higher if compared to hard gelatin capsules.

Animal studies

Pharmacokinetics of hyperforin after administration of an ethanolic SJW extract (WS 5572) to rats were investigated by Biber et al. Maximum plasma levels of approximately 370 ng/mL (approximately 690 nM) were reached after three hours. Estimated half-life and clearance values were six hours and 70 mL/min/kg, respectively.

GARLIC

Garlic (*Allium sativum* L., Alliaceae) is a commonly used food and botanical supplement. Garlic is stated to possess diaphoretic, expectorant, antispasmodic, antiseptic, bacteriostatic, antiviral, hypotensive, and anthelmintic properties. Traditionally, it has been used to treat chronic bronchitis, respiratory catarrh, recurrent colds, bronchitic asthma, influenza, and chronic bronchitis. Modern use of garlic and garlic preparations is focused on their reputed antihypertensive, antiatherogenic, antithrombotic, antimicrobial, fibrinolytic, cancer preventive, and lipid-lowering effects. Garlic contains

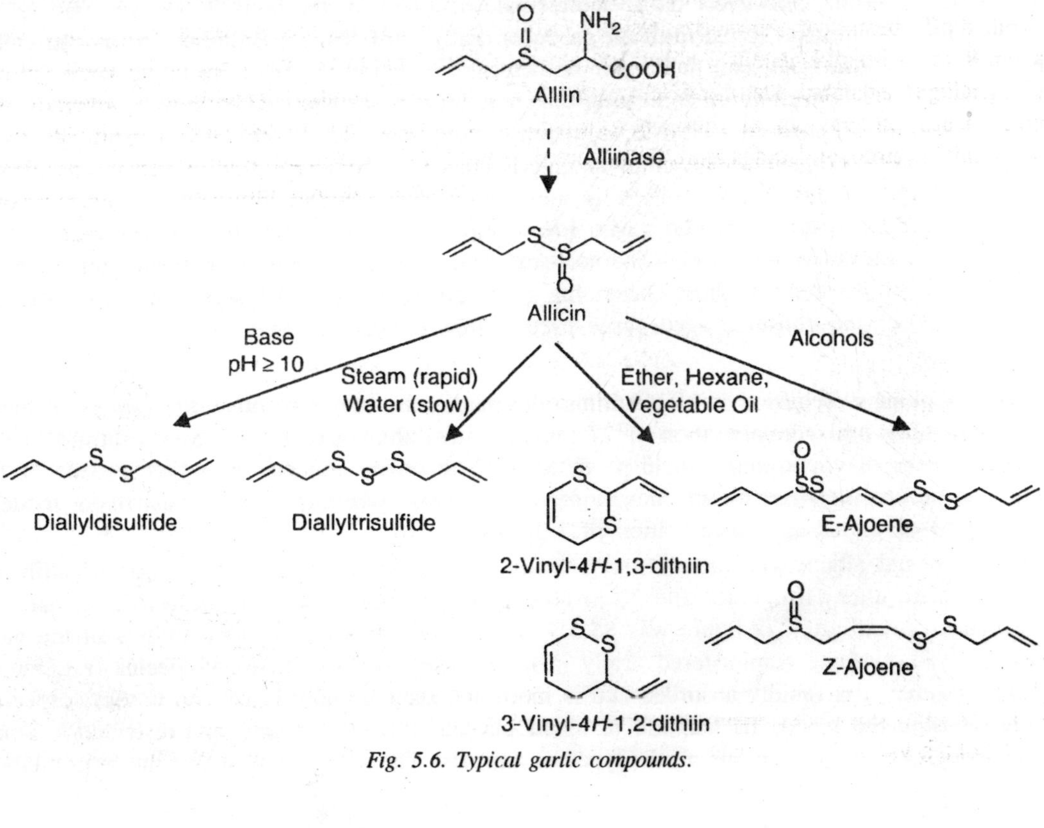

Fig. 5.6. Typical garlic compounds.

a large number of biologically active constituents. The constituents of garlic can be simply divided into two groups: sulfur-containing and non–sulfur-containing compounds. Most of the medicinal effects of garlic are referable to the sulfur compounds and the alliin-splitting enzyme alliinase, which converts alliin into allicin. This enzymatic reaction occurs when fresh garlic is chopped or crushed and alliin comes into contact with alliinase (enzyme and substrate are located in different compartments in the garlic bulb). Allicin is responsible for the characteristic garlic odor but it is unstable in aqueous and oily solution, and within a few hours it degrades into vinyldithiins and ajoenes. Depending on the chemical nature of the solvent, the extract can contain a spectrum of different compounds. As a result, garlic is available in the form of different pharmaceutical preparations, such as dry powder products, oil-macerates, volatile garlic oil (obtained by water vapor distillation), or juices of fresh garlic. Most clinical studies have been mainly performed with the dry powder preparations and some volatile oil macerates.

Pharmacokinetics

There are only a few reports on the absorption, metabolism, and excretion of garlic's sulfur compounds available. Further, until now it is not known what metabolic form of allicin actually reaches the target cells, and it is still unknown how garlic compounds might function in the body.

Allicin

Human clinical studies

Allicin is well absorbed, as indicated by a persistent garlicky odor on the breath, skin, and amniotic fluid of persons after consumption of fresh garlic. Because oral comsumption of pure allicin has been shown to significantly increase overall body catabolism of triglycerides, a substantial absorption of allicin is assumed to occur. However, the metabolic fate of allicin in the body is not well understood. Neither allicin nor its common transformation products diallyl sulfides, vinyldithiins, or ajoene can be found in the blood or urine, nor can their odor be detected in the stool after consuming large amounts of garlic (up to 25 g) or pure allicin, indicating that it is rapidly metabolized to new compounds. In a recent study, Rosen et al. used GC–MS/MS as major techniques to determine various metabolites after consumption of dehydrated granular garlic and an enteric-coated garlic preparation, in breath and plasma. The authors found that methyl allyl sulfide is the main volatile metabolite on the consumption of dehydrated dry garlic and enteric-coated garlic formulations. Hydrogen sulfide was observed but not quantified due to its extremely low levels. The non–sulfur-containing compounds limonene and *p*-cymene were also observed in the breath in those consuming garlic preparations. S-Allylcysteine can be observed in the blood of those who consume aged garlic preparations (so-called "*Kyolic*").

Animal studies

The pharmacokinetic behavior of vinyldithiins, the main constituents of oily preparations of garlic, was investigated after oral administration of 27 mg 1,2-vinyldithiin and 9 mg 1,3-vinyldithiin to rats. In serum both forms of vinyldithiins could be detected. The serum concentration–time profile of 1,2-vinyldithiin can be characterized by an one-compartment model, whereas a two-compartment model is used as a best fit of the serum concentration of 1,3-vinyldithiin.

In rats, alliin and allicin were administered orally at doses of 8 mg/kg. Absorption of alliin and allicin was complete after 10 minutes and 30 to 60 minutes, respectively. The mean total urinary and fecal excretion of allicin after 72 hours was 85.5% of the dose. Pharmacokinetic studies of the garlic constituent S-allyl-L-cysteine administered orally in large doses to three different species (rat, mouse, and dog) showed that it is rapidly absorbed and is more abundant initially in several tissues, especially kidney, than it is in the blood. Its half-life in blood plasma (0.8–10.3 hours) and distribution among urinary metabolites varied greatly among the types of animals. The study showed that the bioavailability of S-allylcysteine decreased linearly with decreased dose, from 98% at 50 mg/kg body weight to 77% at 25 mg/kg and 64% at 12 mg/kg. Recently, Germain et al. studied the in vivo metabolism of diallyl disulfide (DADS), a garlic compound claimed to have anticarcinogenic effects. After oral administration of a single dose of 200 mg/kg, metabolites were measured in the stomach, liver, plasma, and urine by GC coupled with MS over 15 days. DADS was detected in almost all analyzed tissues within the first hours. In addition, the metabolites allylmercaptan (AM) and allyl methyl sulfide (AMS) were detected. The C_{max} of the metabolites were higher than that of DADS (1.46 μg/mL). The t_{max} for DADS was estimated to be less than one hour, whereas this time increased to 24 hours for AM and AMS.

Willow Bark

The willow family includes a number of different species of deciduous trees and shrubs native to Europe, Asia, and some parts of North America. Some of the more commonly known are white willow/European willow (*Salix alba*), black willow (*Salix nigra*), crack willow (*Salix fragilis*), purple willow (*Salix purpurea*), and weeping willow (*Salix babylonica*). The willow bark sold in Europe and the United States usually includes a combination of the bark from white, purple, and crack willows. Willow

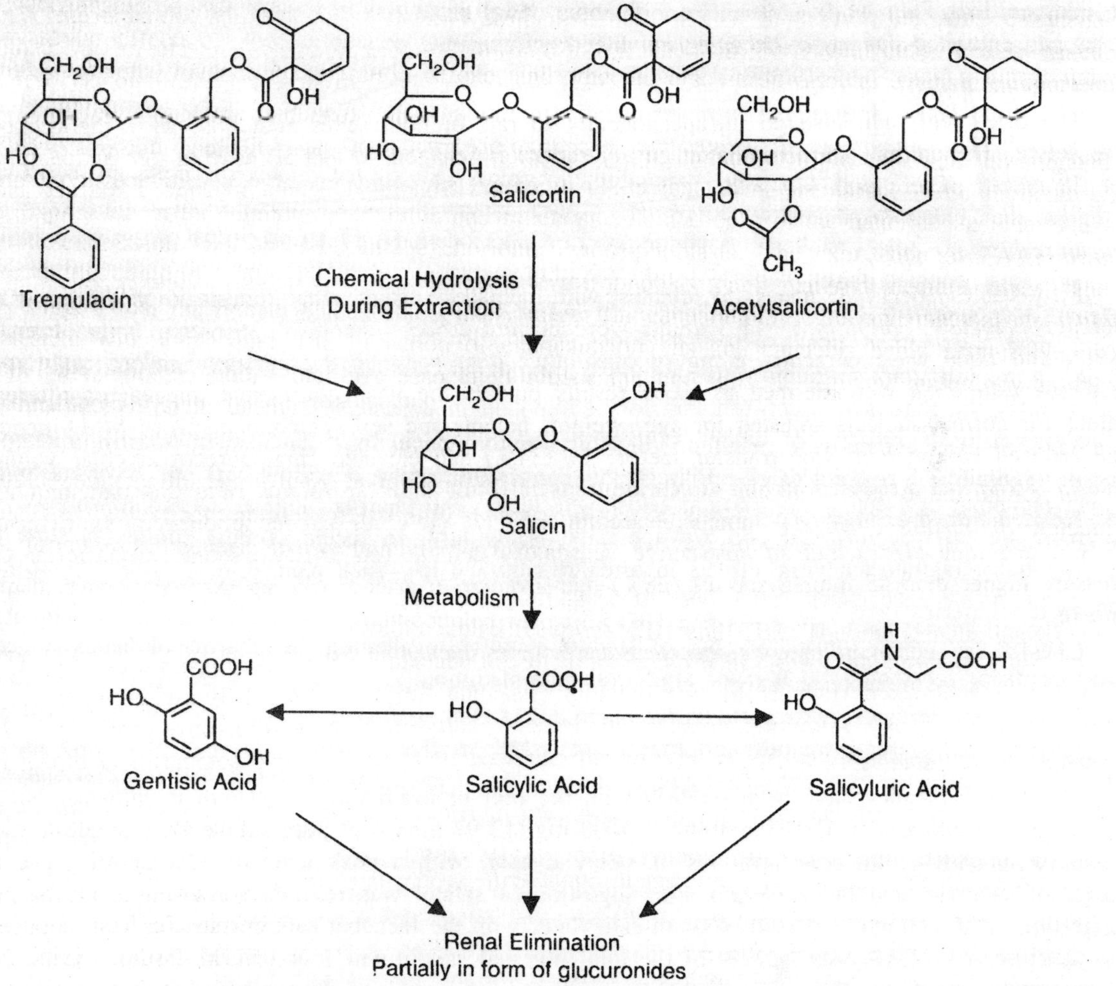

Tremulacin

Salicortin

Acetylsalicortin

Chemical Hydrolysis During Extraction

Salicin

Metabolism

Gentisic Acid

Salicylic Acid

Salicyluric Acid

Renal Elimination
Partially in form of glucuronides

Fig. 5.7. Main constituents of willow bark and pharmacokinetics of salicin in humans.

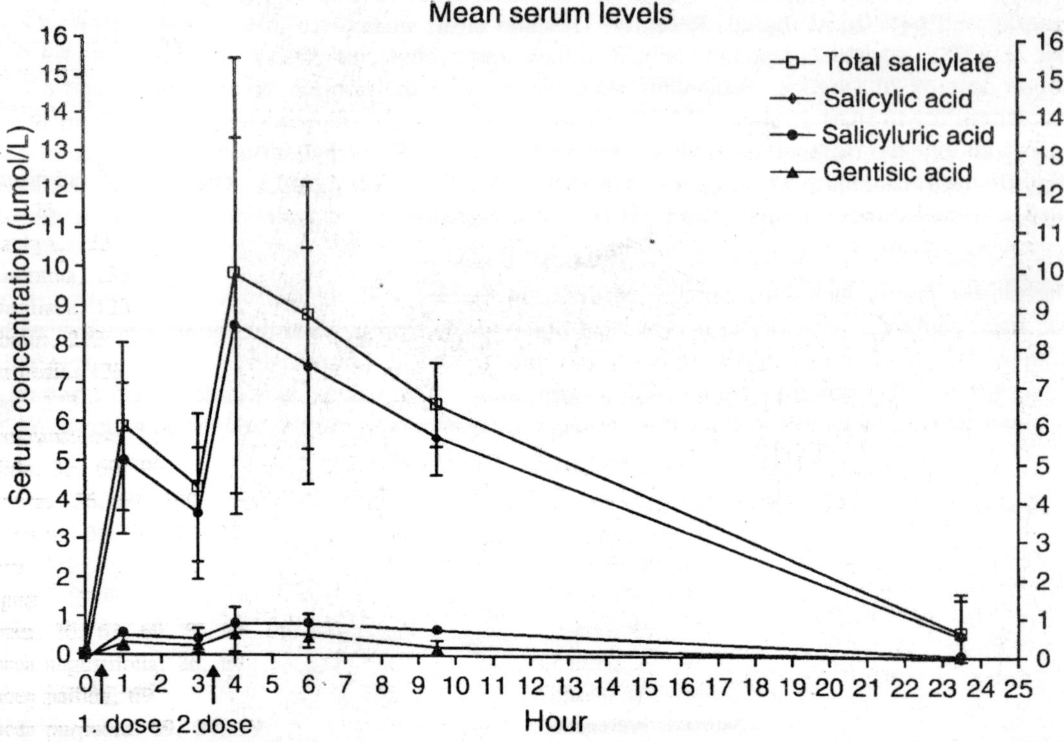

Fig. 5.8. Mean serum levels of salicylic acid, gentisic acid, and salicyluric acid of 10 volunteers. "Total salicylate" represents the sum of the three individual compounds.

bark's most important medicinal qualities are its ability to ease pain and reduce inflammation. Salicin is probably the most active anti-inflammatory compound in willow; it is metabolized to salicylic acid. The enzymatic degradation of salicin, salicortin, and tremulacin by β-glucosidase and by esterase has been investigated. Studies have identified several other components of willow bark that have antioxidant, fever-reducing, antiseptic, and immune-boosting effects. The pharmacological actions of salicylates in humans are well documented and are applicable to willow. In recent clinical studies, willow bark extract had moderate analgesic effects in osteoarthritis and low back pain.

Pharmacokinetics

There are only a limited number of studies available evaluating the pharmacokinetics of salicin and its major metabolites in humans after oral administration.

Salicin

Human clinical studies

Two early studies have been published on the oral bioavailability of salicin in humans, but they showed different results. Oral ingestion of 4000 mg (13.97 mmol) of pure salicin by a single volunteer resulted in high serum concentrations of salicylic acid, with a peak level of 110 μg/mL. The peak level of salicylic acid in the serum after ingestion of salicin was reached somewhat later than after ingestion of an equimolar dose of sodium salicylate, with the fact that salicin must be first metabolized to salicylic acid. Eighty-six percent of the total ingested salicin was found in the 24-hour urine in the form of the usual salicylic acid metabolites, indicating a good oral bioavailability of pure salicin. In contrast, oral administration of a willow bark extract showed a low bioavailability of salicin. After ingestion of commercial sugar-coated tablets containing willow bark extract corresponding to a total amount of 54.9 mg (0.192 mmol) salicin, the serum of 12 volunteers showed a peak concentration of only 0.13 μg/mL salicylic acid; this is only 5% of the serum level expected after oral intake of an equimolar amount of synthetic salicylates. In a recent pharmacokinetic study of willow bark, Schmid et al. carried out a pharmacokinetic study on the oral bioavailability of salicylates from willow bark extract in 10 healthy volunteers. A chemically standardized willow bark extract was used in the form of coated tablets. Willow bark extract was given in two equal doses corresponding to 240 mg salicin at times zero and three hours. Over a period of 24 hours, urine and serum levels of salicylic acid and its metabolites, gentisic acid and salicyluric acid, were determined. Peak plasma levels of salicylic acid were on average 1.2 μg/mL and were reached less than two hours after oral administration. Salicylic acid was the major metabolite of salicin in the serum (86% of total salicylates), besides salicyluric acid (10%) and gentisic acid (4%). After 24 hours, 15.8% of the orally ingested dose of salicin was detected in the urine on average. The AUC of salicylate obtained in this study was equivalent to that expected from an intake of 87 mg of acetylsalicylic acid. Taken together, in the study by Schmid et al., willow bark in the dosage of 240 mg led to much lower serum salicylate levels than observed after analgesic doses of synthetic salicylates. The authors conclude that the formation of salicylic acid alone is therefore unlikely to explain analgesic or antirheumatic effects of willow bark.

Horse Chestnut

The horse chestnut, *Aesculus hippocastaneum* (Hippocastanaceae), was introduced into the northern Europe from the Near East in the 16th century. Extracts from horse chestnut seeds were already being used therapeutically in France in the early 1800s. Several French works published between 1896 and 1909 reported successful outcomes in the treatment of hemorrhoidal ailments. Traditionally, horse chestnut has been used for the treatment of varicose veins, hemorrhoids, phlebitis, diarrhea, fever, and

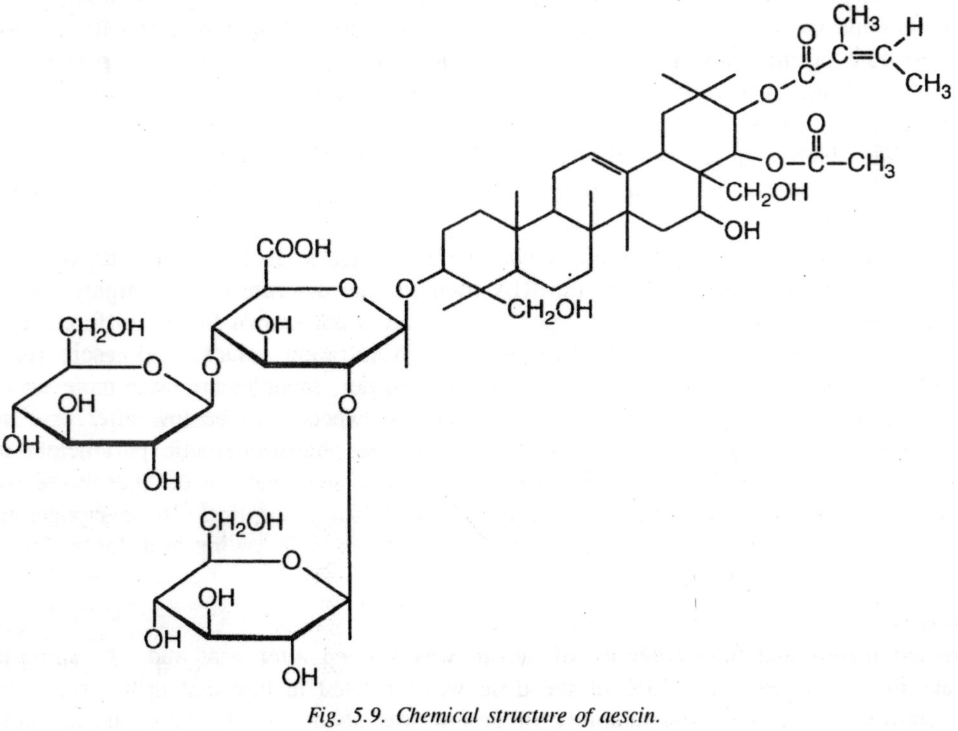

Fig. 5.9. Chemical structure of aescin.

enlargement of the prostate gland. The German Commission E approved its use in the treatment of chronic venous insufficiency in the legs. The seeds of horse chestnut contain mainly aescin, which is a complex mixture of various chemically very similar triterpene glycosides having saponifying activities. The aglycones of these saponins are barringtogenol C and protoescigenin. Former investigations of the saponin mixture differentiated between three aescin subtypes, including the slightly soluble β-aescin, a C-21 and C-22 diester, readily water soluble crypto-aescin, which is created by spontaneous migration of the acetyl group from C-22 to C-28, and α-aescin, an equilibrium mixture of these two isomer diesters. A number of other compounds have been isolated from the chestnut seeds, i.e., flavonols (kaempferol and quercetin), flavonol glycosides (astragalin, isoquercitrin, and rutin), and coumarins (aesculetin, fraxin, and scopolin). However, all of these compounds can be found in larger amounts from other sources and, furthermore, during 1960, Lorenz and Marek concluded that the antiedemigenous, antiexudative, and vasoprotective activities of horse chestnut extracts are mainly due to aescin.

Pharmacokinetics

Pharmacokinetic studies with horse chestnut focus on the absorption, metabolism, and excretion of the main constituent—β-aescin.

Aescin

Human clinical studies

The significant advances in understanding of bioavailability and kinetics of β-aescin should be attributed to the development of a highly specific radioimmunoassay (RIA) allowing the detection of concentrations in the nanogram per milliliter range. The relative oral bioavailability of beta-aescin from a sugar-coated tablet formulation was compared to a reference preparation available in capsule form to 18 healthy, male volunteers over a 48 h period. A large variation in absorption parameters for beta-aescin was measured. Maximum concentration (Cmax) after a dose containing 50 mg aescin varied from 0.19 to 45.1 ng/mL, time for maximum concentration (t_{max}) varied from 0.73 to 8.5 hours, and the AUC varied from 24.6 to 389 ng/hr/mL. The second study, also on two solid-dose preparations (one with sustained release), using 24 volunteers found more consistent results. Parameters of the sustained-release tablet were superior. For example, after a dose containing 50 mg aescin, C_{max} for the sustained-release tablet 9.81 ± 8.9 ng/mL, t_{max} was 2.23 ± 0.9 hours, and AUC averaged 187.1 ng/hr/mL.

The half-life time for both preparations was about 20 hours. However, the data of both studies have to be evaluated with care, because the RIA used for the determination is highly specific and its potential of cross-reactivity with different types of β-aescin is not known for the different extracts used in the different preparations. Thus, absolute plasma concentration values of β-aescin from a single preparation cannot be compared with other preparations. Further, saponins are large molecules containing highly polar groups and their intact bioavailability can be expected to be low after oral doses. This was confirmed in the above-mentioned studies, because the pharmacokinetic parameters indicate an absorption less than 10% of the administered dose. However, saponins can be hydrolyzed by intestinal flora, leaving the less polar aglycone or sapogenin available for absorption. These sapogenins, or their hepatic metabolites, may in fact be the main active form of aescin following oral doses. Further studies are needed to clarify this issue.

Animal studies

Pharmacokinetics and bioavailability of aescin was studied after oral and i.v. administration of tritiated aescin. About 66% and 33% of the dose was excreted in bile and urine, respectively, after i.v. administration. The oral bioavailability of aescin was about 12.5%. Percutaneous absorption of aescin was studied in mice and rats. The amounts of aescin in muscle were greater than in other organs. These results indicate that percutaneous administration of aescin could be beneficial.

GINSENG

Ginseng, a commonly used natural product, has a reputation as a "*herb of eternal life.*" A survey of herbal-based over-the-counter products indicates that 28% of them contain ginseng. It has been estimated that ginseng comprises 15% to 20% of the total annual sales of botanical products in the United States. Botanical remedies known as "ginseng" are based on the roots of several distinct species of plants, mainly Korean or Asian ginseng (*Panax ginseng*), Siberian ginseng (*Eleutherococcus senticosus*), and American ginseng (*Panax quinquefolius* and *Panax notoginseng*). All of these species belong to the Araliaceae family, but each of these different species has specific pharmacological effects. *P. ginseng* is the most commonly used and highly researched species of ginseng. This species, which is native to China, Korea, and Russia, has been an important botanical remedy in traditional Chinese medicine for thousands of years, where it has been used primarily as a treatment for weakness and fatigue. The main constituents of *P. ginseng* are the so-called ginsenosides, a complex mixture of saponins from the tetracyclic dammarane type (sapogenins protopanaxadiol and protopanaxatriol) and a pentacyclic triterpene from the oleanolic acid type. The ginsenosides can be divided into two classes: the protopanaxatriol class, consisting mainly of Rg_1, Rg_2, Rf, and Re, and the protopanaxadiol class, consisting mainly of Rc, Rd, Rb_1, and Rb_2. Further, ginseng constituents are polysaccharides, essential oil constituents, polyacetylenes, peptides, and other lipids. Ginseng is included in the pharmacopeias of several countries such as China, Germany, and the United Kingdom. Modern therapeutic claims refer to vitality, immune function, cancer, cardiovascular diseases, and sexual function.

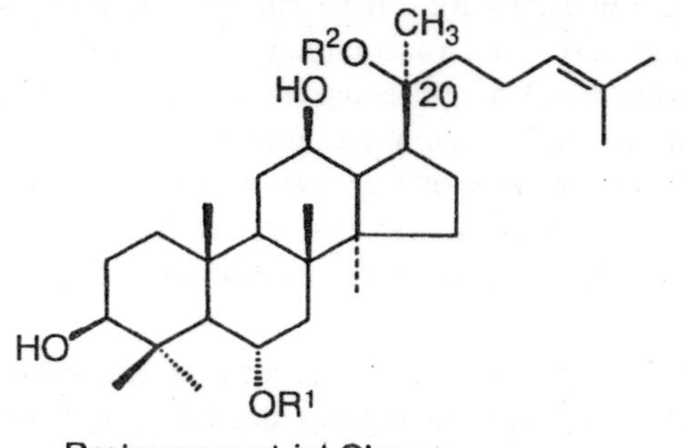

Protopanaxatriol Class

	R^1	R^2
20(S)-Protopanaxatriol	H	H
Ginsenoside Re	glc(2-1)rha	glc
Ginsenoside Rf	glc(2-1)glc	H
Ginsenoside Rg_1	glc	glc
Ginsenoside Rg_2	glc(2-1)rha	H

Protopanaxadiol Class

	R^1	R^2
20(S)-Protopanaxadiol	H	H
Ginsenoside Re	glc(2-1)glc	glc(6-1)glc
Ginsenoside Rf	glc(2-1)glc	glc(6-1)ara *p*
Ginsenoside Rg_1	glc(2-1)glc	glc(6-1)ara *f*
Ginsenoside Rg_2	glc(2-1)glc	glc

Fig. 5.10. Chemical structures of ginsenosides.

Pharmacokinetics

Although ginseng is commonly used as an adaptogenic and immunomodulatory drug, and its efficacy has been shown in pharmacologic assays as well as in clinical trials, the number of pharmacokinetic studies is limited. This might be due to the chemical nature of the ginsenosides—as saponins, they are large molecules containing highly polar groups. Thus, their intact bioavailability can be expected to be low after oral intake.

Ginsenoside

Human clinical studies

Analysis of urine samples from Swedish athletes who had consumed ginseng preparations within 10 days before urine collection showed that out of 65 samples analyzed, 60 were found to contain the sapogenin 20(S)-protopanaxatriol. The concentrations of 20(S)-protopanaxatriol varied from 2 to 35 ng/mL urine. The results after intake of oral doses of ginseng preparations demonstrated a linear relationship between the amounts of ginsenosides consumed and the 20(S)-protopanaxatriol glycosides excreted in the urine. About 1.2% of the dose was recovered in the glycosidic form over five days. The main metabolites of ginsenosides Rb_1, Rb_2, Rc, Re, and Rg_1 after anaerobic incubation with fecal flora were identified as prosapogenins and sapogenins, although the metabolic rate and mode were affected by fermentation media. Further, prosapogenins and sapogenins were detected in blood (0.3–5.1 μg/mL) and in urine (2.2– 96 μg/day) after the oral administration of ginseng extract (150 mg/day) to humans. One organism in the human fecal flora hydrolyzing ginsenosides was *Prevotella oris*.

Animal studies

Studies in rats focus on the pharmacokinetics of ginsenosides Rg_1, Rb_1, and Rb_2. Ginsenoside Rg_1 was absorbed rapidly from the upper parts of the digestive tract (accounting for 1.9–20.0% of the dose of Rg_1 administered orally, 100 mg/kg). The serum level of ginsenoside Rg_1 reached its peak at 30 minutes, and the maximum levels of ginsenosides Rg_1 in tissues were attained within 1.5 hours. However, Rg_1 was not found in the brain. About 80% of the dose of Rg_1 was excreted into urine and bile after i.v. administration to rats, showing that Rg_1 is hardly metabolized in rat liver. The authors could further show that degradation and/or metabolism of Rg_1 occurs in the stomach and large intestine of rats.

Only a small amount of Rb_1 was absorbed from the digestive tract after oral administration (100 mg/kg) to rats. The serum level of Rb_1 in rats after i.v. injection (5 mg/kg) declined biexponentially, a rapid decline (α-phase) followed by a slow decline (β-phase). The half-lives of Rb_1 were 11.6 minutes for the α-phase and 14.5 hours for the β-phase. The persistence of Rb_1 in serum and tissues in rats for long after i.v. administration was assumed to correlate with the high activity of plasma protein binding.

In one study, the degradation of ginsenoside Rg_1, Rb_1, and Rb_2 was studied in further detail. Rg_1 was decomposed to its prosapogenin in both the rat stomach and diluted hydrochloric acid, whereas Rb_1 and Rb_2 were little degraded in rat stomach but were easily converted to their prosapogenins by diluted hydrochloric acid. The ginsenosides were also metabolized to several prosapogenins by gut bacteria and enteric enzymes. The amount of Rg_1, Rb_1, and Rb_2 absorbed from the gastrointestinal tract (GI) of the rat were 1.9%, 0.1%, and 3.7%, respectively. Ginsenoside Rg_1 was excreted into rat urine bile in a ratio of 2:5. Rb_1 and Rb_2 were mainly excreted into the urine.

MILK THISTLE

Milk thistle (*Silybum marianum*) is an annual to biennial plant of the Asteraceae family. It is native principally to southern Europe and northern Africa. The crude drug consists of the ripe fruits from which the pappus has been removed. Milk thistle fruits contain 15% to 30% proteins. The main

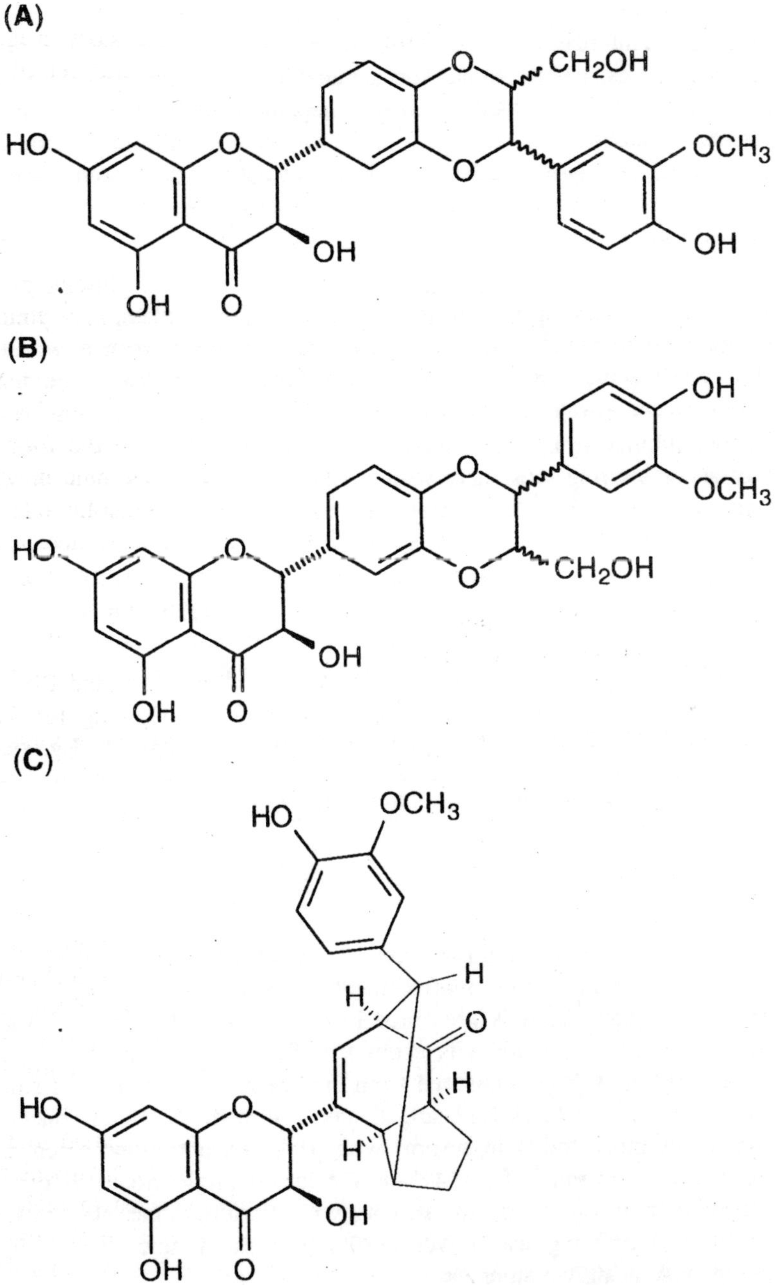

(A)

(B)

(C)

Fig. 5.11. Structures of (A) silibinin, (B) isosilibinin, and (C) silidianin.

active compounds constitute only about 2% to 3% of the dried fruits. The active principle is a mixture of flavolignans called silymarin. Silymarin, a polyphenolic extract isolated from the seeds of milk thistle, is composed mainly of silybin (50–70%), with small amounts of other silybin structural isomers, namely isosilybin, silydianin, and silychristin. The highest concentration of silymarin is found in the ripe fruits. Silibinin is the main compound, also considered to be the most active one in several paradigms. Traditionally, milk thistle fruits have been used for disorders of the liver, spleen, and gall bladder, such as jaundice and gall bladder colic. Milk thistle has also been used for nursing mothers for stimulating milk production, as a bitter tonic, for hemorrhoids, for dyspeptic complaints, and as a demulcent in catarrh and pleurisy. It is stated to possess hepatoprotective, antioxidant, and choleretic properties. Current interest is focused on the hepatoprotective activity of milk thistle and its use for the treatment of liver, spleen, and gall bladder disorders. Recently it has been shown that silibinin reduced prostate-specific antigen levels in prostate carcinoma cells lines, indicating a possible role of silibinin in human prostate cancer.

Pharmacokinetics

Studies of the pharmacokinetics of silymarin and of a silibinin–phosphatidylcholine complex preparation (IdB 1016; silipide) in humans as well as rodents have been performed. Because silibinin is the main compound, pharmacokinetic parameters of silymarin and the active principle of any silymarin-containing products are always referred to, and standardized, as silibinin.

The bioavailability of silibinin from the extract is low and seems to depend on several factors such as (i) the content of accompanying substances with a solubilizing character such as other flavonoids, phenol derivatives, aminoacids, proteins, tocopherol, fat, cholesterol, and others found in the extract and (ii) the concentration of the extract itself. The systemic bioavailability can be enhanced by adding solubilizing substances to the extract. The bioavailability of silibinin can also be enhanced by the complexation with phosphatidylcholine or β-cyclodextrin, and possibly by the choice of the capsule material. The variations in content, dissolution, and (oral) bioavailability of silibinin between different commercially available silymarin products—despite the same declaration of content—are significant.

Therefore, comparisons between studies should be carried out with caution and consider the differences between the analytical methods used and whether free, conjugated, or total silibinin is the object of measurement. Systemic plasma concentrations are usually measured, even though the site of action of silymarin is the liver, because they provide an estimate on the quantity of the drug being absorbed from the GI tract.

Silibinin

Human clinical studies

In male volunteers, after single oral administration of a standardized dose of silibinin 100 to 360 mg, plasma silibinin Cmax is reached after approximately two hours and ranges between 200 and 1400 ng/mL, of which approximately 75% is presented in the conjugated form. For total silibinin, an elimination half-life of approximately six hours is estimated. Between 3% and 8% of an oral dose is excreted in the urine, while 20% to 40% is recovered from the bile as glucuronide and sulfate conjugates. The remaining part is excreted via the feces (unchanged, not absorbed). Silibinin concentrations in bile reach approximately 100 times those found in serum, with peak concentrations reached within two to nine hours. Biliary excretion continues for 24 hours after a single dose. After multiple dose administration, no accumulation is observed. In patients with cirrhosis, the plasma C_{max} of silibinin after a single dose of silibinin 360 mg was lower (120 ng/mL) and time to C_{max} (t_{max}; 2.6 hours) slightly delayed compared with healthy volunteers.

Animal studies

The comparative pharmacokinetics of silipide (IdB 1016, a silybin–phosphatidylcholine complex) and silybin were investigated by measuring unconjugated and total plasma silybin levels as well as total biliary and urinary silybin excretion in rats following administration of a single dose (200 mg/kg as silybin). Mean peak levels of unconjugated and total silybin after IdB 1016 were 8.17 and 74.23 mg/mL, respectively. Mean AUC (0–6) values were 9.78 and 232.15 μg/hr/mL, indicating that about

94% of the plasma silybin is present in a conjugated form. Cumulative biliary (zero to two hours) and urinary (0 to 72 hours) excretion values after administration of IdB 1016 accounted for 3.73% and 3.26% of the administered dose, respectively. After silybin administration, the biliary and urinary excretion accounted for only 0.001% and 0.032% of the dose, respectively.

The use of HMPs, including their use in addition or instead of conventional drugs, is continuing to increase. Unfortunately, only limited information is available regarding the pharmacokinetics and bioavailability of herbal medicines but the awareness of this issue is increasing. In the present chapter, we focused on the bioavailability and pharmacokinetics of some of the top selling botanical products on the U.S. and European market. However, some of the most popular medicinal plants were not further discussed in this chapter, because pharmacokinetic data are not available yet (e.g., Chasteberry, Saw palmetto, or Feverfew). A reason for the lack of pharmacokinetic data could be that the active compounds of these plants are still unknown. This issue points to the fact that the determination of herbal pharmacokinetics is a unique field, which is extremely complex. Thus, the question arises—Is studying herbal pharmacokinetics of any value in the therapeutic use of the plant, especially when the drug has been used therapeutically without this information for centuries?

However, the information derived from a detailed pharmacokinetic study will help to anticipate potential botanical product–drug interactions, to optimize the bioavailability, the quality, and hence the efficacy of herbal medicines, to support evidence for the synergistic nature of herbal medicines, and to better appreciate the safety and toxicity of the plant. Because pharmacokinetic studies with herbal medicines are often complicated by their chemical complexity and by the fact that the active compounds are often unknown, it could be one future issue to assess bioavailability by measuring surrogate parameters in plasma or tissue instead of directly assaying putative active compounds in the blood. In summary, to use HMPs in an evidence-based approach and to achieve the status "*rational phytomedicine*," more experimental studies are needed to characterize the bioavailability and pharmacokinetics of botanical products.

Plant Products: Garlic, Ginkgo and Ginseng

Garlic, ginkgo, and ginseng are, respectively, the second, first, and fourth top-selling botanical supplements in U.S. retail outlets. Their retail sales are impressive, totalling US $35, $46, and $31 million, respectively. These 2002 figures look impressive but actually represent a substantial decline in comparison to those of 2001 (-17%, -35%, and -33%, respectively). Given this nevertheless huge popularity, it is important for health care professionals to advise patients responsibly about the proper use of these products.

Garlic (Allium sativum L.)

Fresh garlic bulb, dried and powdered extract, or oil extracted from the bulb have been used for medicinal purposes. The active constituents include alliin, allinase, diallyldisulfide, ajoens, and others. Alliin is enzymatically converted to allicin, the major garlic component, which is also responsible for its characteristic, sulfur-like smell. Although the best-researched pharmacological property of garlic is that of lowering total serum cholesterol levels, probably via inhibition of hepatic cholesterol synthesis, multiple additional pharmacological actions of garlic, including antibacterial, antiviral, antifungal, antihypertensive, hypoglycemic, antithrombotic, antimutagenic, and antiplatelet activities, have been described. The recommended dose is about 4 g of fresh garlic daily, which is equivalent to approximately 8mg garlic oil or 600 to 900 mg garlic powder preparations standardized to 1.3% alliin content. Adverse effects of garlic are usually mild and transient; they include breath and body odor, allergic reactions, nausea, heartburn, and flatulence. Garlic has been reported to inhibit platelet aggregation, and patients with bleeding abnormalities should be cautioned about the uncontrolled use of garlic supplements. It is recommended that garlic supplements be discontinued before major surgery. The following section describes the available evidence of pharmacodynamic and pharmacokinetic interaction between garlic and prescription drugs.

Interactions

Pharmacodynamic interaction

The primary garlic metabolite allicin has been shown to possess antiplatelet activity. Bordia showed that administration of essential oil of garlic 25 mg daily for five days resulted in significant inhibition of platelet aggregation. A case report of spontaneous epidural hematoma in an 87-year-old male was attributed to excessive garlic consumption. Because the patient was not taking any prescription medications at the time of the bleeding episode, and all laboratory parameters, including clotting factor profile, were normal, the clinicians believed that the only probable explanation for the occurrence of the hematoma was the patient's daily ingestion of four cloves (approximately 2 g) of garlic for an unspecified time period. Another case reporting bleeding disorders associated with garlic use described a 72-year-old male patient admitted to the hospital with acute urinary retention and scheduled to undergo a transurethral resection for benign prostrate hyperplasia. He was not taking any medications on admission except for years of garlic tablets consumption for "*medicinal purposes.*" However, no information regarding the strength and amount of garlic was provided. The patient experienced hemostasis and hemorrhage at the site of resection during and after surgery. The patient had a full recovery with four units of blood transfusion.

Platelet aggregation test was not done during the hospitalization. However, three months after resumption of garlic use, the patient returned to clinic and blood tests were reported to show abnormal platelet aggregation. Therefore, this pharmacological effect suggests that use of garlic could potentially increase the effect of anticoagulants. Two brief cases described that patients who had been stabilized on warfarin experienced a doubling of international normalized ratio (INR) after they took garlic products, but there were no information provided regarding the strength of garlic preparation and the duration of use, the INR values, description of symptoms, and clinical outcomes. Other than these two anecdotal cases, there are no literature reports of interaction between garlic and anticoagulants such as warfarin. Despite this lack of clinical data, especially from pharmacokinetic studies, the potential for irreversible platelet function inhibition has prompted the suggestion to discontinue garlic use at least one week prior to surgery, so as to minimize the risk of postoperative bleeding.

Garlic, as a component of curry, had also been suggested to enhance the hypoglycemic effects of the oral hypoglycemic agent chlorpropamide in a 40-year-old Pakistani woman. The estimated amount of garlic consumed by the patient was not provided. In addition, because the food product also contains karela, another ingredient reported to also possess hypoglycemic effect, it is impossible to conclude from this brief report that a cause–effect relationship exists for garlic. Since the publication of this report there have been no additional evidence to confirm the clinical observation or formal study to evaluate the likely mechanism.

Pharmacokinetic interaction

Garlic is one of the most common botanical remedies used by patients with human immunodeficiency virus (HIV), probably because of to the antiviral claim associated with its consumption, as well as the possibility of lowering total serum cholesterol, which could counteract the common side effect of hypercholesterolemia associated with the use of antiretroviral drug regimens. Conflicting results have been reported in vitro and in animals regarding the effect of garlic on drug metabolism. Piscitelli et al. investigated in human volunteers the effect of garlic supplements on the pharmacokinetics of saquinavir. Because saquinavir has negligible inhibitory or induction effect on drug-metabolizing enzymes, its use as a study drug by the investigators would minimize the potential of confounding the effect of garlic. Ten healthy volunteers participated in a three-period, single-sequence interaction study. In period 1, they received 1200mg of saquinavir three times daily for three days, with blood sampling over eight hours after administration of the 10th dose on day 4. All subjects then entered phase 2 of the study, in which they received garlic caplets twice daily for 20 days (days 5–24). In addition, saquinavir was administered concurrently for three days (days 22–24), with blood sampling over eight hours after administration of the 10th saquinavir and 25th garlic dose on day 25. Both garlic and saquinavir were discontinued for 10 days, after which saquinavir was administered in period 3 for 10 doses with blood sampling as in period 1. Adherence assessment was based on interview at each study visit and dosing calendars kept by the subjects.

Compared to baseline saquinavir pharmacokinetic parameters obtained in period 1, the use of garlic reduced the mean saquinavir area under the concentration–time curve (AUC) by 51%, and the maximum (C_{max}) and minimum (C_{min}) saquinavir concentrations by 54% and 49%, respectively. After a 10-day washout, the AUC, C_{max}, and C_{min} values were within a range of 60% to 70% of baseline values. The magnitude of the decline in concentration might result in therapeutic failure and viral rebound in patients with HIV. Based on the pharmacokinetic parameters obtained in period 3, it also appears that garlic might have a prolonged, albeit lesser, effect on saquinavir exposure. The effects of combined treatment with other protease inhibitors that are also potent cytochrome P-450 (CYP) enzymes modulators need to be further evaluated.

Although this study was not designed to address the mechanism of the interaction, the use of garlic clearly resulted in reduction of saquinavir bioavailability, possibly via induction of CYP enzymes, specifically the CYP3A4 isoform that is primarily responsible for metabolism of saquinavir. Therefore, it is likely that other drugs with significant CYP3A4-mediated metabolism could also be affected. Other mechanisms could include induction of P-glycoprotein and/or impairment of absorption. The results from this study also highlight several problems associated with interpretation of botanical product–drug interaction data from different studies. First, the study results were consistent with that of Dalvi, who showed a significant increase in CYP enzyme activity after five days administration of garlic in rats, and provided further evidence that in vivo studies employing short-term or single-dose administration and in vitro microsomal studies could provide contradictory results that might not be observed with prolonged use in clinical setting.

On the other hand, Gurley et al. showed that a four-week administration of 500mg garlic oil three times daily resulted in no significant change in phenotypic ratios of probe drugs for several CYP enzymes, including CYP3A4. While the specific constituent(s) responsible for the effect on drug metabolism is not known, Borek suggested that allicin is converted to an intermediate by-product that induces production of CYP enzymes. To minimize product variability in content, investigators from both studies used single lot of the supplement from the same manufacturer. The study of Piscitelli et al. used garlic preparation that is supposedly standardized according to allicin content, whereas product information regarding allicin content were not provided by Gurley et al. If allicin is responsible for garlic's effect on metabolism, garlic preparations that contain minimal or no allicin content might have a different metabolic effect compared to one with maximum allicin content.

Finally, even if all garlic preparations were standardized to allicin content, currently the "standardization" practices and therefore the standardized content can vary significantly from one manufacturer to another, while product inconsistency has not been demonstrated for garlic preparations, there is literature data on the disparity of constituent content among different *echinacea* products and ginseng products. As such, the choice of a specific botanical product or preparation may make a difference in the presence and magnitude of a botanical product–drug interaction.

Ginkgo (Ginkgo biloba L.)

Medicinal ginkgo products are made from the leaves of the plant, the main pharmacological constituents of which include ginkgolides A, B, C, J, bilobalide, and flavonoids. Ginkgo leads to an increase in microcirculatory blood flow, inhibition of erythrocyte aggregation, platelet-activating factor antagonism, free radical scavenging, and edema protection. These actions suggest that there is no single mechanism of action but that a complex interaction of a multitude of effects could be responsible for its many therapeutic claims, including intermittent claudication, dementia, and tinnitus. The recommended dosages of an oral standardized dry extract of ginkgo (24% ginkgo flavonol glycosides and 6% terpene lactones) are 120 to 240 mg daily for dementia and memory impairment, and 120 to 160mg daily for intermittent claudication and tinnitus. Adverse effects include gastrointestinal disturbances, diarrhea, vomiting, allergic reactions, pruritus, headache, dizziness, and nose bleeds.

Interactions

Pharmacodynamic interaction

The most frequently cited potential interaction associated with the use of ginkgo is the potentiation of anticoagulants. This is biologically plausible considering the well-documented antiplatelet effects of the various ginkgolides of ginkgo, which have been associated with cases of postoperative bleeding, spontaneous hyphema, and spontaneous intracranial bleeding.

While it is not known whether these case reports are just coincidence or actually have a cause–effect relationship, it does establish the bleeding potential of ginkgo and therefore the caution regarding additive pharmacodynamic effect with concurrent use of aspirin or anticoagulants, although currently there is little supporting data. A 78-year-old female patient who had been stabilized on warfarin for five years experienced an intracerebral hemorrhage after taking an unknown regimen of ginkgo for two months. The prothrombin time and partial thromboplastin time were 16.9 and 35.5 seconds, respectively, when the hemorrhage was discovered. Discontinuance of both warfarin and ginkgo resulted in no further bleeding episode and no vitamin K administration was required.

A 70-year-old man developed a spontaneous bleeding of the iris into the anterior chamber of the eye (hyphema) after ingesting concentrated ginkgo extract 40mg twice daily for one week. His only other medication was aspirin taken for three years at a dosage of 325 mg without any adverse bleeding event. After the bleeding episode, the patient continued the aspirin regimen but not the ginkgo supplement, and there were no additional bleeding events over the next three months. The clinicians attributed the bleeding event to an interaction between aspirin and ginkgo. Based on these and reports of spontaneous bleeding with ginkgo alone, ginkgo has also been recommended to be discontinued before major surgery. In addition, concurrent use of ginkgo with aspirin or other nonsteroidal anti-inflammatory drugs might pose an additive risk of bleeding. Further studies are necessary to confirm the potential of interaction between ginkgo and warfarin or aspirin.

An interaction between *G. biloba* administered as 80 mg leaf extract twice a day and low-dose trazodone (20 mg twice daily) was suspected in a patient with Alzheimer's disease, who took the two products together. It is postulated that a pharmacodynamic (increased gamma-aminobutyric acid-ergic activity) and pharmacokinetic mechanisms [increased metabolism of trazodone to *m*-chlorophenyl-piperazine (*m*-CPP), which acts on the benzodiazepine-binding sites and releases gamma-aminobutyric acid] contribute to the observed effect.

Pharmacokinetic interaction

Gurley et al. evaluated the effect of *G. biloba* standardized to contain 24% of flavone glycosides and 6% terpene lactones on phenotypic markers of CYP1A2, 2D6, 3A4 and 2E1. Twelve healthy individuals (six males and six females) received 60 mg of the standardized G. biloba preparation four times per day for 28 days. The four CYP phenotypes were assessed before and at the end of the 28-day study period. Although there was a trend of increased CYP2E1 activity by 23%, the effect was not statistically significant. *G. biloba* produced no significant changes in phenotypic ratios of the other three CYP isoforms. The results from this study suggested that standardized *G. biloba* preparation containing 24% of flavone glycosides and 6% terpene lactones have minimal effect on the activity of the four CYP isoforms, and therefore the metabolism of drugs mediated by these enzymes.

In general, the study results also are consistent with that of Duche et al., who reported that administration of a 13-day regimen of *G. biloba* in healthy volunteers did not affect the pharmacokinetics of antipyrine, a nonspecific marker of overall CYP enzyme activity. However, there was no evaluation of CYP2C9 activity in both studies, so it remains unknown whether any reports or concerns of interaction with warfarin could have a pharmacokinetic basis as well. Separately, Smith et al. reported in an abstract that *G. biloba* administered over an 18-day period produced a 53% increase in concentration of nifedipine, a CYP3A4 substrate. However, the amount of flavone glycosides and terpene lactones in the botanical supplement were not known, and nifedipine might not reflect CYP3A4 activity in a manner similar to that with midazolam, which is a well-recognized, specific marker for CYP3A4. In another study, 14 patients with Alzheimer's disease received the acetycholinesterase inhibitor donepezil 5 mg per day for at least 20 weeks with steady-state donepezil concentrations of 22.7 ± 10.3 ng/mL and a Mini-Mental Scale Examination (MMSE) score of 9.0 ± 7.8. Concurrent administration of 90 mg/day of *G. biloba* for 30 days did not alter the concentration (24.4 ± 12.6 ng/mL) and MMSE score (8.7 ± 7.7), suggesting that there is no adverse interaction between the *G. biloba* and donepezil when used together in patients with Alzheimer's disease. However, the amount of flavone glycosides and terpene lactones in the botanical supplement were also not known.

Asian Ginseng (Panax ginseng)

There is considerable confusion about terminology for the different species known collectively to the consumer as ginseng; Asian ginseng is also sometimes called Chinese ginseng, Korean ginseng, ninjin (Japanese), or true ginseng. It is often confused with Siberian ginseng (*Eleutherococcus senticosus* Maxim), which belongs to the same family (*Araliaceae*) but is a different genus. Another popular ginseng product is American ginseng or Canadian ginseng (*Panax quinquefolius* L.)

The dried roots of Asian ginseng are used for medicinal purposes and its main constituents are triterpene saponins known as ginsenosides or panaxosides. The pharmacologic actions of ginseng include immunomodulatory, anti-inflammatory, antitumor, smooth muscle relaxation, stimulant, and hypoglycemic effects. The recommended dosage is 200 mg daily of standardized extract (4% total ginsenosides). Reported adverse effects include insomnia, diarrhea, vaginal bleeding, mastalgia, swollen tender breasts, increased libido, manic episodes, and a possible cause of Stevens–Johnson syndrome. A "*Ginseng abuse syndrome*" (consumed dose approximately 3 g daily) has been described with symptoms such as hypertension, sleeplessness, skin eruptions, morning diarrhoea, and agitation. Doses of 15 g daily and over were associated with depersonalization, confusion, and depression.

Interactions

Pharmacodynamic interactions

The most frequently reported interactions are those with monoamine oxidase inhibitors (MAOIs). There were reports in the literature of a potential interaction between ginseng and the MAOI phenelzine. The symptoms described included insomnia, headache and tremulousness in a 64-year-old woman when ginseng was added to her phenelzine regimen. Jones and Runikis reported that the use of ginseng in a 42-year-old woman treated with phenelzine was associated with manic-like symptoms, irritability, tension headache, and occasional vague visual hallucinations. After discontinuing the ginseng, the patient's symptoms resolved with only a few headache episodes thereafter. The clinicians considered that her other medications including lorazepam and triazolam were not contributory factors and the symptoms were mostly likely associated with ginseng–phenelzine interaction. However, ginseng is a stimulant and one of the common side effects of the use of ginseng is insomnia, and MAOI can also cause insomnia and headache; it is difficult to determine whether the described symptoms in these two reports are related to the use of each product alone or are attributed to botanical product–drug interaction.

Ginseng has the potential to interfere with the coagulation cascade and therefore interact with warfarin. However, there is no literature report of increased INR with concurrent use of both drugs. Interestingly, the use of ginseng has been associated with decreased INR. Janetzky and Morreale described a 47-year-old man with a mechanical heart valve, who was stabilized on warfarin with therapeutic INR within the range of 3.0 to 4.0. Two weeks after the patient took an unknown strength of ginger (*P. ginseng*) three times daily, his INR decreased to 1.5. Upon discontinuance of the ginger preparation, his INR returned to 3.3 two weeks later. Fortunately, there were no adverse effects during the two-week period of subtherapeutic INR. On the other hand, a thrombosis of a prosthetic aortic valve was attributed to the use of ginseng in a patient who had been stabilized on warfarin regimen (dosage regimen and INR values not reported). Because there is no human pharmacokinetic study evaluating this potential interaction or metabolic study data showing an effect of ginseng on CYP2C9, the primary CYP isoenzyme responsible for metabolism of the pharmacologically more potent S-isomer of warfarin, the reported paradoxical decrease in INR associated with ginger use is difficult to explain and evaluate. An animal study in rats demonstrated no effect of ginseng on either absorption or elimination of a single dose of warfarin. There were also no changes in prothrombin time after steady-state dosing of warfarin. The confusion about ginseng terminology mentioned above extends to case reports of ginseng–drug interactions, and it is not sure that all of these pharmacodynamic interactions are related to *P. ginseng*. In particular, the intake of concomitant drugs is a potentially significant confounding factor. Any conclusions about causality seem premature at this time.

Pharmacokinetic interactions

Anderson et al. evaluated the effect of *P. ginseng* on the 6-β-hydroxycortisol to cortisol ratio, a marker of CYP3A4 activity. Ten male and 10 female healthy subjects were given 100 mg of *P. ginseng* standardized to contain 4% ginsenosides (Ginsana) twice daily for 14 days. Comparing the 6-β-hydroxycortisol-to-cortisol ratio before and after the 14-day regimen of *P. ginseng* showed no appreciable difference, suggesting that there is no enzyme induction effect on CYP3A4. The result from this study confirms the data of Gurley et al. who administered 500 mg of *P. ginseng* standardized to contain 5% ginsenosides. It is of note that both Asian and Siberian ginsengs contain glycosides with structural similarities to digoxin, and both ginseng products have been reported to interfere with fluorescent polarization immunoassay determination of digoxin concentrations. Even though not an in vivo interaction per se, the digoxin-like immunoreactive substances associated with the use of ginseng cause false elevation of digoxin concentrations and could result in inappropriate dosage adjustment. There is little doubt about the therapeutic potential of some of the herbal medicines. Their safety profile is equally encouraging. The scope for botanical product–drug interactions, however, seems considerable. This is sharply contrasted by the paucity of actual clinical reports of such interactions occurring in practice. There are at least two explanations for this overt discrepancy. Firstly, interactions could indeed be rare. Secondly, the paucity of reports could be the result of underreporting. Under-reporting of adverse effects is significant and, in the realm of herbal medicine, it is likely to be even larger than with conventional drugs. In the absence of sufficient data it is impossible to decide which explanation is correct. What we can say, however, is that the subject of botanical product–drug interactions is potentially important and grossly under-researched. Thus it warrants further systematic study.

6

IN VITRO INHIBITION WITH PLANT PRODUCTS

The nature of plants having secondary metabolites as defensive agents greatly increases the expectation that there will be interactions with other botanical products and drugs. If well-established traditional botanical products are used according to directions, they are likely a "low risk." Risk increases when botanical products are combined with conventional drug therapies and lies in the possibility of unknown natural product–drug interactions. Other risks include product deviation due to misidentification of species, the lack of standardization, or adulteration. Most of the interactions have been reported with cytochrome P-450 (CYP) 3A4, but there are interactions with other metabolism enzymes and transport proteins.

The intent of this review is to provide an understanding of what constitutes a representative product, experimental test conditions, and the presence of contaminants that can affect the interpretation of these interactions using an in vitro assay system. Zou et al. evaluated the effects of 25 purified components of commonly used botanical products and found that many significantly inhibited one or more of the cDNA human P450 isoforms at concentrations of less than 10 μM. These findings are consistent with that of many other botanical components and suggest that there may be a potential for botanical products to affect drug disposition. However, negative findings with such purified biomarkers do not preclude the possibility that the combined total of the plant constituents could have an effect on drug safety and efficacy.

NATURAL PRODUCT VARIATION

Unlike conventional drugs, botanical products are complex mixtures that have inherent variation due to environmental and genetic factors affecting the fresh product, processing and manufacturing conditions, and the possible presence of nonactive components, which need to be converted to the active moiety. In addition, there are individual variations in the amount taken, form, manner of preparation, length of use, combination with other products, genetic characteristics, and health status of the user. Botanical products can be used either fresh or as a formulated single entity or in the blended dosage form. These forms include powder or soft gel liquid-filled capsules, cosmetics, liquids, ointments, tablets, tinctures, or suppositories. Exposure to botanical products may be intentional or fortuitous through their use in beverages, cosmetics, and foodstuffs.

Botanical bulk products may be sourced from several regions or countries and may have unique genotypic and phenotypic characteristics, which can confound interpretation of adverse event reports and product selection for clinical examination. The examination of one or even a few samples may inadvertently lead to the testing of a single chemotype. A chemotype is a variety or population of plants belonging to one particular species, which differ chemically from others of that species. These differences are genetic not phenotypic. Examples include *Melaleuca alternifolia*, volatile oils from single plants of *Thymus serpylloides* ssp. *gadorensis*, and peel and leaf oils of 43 taxa of lemons and limes obtained from fruits and leaves collected from trees under the same climatic and cultural conditions where there are multiple chemotypes. Together, these factors can compromise the testing process.

An interesting example is the report from Fukuda et al. who determined the amounts of three furanocoumarins in 28 white grapefruit juices, and orange, apple, lemon, grape, and tangerine beverages. Considerable differences were observed on the contents among commercial brands and also batches. The contents were determined to be 321.4 ± 95.2 ng/mL GF-I-1, 5641.2 ± 1538.1 ng/mL GF-I-2, and 296.3 ± 84.9 ng/mL GF-I-4 in white grapefruit juices. None was detected in beverages from orange, apple, grape, and tangerine, although trace amounts of GF-I-2 and GF-I-4 were found in lemon juice. The average levels of these furanocoumarins were lower in the juice from red grapefruit than in that from white fruit. This variation may reflect both genetic chemotype and phenotypic differences. The highest level of these components was found in the fruit meat. Sources of variation include the distribution of the constituents into various compartments within the fruit and procedures used in extracting juice from the whole fruit. Grapefruits exposed to freezing temperatures produce more naringin and less limonin.

It was reported that even under stable conditions, temperature and humidity could modulate naringin concentrations. Naringin, limonin, and nomolin reach peak levels in the early development stage of the fruit and decline as the fruit matures. Processing factors can include the pressure used to extract the juice, removal of bitter components, and the balancing of the final juice product by adding back the volatile essential oils and pulp. The clinical effects of the products containing any of the above chemotypes may vary; hence, it is difficult to attribute an effect without additional studies or set limitations on how the data can be extrapolated.

Labeling Information

Currently, there is little consistency in the information provided on botanical product labels in Canada. Some labels are clear as to the amount of botanical product and excipients, stated indications, contraindications, and warnings. In some cases, the information is confusing or difficult to understand, confounding the comparison of products. Examples of confusing or potentially misleading information on some valerian, milk thistle, St. John's wort (SJW), and echinacea product labels are provided. Three of these examples demonstrate the confusion created by or within the industry on chemical names. Information printed on some valerian root product labels stated that the products were standardized to valerenic acid and valeric acid. Although the names are similar, valeric acid is a five-carbon molecule that is not related chemically or pharmacologically to the larger C15 valerenic acids.

Silymarin is considered the active constituent of the milk thistle seed, but it is not a single compound but a descriptive term for several flavonolignans. Constituent analysis of five milk thistle products identified six constituents (representative amount): taxifolin (3.3%), silichristin (23.6%), silidianin (5.3%), silybin A (20%), silybin B (30.7%), and isosilybin (17.3%). The total amounts of silybin A and B in the five different products analyzed ranged from 45.7% to 61%. The biological effect of each constituent is not known; hence, spectrophotometric analysis would not provide sufficient information for a critical comparison of these products.

Many SJW products are standardized to either 0.3% hypericin or 4% hyperforin. Some product labels stated standardization to hypericins. Together hypericin and pseudohypericin have been referred to as total hypericins. Other biosynthetic precursors such as isohypericin, protohypericin, and proto-pseudohypericin may also be present, which are indistinguishable when analyzed spectrophotometrically.

The perceived message from this standardization is that the remaining 96% to 99.7% of the botanical material has no pharmacological effect. This is particularly disturbing with regard to SJW where pharmacological activity has been attributed to at least two dozen constituents or groups of compounds present in *Hypericum* extracts (8,9), including quinones such as hypericin and pseudohypericin, flavonoids such as hyperoside, and phloroglucinols such as hyperforin (about 8%), and water-soluble components such as organic acids.

Hence, standardization to one or two SJW constituents and testing of one to two products should be viewed as a starting point but not the end of the comparative process. In the end, simple unit weight was used to prepare solutions to determine the potential of each sample to inhibit metabolism of the test substrate. There are confounding issues with such an approach, but it does provide a simple quantitative basis for comparison of products. Hypericin, pseudohypericin, and hyperforin levels in the products examined varied widely within and between products. The results obtained from this study showed wide variation in inhibitory potential of teas and tablets, which does not correlate with the constituent levels. The final example of potentially misleading or confusing product labels is *Echinacea* (Asteraceae). The taxonomy has been revised by morphometric analysis to four species with distinct varieties. Two *Echinacea* species are widely used as botanical medicines: *Echinacea angustifolia* (syn *Echinacea pallida* var. *angustifolia*) and *Echinacea purpurea*. A third species, *E. pallida* (*E. pallida* var. *pallida*), has been widely used in Europe. Over 70 compounds were identified in the headspace volatile components of roots, stems, leaves, and flowers of *E. angustifolia*, *E. pallida*, and *E. purpurea* analyzed by gas chromatography/mass spectrometry (MS).

The constituents of echinacea include alkamides, caffeic acid, glycoproteins/ polysaccharides, and ketoalkenynes. In our study, label information indicated that two products contained 4% phenols and three products (NRP 70, 71, 73) were standardized to contain 4% echinacoside. Echinacoside is not, however, a good marker for the genus Echinacea. This is also indicative of the confusion in the botanical product industry, because echinacoside is a marker only for *E. angustifolia*. Echinacoside was not detected in a number of products. Where detectable, the level ranged from above detection up to 32.4 mg/g. Echinacea Special tea is an example of a blended tea, which is not readily evident from the label. The Special tea tested contained multiple ingredients such as lemon grass, peppermint leaf, spearmint leaf, triple echinacea root (*E. angustifolia*, *E. purpurea*, and *E. pallida*), liquorice root, ginger root, wild cherry bark, cinnamon bark, fennel seed, astragalus root, cardamom seed, rose hips, elder berry, burdock root, mullein leaf, clove bud, black pepper, and standardized E. purpurea root extract (4% phenols). All echinacea extracts markedly inhibited CYP-mediated metabolism. The findings with aliquots of the soft gel product extracts were variable. Inhibition was moderate to high toward CYP2D6 and 3A4, but only NRP 69 and 72 had an inhibitory effect against CYP2C9. In addition, NRP 71 did not inhibit CYP2C19-mediated metabolism.

Manufacturing and Storage

Echinacea products provide an example of the large variation in manufactured products. The whole plant has been used for therapeutic purposes, but single-entity product can consist of *E. purpurea* herb extracts, combination root and herb extracts, and teas, or they could be blended as root extracts of the three species, herbal teas with other botanicals, and blends of *E. purpurea* and *E. angustifolia* leaves, stems, and flowers plus a dry extract of *E. purpurea* root. The relative amounts of aerial and root stock materials in formulated products vary widely in composition. All plant tissues, irrespective of the species, contained acetaldehyde, dimethyl sulfide, camphene, hexanal, β-pinene, and limonene (12). The main headspace constituents of the aerial parts of the plant are β-myrcene, α-pinene, limonene, camphene, β-pinene, *trans*-ocimene, 3-hexen-1-ol, and 2-methyl-4-pentenal. The major headspace components of the root tissue are α-phellandrene (present only in the roots of *E. purpurea* and *E.*

angustifolia), dimethyl sulfide, 2-methylbutanal, 3-methylbutanal, 2-methylpropanal, acetaldehyde, camphene, 2-propanal, and limonene. Aldehydes, particularly butanals and propanals, make up 41% to 57% of the headspace of the root tissue, 19% to 29% of the headspace of the leaf tissue, and only 6% to 14% of the headspace of flower and stem tissues. Terpenoids including α- and β-pinene, β-myrcene, ocimene, limonene, camphene, and terpinene make up 81% to 91% of the headspace of flowers and stems, 46% to 58% of the headspace of the leaf tissue, and only 6% to 21% of the roots. The relative amounts of these products will vary greatly in the manufactured and fresh product making it difficult to choose a representative product.

A second example is garlic products. As with other botanicals, garlic can be processed by different procedures including drying or dehydration without enzyme deactivation, aqueous or oil extraction, distillation, and heating, including frying and boiling, processes that contribute to the variability of the extracts and complex nature of these products. These garlic preparations can then be formulated into single and blended products as oils of steam-distilled garlic, aged garlic, garlic macerated in vegetable oils, garlic powder, and gelatinous suspensions. Lawson et al. conducted an extensive phytochemical analysis of organic sulfur compounds of representative fresh and commercially available garlic products and found a wide variation in composition and chemical profile of sulfur compounds. Garlic powders suspended in a gel did not contain detectable amounts of nonionic sulfur compounds. Thiosulfinates were only recovered from garlic cloves and powders. Vinyldithiins and ajoenes were only detected in garlic macerated in vegetable oil. Diallyl, methyl allyl, and dimethyl sulfides were exclusively found in oil of steam-distilled garlic.

Typical steam-distilled garlic oil products contained similar amounts of total sulfur compounds as the total thiosulfinates released from freshly homogenized cloves; however, oil-macerated products contained about 20% whereas the garlic powders varied from 0% to 100%. Garlic is aged to reduce the content of sulfur compounds, such as alliin and the odor commonly associated with garlic. Gel and aged garlic in aqueous ethanol products did not have detectable levels of these nonionic sulfur compounds. Analysis of thiosulfinates from various *Allium* sp. revealed a threefold order of magnitude variation among species. Common garlic (*Allium sativum*) and wild garlic had the highest levels, while Chinese chives and the leek known as elephant garlic (*Allium ampeloprasum*) had intermediary levels. Environmental conditions were also found to influence the total thiosulfinate levels. As would be expected, storage conditions affect constituent content.

Fresh garlic stored at 4EC for two months was found to have decreased levels of γ-glutamly-S-allylcysteine with increased levels of alliin and allicin. Lawson et al. considered the increase in alliin and allicin contents in stored garlic to be a result of sprouting. In a study considering four product classes, there were marked differences in how the various extracts inhibited CYP-mediated metabolism. The effect was lowest against CYP2D6. Extracts from common garlic exhibited a similar inhibitory effect on all CYP3A isoforms. Chinese and elephant garlic had a lesser inhibitory effect on 3A7; Chinese garlic extracts also had a lesser inhibitory effect on CYP3A5-mediated metabolism. These findings were confirmed in a broader study with more products.

Extracts from three fresh garlic varieties were screened for their effect on CYP 2C9*1–, 2C9*2–, 2C19–, 2D6–, and 3A4–mediated metabolism. All three varieties had a slight inhibitory effect on 2C9* 1-mediated metabolism, but highly stimulated metabolism of the marker substrate with the 2C9*2 isoform. The extracts had negligible to no effect on 2C19- and 2D6-mediated metabolism. However, all extracts strongly inhibited 3A4-mediated metabolism. The effects of aqueous extracts from aged garlic capsules and the three fresh varieties were examined for their ability to interact with human P-glycoprotein membranes. Relative to 20 μM verapamil as the positive control, the aged, common, and Chinese phosphate buffer extracts had moderate levels of product-stimulated, vanadate-sensitive adenosine triphosphatase activity. Elephant garlic was inactive. Spices were analyzed for their capacity to inhibit in vitro metabolism of drug marker substrates by human CYP isoforms. Aliquots and infusions of all natural product categories inhibited 3A4 metabolism to some extent. Of the spices tested with 2C9, 2C19, and 2D6, most demonstrated significant inhibitory activity. Spices showed species-specific isoform inhibition with cloves, sage, and thyme having the highest activity against the four isoforms examined.

Blended and single-entity herbal teas, and some bulk spices were analyzed for their capacity to inhibit in vitro metabolism of marker substrates. Aliquots and infusions of all natural product categories inhibited 3A4 metabolism to some extent. Of the aliquots tested with 2C9, 2C19, and 2D6, many demonstrated significant inhibitory activity on the metabolism mediated by these isoforms. Herbal tea mixtures were generally more inhibitory than single-entity herbal teas. Single-entity herbal teas showed species-specific isoform inhibition with SJW and goldenseal having the highest activity against several isoforms.

Traditional Chinese medicine (TCM) includes both crude Chinese medicinal materials (plants, animal parts, and minerals) and Chinese proprietary medicine (CPM). The quality of TCM, as with other products, can vary emphasizing that there can be broad differences in these products. In a study undertaken with 12 purported TCM products, one was found to be a CPM containing three drugs. Extracts from most products inhibited at least three of the four CYP450 isozymes examined in a range from 25% to 100%. All liquid samples markedly inhibited the metabolism of all four isozymes. De le ke chuan kang and Rensheng dao were the strongest CYP450 inhibitors. These in vitro findings helped demonstrate that TCMs can inhibit CYP450 2C9–, 2C19–, 2D6–, and 3A4–mediated metabolism. TCMs need to be examined further under clinical settings to determine if potential interactions occur, which affect the safety and efficacy of conventional therapeutic products.

Experimental Factors

Many of the reported interaction studies with botanical products have used a single aqueous or organic extract. However, botanical products contain several major classes of biologically active constituents, with differing chemical characteristics that can directly or indirectly affect drug disposition. A series of solvents ranging from hexane, with high lipophilicity, down to water have been used to sequentially extract different botanical products. In this representative example, capsule material from aged garlic extract, and the two fresh varieties of garlic and one leek were extracted sequentially. Extracts were reduced to dryness and reconstituted into methanol prior to testing for their effect on 3A4-mediated metabolism. Most extract fractions exhibited a high inhibitory activity against the isoforms studied. There was varietal variation. Inhibition results with hexane (136%) and chloroform (116%) extracts suggest the presence of botanical fluorescent quenching substances. This observation is consistent with the concerns for intrinsic fluorescence and quenching as confounding variables in these assays noted by Zou et al.

As was determined with other botanical products, several 100% inhibitions were evident with this simple extraction sequence. A series of nonsequential extracts with these solvents (data not reported) also revealed high inhibitory activity in all extracts. Selective pH extraction of the Chinese garlic bulb showed significant (50–80%) inhibitory activity in the strong acid, weak acid, neutral, and basic fractions against CYP3A4. Because differences in the inhibitory effects of aqueous and methanolic extracts of fresh and aged garlic cloves on CYP3A4-mediated metabolism were previously noted, the three varieties were extracted under four different conditions. Results varied with variety, but in general, the distilled water and phosphate buffer extracts gave the strongest overall inhibitory effect on CYP450-mediated metabolism of the marker substrates.

Many plant constituents are conjugates, the main form being glucosides. Seven soybean varieties were analyzed by high-performance liquid chromatography (HPLC) for the isoflavones, daidzein and genistein, and their respective glycoside derivatives to determine if the amounts of these compounds could reliably predict the activity of the variety or year of harvest. Genistein levels ranged from 5.8 to 28.7μg/g and daidzein levels from 0 to 42.6 μg/g. The glycosides daidzin and genistin were present in much larger amounts ranging from 198 to 792 μg/g and from 458 to 1261 μg/g, respectively. The free aglycones accounted for less than 7% of the total in these samples. Neither the concentration of the individual compounds nor their total correlated with the inhibition of CYP3A4-mediated metabolism across genotypes and years. In a comparative inhibition test, the glycones were inactive relative to the corresponding aglycones.

Dissolution

As with formulated pharmaceuticals, dissolution of constituents from teas is an important consideration. Visual examination of the contents from tea bags from several botanical products showed that there were inter- and intraproduct differences in particulate size. Particle sizes ranged from fine powder to substantially intact leaves and stems. Under controlled tea-brewing conditions, there were marked differences in the dissolution of constituents relative to the amount and temperature of the water, degree of agitation, and time. Many of these factors are individualistic, so one individual may be exposed to different amounts of constituents relative to another person. In a representative study, several botanical products were examined as tea bag infusions at different temperatures. In this example, three different patterns were noted. The initial 10-minute values for the echinacea special tea were markedly higher than that for the other botanical products, but the inhibition curve only increased slightly with time. Two products, goldenseal herb, and echinacea and goldenseal, had low initial values, which nearly doubled after a second 10-minute incubation. The third pattern with feverfew showed a linear increase throughout the incubation period.

Westerhoff et al. studied the dissolution characteristics of several SJW products under biorelevant conditions. Components of SJW have a broad spectrum of polarity and solubility, and representative compounds from each group were examined. Although labelling indicates that several of the products studied should be pharmaceutically equivalent, dissolution under biorelevant conditions revealed that they have quite different release profiles and cannot be considered interchangeable. It was concluded that biorelevant dissolution testing can be a powerful tool for comparing botanical products as well as synthetically produced drug products. Jurgenliemk and Nahrstedt examined the dissolution of water-soluble phenolic constituents of Hypericum perforatum from a medicinal tea and a coated tablet formulation and found different dissolution profiles.

In general, the flavonoid glycosides were well dissolved, followed by flavonoid aglycones and hypericin, while hyperforin was only detectable at a very low level. Interestingly, hypericin exhibited much better extraction and dissolution rates than the similarly lipophilic hyperforin. When determining the octanol– water partition coefficient, it became obvious that the solubility of pure hypericin in water increased upon addition of some phenolic constituents typical for Hypericum extracts. Most effective in solubilizing hypericin was hyperoside, which increased the concentration of hypericin in the water phase up to 400-fold in this model.

Stability

As with drugs and purified biomarkers, thermal- and photostability of botanical products are the factors that must be considered. Commercial dried extract and capsules of SJW were evaluated under harmonized test conditions. Photostability testing showed all the constituents to be photosensitive in the tested conditions. However, different opacity agents and pigments influenced the stability of the constituents. Amber containers had little effect on the photostability of the investigated constituents. Long-term thermal stability testing showed a shelf life of less than four months for hyperforins and hypericins, even when ascorbic and citric acids were added to the formulation.

Photostability affected several botanical products. To overcome this confounding factor, samples of all products were extracted and tested under reduced lighting conditions, frequently under F40 gold fluorescence lighting.

Marked variations in the stability of 21 tinctures and 13 related single- entity plant compounds were noted. Bilia et al. investigated the stability of 40% and 60% v/v tinctures of artichoke, SJW, calendula flower, milk thistle fruit, and passionflower. The investigation showed a very low thermal stability of the constituents from accelerated and long-term testing as determined by HPLC–diode array detector and –MS analyses. Stability was related both to the class of flavonoids and water content of the investigated tinctures. Shelf life at 25°C of the most stable tincture (passion-flower 60% v/v) was about six months, whereas that of the milk thistle tinctures was only about three months. The stability of artichoke and SJW tinctures also were shown to be variable and seem to be related to the water content of the preparations.

Test Conditions

The first step with all botanical products is authentication to confirm identity. Bulk single-entity products can occasionally be inspected visually and authenticated by comparison to reference materials. However, most products require authentication through phytochemical analysis with comparison against authentic marker substances, preferably with a chromatographic stage to separate constituents for individual assessment. Authentication generally confirms the presence of a botanical product but does not exclude the possibility of the presence of other botanical products, adulterants, or contaminants. In some cases, historical information may suggest that samples be examined in depth for potential contaminants. Ideally, test samples of a product should be prepared from a minimum of five units mixed together to provide a representative sample of a particular product or lot. Test samples should be reduced to a consistent size using a mortar and pestle, ball mill, or blender. In our studies, we routinely begin with either a 100 mg/mL aqueous suspension or a 25 mg/mL ethanolic suspension that is then reduced further to constant particle size in a polytron to facilitate reproducible extraction for one minute. Standardization of this phase of testing is critical to ensure intra-and interday reproducibility in testing. Soft liquid–filled gel capsules are cut open, and the contents emptied into a 1.5 mL microfuge tube and extracted. The mixtures are centrifuged for 18 minutes in a microcentrifuge at a high setting to give a particulate-free stock solution.

Aliquots of aqueous or organic extracts are screened for their ability to inhibit the major human cDNA–metabolizing CYP isozyme CYP 2C9, 2C19, 2D6, and 3A marker substrates using an in vitro fluorometric micro- titer plate assay (4), modified from the one reported by Crespi et al., with balanced amounts of specific activity and protein content. Despite the inherent limitations of these test substrates, these probes provide a quantitative basis for additional studies. Briefly, assays are performed with either a 2 to 4 μL of an organic extract or up to 10 μL of an aqueous extract in a total volume of 200 μL in 96-well, clear-bottom, opaque-walled microtiter plates. The complex nature of the extracts requires blank and test controls to evaluate the effects of intrinsic fluorescence and quenching as confounding variables in these assays. We include controls for both the blank and test product using denatured enzyme with the extraction solvent. Where possible, studies should include one or more test substrates representative of the isozyme as positive control(s).

The effect of botanical products on the expression of drug-metabolizing enzymes or transport proteins can be examined in cell culture with established cell lines or primary human hepatocytes. The cells are incubated under standard conditions and treated with the blank, positive, or negative controls and the botanical extracts. Vehicle use should be consistent in all cultures. Multiple time points should examine the immediate and prolonged effect of these treatments. After treatment, total RNA and/or microsomes are prepared from the harvested cells using standard methodologies.

All assays should be performed under reduced or F40 gold fluorescence lighting to minimize the potential for photodecomposition or activation. Assays are run in triplicate to determine percent inhibition. The tests are repeated at least once with a freshly prepared sample. If there is greater than 15% coefficiency of variation, the samples are run at least one additional time. When the reaction mixture is incubated within the plate reader, readings are taken immediately and at set times throughout the prescribed incubation period as established by the microsome supplier. For assays incubated outside of the plate reader, reactions were stopped in accordance with the product test procedure.

In studies where either intrinsic fluorescence or quenching is a confounding variable, the botanical product should be examined in an assay using a chromatographic separation step with a representative probe substance for the isozyme being examined.

Product Selection for Clinical Studies

In vitro testing with cell-free systems can provide only qualitative information on the inhibitory potential of the particular extract from a specific sample to affect the isozyme-mediated metabolism of a test substrate. There is no a priori basis to extrapolate either positive or negative in vitro inhibitory results to acute or chronic clinical exposure. Despite this caveat, there were, however, clinical reports with echinacea, garlic, and SJW that these botanical products can affect drug pharmacokinetics. The explanation being that in some cases, prolonged exposure to an inhibitory product led to reduced plasma levels of a probe substance presumably due to induction of a transport protein or metabolic enzyme. Negative in vitro findings are limited to the extract and the inherent weakness of these probes' substrates; only further testing with additional extracts and test products can truly demonstrate the potential of these botanical products to cause interactions. In addition, there is no a priori basis to extrapolate in vitro findings from an single active ingredient (SAI) to the complex botanical product.

The number of fresh varieties, dosage forms, and formulations in combination with the variability in botanical material make it impossible to evaluate all of these products in animal models or clinical trials. As a minimum, several products used by the patient community should be obtained and authenticated. The testing and selection criteria should include multiple-lot testing, cost, and product availability, and take into consideration how these products are used. Drug combinations are being examined increasingly in comparative clinical trials with a goal of enhancing efficacy with the same or fewer adverse events. In many instances, the amount of drug exposure, the total drug load, is a major contributing factor to the safety of the combinations.

Four SJW products with similar CYP3A-inhibitory activity were evaluated for their effects on cell viability, the potential of such preparations to modulate induction of nitric oxide and CYP1A1/2-mediated ethoxyresorufin *O*-deethylase (EROD) activity in glial cell cultures. SJW A, B, and D had little effect on EROD activity. SJW C had the highest inductive effect on EROD activity. SJW B and C treatment resulted in the highest nitric oxide levels, raising concern for potential central nervous system toxicity. SJW A and D produced significant lactic dehydrogenase–released cell toxicity. Which product should be studied? The difficult decision is whether to choose an average or a superior product because the results of the study will subsequently be viewed as representative of all related products.

Synergistic interactions are of vital importance in phytomedicines and under-pin the philosophy of herbal medicine. Spinella emphasizes that, in addition to searching for more potent mechanisms, one must consider the additive and supra-additive effects of a plant's multiple constituents. Synergy may occur through pharmacokinetic and/or pharmacodynamic interactions. Synergistic interactions are documented for constituents within a total extract of a single botanical product, as well as between different botanical products in a formulation. Thus interactions with pharmacologically active secondary metabolites are not unexpected because these constituents are part of the plant defensive mechanisms.

In vitro studies can help determine the potential for adverse effects associated with botanical product–drug interactions. Accumulated findings from many studies have confirmed our earlier observations that there is seldom a direct correlation between levels of the purported active ingredient biomarkers and the potential for these extracts to affect P450-mediated metabolism. At best, in vitro studies with an inhibitory finding in these cell-free extracts can only provide a qualitative basis for further studies. A negative finding, particularly with an SAI, can eob oe interpreted to mean that there is no activity under the stated test conditions. Unfortunately, some negative findings have been erroneously taken to mean that related botanical products containing this SAI would not affect drug disposition. The dilemma for all health care professionals and consumers is that what is apparently safe with one botanical product and pharmaceutical, or another botanical product, may be neither safe nor effective in another combination or patient population.

7

Drug Interactions with Plant Products

The use of botanicals by consumers in North American and European countries has significantly increased over the last decade, with one survey showing an almost 10% increase in usage from 1990 to 1997. Although the efficacy of some botanicals has been documented, there is concern regarding the perceived safety of these products, particularly with respect to the lack of research and knowledge on botanical-drug interaction potential and significance. As more consumers use botanicals for various purposes, the likelihood of concurrent use of botanicals with prescription and/or over-the-counter medications, as well as the potential of pharmacokinetic and/or pharmacodynamic botanical-drug interactions will increase. The survey conducted by Eisenberg et al. reported that as many as 15 million adults in 1997 took botanical supplements concurrently with prescription drugs. Over the subsequent years, there has been no change in this usage pattern, with as many as 16% of consumers surveyed indicating concurrent use of botanical dietary supplements and prescription drugs. This continued trend of concurrent use of drug and botanical supplements, together with an underreporting of such use and a general lack of knowledge of the interaction potential, poses a challenge for the health care professionals and a safety concern for patients and/or consumers. Indeed, several clinically important botanical-drug interactions have been reported and some have resulted in altered efficacy and/or toxicity of the drug. The purpose of this chapter is to present an overview of common mechanisms of botanical-drug interactions, and using specific literature examples, discuss challenges associated with the interpretation of available study data or reports, and the of prediction of botanical-drug interactions.

Mechanism of Plant-drug Interactions

In essence, interactions between pharmacologically active botanicals and drugs involve the same pharmacokinetic and pharmacodynamic mechanisms as drug-drug interactions. Pharmacokinetic interactions may involve alteration in absorption, distribution, metabolism, or excretion of the affected drug or botanical. Pharmacodynamic interactions, on the other hand, alter the relationship between the drug concentration and the pharmacological response for a drug or botanical. Although most pharmacodynamic interactions reported in the literature and reviewed in this chapter focus on adverse effects as an outcome, not all pharmacodynamic botanical-drug interactions result in an undesirable effect. Animal studies have shown that the combination of an extract of the Chinese medicinal plant, *Tripterygium wilfordi*, and cyclosporine significantly increased the heart and kidney allograft survival compared to cyclosporine administered alone.

The effective cyclosporine dose required for 100% kidney allograft survival was reduced by 50% to 75% in the presence of the botanical extract. The immunosuppressive activity associated with the use of the botanical extract needs to be studied further in humans in order to explore the clinical potential of their combined use, perhaps by a mechanism similar to that for the ketoconazole and cyclosporine interaction. Most pharmacokinetic and pharmacodynamic botanical–drug interaction studies and clinical cases in the literature evaluated the quantitative effect or reported the consequence of adding a specific botanical to a drug regimen, and not the other way around. This likely represents the challenge of not knowing the identify and constitution of the botanical or botanical product, the difficulty of measuring concentration of a specific botanical or its active ingredient(s), and the more common scenario of patients using botanical preparations on a sporadic basis, while being stabilized on a drug regimen. Nevertheless, the pharmacokinetic profiles of different botanical products are currently being investigated. A better understanding of botanical pharmacokinetics in humans is needed if the prediction of botanical–drug interactions is to be successful.

ALTERED PHARMACOKINETICS

Drug Absorption

While reduction in the extent of drug absorption can potentially occur as a result of increased intestinal transit time, secondary to the use of botanicals containing anthranoid laxatives (e.g., aloe; *Aloe* spp.) or as a result of complex formation between botanical constituents (e.g., polyphenols in green tea), clinical cases of these types of interaction have not been reported. Nevertheless, based on the well-documented chelation of fluoroquinolones by different divalent and trivalent cations (e.g., sucralfate and didanosine) with the resultant significant decrease in fluoroquinolone concentrations and potential treatment failure, there remains the possibility that natural product supplements containing cations might also exert the same undesirable effect. Indeed, concurrent administration of an aqueous extract of fennel, the fruit of *Foeniculum vulgare*, in rats was shown to reduce maximum blood concentration, area under the concentration time curve (AUC), and urinary recovery of ciprofloxacin by 83%, 48%, and 43%, respectively. None of the phenolic or terpene constituents were reported to have an interacting effect, and the most likely mechanism is chelation of ciprofloxacin by metal cations present in the extract. The dose of the extract employed (2 g/kg) is unlikely to be consumed by humans, but the potential of impaired absorption of ciprofloxacin and other fluoroquinolones needs to be considered when patients take concurrent botanical products containing large amount of inorganic materials; staggering administration times of the two products should be considered when such physical interactions are possible.

Interestingly, there was a report of an interaction between aspirin and tamarind, an Asian fruit used not only as an Ayurvedic medicine, but also as a flavoring ingredient for cooking. In six healthy volunteers, tamarind significantly increased the extent of absorption of a single 600mg dose of aspirin, which might result in toxicity if a large amount of acetylsalicylate was ingested concurrently with tamarind. The more significant botanical–drug interaction resulting in altered extent of drug absorption involves modulation of P-glycoprotein within the gastrointestinal tract. Originally discovered by Juliano and Ling, P-glycoprotein has been primarily known for its association with drug resistance to chemotherapeutic agents. However, P-glycoprotein also possesses a physiological protective role by transporting toxic xenobiotics or metabolites out of normal cells. In humans, P-glycoprotein is also expressed in several tissues including the gastrointestinal tract, liver, and blood-brain barrier. The presence of this efflux transporter on the luminal surface of the intestinal mucosa suggests a possible role in limiting drug bioavailability after oral administration.

In early 2000, several reports publicized the now well-recognized interaction between St. John's wort (*Hypericum perforatum*) and commonly used drugs such as cyclosporine, some with significant clinical consequences, e.g., organ transplant rejection. Because cyclosporine is primarily metabolized by cytochrome P-450 3A4 (CYP3A4), induction of CYP3A4 was originally thought to be the primary mechanism of the interaction. However, there is substantial overlapping drug selectivity between CYP3A4 and P-glycoprotein, and it has been demonstrated that St. John's wort is an inducer of P-glycoprotein. P-glycoprotein induction results in reduced oral absorption and at least partially accounts for the reduced systemic concentration of cyclosporine when St. John's wort is coadministered. Indeed, demonstration of correlation between pharmacokinetic parameters of cyclosporine and intestinal P-glycoprotein level in kidney transplant recipients suggests a significant role of P-glycoprotein in reducing cyclosporine absorption after oral administration. St. John's wort has also been reported to reduce concentration of other P-glycoprotein substrates such as digoxin. Current evidence strongly indicates that long-term administration of St. John's wort (longer than 14 days) induces both intestinal CYP3A4 and P-glycoprotein, secondary to activation of the nuclear factor pregnane X receptor (PXR) by the hyperforin component of St. John's wort.

Based on the same principle of modulation of drug absorption, other less known botanicals could produce similar or different effects compared to St. John's wort. Rosemary (*Rosemarinus officinalis* Labiatae) is a commonly used dietary botanical that has been found to have a chemopreventive effect. Furthermore, in drug-resistant MCF-7 human breast cancer cells expressing P-glycoprotein, methanol extracts of Rosemary at two concentrations (16.5 and 85 μg/mL) inhibited the efflux and increased intracellular accumulation of doxorubicin and vinblastine, two chemotherapeutic drugs that are known substrates of P-glycoprotein. Treatment of drug-resistant cells with the extracts also increased the cytotoxic effects of doxorubicin. On the other hand, in wild-type MCF-7 cells that do not express P-glycoprotein, the extracts did not affect accumulation or efflux of doxorubicin. Binding of azidopine, an analog of vinblastine, to P-glycoprotein was also reduced by the extract. The investigators concluded that Rosemary extracts appear to exert an inhibitory effect on P-glycoprotein activity via inhibition of drug binding to P-glycoprotein, with the responsible constituent(s) yet to be identified. Therefore, despite no reported interaction with cyclosporine, this botanical has the potential to increase plasma concentration of cyclosporine via an increase in its oral bioavailability.

Similarly, green tea (*Camellia sinensis*), a commonly consumed dietary supplement in many Asian countries, contains catechins, which have been shown to inhibit the activity of P-glycoprotein and the efflux of doxorubicin by a carcinoma cell line. Although currently there is no literature report of an interaction between green tea and prescription or over-the-counter drug based on modulation of P-glycoprotein, a potential interaction between green tea and warfarin was reported.

Drug Distribution

Changes in distribution of drugs resulting from altered protein binding of highly protein-bound drugs have been studied intensively. However, the clinical significance of interactions based on this mechanism is usually minor and transient, unless accompanied by impaired metabolism and/or excretion that almost always result in persistently elevated blood concentrations of the affected drug. Examples of botanical–drug interactions involving changes in drug distribution and/or protein binding have not been reported in the literature.

Drug Metabolism

Inhibition

More than half of the drugs in current use or in development are eliminated primarily by metabolism, and the most common cause of clinically significant drug–drug interaction is a result of drug-metabolizing enzyme inhibition or induction. Botanicals can have similar effects on drug-metabolizing enzymes, and therefore it is not surprising that most of the reports of botanical–drug pharmacokinetic interactions involve altered drug metabolism. The most common pathway of drug metabolism is oxidation by the cytochrome P-450 (CYP) super family of enzymes located in the endoplasmic reticulum of the hepatocytes. Although there are many subfamilies in the human CYP superfamily, only three are responsible for the majority of drug oxidations in humans, namely CYP1, CYP2, and CYP3. Nine individual CYPs make contributions to drug oxidation in humans: CYP1A2, CYP2B6, CYP2C8, CYP2C9, CYP2C19, CYP2D6, CYP2E1, CYP3A4, and CYP3A5. However, a simpler outlook is often useful because the majority of drug interactions are seen with substrates of just four enzymes. CYP2C9, CYP2D6, and CYP3A4/5 metabolize 15%, 20%, and 60%, respectively, of the drugs that are principally eliminated by metabolism.

Considerable effort has been focused on understanding and predicting the inhibition of CYPs in vivo. Over the past decade, an impressive arsenal of gene- and protein-based tools has been brought to bear on this issue and significant advances have been made in the use of in vitro data to identify the specific CYPs involved in a given biotransformation and to predict clinically important drug interactions. These techniques have recently been extended to characterize botanical–drug interactions. All new drugs are required by the Food and Drug Administration to have the extent of metabolism defined, the CYPs responsible for major metabolite formation to be identified, and the potency of CYP inhibition to be quantified. These regulatory requirements are in part a response to the need to withdraw several drugs from the marketplace due to an unacceptable level of adverse events that stemmed from drug interactions. Whenever two substrates are cometabolized, there is the potential for a metabolic drug interaction, but in most cases a clinically important event does not occur because sufficient systemic blood concentrations of inhibitor are not achieved.

In some cases, drug interactions occur in the wall of the small intestine as well as in the liver. To date, this has only been described for drugs metabolized by CYP3A4/5, because these are the only enzymes expressed at a high level in the gut wall. The balance between the rates of absorption through the intestinal epithelium and the rates of metabolism will determine the net availability at the gut wall. Thus a CYP3A substrate that is either rapidly absorbed or not efficiently metabolized will not experience significant gut wall metabolism, e.g., alprazolam. For a drug such as midazolam, the complete inhibition of intestinal CYP3A4/5 alone could increase the oral AUC of midazolam by 2.5-fold. However, it has been speculated that the remarkable sensitivity of some CYP3A substrates, such as lovastatin, simvastatin, and buspirone, to drug interactions reflects a very low gut wall availability. The clinically important inhibitors of CYP3A4/5 share the capability to completely inhibit the enzymes in the intestinal wall, as illustrated by the high intestinal wall availability of oral midazolam in the presence of clarithromycin and ketoconazole. This is not unexpected because there are high concentrations of inhibitor at the gut wall during absorption. A similar pattern should be anticipated for botanical products that contain strong inhibitors of CYP3A enzymes.

We often rationalize drug interactions as reflecting the reversible competition of two substrates for an active site. However, it is becoming increasingly clear that other mechanisms of inhibition are operational in vivo. For example, some mechanism-based inhibitors are activated during metabolism and form a complex with the heme of CYP3A, known as a metabolite intermediate complex, or make a covalent modification of enzymes and result in irreversible loss of enzyme activity. These irreversible mechanisms appear to contribute to the inhibition of CYPs that occurs following exposure to bergamottins (in grapefruit juice), capsaicin (in chili peppers), glabridin (in licorice root), isothiocyanates (from cruciferous vegetables), oleuropein (from olive oil), diallyl sulfone (from garlic), and resveratrol, a red wine constituent. An important consequence of this irreversible inhibition is that interactions take one to two weeks to resolve on termination of the drug because this is how long the CYP3A takes to resume its predrug steady state. This is one reason why a good medical history should include questions about drugs and botanical products that have been discontinued in the past two weeks, when addressing possible botanical–drug interactions.

Induction

The term "*induction*" has evolved to include any mechanism that results in increased tissue concentration of catalytically active protein involved in drug metabolism. This increased enzyme activity results in greater systemic clearance and lower bioavailability of extensively metabolized drugs. The resulting lower drug concentrations often result in therapeutic failure. For example, it is well known that oral contraceptive pills become ineffective when rifampin is coprescribed. In general, induction may result from enhanced gene transcription rates, increased mRNA stability or translational efficiency, and protein stabilization induced by substrate binding or posttranslational modifications. However, the most common mechanism of induction is binding to and activation of discrete nuclear factors that act in the form of protein heteromers to enhance rates of gene transcription. It is clear that a single nuclear factor may modulate the expression of numerous genes and this mechanism of induction most likely applies to all drug-metabolizing enzymes but the extent of induction, tissue selectivity, and ligand selectivity vary widely between genes. Some degree of predictability has arisen from the discovery of the nuclear factors primarily responsible for the effects of the clinically important inducers. It is worth noting that some inducers, such as ritonavir for CYP3A4, are also potent inhibitors of at least some of the enzymes induced. Therefore, despite greater concentrations of enzyme, the net interaction maybe inhibition prior to full induction, followed by induction or even no effect.

The transcriptional regulation of drug-metabolizing enzymes is commonly cell-type and tissue selective. Thus, tissues that express low concentrations of nuclear factors do not experience significant induction. In contrast, both liver and intestines express significant concentrations of nuclear factors such as the PXR and experience profound induction in the presence of its ligands. The best example of a botanical product altering drug metabolism efficiency is that of St. John's wort, which is a potent inducer of CYP3A4 and causes accelerated metabolism of cyclosporine and indinavir. Of equal importance but less studied is the effect of botanicals on oral contraceptive disposition, which can potentially affect a large number of subjects. Recently, Hall et al. reported that St. John's wort induced the metabolism of both ethynyl estradiol and norethindrone in 12 healthy women via enhanced CYP3A4 activity.

The incidence of breakthrough bleeding was higher with concurrent use of St. John's wort (seven subjects) than without (two subjects), and subjects with breakthrough bleeding had a higher CYP3A4 activity, as measured by midazolam clearance. Therefore, this study provides supportive evidence and explanation for case reports of unexpected menstrual bleedings in women taking concurrent oral contraceptive and St. John's wort. Although Hall et al. did not find evidence of loss of oral contraceptive efficacy or ovulation, breakthrough bleeding is well known as a contributory factor for discontinuance of oral contraceptive use that may lead to a higher incidence of pregnancy. Schwartz et al. reported the loss of contraceptive efficacy associated with the use of St. John's wort with resultant unwanted pregnancies. This issue of oral contraceptive–botanical interaction requires further studies.

The effect of garlic (*Allium sativum*)–containing botanicals on drug metabolism has also been studied in vitro, and in vivo in both animal and human studies. Using human liver microsome as an in vitro drug- metabolism model, Foster et al. showed that raw garlic constituents inhibit CYP3A4-mediated drug metabolism. In rats, acute administration of a single dose of garlic oil produced significant reduction in the activity of several enzymes, including CYPs. However, chronic administration for five days produced the opposite effect—a significant increase in CYP activity. Gurley et al. reported that chronic administration of garlic oil for 28 days in humans reduced CYP2E1 activity by 39%, possibly a result of inhibition of the CYP by diallyl sulfone, a metabolic product of alliin, the major component of garlic. A pharmacokinetic study in healthy volunteers showed that a three-week course of garlic tablets taken twice daily resulted in a 51% reduction in AUC of the protease inhibitor saquinavir, a CYP3A4 substrate. In view of the multiplicity of CYPs and the many possible botanical–drug interactions, highly efficient clinical study designs using CYP probe cocktails have been explored. Following successful application to St. John's wort, other botanicals that have been evaluated in this fashion include echinacea, saw palmetto, garlic, peppermint oil, and ascorbyl palmitate. Curbicin, a botanical remedy taken by patients for the management of prostate enlargement, contains saw palmetto as one of the ingredients. Elevated international normalized ratio (INR) values were reported in two patients taking curbicin, and one of the patients also took warfarin. Cheema et al. also reported a patient who suffered from severe intraoperative hemorrhage with doubling of the bleeding time value after taking saw palmetto. The prolonged bleeding times were normalized after the botanical use was discontinued.

Although the effect of saw palmetto on CYP2C9, the enzyme responsible for metabolism of the active S-isomer of warfarin, has not been studied, it is of note that the prothrombin time (PT) and activated partial thromblastin time were both within normal limits before, during, and after the surgical procedure. It is also not known whether the bleeding abnormality observed in the patient might be related to the reported inhibitory effect of the botanical on cyclooxygenase in animal studies. The differential effect of echinacea on CYP3A4 will be discussed later in this chapter. With the exception of garlic, at present there are no reported drug interactions with the other three botanicals, but based on available data, potential interaction, especially with drugs metabolized by CYP3A4, could be expected.

Drug Excretion

While theoretically it is possible that botanicals with diuretic effects can increase drug excretion, most botanical diuretics are not as potent as furosemide and are unlikely to result in significant interactions. Most botanicals also do not affect urinary pH significantly, and hence are unlikely to affect renal tubular reabsorption of drugs. Nevertheless, lithium toxicity was thought to be related to the use of a botanical diuretic mixture in a patient. If the toxicity indeed is related to the use of the botanical diuretic, the mechanism of action or the responsible constituent(s) is not known.

Altered Pharmacodynamics

In addition to pharmacokinetic botanical–drug interaction, pharmacodynamic interactions can also occur, resulting in either an augmented or attenuated response. These effects can occur without any significant changes in either the systemic or tissue concentrations of the affected drug or botanical, and generally are more difficult to predict. In addition, unlike pharmacokinetic botanical–drug interactions, most pharmacodynamic interactions reported in the literature are mostly based on patient cases or clinicians' experience and seldom involve clinical or experimental study. For example, combining St. John's wort and selective serotonin reuptake inhibitors have been reported to result in an additive pharmacological effect and possibly serotonin syndrome, but there were no clinical studies or literature reports of changes in the pharmacokinetics of the selective serotonin reuptake inhibitors when combined with St. John's wort. The most commonly reported pharmacodynamic botanical–drug interactions primarily involve anticoagulants and antiplatelet agents.

Augmented Pharmacological Effect

Warfarin

Most literature reports of pharmacodynamic botanical–drug interaction involve the anticoagulant warfarin, likely because it has therapeutic end points such as the INR and PT, which are routinely closely monitored. In addition, most botanicals possess anticoagulant and/or antiplatelet activities, and their combined use with warfarin provides a good example of pharmacodynamic interaction with additive pharmacological effect. Botanicals such as garlic can inhibit platelet aggregation, likely accounting for episodes of spontaneous spinal epidural hematoma and postoperative bleeding reported in the literature. Currently, there are no reports of an interaction between garlic and warfarin, but based on the inhibitory effect of garlic on platelet aggregation, one would expect that there is at least a risk of additive pharmacological response to warfarin when the two compounds are taken concurrently. In fact, such interactions have been reported with other botanicals that also inhibit platelet aggregation, including ginkgo and the traditional Chinese medicines, dong quai (*Angelica sinensis*) and dan shen (*Salvia miltiorrhiza*).

In a patient who had been stabilized on warfarin for five years, recent use of ginkgo was reported to result in intracerebral hemorrhage. In an in vitro model using human liver microsomes, the activity of CYP2C9, which metabolizes the active S-isomer of warfarin, was inhibited by commercial ginkgo extracts. Therefore, the interaction between ginkgo and warfarin potentially involves both pharmacokinetic and pharmacodynamic mechanisms. Dong quai also inhibits platelet aggregation, and there have been several reports of increased INR in patients taking concurrent warfarin and dong quai. Despite the elevated INR, the patient did not experience any bleeding episodes. It is not known whether the lack of clinical consequence in this report is a result of intersubject variability in the magnitude of interaction or the absence of a pharmacokinetic component, as an animal study demonstrated that warfarin pharmacokinetics was unchanged by dong quai.

Antiplatelet drugs

Although no pharmacokinetic antiplatelet drug–botanical interactions have been reported in the literature, there is the potential of an additive pharmacodynamic effect with concurrent use of antiplatelet drugs or botanicals that possess antiplatelet activity or contain salicylates, such as willow bark (*Salix* spp.) and meadowsweet (*Filipendula ulmaria*). The ginkgolide constituents, found in ginkgo, are known to exhibit platelet-activating factor antagonistic activity. In an elderly patient who was prescribed aspirin therapy after a coronary bypass surgery, self-initiation of ginkgo use resulted in spontaneous bleeding within the eye and blurred vision. On cessation of ginkgo use, the bleeding stopped and the visual changes resolved.

Drugs Acting on the Central Nervous System

Selective serotonin reuptake inhibitors

The similar pharmacological profile of selective serotonin reuptake inhibitors and St. John's wort would suggest the potential of a pharmacodynamic interaction due to an additive effect. A case of concurrent use of sertraline and St. John's wort, resulting in mania, was reported for a patient with a history of depression who was prescribed sertraline and who also took St. John's wort against medical advice. A similar potentiation of serotonergic effect was reported by Gordon.

Benzodiazepines

Kava (*Piper methysticum*) is a popular botanical product used for management of anxiety and insomnia. Almeida and Grimsley reported a case of potentiation of the central nervous system (CNS)-depressant effect of alprazolam by kava extract and/or kavalactones in a 54-year-old, male patient who became lethargic and disoriented after taking kava for three days. Kava ingestion was concurrent with his usual medications, including alprazolam, cimetidine, and terazosin, but the patient denied overdose of any of his medications. The physicians attributed the patient's mental state to a kava–alprazolam interaction. Both kava extract and kavalactones have been shown in vitro to inhibit several CYPs, including CYP3A4. However, it is possible that in this case, the enhanced effect involves not just a pharmacokinetic component but also a pharmacodynamic basis secondary to synergistic activity at the gamma-aminobutyric acid (GABA) receptor.

Miscellaneous central nervous system acting drugs

Ephedra (ma huang) is a popular botanical incorporated into a variety of formulations for weight loss, "energy" or "performance" enhancement, and symptomatic control of asthma. A pharmacodynamic interaction leading to a fatality has been reported with concurrent use of caffeine and ephedra, possibly as a result of additive adrenergic agonist effect of the ephedrine alkaloids and caffeine on the cardiovascular system and the CNS. Ephedra was recently withdrawn from the market. A botanical–drug interaction postulated to have both pharmacokinetic and pharmacodynamic mechanisms was reported in an elderly Alzheimer's patient, who developed coma likely as a result of concurrent use of ginkgo leaf extract 80mg twice daily and the antidepressant trazodone 20 mg twice a day. The pharmacodynamic mechanism was suggested because the coma was reversed by flumazenil, indicating increased activity at GABA-activated receptors; ginkgo flavonoids possess GABA agonist activity on the benzodiazepine receptor. A pharmacokinetic basis of the interaction was also proposed to be a result of CYP3A4 induction, and subsequent increased conversion of trazodone, to *m*-chlorophenylpiperazine, an active metabolite with GABA agonist activity.

Digoxin

The use of botanicals containing laxatives has not been reported to result in altered drug absorption to date. However, excessive use of laxative-containing botanicals such as cascara (*Rhamnus purshiana*), senna leaves, and/or pods from *Cassia senna* can potentially decrease serum potassium and other electrolyte concentrations, and therefore enhance toxicity of digoxin. To date, no clinical interactions have been reported between digoxin and these botanicals, but given the narrow therapeutic range of digoxin, it would be prudent to monitor for signs and symptoms of digitalis toxicity with long-term, excessive use of these botanical laxatives. The concurrent administration of these botanicals with prescription diuretics should be approached with caution.

The narrow therapeutic index of digoxin necessitates the monitoring of serum digoxin concentration as an aid for optimizing drug therapy in patients, and an in vitro laboratory interaction between digoxin and several botanicals such as danshen and ginseng products have been reported in the literature. These "*cardioactive*" botanicals possess active constituents with structures similar to digoxin, and therefore can demonstrate digoxin-like immunoreactivity. Chow et al. reported that small amounts (2–5 mL) of aqueous extracts of these "*cardioactive*" botanicals interfered with immunoassays used to determine digoxin concentration, both in vitro and ex vivo. Patients taking digoxin might also take these "*cardioactive*" botanicals and this laboratory interference could result in falsely elevated digoxin concentrations in patients. McRae reported a 74-year-old man who had been stabilized on digoxin for about 10 years with therapeutic concentrations between 0.9 and 2.2 ng/mL. Ingestion of Siberian ginseng resulted in a serum concentration of 5.2 ng/mL, even though the patient was asymptomatic with no electrocardiographic changes. The digoxin concentration returned to normal after the patient stopped taking the ginseng product.

Oral Hypoglycemic Agents

In a brief report, a potential interaction between curry and chlorpropamide, leading to reduction in chlorpropamide dose in a 40-year-old woman was attributed to the garlic and karela components of this complex mixture. Garlic reportedly can lower blood glucose. However, there was no information provided regarding the estimated amount of garlic intake in this patient. To date, there are no formal studies that confirm the initial clinical observation or evaluate the likely mechanism.

Antagonistic Pharmacodynamic Effect

While dong quai and possibly garlic have an additive effect on the pharmacological action of warfarin, an antagonistic interaction between warfarin and coenzyme Q_{10} had been reported. Spigset reported three elderly patients who were all stabilized on different warfarin dosage regimens. All experienced a decrease in INR to values below 2 after taking ubidecarenone (coenzyme Q_{10}). The dose of coenzyme Q_{10} was documented as 30 mg/day in two of the patients. In both patients, the warfarin dose was temporarily increased and coenzyme Q_{10} discontinued. The INR returned to the patients' previous stabilized values prior to taking the coenzyme Q_{10}. Because coenzyme Q_{10} is structurally similar to vitamin K_2, the authors suggested that one potential mechanism might be related to the enhanced coagulation effect of coenzyme Q_{10}. Animal data showed that antagonism of coenzyme Q_{10} resulted in increased PT, suggesting that coenzyme Q_{10} might have an opposite pharmacological effect to that of warfarin.

Two patients stabilized on a phenytoin regimen suffered a loss of seizure control after taking shankhapushpi, an Ayurvedic antiepileptic medicine, three times a day. There was also a significant decrease in serum phenytoin concentration from 9.6 to 5.1 mg/L. To investigate the possible mechanisms, multiple doses of shankhapushpi were administered to rats and resulted in decreased plasma phenytoin concentrations, whereas single-dose administration was reported to interfere with the antiplatelet effect of phenytoin, thereby implying both a pharmacokinetic and pharmacodynamic basis for the interaction.

There are several botanicals that have purported immunostimulating effects. These include *Panax ginseng* and *Echinacea purpurea*, which have both been used as an immune stimulant. Any potential adverse effect on the pharmacological activity of immunosuppressants has not been reported in patients or evaluated in clinical studies. Given the lack of data, it would be prudent to advise against concurrent intake of these botanicals, and closely monitor changes in efficacy in patients who self-administer these botanicals.

Evaluating Plant-Drug Interaction

Overall, an accurate assessment of the reliability of reported botanical–drug interactions with a pharmacodynamic basis or mechanism is usually more difficult than the assessment of those with a pharmacokinetic basis. This likely reflects the fact that the former reports are usually case reports, whereas the later reports are often accompanied with objectively measured end points. Also, despite a common belief that botanical–drug interactions are underreported, the overall incidence of this phenomenon is difficult to define. This partly reflects the lack of a mechanism for reporting the interactions, and difficulty in obtaining reliable information to assess clinical relevance or to establish a definitive causality relationship. For example, even with the evidence of St. John's wort increasing the metabolism of oral contraceptive hormone and possibly contributing to reports of break-through bleeding and pregnancy, it is well established that pregnancy can occur with oral contraceptive used alone or with other drugs, and a definitive causality relationship has not been established. Nevertheless, there is sufficient clinical evidence that interaction involving commonly used drugs such as cyclosporine and protease inhibitors with St. John's wort can be serious and sometimes life threatening. In addition, the lack of fatalities resulting from the various reports of botanical–anticoagulant interaction likely reflects close clinical and laboratory monitoring with appropriate dosage adjustment, if necessary, in the patients.

Challenges of Predicting Botanical–Drug Interaction

While defining the overall pharmacokinetic or pharmacodynamic basis of botanical–drug interactions may be relatively straightforward, attempts to explain the underlying mechanism of altered drug concentrations or to predict the magnitude and significance of the interaction is certainly not easy. There are several factors that contribute to this difficulty, and they are briefly discussed below.

Lack of Definition of Active Constituents

First and foremost, it must be emphasized that botanicals or botanical preparations are not pure synthetic molecules but are composed of many constituents, sometimes from multiple botanicals, and some or many of them can be biologically active. Although altered drug concentration can be caused by induction or inhibition of intestinal and hepatic drug-metabolizing enzymes as well as P-glycoprotein, the identity of the biologically active constituent(s) that is responsible for these effects is usually not known. Without this knowledge, most investigations are restricted to studying the commercially available products containing multiple constituents with potentially different modulating effects on these proteins. Commercial preparations of St. John's wort used in most clinical and interaction studies are usually standardized to contain specific amounts of hypericin, but it is another constituent, hyperforin, which was shown to be responsible for induction of CYP3A4. Similarly, although administration of milk thistle 175 mg (containing 153 mg of silymarin) three times a day for three weeks resulted in 9% and 25% reduction in AUC and trough concentration, respectively, of indinavir, its differential effect on CYP3A4 and P-glycoprotein needs to be further studied. In addition, while the overall study result suggested a minimal clinical consequence for AIDS patients receiving indinavir, whether botanical constituents other than silymarin would have a greater modulating effect remains unknown.

In addition, very few studies provide information on the content of important constituents of the botanical or botanical preparation. This obviously poses a problem of general applicability in terms of predicting interaction across different preparations with variable content of constituents, or extrapolating the result of one study to the overall interaction potential. For example, one of the active constituents in garlic is allicin, which gives garlic its specific, well-known odor. Although allicin has been suggested to enhance production of CYP, there is no data to confirm or refute the possibility, let alone the identity of the specific enzyme that is induced. It is clear, however, that commercial garlic preparations have highly variable contents ranging from no allicin to maximum standardized allicin content used in the garlic-saquinavir study described above. In the study by Gurley et al. garlic oil 500 mg did not result in appreciable differences in the 1-hydroxymidazolam/midazolam phenotypic ratio for CYP3A4, the enzyme that mediates the metabolism of saquinavir. Both studies administered the garlic preparations for at least three weeks, and therefore it is unlikely that the duration of therapy would account for the difference between studies. On the other hand, if allicin content is a critical issue, it may be that one of the reasons for the conflicting results between the two studies is potential variability in this active constituent in the two garlic preparations used.

Lack of Standardization of Known Active Constituents

Even though the active constituent responsible for the interaction has been identified, conflict in study results can still occur due to variable content of the known active constituent(s). In the study by Piscitelli et al., the investigators took extra effort in analyzing the allicin and allin content of the commercial garlic caplets administered to the subjects. They reported that the allicin and allin contents were 4.64 and 11.2 mg per caplet, which were different from the labeled content. This study highlights the challenge associated with evaluating any aspect of pharmacology or therapeutic use of dietary supplement, including botanicals. While consumers increasingly are aware of the fact that dietary supplements and botanical products do not have to be proven to be efficacious or to be safe, they are less aware of the lack of standardization among products.

Although dietary supplements and botanical products are required to state exactly the content of active ingredients and their amounts on the label, the manufacturers do not necessarily comply. More importantly, the labels are not routinely checked for compliance by any government agency. In addition, unlike prescription drugs, dietary supplement and botanical products are not required to be manufactured under standardized conditions. This has led to substantial variability in the amount of active constituent(s) between batches. Prime examples of this include *echinacea* and ginseng products. Gilroy analyzed different single botanical *echinacea* preparations purchased from retail stores and reported that only 10 of 19 preparations (53%), labeled as standardized, had an assayed content consistent with the labeled content. There were only weak correlations between labeled milligram content of *echinacea* versus measured milligram for the standardized preparations ($r = 0.49$, $p = 0.02$) and the correlation was even lower for nonstandardized preparations ($r = 0.21$, $p = 0.28$). Similar discrepancies in content were reported in a study conducted by the Consumer Unions, in which they tested the content of 10 marketed ginseng preparations, and found significant differences in the amount of the active constituent ginsenosides (range: 0.4–23.2 mg). Of particular concern is that this inconsistency in product and active constituent occurs even within the same batch. As part of a clinical study with St. John's wort, Hall et al. analyzed 10 capsules of St. John's wort from the same lot and found the mean total weight to be 444 mg (4.6% CV) versus 300 mg as stated on the label. In addition, the dosage form was supposed to be standardized to contain 900 mg of hypericin, but the mean content was found to be 840 mg (6.6% CV). There was also variability of the hyperforin content (mean 11 mg and 5.7% CV), which was not stated on the label. Our experience with two random capsules from one batch of kava-kava also showed the same extent of undesirable variance: the total content of the pharmacologically active kavalactone was 47.3 mg in one capsule and 39.4 mg in the second one.

Confounding Issue Related to Study Design

Extrapolation of Result from In Vitro Study

Similar to evaluation of potential inhibitory effect of different drugs on the CYPs, in vitro preparations such as human liver microsomes have also been used to evaluate the potential of a botanical to cause interaction. Nevertheless, there are numerous reasons why in vitro results based on human liver microsomes do not necessarily agree with in vivo study results. One reason is the inability of human liver microsomes to evaluate and predict enzyme induction. St. John's wort serves as an excellent example to illustrate this limitation. In vitro, St. John's wort has been shown to inhibit CYP2C9, CYP2D6, and CYP3A4. However, as discussed above, St. John's wort has been shown in numerous human studies to induce CYP3A4. This may be due to the finding of hyperforin, a constituent of St. John's wort, binding to the PXR and upregulating CYP3A4 gene expression. Importantly, microsomal preparations lack the capability to synthesize new protein and cannot be expected to provide any insight into the potential for induction to occur in vivo.

Differential Effect on Intestinal and Hepatic CYP3A4

Another confounding issue specific to CYP3A4 would be the potential differential effect of a specific botanical or botanical constituent on intestinal and hepatic CYP3A4. The clinical study by Gorski et al. elegantly showed that, consistent with in vitro inhibition of CYP3A4 by *echinacea* tinctures, administration of echinacea 400 mg four times a day for eight days in healthy volunteers inhibited intestinal CYP3A4 and resulted in an 85% increase in systemic bioavailability. However, hepatic CYP3A4 activity, as measured by systemic clearance of midazolam after intravenous administration, was increased by 34%. Therefore predicting potential interaction between *echinacea* and CYP3A4 substrate would depend on whether the substrate has high oral bioavailability, in which case the likely pharmacokinetic and clinical outcome would be increased clearance secondary to hepatic CYP3A4 induction and lower serum drug concentration versus substrate with a low bioavailability secondary to extensive intestinal first-pass effect, in which case the likely pharmacokinetic and clinical outcome would be decreased oral clearance secondary to intestinal CYP3A4 inhibition and increased serum drug concentration. One can only imagine the difficulty of predicting the potential and extent of interaction between echinacea and CYP3A4 substrates if a patient who is receiving CYP3A4 substrate for medical conditions also treats a cold at the same time by taking *echinacea* and drinking grapefruit juice, which potently inhibit intestinal CYP3A4.

Single-Dose Administration vs. Multiple Dosing

Results from single-dose studies could be different from chronic dose administration. Although St. John's wort administered as a single 900mg dose to healthy volunteers was found to increase the maximum plasma concentration and decrease the oral clearance by 45% and 20%, respectively, of the P-glycoprotein substrate fexofenadine, the opposite effects (35% decrease in maximum plasma concentration and 47% increase in oral clearance) were observed after daily administration of the same dose of St. John's wort for two weeks. Similar differential effects between single versus chronic dose administration have been shown before with CYP3A4: single-dose ritonavir caused inhibition of CYP3A4 and chronic administration resulted in CYP3A4 induction.

Drug Interaction with St. John's Wort

Botanical use is prevalent throughout the world with between 10% and 30% of individuals residing in the United States using complementary and alternative medicines routinely. Of particular concern is the finding that up to 30% of individuals taking prescription medicines have also used botanical remedies concurrently within the past year. The number of individuals consuming St. John's wort on a daily basis has been estimated at more than 11 million and approximately one-third of these are using St. John's wort to treat self-diagnosed depression. In the United States, St. John's wort is one of the top-selling botanical preparations with sales ranking second in 1999 and seventh in 2002. Although botanical preparations are widely considered by the public to be without adverse effect or a source of drug interactions, this is not the case. The report of Ruschitzka et al. clearly illustrates the danger of coadministering botanical products (i.e., St. John's wort) with prescription products and demonstrates that, despite popular belief, the indiscriminant use of botanical products does involve risk. This chapter will review the historical indications, formulations, pharmacology, and interactions between St. John's wort, echinacea, and other medicines.

Indications

St. John's wort (*Hypericum perforatum*) is a perennial wildflower indigenous to Europe, North Africa, and western Asia and has been used for medicinal purposes for over two millennia. As far back as the early 16th century, St. John's wort was used primarily to treat anxiety, depression, and sleep disorders. In the late 20th and early 21st century, St. John's wort has been recommended for the treatment of mild to moderate depression. In support of its use for the treatment of mild to moderate depression, a number of clinical trials have demonstrated that St. John's wort has comparable efficacy to the tricyclic antidepressants (i.e., imipramine) and selective serotonin reuptake inhibitors (e.g., fluoxetine and paroxetine). It should be noted that these clinical trials are typically conducted within a short time period and thus may not reflect long-term outcomes. The utility of St. John's wort in the treatment of moderate to severe depression has been investigated in large randomized placebo-controlled multi-institutional studies. Some such studies demonstrated efficacy, but others failed to detect a clinically significant effect on the symptoms of the moderate to severely depressed individuals. Gelenberg et al. demonstrated a relapse rate of approximately 30% in moderate to severely depressed individuals who initially responded to St. John's wort therapy as would be expected from experience with prescription antidepressants. Other conditions in which St. John's wort has been advocated include neuralgia, anxiety, neurosis, dyspepsia, and external treatment of wounds, bruises, sprains, myalgia, and first-degree burns. In vitro studies conducted in the late 1 980s and early 1990s suggested that components of St. John's wort (e.g., hypericin) may have antiviral properties. However, an open- label clinical trial demonstrated that the intravenous or oral administration of the St. John's wort constituent, hypericin, provided no clinical benefit, as reflected by increasing CD4 counts or decreasing viral load in a group of HIV-infected individuals and resulted in significant adverse events necessitating discontinuation of therapy.

Dosage Forms

St. John's wort and some individual constituents of the preparations have been administered orally, topically, and intravenously in various pharmaceutical formulations, including tinctures, teas, capsules,

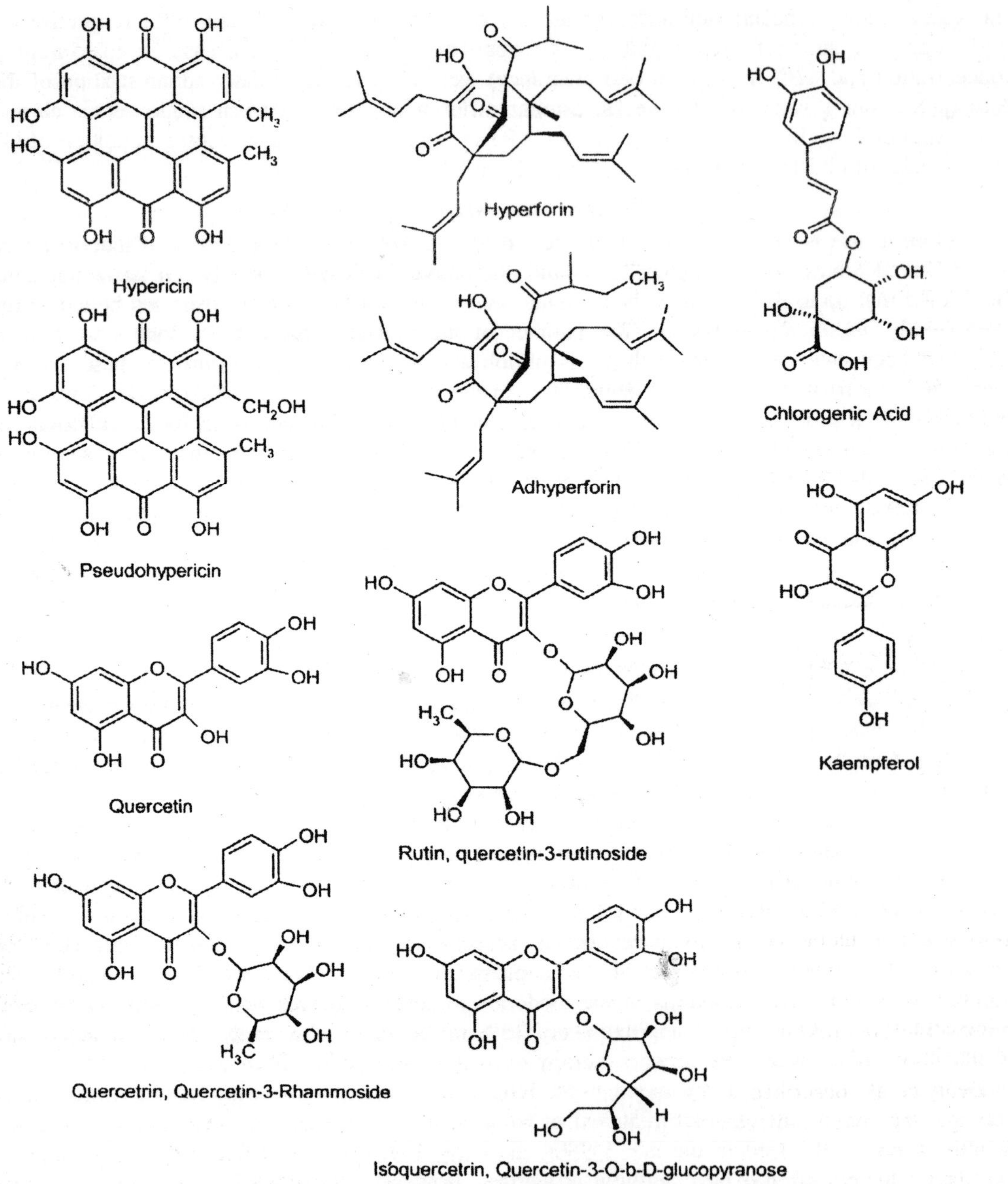

Hypericin

Hyperforin

Chlorogenic Acid

Pseudohypericin

Adhyperforin

Kaempferol

Quercetin

Rutin, quercetin-3-rutinoside

Quercetrin, Quercetin-3-Rhammoside

Isoquercetrin, Quercetin-3-O-b-D-glucopyranose

Fig. 7.1. Chemical structures of common phytochemicals found in SJW.

purified components, and tablets. These botanical preparations of St. John's wort are prepared from plant components (i.e., flowers, buds, and stalk) whose content of the wide array of structurally diverse bioactive constituents may differ. Many commercial tablet and capsule formulations of St. John's wort are standardized using the ultraviolet absorbance of the naphthodianthrones, hypericin, and pseudohypericin, to contain 0.3% "*hypericin*" content. Thus, a 300 mg dose of St. John's wort contains approximately 900 μg "*hypericin*" per dose.

Despite the standardization of dosage forms on hypericin content, the principal active ingredient is thought to be a phloroglucinol, hyperforin. As a result of inappropriate standardization on an ingredient that has limited pharmacological activity, the concentration of hyperforin varies greatly among commercial preparations. Draves and Walker assessed the hypericin and pseudohypericin (naphthodianthrones) content in 54 commercially available St. John's wort products and determined that only two of the products were within 10% of the labeled claims for "*hypericin*" content. Likewise, Wurglics et al. assessed hypericin and hyperforin content and inter-batch variability in eight German St. John's wort products. Pronounced interbatch variability was observed for some products whereas others demonstrated consistent hyperforin and hypericin content. In addition, the expected naphthodianthrone (*hypericin*) content in the preparations also demonstrated considerable variability. It is clear from the reports of a number of investigators that there is wide inter- and intraproduct variability in hyperforin and hypericin content. The lack of consistent phytomedicinal (hypericin and hyperforin) content across and within products is not limited to St. John's wort preparations but is seen with many other botanical medicines. The administration of St. John's wort via tea is no longer recommended because the efficacy of this preparation is questionable; however, the drug interaction potential of St. John's wort in this formulation appears to be maintained.

The preparation used in many of the described interactions between St. John's wort and conventional pharmaceutical products is the product manufactured by Lichtwer Pharma GmbH. This product is marketed under the trade name, Jarsin (LI 160) in Germany and marketed in the United States under the trade name Kira. St. John's wort may also be sold in combination products with vitamins and other botanical preparations. The drug interaction potential between these combination products and cytochrome P450 (CYP) 3A and P-glycoprotein substrates has not been investigated, but should be assumed to be no different than single-agent St. John's wort products.

Adverse Effects and Pharmacodynamic Interactions

It is a reasonable expectation that, as observed with other pharmacotherapies, the administration of St. John's wort will result in adverse effects. In a study examining the efficacy of St. John's wort for mild to moderate depression, dry mouth was the most common adverse effect occurring in 8% of patients (13/157), and other adverse events including headache, sweating, asthenia, and nausea occurred in 3% or less of the participants. In addition, only four individuals withdrew from the trial compared to 26 individuals who withdrew while taking the comparator drug, imipramine. Likewise, Woelk et al. reported a low incidence of adverse events in a group of 3250 (76% women) patients receiving St. John's wort three times daily (LI 160) for the treatment of depression (33). The most frequently recorded adverse events were gastrointestinal irritation (0.6%), allergic reactions (0.5%), tiredness (0.4%), and restlessness (0.3%). Other adverse effects associated with St. John's wort intake include sedation, anxiety, and dizziness. It is clear from these reports that St. John's wort is well tolerated.

Dean et al. described a 58-year-old postmenopausal woman who experienced nausea, anorexia, retching, dry mouth, dizziness, thirst, cold chills, weight loss, and extreme fatigue following the discontinuation of St. John's wort (1800 mg three times daily for 32 days). The symptoms peaked three days after cessation of St. John's wort for suspected photosensitivity reaction and resolved within eight days. The reported symptoms and the temporal relationship to the discontinuation of the St. John's wort dosing were considered by Dean et al. to be consistent with "*withdrawal syndrome.*" Additionally, the high dose of St. John's wort administered was considered to be a contributing factor in the patient adverse-event profiles.

In studies examining the antiviral activity of synthetic hypericin following oral and intravenous administration for the treatment of HIV infection, a dose-limiting toxicity was moderate to severe photosensitivity, including the erythema, numbness, pain, and temperature sensitivity. There is a case report of hypertensive crisis in a 41-year-old male, following the ingestion of St. John's wort for approximately one week, and the consumption of tyramine-rich foods (aged cheese and red wine). Although the interaction between monoamine oxidase inhibitors and the eating of tyramine-rich foods is well recognized, alcoholic extracts of St. John's wort have been shown to weakly interact with monoamine oxidase receptors A and B. Thus, the mechanism of the observed hypertensive crisis is unclear. Other serious adverse effects attributed to St. John's wort due to drug–drug pharmacokinetic or pharmacodynamic interactions include cardiovascular collapse, mania in patients with bipolar depression, and photosensitivity.

Mechanisms of St. John's Wort–Mediated Drug Interactions

In vitro

Using crude extracts and isolated constituents, Obach demonstrated that St. John's wort was capable of inhibiting cDNA-expressed CYP-mediated metabolism. cDNA-expressed CYP2C9-, CYP2D6-, and CYP3A4-mediated biotransformations were inhibited by purified hyperforin. Likewise, I3, II8 biapigenin was shown to competitively inhibit CYP1A2-, CYP2C9-, and CYP3A4-mediated phenacetin *O*-deethylation, diclofenac 4-hydroxylation, and testosterone 6β-hydroxylation, respectively. The results demonstrated that constituents of *H. perforatum* were capable of inhibiting biotransformations mediated by both CYPs, CYP2D6 and CYP3A. Likewise, Budzinski et al. demonstrated that commercial tinctures of St. John's wort and hypericin, a principal component of these tinctures, were capable of inhibiting cDNA-expressed CYP3A4- mediated metabolism of 7-benzyloxyresorufin. Although the crude extracts and purified constituents of St. John's wort were relatively good inhibitors of CYP3A in vitro, the subsequent in vivo studies failed to confirm these observations. It is clear from the current body of literature that the coadministration of St. John's wort with many therapeutic agents, especially those that are CYP3A substrates, results in reduced serum concentration and diminished drug efficacy. These observations are consistent with increased drug elimination.

CYP3A4 and P-glycoprotein are transcriptionally regulated by an orphan nuclear receptor designated as the pregnane X receptor (PXR). Small molecule ligands such as rifampicin bind to PXR and encourage heterodimerization of PXR with the retinoid X receptor. mRNA synthesis of numerous target genes is stimulated after this complex undergoes trans- location to complimentary sequences in the regulatory region of the genes. Moore et al. examined the effect of extracts of commercial St. John's wort preparations on CYP3A4 mRNA expression in cultures of human hepatocytes. CYP3A4 mRNA expression was induced in human hepatocytes treated for 30 hours with either extracts of commercial St. John's wort preparations or purified hyperforin. Additional experiments conducted by this group employing CV-1 cells transiently transfected with both a PXR expression vector and a human chloramphenicol acetyltransferase reporter system containing a PXR-binding site demonstrated that hypericum extract and hyperforin, but not hypericin, induced CYP3A4 mRNA expression via activation of the PXR. Hyperforin has an EC_{50} for the activation of PXR of around 20 nM and is one of the most potent inducers discovered to date.

In vitro studies indicate that other constituents of St. John's wort, such as hypericin, kampferol, pseudohypericin, and hyperoside, are not PXR ligands and thus do not contribute to the enhanced CYP3A4 mRNA expression. Likewise, Wentworth et al., using a reporter gene construct containing the ligand-binding domain of the CYP3A promoter, determined that hyperforin but not hypericin was capable of activating CYP3A transcription when coexpressed with the steroid X receptor, which is synonymous with PXR. Hyperforin but not hypericin interacts directly with the receptor ligand–binding domain of PXR and contributes to the recruitment of steroid receptor coactivator-1 with an efficiency that is comparable to that of rifampicin. In addition, hyperforin has been shown to induce other PXR-responsive genes, such as CYP2C9, CYP2C19, CYP2B6, and p-glycoprotein (MDR-1), by mechanisms that may involve both the PXR- and the constitutive androstane receptor (CAR)-responsive elements, but the extent of induction of these genes is modest compared to that of CYP3A. In the case of the CYP2C9 gene, a PXR-responsive element was identified −1839/−1 824 base pairs upstream from translation start site at the same location as the CAR-responsive element. Komoroski et al. reported increased mRNA and protein expression and catalytic activity following exposure of human hepatocytes to hyperforin, confirming the in vivo observations (*vide infra*) concerning CYP2C9 and CYP3A4 induction by St. John's wort. The treatment of human hepatocytes with hyperforin did not alter CYP1A2 expression (mRNA and protein) or catalytic activity. In vitro, studies using LS-180 cells have demonstrated that hyperforin was capable of inducing the PXR-dependent expression of P-glycoprotein by western blot analysis, and functionally reduced the cellular uptake of the P-glycoprotein substrate, rhodamine 123.

Role of CYP3A in Drug Interactions

The most abundant CYPs in humans belong to the CYP3A subfamily, which accounts for up to 60% of total hepatic and up to 90% of total intestinal CYP. Like other CYPs, CYP3A family members are heme-containing proteins that along with the conjugating enzymes, such as the sulfotransferases (SULTs) and glucuronosyltransferases (UGTs), are instrumental in metabolizing a wide variety of endogenous and exogenous agents. The human CYP3A subfamily includes four members, namely CYP3A4, CYP3A5, CYP3A7, and CYP3A43. CYP3A4 is abundantly expressed in all adults and is responsible for the metabolism of a wide variety of structurally diverse chemicals including macrolide antibiotics, 3-hydroxy-3-methylgluatryl coenzyme A (HMG-CoA)–reductase inhibitors, HIV protease inhibitors, benzodiazepines, and immunosuppressants. It has been estimated that approximately 40% to 50% of drugs requiring metabolism for elimination undergo biotransformation by CYP3A4. The CYP3A5*1 gene product is detected in about 30% of Caucasian and 70% of African-American human livers and intestines and has comparable catalytic activity and substrate selectivity to CYP3A4, although there are important exceptions to this generalization. CYP3A7 is expressed only in fetal tissue and the level of expression and catalytic activity of CYP3A43 are extremely low and consequently, an important role for these enzymes in drug metabolism is not anticipated.

The expression of CYP3A4 and CYP3A5 at both the intestine and liver results in a greater first-pass removal of CYP3A substrates than would be predicted if the liver was the sole organ of removal. For example, the CYP3A substrates cyclosporine, nifedipine, midazolam, and verapamil exhibit low oral bioavailability because of the substantial contribution of both intestinal wall and hepatic metabolism to their first-pass elimination in man. In view of the broad substrate selectivity, along with expression in both the enterocyte and hepatocyte, it is not surprising that modulation of CYP3A expression and activity by environment, disease, and other drugs, such as St. John's wort, is a significant public health issue with implications in regard to drug safety and efficacy. The remaining portion of the chapter will review the reported interactions between St. John's wort and prescription medications.

Interactions with CYP3A Substrates

Anticancer agents

Irinotecan is a topoisomerase-I inhibitor, which is used in the treatment of colorectal and non-small cell lung cancer. Individuals diagnosed with cancer routinely become depressed and may require pharmacotherapy with prescription or botanical antidepressants. Mathijssen et al. reported that the disposition of 7-ethyl-10-hydroxycamptothecin (SN-38), the active metabolite of irinotecan, was altered following coadministration of St. John's wort to five individuals. SN-38 levels were reduced 43%, while plasma concentrations of irinotecan remained unaltered. The formation of a CYP3A-mediated metabolite of SN-38, 7-ethyl-10-[4-N-(5-aminopentanoic acid)-1-piperidino]-carbonyl-oxy-camptothecin

(APC), did not appear to be significantly altered, although the APC/irinotecan serum ratio was reduced by 28%. Likewise, SN-38 glucuronidation was not altered by St. John's wort. The investigators concluded that St. John's wort and irinotecan should not be coadministered.

Imatinib is an inhibitor of the protein tyrosine kinase involved with platelet-derived growth factor (Bcr-ABL). A loss of cellular control of this tyrosine kinase has been identified as a key mechanism for malignant cell growth. The ability of imatinib to inhibit Bcr-ABL provides a rationale for its use in the treatment of human cancers such as Philadelphia chromosome–positive chronic myologenous leukemia. CYP3A4 plays a principal role in the biotransformation of imatinib. The effect of St. John's wort on imatinib disposition was investigated in 12 healthy volunteers using a two-period, open-labeled, fixed-sequence study by Frye et al. Imatinib (400 mg) was administered before and after the administration of St. John's wort [300 mg; Kira (LI 160), Lichtwer Pharma AG, Berlin, Germany] three times a day for 14 days. The administration of St. John's wort resulted in 30% reduction in imatinib exposure from 34.5 ± 9.5 to 24.2 ± 7.0 μg hr/mL. A corresponding 43% increase in the oral clearance of imatinib was observed following St. John's wort dosing. Frye et al. concluded that the imatinib–St. John's wort (drug–botanical product) interaction is clinically significant and may result in a loss of imatinib efficacy.

Anticonvulsants

Carbamazepine is a dibenzazepine carboxamide derivative that is used to treat epilepsy and other neurologic conditions. A substrate of CYP3A, carbamazepine, is also recognized as a potent in vivo and in vitro inducer of CYP3A4 (74–77). Induction of CYP3A4 by carbamazepine is mediated at least in part through activation of PXR, although other mechanisms such as glucorticoid receptor–activation have been proposed. The effect of St. John's wort administration (300 mg t.i.d. × 14 days) on carbamazepine disposition at steady state was examined in eight healthy adults. The oral clearance of carbamazepine (2.8 ± 0.3 L/hr) was not significantly altered by St. John's wort (2.9 ± 0.6 L/hr) administration. Likewise, the area under the plasma concentration–time curve (AUC) at steady state of the CYP3A4-mediated metabolite carbamazepine 10,11-epoxide was not altered by St. John's wort dosing (37.5 ±7.4 vs. 41.9 ± 10.9 mg hr/L). These data indicate that a 14-day course of therapy with St. John's wort does not enhance the elimination of carbamazepine. This may reflect a lack of influence of intestinal CYP3A4 on carbamazepine disposition, given that the oral availability of carbamazepine approaches unity. In addition, the product used in this study may have lacked sufficient quantities of hyperforin to induce hepatic CYP3A4 activity. Also, many anticonvulsants (phenobarbital, carbamazepine, and phenytoin) are CYP3A inducers and modulate their own pharmacokinetics via enzyme induction. The lack of effect of St. John's wort on carbamazepine disposition may therefore reflect the possibility that the enzyme system (CYP3A4) is already close to maximal induction.

Antihypertensives

Nifedipine is a dihydropyridine calcium channel modulator, often used in the treatment of hypertension and angina. CYP3A4, with a minor contribution from CYP3A5, is the principal enzyme involved in the metabolism of nifedipine. Smith et al. examined the effect of St. John's wort (900 mg/day for 18 days) on nifedipine disposition by examining changes in C_{MAX} in 22 healthy volunteers. St. John's wort coadministration reduced the maximum nifedipine plasma concentration obtained by approximately 50%, following a 10 mg oral dose. It is to be expected that other dihydropyridine calcium channel blockers that rely on CYP3A for their metabolism (e.g., isradapine and nimodipine) will be similarly affected by St. John's wort administration.

Verapamil is a diphenylalkylamine calcium channel modulator that is widely used in the treatment of hypertension, angina, and cardiac arrhythmias. Verapamil is extensively metabolized by the CYP3A enzymes. Tannergren et al. examined the effect of St. John's wort on the jejunal transport and presystemic extraction of single-dose verapamil in eight healthy male volunteers using a fixed-order design (control–treatment). St. John's wort 300 mg was administered three times daily for 14 days. The administration of St. John's wort did not alter the cellular permeability of verapamil, but did increase the excretion of the CYP3A-mediated metabolite norverapamil into the intestine. Furthermore, jejunal transport of the verapamil enantiomers was not altered by St. John's wort pretreatment. St. John's wort administration resulted in an 89% reduction in R- and S-verapamil plasma concentrations. Verapamil has also been shown to inactivate CYP3A4 through the formation of a metabolic intermediate complex and is also a modest inducer of CYP3A4. The effect of St. John's wort on the disposition of verapamil at steady state has not been assessed and is not readily predictable from single-dose data. Unless proven otherwise, it would be prudent to expect that the coadministration of St. John's wort with verapamil will result in decreased verapamil concentrations and possibly efficacy.

Antiretroviral agents

Indinavir is a protease inhibitor used in the management of HIV infection. CYP3A4 mediates the biotransformation of indinavir in vitro, and in vivo, indinavir has been shown to be a potent competitive and mechanism-based inhibitor of CYP3A4. Piscitelli and coworkers examined the effect of St. John's wort (300 mg t.i.d. × 14 days) administration on indinavir (800 mg q.i.d. × 8 hr × four doses) exposure in eight healthy volunteers (two females). The administration of St. John's wort for 14 days resulted in a significant 54% reduction in the indinavir eight- hour area under the concentration–time curve, from 35.8 ± 13.0 to 15.6 ± 5.8 μg × hr/mL. The authors conclude that the magnitude in the reduction in indinavir concentrations may result in the development of antiretroviral resistance and subsequent treatment failure.

Nevirapine is a non-nucleoside reverse transcriptase inhibitor used in the treatment of AIDS. Elimination of nevirapine from the body occurs via P-glycoprotein, and it is extensively metabolized by the CYPs. In addition, nevirapine dosing is known to increase CYP3A and CYP2B6 enzymes by approximately 25%. This induction appears to be mediated by the orphan nuclear factor PXR. de Maat et al. reported data from five HIV-1-infected individuals who were treated with nevirapine and coadministered St. John's wort for several months. The median oral clearance of nevirapine for all patients (n = 171) was 3.2 L/hr (range 2.7–3.9); for the five individuals taking St. John's wort, the median oral clearance of nevirapine on St. John's wort was 4.3 (range 3.8–4.7 L/hr), whereas the median oral clearance without St. John's wort coadministration was 3.3 L/hr (range 3.2–4.2 L/hr). The authors concluded that the coadministration of St. John's wort resulted in a 35% increase in median oral clearance of nevirapine and that dose adjustment is indicated.

Benzodiazepines

Midazolam

Midazolam is a 1,4-imidazobenzodiazepine that is widely employed therapeutically as a sedative/hypnotic in major and minor surgical procedures. In humans, midazolam is primarily eliminated from the body by CYP3A-mediated metabolism to the major primary metabolite, 1-hydroxymidazolam, and to a much lesser extent to 4-hydroxymidazolam. Midazolam is widely used as a selective metabolic probe for assessing CYP3A activity in vivo, because it is not a substrate for the P-glycoprotein efflux transporter. Following intravenous administration, less than 1% of the dose is excreted unchanged in the urine. Consequently, the clearance of midazolam following intravenous administration has proven to be an effective index of hepatic CYP3A activity in vivo. Up to 75% of the first-pass loss of midazolam following oral administration occurs in the intestinal wall using the simultaneous administration of oral and intravenous drug. Additionally, approximately 90% of the variability in oral availability was accounted for by variations in intestinal availability alone.

Wang et al. demonstrated that multiple-dose St. John's wort dosing resulted in a 50% reduction in the midazolam oral AUC and maximum serum drug concentration and a corresponding doubling of the oral clearance, from 122 ± 71 to 255 ± 128 L/hr. In contrast, the systemic clearance of midazolam increased from 34.3 ± 10.8 to 43.6 ± 15.8 L/hr, but this change was not significant. The oral bioavailability demonstrated a significant decrease from 0.28 ± 0.15 to 0.17 ± 0.06, but changes in hepatic and intestinal availability were not significant. Similar results were observed following St. John's wort administration for eight weeks in a group of 12 women. These changes are in good agreement with the observation of Du¨rr et al. who reported a 1.5-fold increase in intestinal CYP3A4 expression and a 1.4-fold increase in erythromycin breath test.

Dresser et al. administered St. John's wort (LI 160 300 mg t.i.d.) to 20 ethnically diverse individuals and observed a 44% increase in the systemic clearance of midazolam. In contrast, the oral clearance of midazolam was increased 1.7-fold. The combined changes in midazolam disposition resulted in a significant reduction in the oral bioavailability of midazolam. Gurley et al. examined the one-hour 1-hydroxymidazolam-to-midazolam serum ratio and concluded that St. John's wort administration for 28 days resulted in a significant increase in the ratio, which is indicative of CYP3A4 induction.

Alprazolam

Markowitz et al. initially reported that St. John's wort administration did not alter the disposition of alprazolam following oral dosing. It was subsequently determined that the duration of St. John's wort administration (three days) was insufficient to demonstrate the inductive effects of this botanical medicine on CYP3A4. In a follow-up study in which St. John's wort was administered for 14 days, there was more than a doubling of the oral clearance of alprazolam, from 3.7 ± 0.9 to 8.4 ± 3.2 L/hr. There was also a corresponding reduction in the elimination half-life by approximately 50%, from 12 ± 4 to 6.0 ± 2.0 hours. However, the maximum plasma concentration and the time to maximum concentration were not significantly different before and after St. John's wort dosing. The change in oral clearance is consistent with a change in the systemic elimination but not first- pass elimination of alprazolam, because the maximum serum alprazolam concentration achieved was not significantly different before and after St. John's wort administration. This is to be expected, considering that the oral bioavailability of alprazolam is high (>0.8), which is consistent with alprazolam being a low-affinity substrate for CYP3A4.

HMG-CoA–reductase inhibitors

Sugimoto et al. examined the effect of St. John's wort administration (300 mg three times a day for 14 days) on the disposition of simvastatin and pravastatin in 16 healthy male Japanese subjects in a double-blind crossover study. The administration of St. John's wort significantly reduced the mean maximum plasma concentration from 3.6 ± 1.0 to 2.5 ± 0.7 after oral simvastatin (10 mg) dosing and a corresponding 48% reduction in the mean systemic exposure to simvastatin, from 11.1 ± 3.7 to 5.8 ± 1.8 ng hr/mL. The authors reported similar results for the active metabolite (simvastatin hydroxy acid) of simvastatin. In contrast to the significant changes observed with simvastatin, St. John's wort administration (300 mg three times a day for 14 days) did not significantly alter the mean maximum plasma concentration achieved following pravastatin (20 mg) administration (36.5 ± 5.7 ng/mL before vs. 30.8 ± 5.2 ng/mL after). Likewise, significant differences in the mean systemic exposure to pravastatin were not observed following placebo (109.4 ± 17.4 ng hr/mL) and St. John's wort (96.6 ± 13.4 ng hr/mL) dosing. The differences reflect the fact that simvastatin is a substrate for CYP3A4 and P-glycoprotein, whereas pravastatin is not a substrate for either CYP3A or P-glycoprotein (MDR1). Similar effects are expected for other HMG-CoA–reductase inhibitors, such as lovastatin, cerivastatin, and atorvastatin, which rely on CYP3A4 and P-glycoprotein for their distribution and elimination. In the case of the CYP2C9 substrate, fluvastatin, a drug interaction between St. John's wort and fluvastatin is expected to be at most modest, even though there is evidence that St. John's wort alters CYP2C9 expression in vitro. This is because Wang et al. did not observe an alteration in the disposition of the prototypic CYP2C9 probe drug, tolbutamide, in a group of 12 healthy volunteers.

Immunosuppressants

Cyclosporine is a calcineurin-inhibitor immunosuppressant that is in part metabolized by CYP3A4/5 and transported by P-glycoprotein (MDR1). Coadministration of St. John's wort with cyclosporine has resulted in significant reduction in circulating cyclosporine concentrations, which has led to graft rejection. Breidenbach et al. reported a series of 30 renal transplant recipients who were stabilized on cyclosporine and subsequently administered St. John's wort. Following initiation of St. John's wort therapy, blood cyclosporine concentrations were reduced 47% (range: 33–62%) and the corresponding cyclosporine doses were increased on average 47% (15–115%), to maintain therapeutic cyclosporine blood concentrations. Cessation of St. John's wort dosing resulted in a 187% (84–292%) rise in blood cyclosporine concentration, which required subsequent cyclosporine dose adjustment. These changes

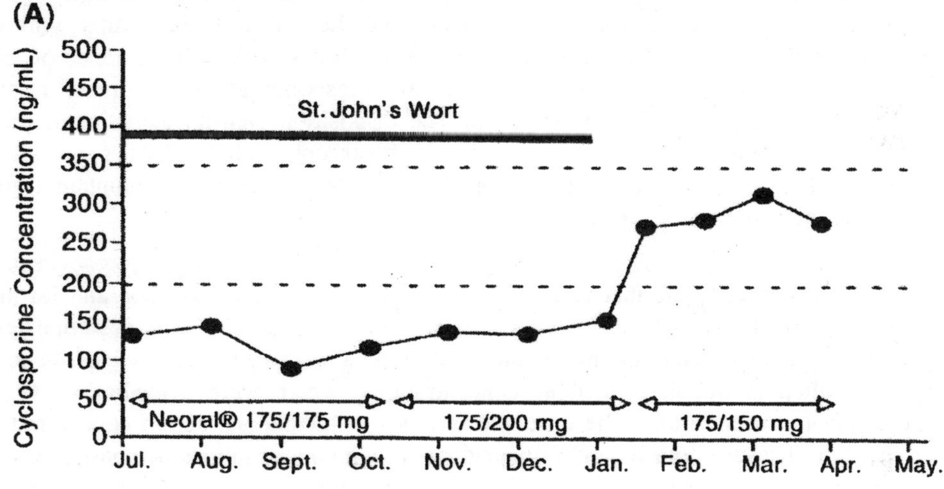

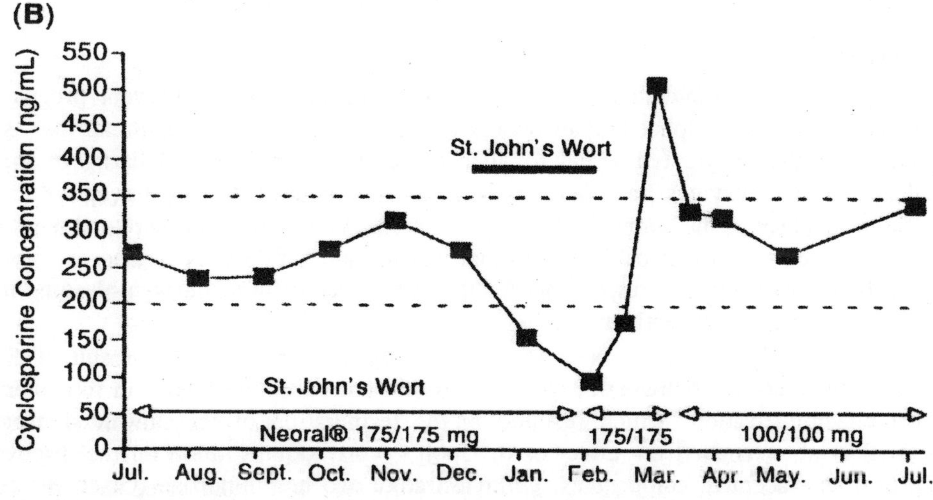

Fig. 7.2. A–Chronology of CSA through concentrations in patient 1 self-mediating with SJW (dotted lines). B–Chronology of CSA trough concentration in patient 2 self-medicating with SJW (dotted lines)

have been confirmed by others. Barone et al. reported the occurrence of acute graft rejection in two kidney transplant patients and Ruschitzka et al. reported a similar loss of immunosuppression in cardiac transplant patients. Bauer et al. also examined the effect of St. John's wort administration on the disposition of cyclosporine and its metabolites in 11 renal allograft recipients. St. John's wort was administered for 15 days and cyclosporine plasma concentrations were adjusted every four days by assessing trough concentrations. The investigators demonstrated that St. John's wort administration resulted in a 45% decrease in cyclosporine exposure compared to baseline. Likewise, metabolite exposure was altered significantly following St. John's wort administration with metabolites AM1c and AM1, demonstrating a 60% decrease after dose correction, but exposure to the metabolites AM9 and AM19 was not affected. In addition, the interaction between cyclosporine and St. John's wort was confirmed by Dresser et al., with the oral clearance increasing 63% from 728 ± 195 to 1155 ± 236 mL/min.

Tacrolimus is a calcineurin-inhibitor immunosuppressive used to prevent organ rejection following kidney and liver transplantation. The disposition of tacrolimus, like cyclosporine, is dependent on both CYP3A activity and P-glycoprotein activity. Circulating tacrolimus concentrations were reduced following the administration of St. John's wort. Hebert et al. examined the effect of St. John's wort (300 mg t.i.d. × 18 days) coadministration on the oral disposition of tacrolimus. The oral clearance of tacrolimus increased from 349 ± 126 to 586 ± 275 mL/hr/kg and a corresponding 35% decrease in the AUC from 307 ± 176 to 199 ± 140 μg hr/L was observed. The disposition pharmacokinetics of the adjunct agent, mycophenolic acid, was not altered following coadministration with St. John's wort. It is clear from these reports that individuals who require immunosuppressive therapy to maintain transplanted organ function should not receive St. John's wort.

Opioids

Methadone is a long-acting opiate that is used in the treatment of opiate addiction and for analgesia. Eich-Hochli et al. described four addicts in whom St. John's wort (Jarsin) was coadministered with three daily doses of methadone. The administration of St. John's wort (900 mg/day) for 14 to 47 days (median 31 days) resulted in trough methadone concentrations, which were a median of 47% (range 19–60%) of the original concentration. The observed changes in methadone serum concentrations were not enantiomer selective, because both R- and S-methadone trough concentrations demonstrated reductions of similar magnitude. Two female patients reported symptoms suggestive of withdrawal and requested increases in their methadone dose.

Oral contraceptives

Oral contraceptives are combination products that are typically used to prevent pregnancy. The combination of an estrogen (17-alphaethinylestradiol) and a progestin (e.g., norethindrone) is used to prevent the release of the oocyte (egg) and to alter the cervical mucous and lining of the uterus. Drugs that induce CYP3A enzymes have been associated with reduced oral contraceptive efficacy or even failure. The metabolism of the components of oral contraceptives, ethinylestradiol and norethindrone, is thought to be catalyzed at least in part by intestinal and hepatic CYP3A. A number of reports have indicated that St. John's wort may be responsible for the occurrence of breakthrough bleeding in women formerly stabilized on oral contraceptives.

In addition, "*miracle babies*" have been identified in the lay press to be a result of St. John's wort consumption. Furthermore, Schwarz et al. reported oral contraceptive failure in four women after St. John's wort coadministration, which resulted in the termination of the unwanted pregnancies. Subsequently, Hall et al. examined the effect of St. John's wort administration on the disposition and efficacy of the oral contraceptive components, ethinylestradiol and norethindrone (Ortho-Novum 1/35), in 12 healthy females. St. John's wort (Sundown Herbals) was administered three times a day for eight weeks. The pharmacokinetics of ethinylestradiol and norethindrone (CYP3A substrates) were assessed before and six weeks after the start of the St. John's wort dosing. St. John's wort significantly ($P \leq 0.05$) increased the oral clearance of norethindrone from 8.2 ± 2.7 to 9.5 ± 2.4 L/hr, with a corresponding decrease in the peak serum concentration of norethindrone (from 17.4 ± 5.1 to 16.4 ±5.2 ng/mL; $P < 0.05$). Likewise, the elimination half-life of ethinylestradiol was significantly reduced from 23 ±20 to 12 ±7 hours.

Furthermore, the incidence of breakthrough bleeding increased with the duration of St. John's wort administration with 7 of 12 individuals having breakthrough bleeding compared to two individuals prior to initiation of St. John's wort dosing. In good agreement with the observation of Hall et al., Pfrunder et al. reported that St. John's wort given twice daily or three times a day resulted in a greater incidence in break-through bleeding, 13/17 or 15/17, respectively, compared to oral contraceptive (20 μg ethinylestradiol and 150 μg desogestrel) dosing alone. Although, pharmacokinetic changes were not observed for ethinylestradiol, the maximum plasma concentration and the AUC of 3-ketodesogestrel decreased 18% and 44%, respectively, during twice-daily dosing of St. John's wort. It is clear that the combination of St. John's wort with oral contraceptive has resulted in the induction of norethindrone clearance, increased incidence of breakthrough bleeding, and reports of unplanned pregnancy and resultant termination. Thus, the coadministration of St. John's wort in women taking oral contraceptives is contraindicated and should be discouraged. To prevent this interaction, it is the author's opinion that all St. John's wort products must clearly carry warning labels concerning the potential for St. John's wort to alter the efficacy of oral contraceptives and of many other prescription products.

Interactions with Substrates of Other P450s

Theophylline

Theophylline is a bronchodilator that is commonly used to treat the symptoms of chronic asthma. The principal enzyme involved in the biotransformation of theophylline is CYP1A2. In a case report, Nebel et al. described an individual who required theophylline dosage adjustment following the initiation and cessation of St. John's wort pharmacotherapy. The dose of theophylline (Theodur) was increased from 300 mg twice daily to 800 mg twice daily following the initiation of St. John's wort intake. The resultant steady-state theophylline concentration was 9.2 μg/mL. Subsequently, termination of St. John's wort resulted in a twofold increase in serum theophylline concentration and necessitated dose reduction. Preliminary in vitro experiments conducted suggested that hypericin and pseudohypericin were capable of activating the xenobiotic response element, which is responsible in part for CYP1A2 induction. However, it should be noted the individual described in the case report by Nebel et al. was a smoker taking a multitude of other medications, including furosemide, morphine, zolpidem, valproic acid, ibuprofen, amitriptyline, albuterol, prednisone, and zafirlukast. Zafirlukast has been shown to be an inhibitor of theophylline both in vitro and in vivo.

Thus, the observed changes in theophylline disposition may be a result of St. John's wort altering the disposition of one of the many concurrent medications. In addition, a study by Morimoto et al. failed to confirm the observation of Nebel et al. Briefly, Morimoto et al. examined the potential for St. John's wort to alter the disposition of theophylline in vivo by conducting a randomized open-label crossover study. The oral clearance of theophylline was determined in 12 healthy Japanese men before and after 15 days of St. John's wort administration (300mg three times daily). St. John's wort did not alter the oral clearance of theophylline (2.3 ± 0.6 L/hr vs. 2.4 ± 0.6 L/hr).

Caffeine

Caffeine is a methylxanthine that is a central nervous system stimulant found in a number of beverages such as coffee, tea, soda (Pepsi, Coke, Mountain Dew, etc.), and over-the-counter products (Vivarin, NoDoz, etc.). Caffeine is principally metabolized to paraxanthine by CYP1A2 and the six-hour plasma ratio (paraxanthine to caffeine) has been used as an index of in vivo CYP1A2 activity. The administration of St. John's wort (300 mg three times daily) for two weeks did not alter the disposition of caffeine. In the same study, Wang et al. reported that a single dose of St. John's wort (900 mg) did not affect the oral clearance of caffeine. Likewise, Wenk et al. examined the effect of 14 days of St. John's wort administration (300 mg t.i.d., $n = 16$) on the in vivo activities of CYP3A4, CYP1A2, and CYP2D6 using 6β-hydroxycortisol-to-cortisol urinary ratio, paraxanthine-to-caffeine salivary ratio, and dextromethorphan-to-dextrorphan urinary metabolic ratio, respectively. The mean values for the salivary estimates of CYP1A2 were not significantly altered by treatment with St. John's wort. This observation is in good agreement with the observation of Morimoto et al. with St. John's wort and theophylline (*supra vide*). In addition, Komoroski et al. noted that hyperforin did not alter the mRNA expression, protein expression, or catalytic activity of CYP1A2 in human hepatocytes. These observations taken together suggest that interactions between CYP1A2 substrates and St. John's wort are unlikely.

Omeprazole

H. perforatum II 300 mg was used in assessing the effect of St. John's wort on the pharmacokinetics of a single dose of omeprazole. A placebo-controlled randomized crossover study was conducted over a five-week period in 12 individuals. Six individuals had CYP2C19 * 1/ *1 and six individuals had either *2/*2 ($n = 4$) or *2/*3 ($n = 2$) genotypes. The sulfoxidation of omeprazole is mediated primarily by CYP3A4 and the 5-hydroxylation of omeprazole is mediated principally by CYP2C19. Administration of St. John's wort 300 mg three times a day for 14 days resulted in a significant reduction in the AUC of omeprazole in both homozygous wild-type individuals and homozygous variant individuals. A corresponding increase in the principal metabolites for both CYP2C19 (5-hydroxyomeprazole) and CYP3A4 (omeprazole sulfone)-mediated biotransformations demonstrated increased AUCs following St. John's wort administration. The authors suggest that the study provides evidence for in vivo CYP2C19 induction by St. John's wort.

Warfarin

Warfarin is an anticoagulant that is administered as a racemic mixture with the S-enantiomer having most of the pharmacologic activity. Warfarin is extensively metabolized in the liver by CYP2C9 with 7-hydroxylation being the principal route of metabolism for the S-enantiomer. R-warfarin is 8-hydroxylated, 6-hydroxylated, and 10-hydroxylated by CYP2C19, CYP1A2, and CYP3A4, respectively. Likewise, additional enzymes are involved in the metabolism of the S-warfarin, namely CYP3A4 and CYP1A2. In light of the overlap between the P450s involved in warfarin metabolism and those affected by St. John's wort, namely CYP2C9, CYP2C19, and CYP3A4, it is clear that an interaction between St. John's wort and warfarin is possible. To determine the potential for interaction, Jiang et al. examined the effect of St. John's wort administration on the pharmacokinetics and pharmacodynamics of warfarin (25 mg) administered to 12 healthy male volunteers.

The oral clearance of S- and R-warfarin was increased 36% and 29%, from 198 ± 38 to 270 ± 44 mL/min and from 110 ± 25 to 142 ± 29 mL/min, respectively. A corresponding reduction in the pharmacodynamic effect was observed with St. John's wort (one tablet three times a day for two weeks) dosing, significantly reducing the area under the effect curve of the international normalized ratio of pro- thrombin time by approximately 20% from 111 ± 49.3 to 88.3 ± 30.7. Although the data quite clearly indicate that St. John's wort and warfarin should not be coadministered, the enzyme(s) responsible for the increased clearance of S- and R-warfarin in vivo cannot be determined, because changes in metabolite formation were not assessed. Thus, it is possible that the observed changes in S- and R-warfarin clearance were a result of a St. John's wort–mediated induction of CYP2C9, CYP2C19, CYP3A4, or some combination of these enzymes.

Role of P-Glycoprotein in St. Johns's Wort Interaction

Fexofenadine

Fexofenadine is a nonsedating antihistamine that has been shown to be transported by P-glycoprotein (MDR1) and organic anion transport poly- peptide in vitro using cell culture models and MDR1 knockout animals. Wang et al. demonstrated that the administration of St. John's wort for 14 days resulted in a significant increase in the oral clearance of fexofenadine observed after a single 900mg dose of St. John's wort, from 62 ± 26 L/hr to 91 ± 32 L/hr. In good agreement with the observations of Wang et al., Dresser et al. observed a significant reduction in the maximum plasma concentration and a 94% increase in the oral clearance of fexofenadine following the administration of St. John's wort (Jarsin 300 three times daily for two weeks). The increase in the oral clearance of fexofenadine reported by these two groups is consistent with PXR-mediated induction of P-glycoprotein (MDR1) by St. John's wort.

Digoxin

Digoxin is a cardiac glycoside that is used traditionally in the treatment of congestive heart failure and is a substrate of the transporter P-glycoprotein. Johne et al. conducted a single-blind, placebo-controlled parallel study in 25 healthy volunteers (12 women). Volunteers were given a 0.25 mg loading dose of digoxin followed by 0.125 mg daily for 10 days. On day 6 of digoxin dosing, a single 900 mg (three tablet) dose of St. John's wort (LI 160) or placebo was administered and on day 15 of digoxin dosing (10 days of St. John's wort or placebo), the pharmacokinetic study was repeated. Single dose of St. John's wort had no effect on the pharmacokinetics of digoxin. In contrast, 10 days of St. John's wort dosing resulted in significant 25% decrease in the AUC from 0 to 24 hours and a corresponding 24% decrease in the maximum plasma concentration. Treatment with placebo did not alter the pharmacokinetic parameters of digoxin.

Likewise, Dürr et al. demonstrated an 18% reduction in digoxin exposure with a corresponding increase in intestinal P-glycoprotein/MDR1 and CYP3A4. In agreement with the above results, Mueller et al. reported that hyperforin-rich St. John's wort products (i.e., LI 160) resulted in a significant 25% reduction in the 24-hour area under the digoxin concentration–time curve. The observation with fexofenadine and digoxin are in good agreement. It is clear that P-glycoprotein, along with CYP3A4, provides a competent barrier to the absorption of xenobiotics. The administration of St. John's wort for a period of two weeks results in a reduction in drug exposure due to the increased efflux activity of P-glycoprotein at the brush border membrane of enterocytes. For drugs that are not substrates of CYP3A, the increased expression of P-glycoprotein results in a reduced bioavailability but no change in the elimination half-life. This pattern of interaction is suggestive of an alteration in first-pass elimination but not systemic clearance.

Echinacea

Echinacea is a widely available over-the-counter botanical remedy used for the treatment of the common cold, coughs, bronchitis, "flu," and inflammation of the mouth and pharynx. It is one of the more popular botanical remedies with a sales ranking of 5 and sales of US $70 million. About 10% to 20% of the adult and child botanical users consume echinacea routinely. Three species of *echinacea* (*Echinacea purpurea*, *E. angustinfolia*, and *E. pallida*) have been used medicinally. However, only the aboveground parts of *E. purpura* and the root of *E. pallida* have been approved for oral administration by the German E Commission.

The beneficial effect of *echinacea* in the treatment of infections appears to be a result of its ability to stimulate the host's immune system. Following exposure to *echinacea*, macrophages and T-lymphocytes demonstrate increased phagocytic activity and release of immunomodulators such as tumor necrosis factor-a and interferons. Although the exact mechanism of echinacea immunostimulatory effect is unknown, controlled studies suggest that the oral administration of echinacea is beneficial in the early treatment of upper respiratory infections. However, this observation is still controversial and the usefulness of long-term echinacea administration to prevent illness appears to be limited. Although echinacea appears to be well tolerated following acute and chronic dosing, the unsupervised self-medication by patients in an effort to cure, ameliorate, and/or prevent sickness provides the potential for a multitude of drug–botanical product interactions.

Budzinski et al. examined the capability of 5% to 10% (v/v) dilutions of marketed *echinacea* tinctures to inhibit the metabolism of 7-benzyloxyresorufin by cDNA-expressed CYP3A4. The relative inhibitory concentration, in relation to the full-strength product for *E. angustifolia* roots, *E. purpurea* roots, *E. angustifolia/purpurea* mixture (1:1), and *E. purpurea* tops was 1.1%, 4.0%, 6.7%, and 8.6%, respectively. The effect of these echinacea extracts on the in vitro catalytic activity of other drug-metabolizing enzymes (e.g., CYP1A2, CYP2C9, and CYP2D6) has not been assessed. However, extracts of teas prepared from combination products containing echinacea plus other botanical products (e.g., goldenseal, lemon grass leaf, spearmint leaf, and wild cherry bark) have inhibited drug metabolism mediated by cDNA-expressed CYP2C9, CYP2C19, and CYP2D6. The effect of extracts of *E. purpurea* on other hepatic and intestinal enzyme systems such as UGTs and SULTs has not been reported.

Although in vitro data suggest that echinacea products may be inhibitory, Gorski et al. observed a mixed effect in vivo. The effect of *E. purpurea* on the in vivo activity of CYP1A2, CYP2C9, CYP2D6, and CYP3A4 was investigated in 12 healthy volunteers (six males). This two-period, open-label, fixed-order study involved the administration of a cocktail of probes (caffeine, dextromethorphan, tolbutamide, and intravenous and oral midazolam) that were administered before and after eight days of *echinacea* (400 mg four times daily) dosing. Briefly, echinacea reduced the oral clearance of caffeine and tolbutamide by 27% and 11%, respectively. Although the change in tolbutamide clearance was statistically significant, the clinical relevance of the observed change is unclear. It appears from this study that echinacea does not alter the in vivo catalytic activity of CYP2D6, as reflected by the absence of change in the oral clearance of dextromethorphan. In contrast to the activity of these enzymes, hepatic and intestinal CYP3A alterations are a little less clear.

The systemic clearance of midazolam, a reflection of hepatic CYP3A activity, was increased significantly. In light of the enhanced systemic elimination of midazolam and considering the "well-stirred" model of hepatic elimination, it is predicted that the oral clearance of midazolam should be increased and midazolam exposure reduced. Or in other words, the oral clearance reflects the contribution of both intestinal and hepatic CYP3A to first-pass elimination and hepatic CYP3A to the systemic elimination of midazolam. However, the oral clearance of midazolam was not significantly altered by *echinacea* dosing. In addition, the oral availability of midazolam (F_{PO}) was significantly increased from 0.23 ± 0.06 to 0.33 ± 0.13. Given the relationship between hepatic (F_H) and intestinal (F_G) availabilities and oral bioavailability ($F_{PO} = F_H \times F_G$), it is possible to examine the effect of *echinacea* on these independent sites of CYP3A expression.

As expected from the change in the systemic clearance, the hepatic availability was reduced significantly from 0.72 ± 0.08 to 0.61 ± 0.016. In contrast to the observed enhanced hepatic extraction ($F = 1 - E$) of midazolam caused by echinacea dosing, the intestinal availability of midazolam was enhanced 85% from 0.33 ± 0.11 to 0.61 ± 38, resulting in an unchanged midazolam exposure (AUC) following oral midazolam administration. This observation suggests that intestinal CYP3A is inhibited. The mechanism(s) of the differential effects of echinacea on intestinal and hepatic CYP3A could be due to one or more of the following: (i) intestinal and hepatic CYP3A induction is mediated by tissue-specific activators; (ii) the inducing component is rapidly absorbed and thus the intestine has limited exposure; (iii) hepatic and intestinal CYP3A are induced, but there is a potent inhibitor of CYP3A which is not systemically available; and (iv) a metabolite of a constituent of the echinacea preparation is responsible for the induction of hepatic CYP3A but not intestinal CYP3A. It is interesting that Gurley et al. reported no effect of echinacea on CYP3A; however, midazolam was only administered as an oral dose. The differential effect of echinacea on intestinal and hepatic CYP3A complicates the predication of drug interactions with other CYP3A substrates. For instance, drugs that undergo minimal first-pass elimination by the intestine and liver may demonstrate an increased oral clearance as expected due to the induction of hepatic CYP3A. However, substrates that undergo high first-pass elimination by the intestine may demonstrate increased serum concentrations due to the inhibitory effect of echinacea on intestinal CYP3A. It is clear from the data presented that caution should be used when echinacea and CYP3A substrates are coadministered.

8

Drugs and Poisons from Plants

In the previous chapter, we discussed drugs that are used medicinally to releive pain or control disease. A large number of same drugs can effect an individual's sense of perception by producing feelings of transquality, invigoration, or otherworldiness. Sometimes people want to escape from, or "alter," reality and thus seek out these drugs, but in excessive quantities, most are toxic. This chapter focuses on plant substances used because of their psychoactive properties, and on the overlapping group of poisons obtained from plants. The psychoactive drugs can be classified on the basis of the chemical nature of the compounds involved, the effects they produce, or thesource from which they are obtained. None of these classification schemes is perfect. The psychological effects of the compounds involved sometimes differ between chemically very similar substances, and a classification based on chemical structure is rather difficult for a nonchemist to remember.

It is difficult to categorises the psychoactive drugs into stimultants or depressants, hallucinogens or narcotics because many drugs can act in a combination of ways. For example, many of the so-called depressants, like the narcotic analgesics (opiates) and the solanaceous alkaloids, can also act as hallucinogens. Stimulants such as nitotine and cocaine will likewise cause vision if taken in sufficient quantities. The word narcotic itself a dangerous and filled with problems because some people use it to refer strictly to those drugs obtained from the opium poppy (as here), while others employ it for any habit-forming drug or even for any drug used illicity. One way of grouping psychoactive drugs by their primary effects, but since this is a botany text, we will follow yet another system and discuss drugs taxonomically in terms of the species from which they are obtained. Caffeine and alcohol these are wildly used. Psychoactive drugs are discussed in next chapter. These are obtined from a variety of plant sources and are economically and culturally important.

Chemistry and Pharmacology of Psychoactive Drugs

Almost all chemicals thathave psychoactive properties contain nitrogen, and most belong to one of the classes of alkaloids. The most notable chemical exception is the active compound of marijuana, Δ-*trans-tetrahydrocannabinol* (THC). Some other alcohols (including ethanol) and terpenes have psychoactive effects and have been designated as hallucinogenic agents. First of all the psychoactive drugs absorbed into the blood stream and transported via circulation to sites where they can exert their effects. Like medicinal drugs, psychoactive drugs can be taken orally, injected into the body, or absorbed through membranes such as those lining the nose, rectum, vagina, or lungs. Once the active compounds enter the bloodstream, they are transported to all parts of the body. A drug may exert its influence in only one place in the the body, but it is still present (at least initially) throughout the bloodstream because blood and the products it carries continuously circulate around the body. As the blood passes through the liver on its cycle, some drugs are degraded in the places where they act, and the breakdown products are picked up by the blood and transported to the kidneys for excretion. In this process of circulation, ending in degradation and excretion, that accounts for the initial "rush" followed by an enventual "wearing off" of a drug's effect.

Most of the psychoactive drugs acts on cell of the central nervous system (the brain and spinal column) and affect very badly as the blood flow to the brain is 10 times that to other tissues. Therefore drugs reach the brain more quickly than tissues. However, blood capillaries around the brain are more tightly packed that those elsewhere and are covered by sheaths that make passage of compounds from them into the brain difficult. Thus the actual amounts of drugs that enter the brain are less than the blood flow to it might suggest. Once they have reached the central nervous system. Psychoactive drugs disturbed the natural interactions between nervous, or sensory cells. These cells transmit information by chemical signals.

Neurons release chemical substances called *neurotransmitters* in response to stimulation. Once released, these transmitters flow across the space, or *synapse*, between a transmitting neurons and a receiving neuron. Specific sites on the receptor recognize the transmitted compound and bind briefly with it. Once the compound has been accepted, it triggers a response in the receptor neuron. Among the receptor neurons involved in psychological reactions are those responsible for our perception of pain and emotion, as well as interpretations of audio and visual stimuli. Psychoactive drugs, for the most part, alter or mimic the behaviour of four kinds of natural neurotransmitters: acetylcholine, norepinephrine, serotonin, and neuropeptides. The way in which THC works is still unclear.

Caffeine appears to act as a stimulant by activating intracellular metabolism,and alcohol is a general central nervous system depressant. Acetylchonine is released by neurons of the brain and the peripheral nervous system. It is a neutral transmitting chemicals. It causes muscle contractions and also speed up the heart beat. The compund is synthesized within the transmitting neuron and stored until the neuron is triggered. Once stimulated, the neuron releases acetylcholine into the synaptic region. After a neighbouring neuron has received the chemical at specific places on its dendrite and reacted, the acetylcholine is broken down by an enzyme so that it can no longer produce an effect.

Drugs can effect this sequence by blocking the transmission of acetylcholine (atropine and scopolamine), preventing the breakdown of the chemical, or by mimicking its action (nitotine). The drugs prevent the action of acetylcholine or neurotransmitter, prevent rapid muscle reaction and thus produce a relaxed sensation.

Compounds that prevent acetylcholine breakdown, or mimic its effects, act as stimulants because they cause the receptor neurons to fire at an increased or continuous rate. Norepinephrine is another of the brain's endogenous (made within the body) neurotransmitters. Synthesis of norepinephrine takes place in transmitting neurons and released ofter receiving stimulus. However, after it exerts its effect, norepinephrine is reabsorbed, not broken down. Apparently, the same molecules are reused over and over. Some drugs (e.g., reserpine) deplete norepinephrine. Others such as cocaine prevent reabsorption of norepinephrine or mimic its action (mescaline, myristicine, and elemicin).

Serotonin is a brain neurotransmitter and it acts on the cells regulating body temperature, sensory perceptin, and sleep. Some psychoactive drugs such as the LSD-type compounds appear to alter the functioning of the neurons that transmit serotonin, thereby producing illusions of strange images. Finally, a recently didsovered group of neurotransmitter are peptides, polymers of amino acids. These include enkephalin, beta endorphin, and oxytocin. These chemicals are produced in minute quqntities and are received by very specific receptors. Some of these appear to act like the body's own pain killers. It is believed that opiates effect the same receptor sites as these endogenous pain killers and thus produce a dull, relaxed sensation. The evidence for this hypothesis isi that opiates are very specific in their effects, effective in low concentrations, and blocked by antagonistic agents. Opiates influence receptor cells in the brainstem, the medial thalamus, and at several places in the spinal cord.

History of Drug Use

We tend to think of our society as "the drug culture," but the use of mindaltering drugs is very ancient. What perhaps distinguishes our culture from previous ones is the modern dissociation of drugs from formal cultural or religious customs. It is easy to understand why humans would appreciate a substance that alleviated pain or reduced hunger and fagigue. At this stage the psychoactive properties of drugs merge into medicinal usage. Nevertheless, many drugs have always been taken in excessive quantities when ther was no physical need for them. These doese led to audio and visual hallucinations. The original appeal of these kinds of psychological effects was quite different from the modern enjoyment of the sensations as recreational experiences. Primitive peoples were unknown about hte natural phenomena and other events that was going around them and they used to call the action of supernatural agents.

Under normal conditions, people cannot see or speak with such powers, but under th einfluence of psychoactive agents, they could be transported to another world where communication with gods, demons, or even the dead was possible. In most cultures, only specific individuals were allowed to ingest psychedelic substances. Such people, usually men, were variously called healers, medicine men, witches, or shamans. Shaman, originally a Siberian word, has been adopted by anthropologists for this group of diviners because it lacks the connotations attached to most of the other designations. The initiation of an individual into the role of shaman was often painful and dangerous. Initiation rites sometimes involved starvation, some form of self-mutation, and consumption pf psychoactive drugs.

The combined effect produced a physical condition that enhanced the illusion of visiting the supernatural world and acquiring the insights necessary to serve as the intermediary between the earthly and spiritual worlds. The shaman took the drugs repeatedly whenever he needed the advice of ancestors or divine guidence to solve a problem. Anthropologists and botanists who have studied psychoactive drug use have found that New World peoples employed many more species of plants for their psychoactive properties than their counterparts in the Old World (40 versus 6 species).

It is assumed that the distribution of psychoactive drug plants is uniform in the two hemispheres, some authors have suggested that the explanation for the use of few plants in the Old World lies in the cultural superiority of Eurasian civilizations. This explanation assumes that shamanism is a primitive trait and that the cultural environments of the ancient European and Asian civilizations provided a milieu that replaced the need for spiritual intermediaries. Such authors forget, however, the localized distributions of many of the plants used by New World peoples and the overwhelming dominance of two psychoactive drug plants, the opium poppy and marijuana, in the Old World.

Cannabaceae

According to a Neolithic Chinese legend, the gods gave humans once plant to fulfill all needs. The plant was *Cannabis sativa*, ma, marijuana, or hemp. This assertion is not so far-fetched. *Cannabis* assuredly ranks among the world's most remarkable plants. It produces highly durable fibres that can be turned into ropes, fish nets, and clothing, its peeds are highly nutritious, and oil obtained from it, used in lamp or in paints and varnishes. Ten-thousand-year-old pot shards imprinted with twisted hempen fibers are part of the evidence that *Cannabis* was among the oldest cultivated plants. Today, the species is best known for the psychoactive chemicals it produces. Marijuana rivals alcohol, caffeine, and nicotine as the most widely used nonmedical drugs. It is said to be the most profitable of California's agricultural commodities and perhaps the second most lucrative crop in the United States. How it came to occupy this position is a fascinating story, involving cultures spanning almost every continent.

Cannabis, native to central Asia, the disperson of the seeds of *Cannabis* takes place by means of water, wind, birds or animals. Still, it is humans who have spread the species worldwide, often without knowledge of its psychoactive properties. The first to use *Cannabis* were the Chinese, who had such high respect for the plant that they referred to their country as the "land of mulberry and hemp." The first true paper was made by the hemp fibre the T'si Lun in 105 A.D. The ancient Chinese *Book of Rites* instructed that mourners wear hempen clothes to show respect for the dead. The Chinese apparently knew of hemp's psychoactive properties, but they were primarily interested in its medicinal virtues.

The legendary Shen Nung (ca, 3000 B.C.) observed that th female plants contained a greater proportion of the creative (yin) principles than the male (yang), and recommended cultivation of the former and its use in correcting spiritual imbalances. The ancient Hindu texts, the Vedas, describe Siva, the Lod of Bhang, bringing *Cannabis* down from the Himalays for the enjoyment of the Indian people. India was the country in which the hallucinogenic properties of *Cannabis* were first exploited. Chinese grew their plants very closely in order to reduce the branching and in this way they improve their fibre production, the Bengalese (from "Bhangland") developed a cultivation strategy that maximized production of the psychoactive compounds. The Indians also realized that marijuana was dioecious and that female plant swere more potent than males.

The resins exuded on the upper leaves and bracts of the female inflorescences are particularly rich in intoxiating substances. Resins protect the vulnerable part of the plant from desiccation, they are produced in greatest abundance when the plants are exposed to heat and sun. By growing the plants widely spaced and removing the male plants (to prevent fertilization and prolong female flowering) resin producton is enhanced. The stimulation of resin production by dry conditions is the reason why plants grown in semiarid climates are more potent than those from cool temperate regions.

Likewise, sinsemilla (unfertilized, or without seeds) marijuana is one of the strongest forms. In India *Cannabis* is taken in the form of bhang, a milk-based beverage concocted with ground *Cannabis* leaves, sugar, and an assortment of spices. Bhang is widely drunk and is commonly offered as a gesture of hospitality. Indians classify *Cannabis* products into ganja, consisting of the potent female flowers and upper leaves, and hashish (charas), which is relatively pure resin. There are many colourful stories about traditional hashish collection. In Nepal, naked men used to run through the fields of flowering female plants and then scraped off the clinging resin globules. In Persia, the plants were beaten on rugs which wer subsequently washed to dislodge the resins.

Marijuana may have also had a role in Western civilizations, but it is generally believed that the species had not spread farther west than Turkey in ancien ttimes. The history of Arab use of hashish is not documented, but many legends exist to fill this gap. One of them credits an ascetic monk with finding Cannabis growing wild on the hillsides and sharing it with the poor (Sufis). The monk Haydar was so delighted with the plant that he ate nothing else until his death 10 years later in 1221 A.D. The Sufis used hashish to comfort the pain of their cheerless existence. When Marco Polo returned from his travels to the east in 1297 A.D., he brought back a story that was destined to become a classic. He recounted that there was a terrorist, Hasan-ibn-Sabah, the "Old Man of the Mountain," who had a large band of followers willing to do anything, even commit murder or sucide, for their leader. This blind loyalty was attributed to a very clever recruitment strategy. Political candidates were drugged and brought unconscious into an exquisite mountain garden. There, among the exotic flowers and water works, beautiful women were ready to minister to every need.

After experiencing the pleasures of this paradise, the recruit was drugged once again and, upon awakening, told that he could return to the garden in this life or after death only if he swore complete allegiance to the Old Man. The technique was apparently very successful, and the band of loyal followers became known as assassins, a name purportedly derived from hashish, the drug they were supposed to have been given.

There is no evidence to substantiate this story, and much to refute it, but it was compelling enough to be retold countless times and has been misinterpreted even in recent history by antimarijuana zealots who imply that the hashish produced violence rather than to ensure loyalty to the assassins. In the fifteenth and sixteenth centuries, Arab traders introduced marijuana to Africa where its use quickly spread. The drug was commonly given to women in childbirth and fed to babies when they were weaned. The Kafirs called the plant daga, a name later applied to a culture that made widespread use of *Cannabis*. The dried plant part was mixed into beverages which was taken by Africans. The technique of smoking became popular only after the Dutch colonized the continent. By the time of the Crusades, when European contact with the Arabs was reestablished, marijuana use was common throughout Africa and Asia.

The *Tales of the Arabian Nights* (written sometime between 1000 and 1700 A.D.) was widely read in Europe and provided a romantic introduction for many to hashish. "The Tale of the Hashish Eater," in particular, described the effects of the drug ona degenerate who used it to delude himself that he was wealthy and handsome. When Napoleon's army was stationed in Egypt in 1798, recrutis experimented with the local intoxicant and returned to France with stories and samples. The French student visited the northern Africa and returned back with glowing youthful reports of marijuana use. A frecnch psychiarist, Dr. Jacques Joseph Moreau de Tours, read the medicinal uses of hashish in combinating plague and dysentery and decided to test the drug himself. Because he had read that the drug produced drastic change in perception, he thought that he might be able to administer it to sane individuals in order to reproduce the symptoms of psychosis and thereby learn about the causes of mental illness. He asked an artist friend, Theophilie Gautier, to try the experiment of the drug and report his reaction.

Gautier was impressed with the effect of hashish on hsi artistic senses, and he recommended it to his frient, the painter Boissard. Eventually the group expanded, meeting monthly in Boissard's hotel suite where they consumed hashish while being observed by Dr. Moreau. The 'Club des Hachichins' named after the story of the Old Man of the mountains by themselves. The club eventually became a meeting place of the great French artists and writers of the day: Alexander Dumas, Victor Hugo, Eugene Delacroix, and Charles Baudelaire.

Not surprisingly, much of the art and literature of this period reflects the preoccupation with the newly discovered drug. Despite Moreau's eventual conclusions published in "On Hashish and Mental Alienation" (1845) that hashish use had overall deleterious effects on mental health, the use of the drug continued to spread. The fascination with hashish travelled across the English Channel, from the French to the British intellectual community, where W.B. Yeats, Oscar Wilde, and Ernest Dawson experimented with it to see if it enhanced creativity. After introduction of Marijuana into the Americans it was scattered all over the country The first introduction of cannabis to the New World was done by the Spaniards during hemp cultivation in Chile in 1545.

The British, hoping to produce a lucrative attempts failed, *Cannabis* managed to escape cultivation and spread around the island. When African slaves were brought to Jamaica in the middle of the century to harvest sugar cane, they found a local source of their familiar drug already established. Before 1800, the British had sent formal orders for the colonists in North America to produce hemp. Even Thomas Jafferson and George Washington realized the military need for rope-making fibers and started hemp cultivation. An appreciation of the other uses of *Cannabis* did not develop in the United States unti the 1800s. The introduction of hemp's psychedelic effects followed the same pattern as in France and England.

First there was tremendous curiosity, primarily by intellecturals and artists, a rash of drug-inspired art, and then a general spread of marijuana use. In fact, the 1876 Centennial Exposition had a Turkish bazaar which featured hashish smoking as a special attraction. Despite the sensatinalized stories that soon began to appear about decadent hashish dens and violent consumption by the poor in such manner the use of the drug spread. During prohibition, marijuana smoke floated up and down the Mississippi to the tunes of Dixieland. Jazz musicians found "moota" a pleasurable way to enhance their music without experiencing the stupefying effects of alcohol. They expressed their appreciation of the drug with such songs as Benny Goodman's "Sweet Mrijuana Brown," Cab Calloway's "That Funny Feefer Man," "Texas Tea Party," the "Mary Jane Polka," and others. It has been suggested that the desire to suppress the increasing popularity of black and Latin-inspired music began the movement that eventually led to the criminalization of marijuana. However, racism would have been only one of many factors that countered marijuans use.

Harry Anslinger was trying to justify the existence of the newly created Narcotics Bureau of the Government. He also carried out the most effective antimarijuana campaign. He publicized stories of the horrors of marijuana "addiction" and depicted the put smoker as a savage fiend whose aroused sexual desires and violent tendencies led to criminal activities and the use of stronger drugs. The story of the "Old Man of Mountain" was revived by Anslinger but the story was twisted a little so that the drug iteself was the cause of irrational violence. Marijuana was tried by the press and condemned by an emotional, ill-informed public. In the late 1920s and 1930s, marijuana use was banned in Louisiana, Texas, and Illinois.

Other states soon followed, and in 1937 the use of cannabis came under federal jurisdiction withe the passage of the 1937 Marijuana Tax Law which actually heavily taxed, but did not prohibit, the drug. Other countries had previously taken hard step to prevent Cannabis use. In this field South Africa passed the first anti-Cannabis law in 1870, and by 1925, the League of Nations agreed to an international statute against its use. All of these legal sanctions against *Cannabis* had two things in common, they were not based on any substantial medical evidence of severe effects of marijuana use, and they universally failed to halt its spread. Most medical studies beginning in 1893 with the Indian Commission's through report failed to find any detrimental effects of moderate use. Sanctions against marijuana only seem to increase its popularity.

In the 1960s the explosive spread of pot smoking provides the drastic demonstration of such kind of effect. Young people throughout the world, disillusioned with the ways in which the older generation was handling affairs, made marijuana smoking a symbol of their rebellion. While attention was focused on marijuana in the 1960s, the suggested medical virtues as well as the possible detrimental effects were investigated. Once THC was finally isolated in 1965 and measured quantities could be used for testing, it was discovered that the active principal is effective in reducing the pressure exerted against the eyes of glaucoma patients. Cancer patients undergoing radiation or chemotherapy treatment when used such compound, the nausea problem reduced. Since THC dilates the bronchial vessels, it provides relief for asthma sufferers. Because of the useful medicinal qualities, marijuana cigarettes or concentrated THC in the form of pills can now be prescribed by physicians when appropriate. In 1972, a federal government study on the effects of marijuana was released. The report suggested decriminalization of marijuana because there was no evidence of medical or mentla damage from use of the drug and because the drug was so widely used that laws against it were unforceable.

In 1981, the results of another study made by the National Academy of Sciences were released. The academy panel observed the use of marijuana very carefully and came to conclusion that the low or moderate use weakened the sense, sensibility and sensitivity of users. They could find no evidence of addiction or permanent deleterious medical effects with low or moderate use, but heavy use was correlated with several psychological and physiological problems. The use of marijuana is directly linked with low sperm count in males. Since marijuana is usually smoked, heavy and regular user suffers lung disorders similar to those incurred by cigarette smoking. Finally, the Academy found that ther was a high probability that heavy users would turn to "hard drugs." The panel did conclude, however, that marijuana was safer thanits legal, more widespread counterparts, alcohol and tobacco, and suggested that the present legal sanctions against possession and trade be removed.

Papaveraceae

In United States poppies are well known as garden ornamentals or for their culinary seeds, but they rose to fame because of opium. The capsules of the fruits from which the seeds are obtained are rich in an alkaloid-containing latex. The latex has medicinal use but opium and morphine these are the most abundant of the opium alkaloids and are used as mind-altering drugs. The collection of opium from fruits of the opium poppy (*Papaver somniferum*) extends back to at least 3000 B.C. Sumerian tablets from 2500 B.C. refer to opium as the "joy plant" and describe the ingestion of smal balls of the latex to induce sleep and relieve pain. Opium is mentioned in every medical treatise written during ancient Greek and Roman times. The name opium comes from the Greek "opion" for poppy "juice."

The early Greeks realized that the driniking of wine in which latex had been dissolved led to trancelike state, and they therefore associated poppies with several divinities such as Hypnos, the God of Sleep, Morpheus, the God of Dreams, and Thanatos, the God of Death. It is commonly known that Chinese was responsible for introducing opium to the rest of the world, but China was ignorant of the drug until the seventh century A.D. when Arab traders first brought samples to the Orient. In the East, it is used to cure dysentory in some manner that paragoric (tincture of opium) is used for diarrhea today.

In the seventeenth century, the Dutch introduced tobacco smoking to Formosa and began to mix tobacco with opium in their pipes as a treatment for malaria. The practice of smoking opium spread to the mainland where it was rapidly adopted throughout the country. Soon, tobacco disappeared from the mixture, and smokers inhaled vapours from heated balls of latex that were droped into the bowl of a pipe. Opium smoking became so popular that Chinese officials thaught to ban the sale of opium. Their antiopium edicts, issued as early as 1729, were ignored by both the Chinese populace and th Portuguese suppliers. In 1800, the British gained a monopoly over trade rights with China. Since China had little use for of opium dealing because it was one of the few commodities for which the Chinese would exchange coveted products such as silk and spices. Eventually, England even established opium plantations in her Indian colonies to provide opium for trade.

Importation of opium into England was, however, illegal at this time, but distance and the need to maintain international trade seems to have exonerated the practice elsewhere. The Chinese lords tried again and again to halt the drug trafficking because they could see the debilitating effects on their people, but without the support of the British, their efforts failed. The situation came to a head in 1839 when the Chinese confiscated and destroyed British opium supplies in Canton. England retaliated by invading China, touching off the first Opium War. The war ended wiht a treaty that ceded Hong Kong to the British.

A second Opium War broke out in 1865 and the Chinese were again defeated. This time they were forced to agree to British opium importation. Naturally, opium consumption among the Chinese continued to escalate. Trafficking in the drug was finally reduced in 1906 when the last emperor of China managed to get the British to agree to a reduction in importation. Still, use did not drastically drop until the revolution and the establishment of the People's Republic of China in 1949. Opium trading and use are now strictly forbidden. Opium consumption was not common in Europe until 1525 when Paracelsus discovered, or rediscovered, a way to dissolve it inalcohol.

The opium is dissolved in alcohol to obtained the tincture was known as laudanum. At that time laudanum was the most popular drug due to its meaningful medical benefits of the case with which it could be consumed. In 1803, morphine was isolated form opium, producing a purified alkaloid that could be given in measured doses. On a per weight basis, morphine is 10 times as strong as opium (which contains at least 24 other alkaloids andnumerous other compounds). Purified morphin can be administred intravenously to releaf the pain. When the hypodermic syringe was developed in 1853, it presented doctors with a rapid method of introducing this potent pain killer into the bloodstream and morphine use rose. Morphine's effectiveness was both a blessing and a curse. In many cases, the speed with which a drug enters the bloodstream (and thus the intensity of the "rush") is correlated with the level of addiction it can produce.

Only after 45,000 soldiers returned home from the Civil War addicted to thepain killer were the dangers of morphine use recognized. In an attempt to develop a nonaddicting pain killer, scientiets discovered that morphine could be chemically altered by the addiction of two acetyl groups. The end product is a semisynthetic compound known as *heroin*. Heroin is an even more powerful analgesic than morphine, and it, like morphine, was initially explained that heroin contain less addicting qualities of its parent drug. While mark of this idea persisted untill 1905, but very soon become apparent that heroin was even more dangerous than morphine because it crosses the cell membranes very easily and rapidly than its natural counterpart. Because of this fast absorption (and hence cycling around the bloodstream), heroin must be more frequently administed than other opiates. In addition to the effects produced by heroin itself, users of the drug suffer from disruption of the blood flow, infections, diseases such as hepatitis, and collapsed blood veins.

Heroin is very physically addicting and produces pronounced withdrwal symptoms once the habit has become established. Two to three weeks of injecting 30 mg per day is considered enough to create a tenacious habit. The main cause of death for heroin addicts is by overdose. Because of the common usage of adulterantnts in heroin and the use of heroin in combination with other drugs, "overdose" levels for an individual ar enot constant and therfore cannot be adequately gauged. The lifestyle of the heroin addict who no longer seeks the drug for pleasure, but because he or she desperately needs it to avoid withdrawal, tends eventually to lead to overall physical problems. The death rate for heroin users is more than twice the normal rate. In 1914, a law was passed prohibiting the possession of opiates for nonmedical purposes. In 1924, the production of heroin was proved illegal in the United States and in 1956 existing legal supplies of the drug were destroyed completely. Because of its addictive properties, even morphine is seldom used in its pure form as a modern-day medicine.

In contrat, the use of heroin, all illegal, has dramatically increased during the last 30 years. Since heroin is made from morphine, the total production of opium has correspondingly increased. In 1972 most of the opium production ended up in the United States but still grow in Turkey. When Turkey and the U.S. government reached an agreement to limit opium poppy cultivation, th ecenter of productio shifted to southeast Asia. Economically, opium is an extremely lucrative crop. In 1972 10 kg of crude opium which sold for $250 where it was grown, would eventually gross $200,000 when sold on the street. With profits such as these, people are willing to risk the penalties if caught.

Fabaceae

A large number of new wrold Indian tribes used the species of the bean family which is rich in alkaloids, as hallucinogenic agents. Snuff made from powdered pods of cohoba (*Anadenanthera peregrina*) was first reported form Hispaniola in 1496 and was much later found to be used by the Indians of the Orinoco region of Venezuela. In South America, the snuff is called "yopo," and it is often mixed with lime (Calcium carbonate) when taken. The snuffs can be inhaled througha number of devices, one of which is a long, hollow pole. One end of the pole is placed in a nostril and the other inot a pile of snuff on the ground. Mescal, or red bean (*Sophora secundiflora*), a common ornamental shrub or small tree of the American southwest, was used by native proples of this region to induce trances. The plant has received some recent publicity because mescal beans have been confused with mescaline and unknowing youths have been tempted to try them. However, *Sophora* contains cytisine, a much more dangerous alkaloid than mescaline, and one that can easily cause death due to respiratory failure if slight overdoses are taken.

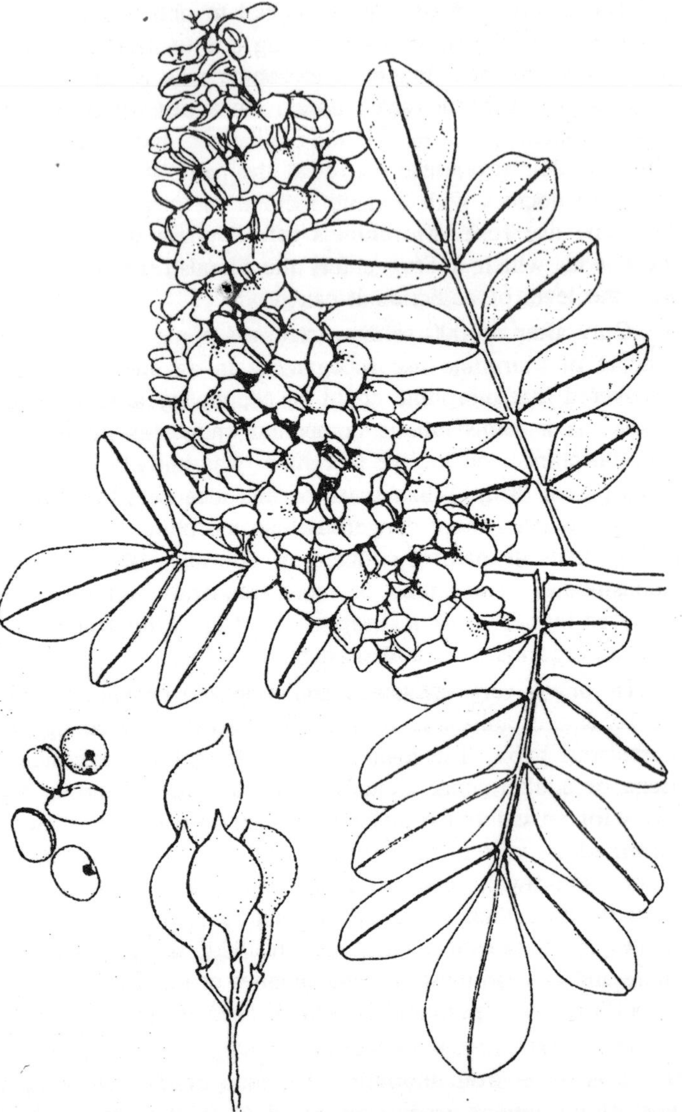

Fig. 8.1. Sophora secundiflora is valued as an ornamental plant because of its fragrant purple flowers and glossy green foliage.

Celastraceae

As Americans, we are accustomed to take a cup of coffee, tea, or other caffeine-containing beverage when we need a lift or freshness, but in other parts of the world the same effect is produced by consuming parts with different but equally effective stimulatory alkaloids. Among these is kat (*Catha edulis*), a native of Arabia but now most widely grown along the eastern edge of Africa. The alkaloids are ingested by chewing wads of freshly cut leaves usually mixed with small amounts of lime. Each lump, or quid, is masticated for about 10 min until all of the juice has been expressed. The remaining cellulose mass is than swalloed.

Erythroxylaceae

The stimulatory alkaloids are extracted by chewing the coca just like kat (varieties of either *Erythroxylum coca* or *E. novogranatense*). Coca is native to the north-central Andes. Discoveries of bags of coca leaves and utensils in 3,000 year old Andean burial sites indicate that it was used by natives of this region long before Europeans discovered South America. By the time Pizarro conqured Peru, coca was an integral part of Inca life and was considered to be a sacred plant. The preparation of coca involves very simple method the leaves are collected, dryed and sometimes allowed to sweet (lightly ferment) in order to render them pliable rather than crisp. Indians normally dip the leaves into lime that they carry in a small bag before chewing them.

As in the case of its use with kat, lime aids in the extraction and absorptin of alkaloids. The masticated residue is either expelled or swallowed. Cocaine, the active principle in coca, affect directly on the central nervous system by blocking reasorbtion of norepinephrine in the brain. This blockage has the result of making the coca chewer feel invigorated and relatively immune to fatigue and hunger, peasants of the Inca Empire depended on coa, as do an estimated 90 percent of their modern descendants, to mitigate the harsh environment of the high elevations of the Peruvian and Bolivian Andes.

The Spanish conquerors tried to prohibit coca use until they realized that the Indians they enslaved would work harder if allowed to chew it. Coca was taken back to Europe by the Spanish, but in leaf form it never received much attention. In 1860, cocaine was isolated and, unlike the homely leaes, it soon became extraordinarily popular. Sigmund Freud publicized the drug in his treatise *Uber Coca*. Among other things, Freud recommended coca as a treatment for alcoholism and morphine addiction. He also lauded it as a local anesthetic and praised its use in psychotherapy to relieve depression. In both Europe and the United States, doctors began to prescribe its use.

An enterprising Italian, Angelo Mariani, developed a coca wine beverage in 1860 which was soon the rage of Europe. Mariáni's wine earned commendations from scores of celebrities including Jules Verne, Ulysses S. Grant, President McKinley, Emile Zola, Henrik Ibsen, and Thomas Edison. Coca cola; which originally consist both caffine-rich extracts from Cola nitida and coca txtracts, was first marketed in 1880 as a headache remedy. Since 1904, however, federal law has prohibited the inclusion of cocaine in nay beverage. Ironically enough, after the Coca Cola Company complied with the federal law, it was sued for misleading advertising on the groudns that the name implied that the beverage contained coca products.

As a result, coca leaves, with the cocaine removed, are now used to flavour the syrup from which the soda is made. As more and more reports of cocaine-induced violence and other unsubstantiated effects of the drug circulated after the turn of the century, the government took steps to curb its use. In 1914, the drug was formally declared illegal by the Harrison Narcotic Act. As a result, cocaine costs escalated, but despite illegality and high costs, its use continues to increase. Status seekers appear delighted to pay exorbitant prices for a brief experience. Indulgers usually chop or grind crystals of cocaine hydrochloride before sniffing ("snorting") a line of the fine powder. Cocaine may also be smoked or injected.

The conversion of cocaine from the salt form to the base form is termed 'Free basing', which alters its solubility (increasing the drug's potency), has proved to be particularly more harmful. Both these latter methods of ingestion produce stronger sensations but are much more dangerous to the user than sniffing. Just how dangerous is cocaine? While researchers have not been able to find any organic damage from prolonged coca use per se among Andean Indians, their findings do nto mean that cocaine is a harmless drug.

Indians used to chew coca leaves and so they absorbs only limited quantities of the drug because the leaves contains alkoloides in relatively small amount (0.65 to 1.25 percent on a dry weight basis). An average native user might consumes 2 oz of leaves per day or about 0.7 grains of cocaine. A person who "snorts" or injects cocaine could be consuming as much as 6–8 grains per day of the drug plus adulterants. 1.2 grams of cocaine is considered as a lethal dose. Cocaine is not physically addiciting, but its use can become a pronounced habit as indicated by the burgeoning number of treatment centers for cocaine abusers.

After becoming accustomed to the elevated feelings it produces, habitual users suffer from depression when denied the drug. While occasional use produces feelings of mood elevation, vitality, mental clarity, sexual stimulation, and reduction in appetite in many users, chronic users can suffer from anxiety, conofusion, insomnia, and impotence. Paranoid psychosis can develop from longterm heavy use. Cocaine ingestion of pregnant animals is harmful to fetal development. This is concluded by recent experiments. As more reserach is conducted, other harmful effects may be discovered.

Malpighiaceae

One of the most important groups of plants used by South American Amazonian tribes as a source of hallucinogenic compounds are members of the genus *Banisteriopsis*, primarily *B. caapi*. Depending on the tribe, this species is called caapi, ayahuasca, yaje, or cipo. Infusions of mashed bark or stems are usually prepared, but stems can also simply be chewed in order to release the active alkaloid. People using ayahuasca describe the perceptions produced as feelings of having experienced death or of experiencing a separation of spirit and body. Sometimes the hallucinations are pleasant, other times apparently terrifying with vivid apparitions of jaguars or snakes. Like many other alkaloid-rich preparations, mixtures of ayahuasca often cause vomiting and diarrhea. Unlike the use of many other psychoactive drugs by native peoples, ayahuasca is often taken in communal rituals.

Fig. 8.2. Banisteriopsis caapi is a tropical vine with stems tha contain the alkaloid hormone, a psychoactive substance known as yaje or ayahuasca by the Amazonian Indians who use it in rituals.

It is thought that this practice of communal ingestion enhances feelings of clairovoyance or telepathy. The group, under the supervision of a yaquero, is led through chants to experience shared hallucinations. This effect was reflected in the name "telepathine" given to the mixture of active chemicals first isolated from *Banisteriopsis caapi*. Since this initial chemical work, the active compounds have been further purified and named harmine and harmaline. In some tribes, the rituals commemorae an act of incest by a deity. In almost all cases, the ceremonies are filled with sexual symbolism, but there is no sexual gratification during the experience.

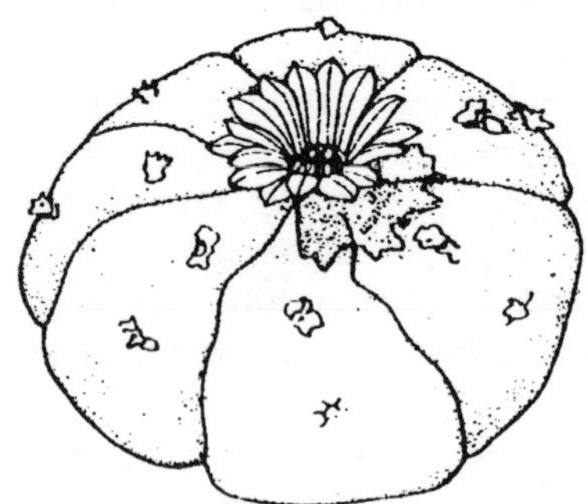

Fig. 8.3. Mescaline is the psychoactive compound in the gray-green peyote buttons that grow natively in southwestern U.S. and adjacent Mexico.

Cactaceae

A large number of species of the Cactaces family are good source of hallucinogenic compounds. The best known is peyote, *Lophophora williamsii*, a small, globose, gray-green cactus native to the Rio Grande Valley bordering the United States and Mexico. Native American Indians appear to have extended its distribution to Arizona and New Mexico centuries ago. No one is sure when peyote was first used, but sixteenth century reports by European explorers describe its use by the Aztecs as a divinatory plant. These accounts refer to peyote as the "diabolic root," because the Spanish observed the Aztecs using the plants ritualistically.

After the collapse of the Aztec Empire, use of peyote survived among a few Mexican Indian tries such as the Huichol, Cora, and Taramara. Inthe United States, the Plains Indians began to use it as late as the 1880s. In India harvesting is done by cutting off the top of the stems and leavign the sturdy taproot for regeneration. The stem tips, or buttons,were eaten fresh, or dried for later consumption. The dried buttons can be swallowed by softening and mastication. The initial reaction after swallowing in nausea. One to two hours later, quesiness disappears and is replaced by kaleidoscopic illusions of vivid colours, sensual hallucinations, and distorted perceptions of time. During this last period, which can last from 5 to 12 hours, the faithful report hearing the voices of their ancestors who help them to diagnose and cure any problems they may have.

Consumption of peyote means ingestion of 30 to 40 different alkaloids and several chemicals at a time, mescaline is one of the most active compound present in the alkaloids. Isolated in 1896, mescaline was the basis for research into psychedelic compounds for the next 50 years. While there is no evidence that peyote or pure mescaline is addicting, repeatrd use may permanently affect the mind. Because of their potentially harmful effects, both the plnat and the compound are illegal to possess or sell in the United States. However, one religious sect, the Native American Church, uses peyote as an integral part of its services. In a case that went before the Supreme Court, the church received an exemption from the Prohibition against peyote use. Originally founded by natives in northern Mexico, the sect holds Christian services modified by the use of peyote and espouses as doctrine, brotherly love, abstinence from alcohol, and high standards of moral conduct. Today, there are over a quarter million members of the church within the United States. Other cacti such as the San Pedro cacti have been widely used by native peoples in their native regions of Peru and Chile. All of these species used mostly contain mescaline and are used to produce hallucinations.

Apocynaceae

While fewer drug plants were used by native Africans than by South and Central American natives, one, iboga (*Tabernathe iboga*) was important as an hallucinogen. Iboga is, in fact, one of the very few psychoactive drugs in the entire periwinkle family. Although th species of the family are rich in alkaloids, most of them are toxic. A few species such as *Rauwolfia serpentina* and *Catharanthus roseus* are used medicinally, but there is no indication that they were used in rituals. In preparing iboga, the bark or roots of the shrubs are powdered or chewed, often with other plant species. Ingestion of the alkaloids, one of which is ibogamine, is said to lead to frightening visions and in large doses can produce severe convulsions, respiratoty arrest, and death.

Convolvulaceae

No species of morning glory family were known by scientists before 1903, which could be used for psychoactive purposes. In 1955, Humphrey Osmond described the use of *Rivea corymbosa* by Mexican natives, and, in 1960. T. MacDougall detailed the use of another member of the family, *Ipomoea tricolor* for altering perceptions. Still more startling was the discovery in 1960 that thc sees of these plants contained D-lysergic acid amide. Prior to this time, lysergic acid

Fig. 8.4. Rivea corymbosa, or ololiuqui a white-flowered vine of the morning glory family contains the psychoactive compound, D-lysergic acid amide.

alkaloids were thought to be restricted in nature to ergot (*Claviceps purpurea*), a rust that infests grains, and to a few other fungi. Synthetically produced LSD had become a part of the Americna counterculture in the 1960s and was regarded as something very modrn and exciting.

Osmond's and MacDougall's discoveries showed that the effects of these kinds of compounds had been recognized and utilized by native peoples, perhaps for thousands of years. Both of the convolvulaceous species now known to be used are called *ololiqui* (pronounced o-low-lee-oo-key) by the peoples that use them. The carefully measured quantities of seeds should be ingested especially by experienced individual because there is a little difference between the quantities seeds that produced vivid hallucinations and the quantities that produce death, there has been little lay experimentation with the seeds, despite the fact that one of the species is a commonly cultivated ornamental plant.

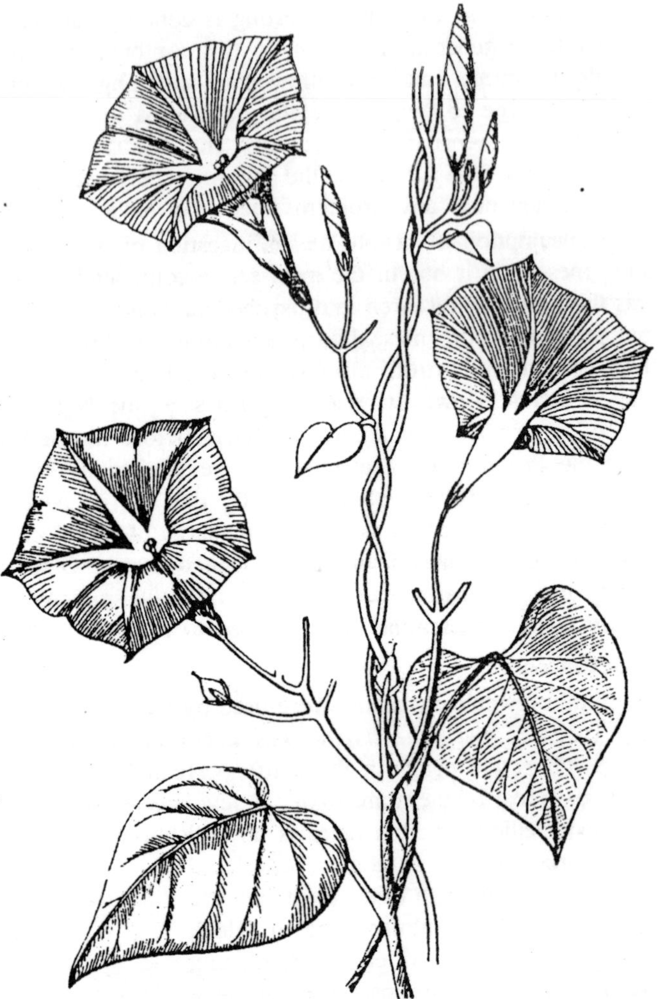

Fig. 8.5. Because of the extremely fine line between effective and lethal doses, the seeds of hallucinogenic morning glories have rarely been used even by primitive peoples.

Solanaceae

The plants of the Solanaceae is highly toxic therefore their use is restricted outside of some Central and South American people rarely used today as source of mind-altering drugs. Undoubtedly, the major reason for their restricted use is their toxicity. At one time, however, various species were used in both Eurasia and the Americas as sources of hallucinogenic compounds. Historically, the most important of these were species of *Datura*, *Hyoscyamus*, *Atropa belladonna*, and *Mandragora officinarum*. Many of these were also used medicinally, and a few have become important modern sources of pharmaceutical compounds. The tropane alkaloids (atropine, scopolamine, and hyoscyamine) which these species contain are effective in controlling smooth muscle action. It is careful measured amount is useful in the treatment of heart irregularities. In larger, but sublethal, doses, the same alkaloids produce hallucinations. Of the plants in this group, only datura is native to the New World.

Datura is particularly rich in scopolamine, which is considered to be the most hallucinogenic of the solanaceous alkaloids. Seeds or roots of various species provides vision upon eating. So it is widely used in Central and South America, but even among people that use datura, it is considered dangerous and too strong for general use. Usually datura was ingested only by boys during puberty rites or by trained shamans. In South America, alcoholic beverages made from datura fruts were given to women and slaves of dead warriors to stupefy them before they were buried alive with their deceased masters.

Infusions of bark and leaves were also used by various tribes. Congeneric species is widely used in the New World while the species of datura native to Europe and Asia were less widely used, but belladonna, mandrake, and henbane assumed great importance in Europe during the Middle Ages.

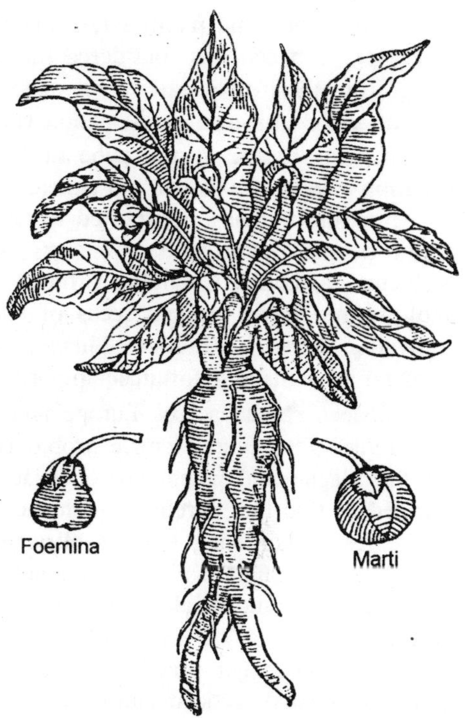

Fig. 8.6. Mandrake, or Mandragora was long thought to be an aphrodisiac because of the human shape of its root.

The ancient use of solanaceous alkaloids in witchcraft and folk medicine led European scientists after the Renaissance to investigate their actionand eventually to the discovery of their medicinal properties. Likewise, their uses as hallucinogens and poisons prompted writers and artists to include them in the writtings and paintings of the time. Belladonna or henbane are the source of Atropine. Atropine is directly absorbed by skin, and during the Middle Ages, witches rubbed ointments containing it on their bodies. The psychological sensations produced give one the feeling of flying. While under the influence of atropine, witches were supposed to be transported to rendezvous with spirits of demons. These meetings, called sabats, are often figured in European art of the period. Our modern portrayal of a witch riding a broom may stem from these ancient rites.

The depiction of the flying comes from the hallucinogenic sensations of being transported, and the broom purportedly represents the stick used to apply atropine-impregnated ointements to vaginal membranes for absorption. Another figure commonly attributed to visions seen while under the influence of solanaceous alkaloids is the werewolf. Similar to the visions of jaguars and snakes produced when South American natives use caapi, wolves were associated with trances induced by henbane, belladonna, and related species. A quite different member of the Solanaceae, however, has had the greates tworldwide impact. N. rustica is believed to be the relative of Nicotiana tabacum. Both of them become important commercial items in almost every country of the world. The genus *Nicotiana* is thought to be native to the New World, but at least one species, *N. suavolens*, which occurs in Australia, was independently domesticated.

Fig. 8.7. Datura acquired its common name jimsonweed in 1705.

At least 1,000 years before Columbus landed in the West Indies, tobacco was smoked, eaten, and snuffed by native peoples throughout the New World. Among American Indians, tobacco was used medicinally to ease childbirth pains and stave off hunger on long hunts. The dried leaves of tobacco become so valuable that it becomes the source of money and incorporated into ritualistic and symbolistic practices such as the smoking of a peacepipe. Tobacco was considered sacred by many American tribes. Mayan priests thought that smoke rising from their pipes carried messages to the gods. Recent studies of Amazonian tribes have shown that the use of tobacco in divination rites continues today.

Shamans of the Warao tribe of Venezuela, for example, fast almost to starvation during their initiation rites and then eat and smoke large quantities of tobacco. The resultant hallucinations produce the sensation of being transported to another world for mettings with the spirits that govern the lives of the people. Later in their careers, shamans frequently turn to tobacco to help them solve tribal problems. Although Columbus brought "cigars" to Queen Isabella after his first voyage, it was not until Andre Thevet brought seeds of Nicotiana tabbacum from Brazil to France in 1556 that tobacco cultivation started in Europe. Linnaeus named the genus after Jean Nicot, the French ambassador to Portugal who made a fortunate importing and popularizing the use of the plant in Paris.

Tobacco spread across Europe and Asia and gained a special place because its medicinal virtues as a remed for several female problems, a snakebite antidote, lung strengthener ulcer remedy, cure for the plague and potent aphrodiasiae. Tobacco use became so widespread that King James I was alarmed to find that purchases of tobacco were depleting England's silve supply. Neither the high cost of the leaves nor King James's blistering attack in his *Counterblaste to Tobacco* (1604) checked the spread of smokingor sniffing. In other countries as well, leader tried to discourage tobacco use. So many hard step and stricked rule were formed in order to prohibit the smoking as one Chinese emperor ordered smokers to be decapitated and Russian Tsar ordered the nostrils of snuffers to be split so that they could no longer practice the habit. Still, tobacco continued to attract followers. Eventually, the British promoted tobacco cultivation in their colonies to ensure a national supply. Virginians started growing tobacco in 1612. Since an acre planted in tobacco yielded four times the revenue of an acre planted to corn, colonists turned more and more exclusively to tobacco farming. The Virginia monopoly on tobacco production granted by the Queen was broken when settlers in Maryland began to cultivate the crop.

Soon, much of the eastern seaborad of the United States was engated in tobacco production. For many years, tobacco was the most important item of trade between England and North America. Primitive method of cultivation of tobacco consisted simply of saving seeds of the previous year's crop and sowing them in a cleared area. After maturity plants are harvested and the leaves dried in the sun. In contrast, modern tobacco farming is quite elaborate and employs various cultivation techniques for the many different cultivars and the types of of tobacco being produced. On modern U.S. farms, tobacco is usually first sown in seed beds and the seedlings transplanted into fields after 2 months of growth. The spacing of plants and fertilizer regime varies with the final type of tobacco to be produced.

Nitrogen-rich fertilizers are applied to encourage the development of large, pliable leaves suitable for cigar wrappers, while the small, more brittle leaves favoured for cigarettes are produced by limiting nitrogen and supplyign phosphorus and potassium at critical stages. While it is growing, the flower stalks and the side shoots are usually removed to prevent the plants from diverting resources to seeds and low branch development. Leaves used for cigar wrappers should be large so the plants are generally grown under cheesecloth where the high humidity and protection from sunscald also promote formation of large, thin, blemish-free leaves.

Two to 4 months after planting, the crop is considered mature Good grades can be picked a leaf at a time, or whole stalks can be harvested. In either case, the leaves or stalks are tied into bunches called "hogsheads" for curing. During the curing process, themoisture content in the leaves is reduced from about 80 to 20 percent. Starches are converted to sugars, and some proteins are broken down enzymatically. Slow drying permits aerobic fermentation to take place but prevents the growht of molds or fungi. Curing can be done by circulating air or by smoking. Occasinally, bundles of leaves are sun-dried. Following curing, the leaves are aged for periods varying between a few months and a year. The tobacco used for the cigarette and cigar filling is remoistered before marketing and the veins and petioles removed. The softened leaves are then cut by machines into strips. Depending on the ultimate tobacco product, are then cut by machines into strips. Depending on the ultimate tobacco product, various materials can be added.

Mositure-retaining substances such as glycerin, cider concentrate, and diethyl glycol are common additives, but honey or sugar, oil of hops, licorice, coumarin, rum, or menthol can be added for flavouring. Additives have been increasingly used in tobacco mixtures since the middle 1970s to compensate for the flavour and aroma lost when manufacturers produce cigarettes with low tar and nicotine. While most of these are drawn from the FDA list of "Additives Generally Recognized as Safe," there is concern that, when burned, many may constitue health hazards. Coumarin, a compound used as additives removed from the FDA list of safe drugs because it has been shown to be carcinogenic, is stull being purchased in large quantities by cigarette manufacturers.

Locorice, likewise, produces carcinogenic compounds when burned. Even the sugars added produce tars when burned. Ironically, the federal government exempts tobacco manufacturers from the labeling requirements that other food and drug producers must observe. While tobacco smoking was widely enjoyed before 1880 change in the curing process make the tobacco smoking more popular after this year. Before 1880, most tobacco was cured by direclty placing it over the hot smoke of a charcoal fire. After 1880, farmers began to use hot air brought to the drying rooms by flues. This indirect drying produced a milder form of tobacco than that produced by smoke drying.

Mild tobaccos were perfect for cigarettes, a relatively "general" way of smoking. However, thee light tobaccos produce an acidsmoke when they are burned. Tobacco cured directly tends to produce an alkaline smoke. Acid tobacco smoke has little physiological effect on the body if it is simply puffed. Consequently, it must be inhaled in orde rto produce an exhilarating effect. The surface of the lungs neutrilized the inhaled smoke and the lung membrane absorbed the nicotine carried in the smoke. Smokers who inhale tend to become addicted to tobacco because they become physiologically dependent on the strong sensory reaction produced when nicotine is absorbed. Pipe and cigar smokers can become accustomed to smoking and dearly miss their habit if they quit smoking, but, in general, there is little physical reaction to the absence of nicotine in their blood since they normally do not inhale.

Nicotine acts as stimulant of the central nervous system as it is a major alkaloid in tobacco. It causes nausea, dizziness, and hallucinations in large doses. It is physiologically addicting, and withdrawal symptoms appear when it is kept from habituates. In pure form, nicotine is a poison and is often used, after treatment with sulfuric acid, as an insecticide. In addition to nicotine, smoking leads to the inhalation of tars. These tars are known carcinogens, and cigarette comapnies have therefore tried to manufacture cigarettes that reduce the intake of tars. However, none of the various filtering or extraction devices is perfect, and some tars are drawn in by smokers who inhale. There is a direct correlations between smoking and lung cancer and heart disease all these are dut to toxicity of nicotine and tars. Smokers age prematurely, die at a younger age than nonsmokers, and tend to be susceptible to emphysema, bronchitis, and other respiratory ailments. Smoking during pregnancy increases the risk of miscarriage and problesm in early infancy. Infants born to mothers who smoke can emerge into the world underweight and addicted to nicotine. Despite the warning of the Surgeon General that smoking is bad for your health, tobacco smoking is a legal form of drugs ingestion. In direct contrast to its stance concerning many less harmful drugs, the government continues to subsidize nicotine drug addiction by providing price supports for tobacco farmers. Cigarette companies also blissfully ignore the effects of tobacco smoking and continue to lure people into smoking with advertisements that portray smokers as healthy, particularly masculine, independent, and/or beautiful.

Plant Poisons

A large number of plants are poisonous to varying degree that an enumeration would be futile. Regional agricultural stations usually publish lists of local plants that are dangerous to humans or livestock, and many horticulture books indicate which ornamental plants can be toxic if eaten. Some of these plants have been used by humans for their poisonous compounds. These are the ones in which we are interested here. As we explain earlier, a large number of plants used for their physiological effects are poisonous if used in large quantities.

It is difficult to know exactly how humans discovered which plants were or were not toxic and the quantities that could be ingested relatively safely. Ancient Greek and Roman medical rocords attest to an early interes in documenting the occurrence and usefulness of plant toxins. Administration of plant poisons as a form of capital punishment is well known from the story of the death of Socrates, after being sentenced to drink the juice of poison "hemlock" (*Conium maculatum*, Apiaceae). In the Middle Ages, "succession powders" were deviously employed to eliminate potential heirs to the throne. The employment of these powders became so common that th wealthy and powerful hired tasters as protection against sudden death from "food poisoning."

Today, scientists often seek out plants known by natives to be poisonous. Many poisonous plants are the sources of drugs or commercial poisons for use as insecticides, herbicides, or fungicides because all these poisons are either alkaloids or steroids. Some modern primitive tribes in the tropics use arrow or fish poisons such as curare or barbasco. Curare is made from extracts of the bark and stems of *Chondodendron tomentosum* (Menispermaceae) and from seeds of species of *Strychnos*, particularly *S. nux-vomica* (Loganiaceae). *Chondodendron*, a vine native to South America, contains tubocurarine. Species of Strychnos occur natively in South America, southeast Asia, and Indonesia and many yield the potent alkaloids curarine, strychnine, and toucine.

Curare alkaloids, primarily tubocurarine chloride, have had limited applications in medicine because they block neuromuscular activity and thus help relax muscles that remain contracted under anesthesia. Initially Strychnine was used medicinally as a central nervous system stimulant but has now been replaced by safer compounds. Barbasco and tuba are fish poisons used by Amazonian peoples. The first is prepared from various species of *Derris*, such as *Derris elliptica*, and the second from *Lonchocarpus nicou* (both Fabaceae).

Recently, species of both these genera have been shown to be effective insecticides because they contain rotenone. This isoflavanoid compound is fatal to insects but comparatively harmless to humans because it leaves no residue. Pyrethrum (*Chrysanthemum coccineum*, *C. cinerariifolium*, and *C. marschallii*) and tobacco acts as the sources of natural insecticides. Insecticidal compounds form species of *Chrysanthemum* are sold as Persian insect powder or Dalmation. Pyrethrins, the active compounds, are esters that break down into acids and alcohols. Since they stun insects, but do not instantly kill them, pyrethrins are usually mixed with other substances that ensure death of the insect. In some regions, notably Kenya, Tanzania, and parts of northeastern South America, pyrethrum is grown commercially for natural compound extraction. Approximately 1,200 plants are use as sources of insecticides. Yet, the fact that these would be natural insecticides does not mean that they are, or would be, less harmful than synthetic compounds. In fact, nicotine sulfate, a commercially available, natural insecticide is one of the most toxic pesticides on the market today.

9

IMPLICATIONS OF INTERACTION

Serious drug–drug interactions have contributed to recent U.S. market withdrawals and nonapprovals of new molecular entities (NMEs). In addition to coadministration of other drugs, concomitant ingestion of dietary supplement or citrus fruit or fruit juice could also alter systemic exposure, leading to adverse drug reactions or loss of drug efficacy.

METABOLISM OF NEW MOLECULAR ENTITIES AND INTERACTIONS WITH OTHER DRUGS

The interactions of concern can be divided into two types. First, other drugs can affect the blood levels of the NME by inhibiting or inducing its absorption, distribution, metabolism, or excretion pathways. Second, the NME can affect the blood levels of other drugs by inhibiting or inducing their absorption, distribution, metabolism, or excretion pathways. Of particular interest is the role of hepatic and intestinal cytochrome P450 (CYP) enzymes. Thus, "Is the drug a substrate for CYP enzymes?" and "Is the drug an inhibitor and/or an inducer of CYP enzymes?" are among the critical questions that need to be addressed when evaluating clinical pharmacology data of NMEs in new drug applications. Recognizing the importance of addressing these questions early in drug development, pharmaceutical companies routinely assess a NME's clearance pathways, including in vitro evaluation of a drug's metabolic pathways and its modulating effects on CYP1A2, CYP2C9, CYP2C19, CYP2D6, and CYP3A activities and subsequent clinical interaction studies based on in vitro data. In addition to CYP enzymes, other enzymes such as glucuronosyl transferases and various transporters also play important roles in drug interactions and changes in systemic exposure.

The clinical significance of altered systemic exposure of coadministered drugs depends on the concentration–response relationships for clinical effects, both effectiveness and toxicity. If the concentration– response relationship is well described, knowledge of the effects of interactions can lead to rational adjustment of dose or dosing interval, or to appropriate warnings and precautions.

Labeling descriptions of drug–drug interactions can be based on interactions observed in clinical studies or projected interactions extrapolated from other studies. If a drug is shown not to inhibit a particular CYP enzyme based on in vitro data, labeling will indicate no interaction with substrates of that CYP. For example, the current ABILIFY labeling states that "aripiprazole and dehydro-aripiprazole did not show potential for altering CYP1A2-mediated metabolism in vitro".

If a drug has been determined to be a sensitive CYP3A substrate (e.g., budesonide, buspirone, eletriptan, eplerenone, felodipine, lovastatin, midazolam, saquinavir, sildenafil, simvastatin, triazolam, and vardenafil) or a CYP3A substrate with a narrow therapeutic range (e.g., alfentanil, astemizole, cisapride, cyclosporine, diergotamine, ergotamine, fentanyl, pimozide, quinidine, sirolimus, tacrolimus, and terfenadine), it does not need to be tested with all strong or moderate inhibitors of CYP3A, to

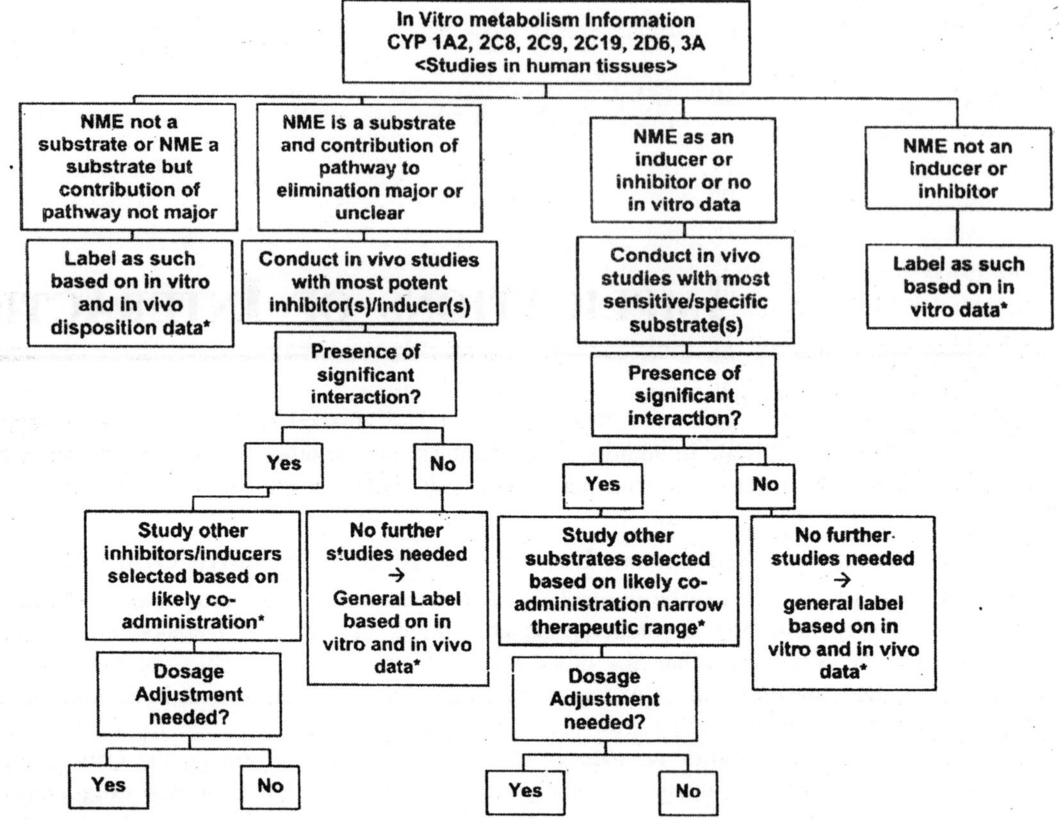

Fig. 9.1. CYP-based drug-drug interaction studies decision tree.

warn about an interaction with "strong" or "moderate" CYP3A inhibitors. For example, the labeling of RELPAX indicates that it should not be taken with strong CYP3A inhibitors such as ketoconazole, itraconazole, nefazodone, troleandomycin, clarithromycin, ritonavir, and nelfinavir. The interactions with itraconazole, nefazodone, troleandomycin, clarithromycin, ritonavir, and nelfinavir have not been evaluated; the warnings are based on data from a clinical interaction study conducted with ketoconazole and on other relevant in vitro data.

Labeling of drugs as "strong" or "potent" CYP3A inhibitors can facilitate, for example, the projection of its interaction with other sensitive substrates. Examples of "strong CYP3A inhibitors" include atanazavir, clarithromycin, indinavir, itraconazole, ketoconazole, nefazodone, nelfinavir, ritonavir, saquinavir, telithromycin, etc. These are drugs that increase the area under the plasma concentration-time curve (AUC) of either orally administered midazolam or other CYP3A substrates by five-fold or greater. Examples of "*moderate CYP3A inhibitors*" include amprenavir, aprepitant, diltiazem, erythromycin, fluconazole, fosaprenavir, (grapefruit juice), verapamil, etc. These are drugs that increase the AUC of orally administered midazolam or other sensitive CYP3A substrates by two-fold or greater, but less than five-fold. The labeling of KETEK indicates that telithromycin is a strong inhibitor of CYP3A, tha its concomitant use with simvastatin, lovastatin, or atorvastatin should be avoided, and that its use is contraindicated with cisapride and pimozide. Note that the warnings in the labeling about interactions with lovastatin, atorvastatin, and pimozide are based on extrapolation from clinical studies with simvastatin and cisapride.

Effect of Dietary Supplements on New Molecular Entities and Interactions

St. John's Wort

FDA has received adverse-event reports that suggested a role of St. John's wort in reducing the effectiveness of a number of drugs, such as cyclosporine (e.g., transplant rejection), oral contraceptives (e.g., breakthrough bleedings and pregnancy), and sildenafil (e.g., loss of efficacy).

Impact on drug labeling

As the effect of an enzyme inducer is reasonably predictable, the current labeling recommendation is that, for drugs that are substrates of CYP3A or P-gp, and when the products' effectiveness would be reduced upon coadministration of St. John's wort, St. John's wort should be listed along with other known inducers, such as rifampin, rifabutin, rifapentin, dexamethasone, phenytoin, carbamazepine, phenobarbital, etc., in the labeling as possibly decreasing the plasma levels. Of these 55 labels, only two labels, CRIXIVAN and INSPRA, are based on actual clinical studies on these drug products. The others either indicate that there are reports of interactions (e.g., NEORAL and TRIPHASIL) or that there are mechanistic reasons and there is the potential for interactions (e.g., AGENERASE, ALESSE, and UNIPHIL).

Several products with teratogenic potential (e.g., ACCUTANE and THALOMID) list interactions not directly related to the drug products, but to the oral contraceptives that need to be taken. Except for ACCUTANE, whose warning about the interaction with St. John's wort is in the "Contraindications" section, all current labelings on St. John's wort appear either in the "Warnings" or the "Precautions" sections, the two sections that will be combined in one section "Warnings/Precautions" under the new labeling rule. A recently published guidance for industry entitled "Labeling for combined oral contraceptives" includes labeling language on drug interactions between oral contraceptives and St. John's wort, such as "Herbal products containing St. John's wort (*Hypericum perforatum*) may induce hepatic enzymes (cytochrome P450) and P-glycoprotein transporter and may reduce the effectiveness of contraceptive steroids. This may also result in breakthrough bleeding". Several St. John's wort products also carry warning language about potential interactions with various drug products.

Effect of Citrus Fruit/Fruit Juice on New Molecular Entities

Grapefruit Juice Effects on CYP3A

FDA has received adverse-event reports that implicated grapefruit juice in the observed exaggeration of pharmacological effects or adverse drug reactions for calcium channel blockers (e.g., resulting in hypotension), statins (e.g., leading to muscle pain), antihistamines (e.g., QT prolongation and arrhythmias), and others. Clinical pharmacology studies have also clearly shown that concomitant grapefruit juice ingestion increased the systemic exposure of orally administered drugs that are CYP3A substrates and have low oral bioavailability due to extensive presystemic extraction contributed by enteric CYP3A.

Impact on drug labeling

The current labeling recommendation is that, for drugs that are primarily substrates of CYP3A with low oral bioavailability due to extensive presystemic extraction, grapefruit juice should be listed along with other CYP3A inhibitors in the labeling as possibly increasing the plasma levels of coadministered drugs. Table 4 lists 28 labeling examples of grapefruit juice–drug interactions. Of these labels, half of them contain actual clinical data (e.g., ADALAT, ADVICOR, BILTRICIDE, CLARINEX, CONERA, CRIXIVAN, INSPRA, KETEK, NOR VASC, PACERONE, LETAL, PROGRAF, SULAR, and ZOCOR). All relevant literature data on grapefruit juice interactions with a particular drug product may be considered in the labeling of the drug product. For example, the latest package inserts for lovastatin and simvastatin include data from two clinical studies; one conducted by the sponsor and the other by an independent researcher.

The results of the two studies differ widely, possibly because of the different study designs (timing and frequency of coadministration), and also because of variables that are difficult to control, such as the source, brand, lot-to-lot variation of the same brand, and the preparation procedure (including the extent of dilution by the consumers when using frozen concentrates) of the grapefruit juice. For both drug products, both sets of data have been included in the labeling. Other labels include those based on theoretical interaction potential (e.g., CENESTIN). Grapefruit juice has been considered as a "*moderate CYP3A inhibitor*" that can be included with other moderate CYP3A inhibitors, such as erythromycin and diltiazem, as appropriate. Unlike most other CYP3A inhibitors, which affect both enteric and hepatic CYP3A, grapefruit juice appears to affect only enteric CYP3A; it would therefore be listed in the labeling only for drug products for oral administration and with low oral bioavailability.

Grapefruit Juice, Apple Juice, and Orange Juice Effects on Transporters

Grapefruit juice has been shown to inhibit P-gp transporter, resulting in increases in plasma levels of drugs that are substrates of P-gp transporter. Limited data have shown that grapefruit juice, as well as apple juice and orange juice, may inhibit organic anion transporting peptides (OATP), leading to decreased systemic exposure of drugs that are substrates of OATP. The overall effect of fruit juices on drugs that are substrates for both transporters may depend not only on the contribution of either transporter and other clearance pathways to the drugs' overall clearance, but also on other variables, including the amount, the type, and frequency of juices being consumed. Until more data are available, current labeling for known substrates of both transporters that are not substrates of CYP3A is to recommend taking these drug products with "water".

Calcium-Fortified Orange Juice Effect on Bioavailability

Chemical complexation of the fluoroquinolones with the calcium ion may play a major role in the reduced absorption and decreased plasma levels when calcium-fortified orange juice was coadministered with ciprofloxacin. The current labeling for Cipro (ciprofloxacin) tablets has warnings against the use of these products with calcium-fortified orange juice.

Future Prospectives

Drug–drug interactions have been a significant cause of adverse drug reactions. Various guidance documents for industry and for reviewers have stressed the importance of evaluating drug–drug interactions during drug development.

With increased understanding of how certain dietary supplements (e.g., St. John's wort) and juices (e.g., grapefruit juice) affect the systemic exposure of drug products, it is possible to anticipate an interaction with a drug based on the drug's clearance pathway and label the drug products accordingly. But while the potential for an interaction can be understood, it is much harder to describe the effect quantitatively or recommend dose modifications or usage. Dietary supplements or juices have multiple unknown components that are not well defined and vary from product to product and are used at very different doses and in a very variable time with relationship to drug use. Current labeling recommendations therefore urge avoidance of the coadministration of a drug and a dietary supplement or food that interacts with it in a clinically significant way, rather than adjustment of the drug dose or dosing interval, a recommendation that is common for dealing with drug–drug interactions.

Despite the increased understanding and documentation of drug interactions in the labeling and in letters to "*dear health care professionals*," adverse reactions resulting from well-recognized drug–drug interactions continue to be reported. The increased use of dietary supplements with significant drug interaction potential increases the propensity for adverse drug reactions.

To better translate information into practice, Center for Drug Evaluation and Research (CDER) has published a final rule on the physician's labeling format. When drug interactions that are significant, or their absence when they are expected to occur, would appear in labeling "Highlights," in addition to being included in the main body of the labeling. In addition, a proposed revision of the 1999 drug interaction guidance includes a proposal to use a classification system for CYP3A inhibitors (including grapefruit juice) in the labeling, in an effort to improve the consistency of labeling language and to highlight key drug interactions. Additional risk management tools have been proposed for particularly serious situations, including use of medication guides and restricted distribution.

With continued improvement in our understanding of the mechanisms of interactions and contributions of additional patient factors (e.g., genetics and gender), the risks associated with these interactions can be better predicted, assessed, and managed to reduce the frequency of clinically significant adverse drug reactions.

10

Utilization of Plant Medicine

Comprehending the use and safety of botanical dietary supplements is challenging largely owing to the lack of regulation and the paucity of data on their utilization, effectiveness, and safety. The literature describing the utilization of botanical products tends to be poorly documented and incomplete and evidence in the form of clinical trials is sparse; safety data are largely derived from anecdotal case reports. Medications from botanical sources have been described as far back as 60 millennia and most of the medications used throughout the world were derived from plants until the early 1900s. It is estimated that 35,000 to 70,000 plants have been used for medical purposes. For example, opium and willow bark have long been used for the treatment of pain. It was not uncommon for over-the-counter medications to contain opium without warnings or legal restrictions. Willow bark may still be purchased over the counter as an extract to relieve pain and many other prescriptions medications are currently derived from botanical sources.

Historical Overview

Prescriptions Derived from Plant Sources

Today, it is estimated that 25% of the Western pharmacopoeia contains chemical entities that were first isolated from plants and another 25% are derived from chemical entities modified from plant sources. In 1999, 121 prescription medicines worldwide came directly from plant extracts and it is now a $10 billion-a-year industry. These medicines are not dietary supplements but rather are botanical products that have passed the more rigorous process of approval to be used as a prescription drug. The World Health Organization estimates that 75% to 80% of the developing world continues to rely heavily on botanicals for medication. However, most products available are considered dietary supplements in the United States.

Plant Dietary Supplements

The use of botanicals in the industrialized world is growing. In the United States, it has been estimated that about 20,000 products are in use, with the top ten botanical products comprising 50% of the commercial botanical market. In China, approximately 80% of medications are obtained from between 5000 and 30,000 types of plants. In the era of increased globalization, many botanical products are available to people all over the world through the Internet, imported for sale by botanical shops catering to high-use ethnic populations, or imported (often illegally) by individuals returning from global travel. Utilization of these products has dramatically increased in the past decade. In 1991, the U.S. Congress passed legislation to establish the National Institutes of Health Office of Alternative Medicine, which later became the National Center for Complementary and Alternative Medicine, to better understand how Americans are embracing the use of unconventional therapies.

Utilization of Plant Dietary Supplements

Although physicians in the United States infrequently prescribe botanicals, they receive little formal training on the benefits and risks of these and other complementary and alternative medications (CAM). This is disturbing because a significant proportion of patients take botanical dietary supplements. More than 37 million Americans utilize botanical remedies and some estimates put forth a much higher. Since the Dietary Supplement Health and Education Act (DSHEA) of 1994, growth of the botanical market has been dramatic. However, the industry is fragmented, with a few large corporations manufacturing the bulk of botanical products and many smaller companies targeting specific herbs. Market research organizations have traditionally avoided analyzing botanical products because the market was too small, but this has changed recently because botanicals are now profitable to analyze. As a result of DSHEA, the public now has many botanical dietary supplements from which to choose. With the increasing number of products competing against one another, corporations have taken action to distinguish their products from one another. As such, dietary supplement manufacturers have taken a page from the pharmaceutical industry and have begun branding botanical products to develop a market following for their product.

Many products also consist of combinations of dietary supplements and at least one of them also uses a nonprescription medication in combination with the botanical dietary supplement. At least one pharmaceutical manufacturer has also entered the branded botanical market. Direct-to-consumer advertising of branded botanical dietary supplements appears to be quite effective, judging from the number of advertisements appearing in the print and electronic media. Many of these products claim to improve conditions that are refractory to conventional medical treatment or they are touted to be natural and, as such, purported to be safer than conventional pharmaceuticals and free of side effects. The public is well aware of dietary supplements, because many of these have appeared on late-night infomercials. Some examples of branded products touted for weight loss include Metabolife, Leptoprin, and Cortislim. Most weight loss products in the United States contained ephedra before the Food and Drug Administration (FDA) banned ephedra-containing dietary supplements. It appears that weight loss products are now being reformulated with other stimulants that have not received the intense scrutiny of the FDA, such as bitter orange (synephrine), green tea extract (caffeine), and guarana (methylxanthines: caffeine, theobromine, and theophylline).

Other branded combination botanical products such as Enzyte and Avlimil are touted for treatment of sexual dysfunction and are advertised in a manner similar to sexual dysfunction pharmaceuticals. Still other formulations are advertised for breast enhancement—Bloussant, hair loss—Avacor, depression—Amoryn, nourishing the brain—Focus Factor, and sleep—Alluna Sleep. All of these contain one or more botanical constituents and are sold under the auspices of DSHEA, and therefore are not regulated by the FDA and the Federal Trade Commission as rigorously as prescription pharmaceuticals or food additives. Sizing up the economics of the botanical dietary supplements market in the United States is challenging because the market is prodigiously dynamic. The market has been estimated to represent a demand between $0.6 and $5.1 billion. It is important to note that each study sampled a different population. Growth in the market occurred rapidly between 1991 and 1998, but recent sales appear to have reached a plateau. Americans usually pay for botanical dietary supplements as well as other CAM therapies out of their own pockets because most health insurance programs do not cover CAM therapies.

In 1997, total CAM out-of-pocket expenses exceeded $27 billion, with the expenditure on botanical products estimated at greater than $5 billion. Insurance coverage that covered CAM therapies would also likely result in growth in the botanical industry. One study found that full insurance coverage for botanical dietary supplements predicted an increase in usage of five-fold and partial insurance coverage

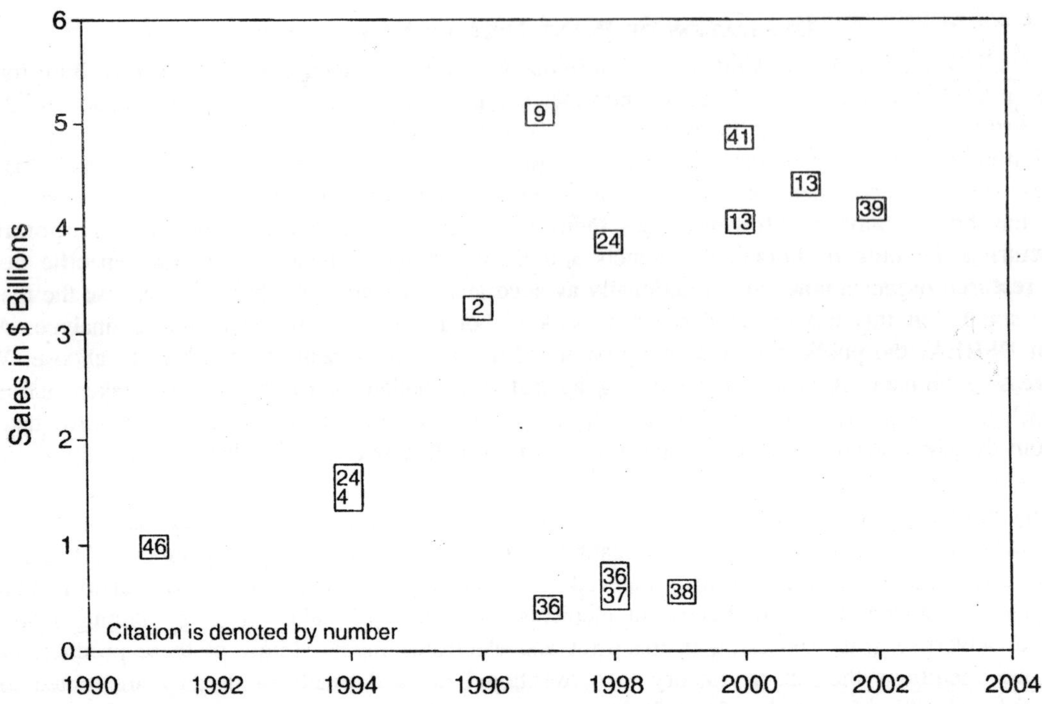

Fig. 10.1. Estimates of retail botanical sales in billions of dollars by year from multiple citations.

predicted a threefold increase in botanical utilization. Rapid growth in the botanical dietary supplement industry occurred within the first four years of DSHEA and there was also a concurrent growth spurt in the U.S. economy in the mid-1990s. DSHEA relaxed regulatory restrictions on dietary supplements, thus lowering the barrier to enter the market. As a result, growth in CAM likely is a result of deregulation by DSHEA and may reflect the disposable income available. This would explain the rapid growth in the mid-1990s and leveling of spending on botanical products at the turn of the century. Also, Eisenberg et al. found that the increase in botanical product utilization between 1990 and 1997 was likely due to an increase in the proportion of the population using botanicals rather than an increase in per patient utilization. In contrast to the growth of botanical products in the mid-1990s reported by Eisenberg et al., growth of the botanical market in early 2000 was reported to be from patients already using sundry botanical products according to the Natural Marketing Institute (NMI). This indicates that botanical dietary supplement market expansion among new patients has moderated.

Market Analyses

Several major surveys of dietary supplement utilization have been conducted recently. The Saskatchewan Nutriceutical Network (SNN), National Nutritional Food Association (NNFA), Consumer Healthcare Products Association (CHPA), Landmark Healthcare, Inc., The NMI, individual investigators, Centers for Disease Control and Prevention (CDC), and FDA have all recently either conducted or contracted market analyses of CAM utilization in the United States, which included botanicals. Each survey is presented individually because the data are so heterogeneous among studies.

Saskatchewan nutriceutical network

The SSN estimated U.S. botanical sales in 1999 to be $4 billion. The network further quantified where consumers buy their botanical products. Forty-seven percent are sold in retail stores, 30% are sold in multilevel distribution systems, 8% are sold by mail order or practitioners, 6% was sold by

Asian herbal shops, and only 1% was purchased on the Internet. Notwithstanding these findings, it is important to note that the Internet was the fastest growing sales market for botanical products, at 150% per year.

National nutritional food association

The NNFA commissioned a telephone survey of 736 adults in October of 2001. The key finding was that women (25%) were more likely to take botanical products than men (15%). The survey emphasizes the importance of accurate labeling. Seventy percent agreed with the statement "*Labels on supplements*' bottles or packages are carefully read by most: they help the majority of older adults choose the right supplement and to determine the correct dosage." Only 22% disagreed with that statement. Fifty-five percent of respondents agreed with this statement: "Labels on dietary supplements help me understand if this is the right supplement for me," while 64% agreed with the following statement: "Labels on dietary supplements help me determine the dosage I need to take." The more educated patients were less likely to agree with this statement.

Consumer healthcare products association

The CHPA commissioned a study entitled "Self-Care in the New Millenium: American Attitudes Toward Maintaining Personal Health and Treatment." They conducted 1505 telephone interviews in January of 2001, using random telephone numbers. African-Americans and Hispanics were over sampled to conduct in-depth subgroup analysis. Of particular interest is the finding that 96% of respondents felt confident that they could take care of their own health. This might explain why so many people want access to pharmacologically active botanicals. These products do not require a prescription and thus allow patients to treat themselves.

Many of these products are being used for specific medical conditions. The top five conditions, in many cases are refractory to conventional medicine, namely menopausal symptoms, colds, allergies/ sinus, muscle/joint/back pain, and premenstrual/menstrual symptoms.

The demographics of utilization in the past six months were reported. Thirty percent of women reported using a dietary supplement and 23% of men used a dietary supplement in the six-month period. Results for the effect of age on utilization have been mixed across studies. Patients who were between 50 and 64 years old had the highest reported use of dietary supplements, and 59% and those who were 18 to 34 years old had the lowest use at 48%. Income may be reflected in the utilization-by-age category. Utilization of dietary supplements by ethnicity was characteristic of other studies. Forty-four percent of African-Americans and 42% of Hispanics reportedly used dietary supplements, as compared to 53% of the general population. Although the study did not report Caucasian dietary supplement utilization rates, we can infer that Caucasians increased the overall utilization rate for the population. Health insurance status was associated with greater dietary supplement use, 56% versus 45%. This likely reflected the fact that patients who had health insurance also had more income. Those with some college education reported the highest utilization rate of 60%. People with college degrees used dietary supplements slightly less, 57%, but those with high school education or lesser educational qualification reported 48% utilization of dietary supplements in the past six months.

Landmark healthcare inc.

In 1997, Landmark Healthcare Inc. commissioned a report entitled "The Landmark Report on Public Perceptions of Alternative Care." They conducted 1500 telephone interviews in November 1997, using random digit selection. The survey included a representative sample of minority patients—85% Caucasian, 8% African-Americans, and 3% Hispanic. The survey found that 17% of the U.S. population used botanical dietary supplements in the past year and even more striking, 75% of the U.S. population was most likely to use botanical products. Eighty-five percent of those reported to have taken a botanical supplement self-prescribed and self-administered the products. Three-fourths of patients who used alternative forms of care did so in conjunction with conventional medicine, yet 15% of patients replaced their conventional treatment with alternative care.

Natural marketing institute

The NMI surveyed by mail 2002 households, July through August 2001. Only 53% of botanical supplement users were satisfied with botanical supplements. Despite the low satisfaction for botanical products, supplement users accounted for most of the increase in the previous year: 46% of botanical users increased utilization while only 10% of the general population increased utilization of botanical dietary supplements. Consumers took botanical supplements primarily for general health benefits, 59% versus 40% for a specific condition. Only 6% took botanicals products for short- term benefits, whereas 80% took them for daily or long-term benefit. Many have recently started, with only 50% having used an herb for more than three years.

Independent investigators

Eisenberg et al. surveyed 1539 adults in 1990 and 2055 adults in 1997. Botanical use in the prior 12 months increased from 2.5% in 1990 to 12.1% in 1997—a 4.8-fold increase. They estimated, in 1997, that 15 million adults took a botanical product or high-dose vitamins with other medications, which represented approximately 18.4% of those taking medications in the United States. Growth in botanicals was found to be from an increase in the percentage of the population taking botanicals and not due to an increase in utilization per patient. More than 60% of patients did not discuss CAM use with their doctor. Patients spent an estimated $5.1 billion on botanical medications. Kaufman surveyed 2590 patients, February 1998 through December 1999. Fourteen percent of the U.S. population reported using botanical supplements. Concurrent use with medication was highest with patients on fluoxetine, 22%; overall, 16% of those taking medication reported using botanical medications.

Centers for disease control and prevention

The Division of Health Interview Statistics, National Center for Health Statistics, CDC conducted a survey entitled "Utilization of Complementary and Alternative Medicine by United States Adults" in 1999. The survey attempted to obtain a representative sample of minorities and also patients without telephones. This is important because these demographic groups tend to report lower utilization of botanicals products than Caucasians and those of higher socioeconomic status. The CDC found that 9.6% of the population took botanical medicines. Hispanics reported the lowest use of CAM followed by African-Americans, and then Caucasians: 19.9%, 24.1%, and 30.8%, respectively. The western part of the United States reported the highest use of CAM.

Food and drug administration

FDA commissioned a study of dietary supplement sales in the United States in 1999. Samples of products were purchased from a representative sample of retail establishments, catalogs, and the Internet. The authors looked at the consistency of botanical products purchased. Forty percent to 46% of botanicals and botanical products were consistent with the ingredients listed on the label. Botanical extracts were even less consistent with the label, only 12% to 24% (depending on where purchased) were found to be consistent with the label. They also gave the mean, minimum, and maximum price paid for dietary supplements by source of purchase. Interestingly, the mean purchased price on the Internet was the most expensive at $23.34, followed by the mean catalog price, $16.40. The mean retail price was less than half the cost of the mean Internet price, at $11.62.

Utilization Summary

Patients who use botanicals tend to have attained higher education, be female, be older persons, have higher incomes, and have a recalcitrant chronic disease unresponsive to conventional medicine.

There is also evidence that cultural differences have a strong impact on the use of botanicals. Certain subpopulations may defy these generalizations to the U.S. population. Asian-Americans have a long history of using botanicals as medication and often consider botanicals a conventional form of treatment. Southern rural poor are also reported to have a higher utilization profile of plant-derived products. Rural poor may treat illness with botanical products while the U.S. population as a whole tends to use botanical products for general health benefits rather than to treat a specific illness.

SAFETY OF PLANT PRODUCTS

As a result of DSHEA, the majority of botanical drug products are used in the United States without medical supervision. Only 8% of those who use botanicals do so under medical supervision and 85% of those who treat themselves with herbs do not seek professional guidance or advice. Even if patients utilizing botanical dietary supplements were medically supervised, adulteration and misbranding are prevalent and so little is known about the supplements that many untoward events could not be prevented or recognized in a timely fashion. Despite the widespread acceptance of CAM by the lay public, clinicians possess little scientific information about the practices of CAM relative to conventional western medicine. This is particularly unsettling because it is estimated that 16% to 18% of prescription medication users took botanical and supplements coincidentally. Medication–botanical interactions are largely unknown. Even more alarming is a report that 14.5% of women used botanical products during pregnancy and 23.5% of children under 16 may be taking botanical products. Neonatal heart failure has been attributed to the use of Blue cohosh during pregnancy.

Up to 60% of patients using alternative therapies are reported to have never informed their physician of their botanical or CAM use. Furthermore, only 40% of physicians ask their patients about alternative therapy. The 60% of physicians who do not ask about the use of botanical supplements and other CAM are unlikely be informed of alternative therapies their patients are using. Clearly, there is a lack of communication between patients and providers. Some patients may fear disapproval by physicians and wish to give socially desirable answers. However, the majority of patients express a lack of concern about their physician's approval, rather they were more concerned with their physician's inability to understand and incorporate CAM into their medical management. Patients are not using alternative therapy because they are dissatisfied with conventional medicine but instead because they value both types of therapy.

Many botanical dietary supplements are potentially unsafe because of adulteration and misbranding. Thirty-two percent of botanical medications collected in California contained an undeclared pharmaceutical or heavy metal. Pharmaceuticals adulterating botanical products are one of the most frequent reasons botanical dietary supplements are placed on the FDA MedWatch site, and this is undoubtedly a small fraction of what actually occurs. Many of these adulterants are not detected until patient illnesses are first detected. Consumers often do not recognize that many imported products, purported to be traditional medications, are actually recognized pharmaceuticals. For example, a "*Mexican asthma cure*" had a claim on the label that said it contained no corticosteroids and was free of adverse effects, but the product was found to contain triamcinolone, a moderately potent corticosteroid with well-documented systemic adverse effects common to all glucocorticoids. In another example, a patient used an illegally imported Chinese medicine; it was reported to last much longer than the medication the physician had prescribed. The label on the Chinese medicine said it contained astemizole, a long-acting antihistamine with-drawn from the United States as a result of its effect of prolonging the cardiac QTc interval. In many cases, patients may not recognize pharmaceuticals that are sold as traditional medicines. In the past, consumers have had difficulty distinguishing between vitamins and botanical products. It is likely no different for botanicals and pharmaceuticals. This may be problematic because corporations are creating proprietary botanical blends and branding them for use in specific medical conditions. Patients could inadvertently assume they are treating themselves with a medication that has undergone the same rigorous clinical testing as other FDA-approved medications. Patients readily read and trust the directions on labels of dietary supplements. In fact 59% of the public incorrectly thought a government body reviewed and approved botanical supplements before they are sold.

There are other risks of contamination to botanical and botanical supplements. Due to stress on the supply of cultivars for botanical supplements, products may vary greatly in their active content. In the era of limited resources, with increasing utilization and decreasing wild production, there is pressure to produce a product. Raw material costs may override the quality and purity of the product. There are few barriers to bringing new products to the market and many newer entrants may lack expertise to prevent quality issues and contamination in their product. This creates the potential for inadvertent poisoning as a result of overdosing or contamination as well as treatment failure through under-dosing. Indeed, a study of botanical consistency found that only 43% of the products tested were consistent for ingredients and dose with the benchmark or recommended daily dose. Twenty percent had the correct ingredient but not the stated dose and 37% were not consistent with either ingredients; dose or the labeling was too vague to draw conclusions. The FDA also found that many botanical products were inconsistent with the ingredients listed on the label and estimated that only 12% to 24% of botanical extracts and 40% to 46% of botanical products contained what was on the label.

Adulteration was found to be a problem in another dietary supplement containing androstenedione; although not strictly a botanical, it is regulated in a similar fashion under the auspices of DSHEA. Ingestion of androstenedione contaminated with trace amounts of 19-norandrosterone resulted in a positive test for 19-norandrosterone, a metabolite used to detect nandrolone. Other samples were also found to be contaminated with testosterone. The FDA has been cautious in its enforcement of DSHEA after its experience with the passage of The Nutritional Labeling and Education Act of 1990. This act severely restricted unproven claims on foods and dietary supplements. Fearful of the loss of the ability to conduct business as usual, the dietary supplement industry responded with forceful lobbying to the Congress, which responded with DSHEA, exempting dietary supplements from the earlier law.

DSHEA severely limited when the FDA could take action to protect the public and what actions could be taken. The burden of proof to show harm is now placed on the FDA. Moreover, dietary supplement manufacturers are not required to report adverse dietary supplement events. In fact, between 1994 and 1999 fewer than 10 of the 2500 adverse events associated with dietary supplements and reported to the FDA were reported by the manufacturer. The Office of Inspector General concluded the spontaneous adverse event reporting "system has difficulty generating signals of possible public health concern" due to "limited medical information, product information, manufacturer information, consumer information, and ability to analyze trends". One weight loss supplement manufacturer is reported to have withheld from the FDA 14,684 complaints of adverse events regarding ephedra, which included heart attacks, strokes, seizures, and deaths.

Recently, the FDA has begun to enforce DSHEA more assertively. Ephedra was banned as a dietary supplement in April of 2004 because ephedra presented an "*unreasonable risk.*" However, this ban does not include foods containing ephedra, approved drugs, or Asian medicines, which are allowed to contain ephedra under the final rule. It appears that FDA may address androstenedione in the near future. In March of 2004, FDA sent warning letters to 23 manufactures or distributors of androstenedione threatening enforcement if they do not immediately cease distribution of androstenedione and within 15 days advise the FDA, in writing, of actions taken. The FDA did this on the grounds that androstene dione was not marketed on October 15, 1994 and as such is not presumed safe under DSHEA. Furthermore, the FDA has stated that androstenedione consumption would be considered an unreasonable risk, given what is now known.

Other botanical products are receiving FDA attention. The acting commissioner of the FDA, Lester Crawford, told members at the American Society for Pharmacology and Experimental Therapeutics in April 2004 that the FDA was compiling data on other botanical products that have been associated with safety issues (64). Kava, used as an anxiolytic, and usnic acid, used for weight loss, have both been associated with liver disease; bitter orange is used as a sympathomimetic in weight loss products to replace ephedra; all the pyrrolizidine alkaloids have the eye of the FDA. There are other products that could receive scrutiny of the FDA in the future. Examples profiled in Consumer Reports include a list of what they call "the dirty dozen herbs listed by risk."

The botanicals are broken down as follows: "*definitively hazardous*": aristolochic acid; "*very likely hazardous*": comfrey, androstenedione, chaparral, germander, and kava; and "*likely hazardous*": bitter orange, organ/glandular extracts, lobelia, pennyroyal oil, skullcap, and yohimbe. These are products with potent pharmacological actions and poorly documented toxicities, and as long as they are available safety will clearly be an issue. As a result of DSHEA, botanical supplements are presumed safe by virtue of being "grandfathered" by the FDA if the product was marketed before October 15, 1994. Products brought to market after that date only require 75-day pre-market notification to the FDA with information that substantiates that the ingredients will reasonably be expected to be safe. FDA cannot take action until patients are injured but it is increasingly clear relatively rare adverse events may not be detected until a significant number of patients are killed or injured.

Safety

With little knowledge of dietary supplements, many physicians do not ask patients about botanical products and patients are also not disclosing the consumption of these products. Some of these products also have substantial pharmacologic activity that interacts with prescription medications and disease states while other are devoid of any biological activity. Many patients may actually think they are taking something that is rigorously tested and regulated by the FDA when in fact some have been reported have serious issues with contaminants. Safety has been presumed as a result of DSHEA despite common misbranding, and adulteration. Several dietary supplements have been linked to cancer, renal and liver failure, and even death.

The vast majority of products are probably safe but many likely have low level undocumented adverse effects. This leaves the possibility most adverse events likely go unrecognized and untreated. Under current practices, the situation is unlikely to change. The profile of the patient who uses a botanical product will likely be someone with higher education, be female, have higher socioeconomic status, have more disposable income, and be older. The market is estimated to be in excess of $5 billion in the United States with an estimated 10% to 20% of the population using botanicals. Utilization of botanical dietary supplements will continue to grow under the deregulation of DSHEA and as they gain acceptance by the public and medical establishment. With increasing stress on the harvesting of wild foliage, corporations must resort to harvesting domestically grown botanical dietary supplements to meet the demand. This should result in a more consistent product base.

By increasing direct-to-consumer marketing and branding of specific products, there will likely be an acceleration of market growth. New ads for branded botanicals have already appeared as this chapter was being published. Products will continue to be imported and Internet sales will continue to grow. As more patients use these products and regulatory issues remain, safety will continue to be a concern and the market will likely be difficult to define. Drug–botanical interactions and disease–botanical interactions are only now beginning to be recognized by health care professionals as a potential source of harm, as the prevalence of botanical dietary supplement utilization increases.

11

EPHEDRA ALKALOIDS

Ephedra, and other medicinal plants have been identified at European neanderthal burial sites dating from 60,000 BCE. Thousands of years later, Pliny accurately described the medicinal uses of ephedra. But thousands of years before Pliny, traditional Chinese healers used ephedra extracts. Chinese texts from the 15th century recommended ephedra as an antipyretic and antitussive. In Russia, around the same time, extracts of ephedra were used to treat joint pain; and though recent laboratory studies confirm that ephedra might be useful for that purpose, additional trials and studies have not been forthcoming. In the 1600s, Indians and Spaniards in the American South-west used ephedra as a treatment for venereal disease. That idea might also have had some merit, as some studies show that ephedra contains compounds with antibiotic activity called *transtorines*. Whether the transtorines will prove to be clinically useful has not been determined.

In 1885, Nagayoshi Nagi, a German-trained, Japanese-born chemist, isolated and synthesized ephedrine. Nagi's original observations were confirmed by Merck chemists 40 years later. Merck's attempts at commercializing ephedrine were unsuccessful, at least until 1930, when Chen and Schmidt published a monograph recommending ephedrine as the treatment of choice for asthma. During the 1920s and 1930s, epinephrine was the only effective oral agent for treating asthma. Epinephrine, which had been available since the early 1900s was (and still is) an effective bronchodilator, but it has to be given by injection, or administered with special nebulizers. Ephedrine was nearly as effective as epinephrine, and could be taken orally.

As a result, ephedrine became the first-line drug against asthma. It was displaced from that front during the late 1970s and early 1980s, when aerosolized synthetic β-agonists were introduced. Unlike most of the other alkaloids contained in ephedra (methylephedrine and cathinone are both psychoactive, but the amount contained in unadulterated ephedra is too low to be of clinical significance), ephedrine is also a potent *central nervous system* (CNS) stimulant. Injections of ephedrine, called *philopon* (which means "love of work") were given to Japanese kamikaze pilots during World War II. A major epidemic of ephedrine abuse occurred in postwar Japan, when stockpiles of ephedrine accumulated for use by the Army were dumped on the black market. Abusers in Tokyo, and other large Japanese cities, injected themselves with ephedrine (then referred to as hirapon), in much the same way that methamphetamine is injected today.

In the Philippines, a mixture of ephedrine and caffeine called *shabu* was traditionally smoked for its stimulating effect. In the late 1980s, shabu smoking gave way to the practice of smoking methamphetamine ("ice"). In what is perhaps a tribute to the past, some "ice" is sold under the philopon name. The chemistry and nomenclature of these compounds are somewhat confusing, and are best understood by reference to the synthetic route used by plants to make ephedrine. All ephedra plants contain phenylalanine-derived alkaloids. Plants use phenylalanine as a precursor, but incorporate only seven of its carbon atoms. Phenylalanine is metabolized to benzoic acid, which is then acetylated and decarboxylated to form pyruvic acid. Transamination, results in the formation of forms (–)-cathinone.

Reduction of one carbonlyl group leads to the formation of either (–)- norephedrine (phenylpropanolamine is the name used to refer to the synthetic mixture of ± norephedrine), or norpseudoephedrine (called *cathine*). N-methylation of (–)-norephedrine results in the formation of (+)-ephedrine. N-methylation of cathine leads to the formation of (+)-pseudoephedrine.

Current Promoted Uses

Physicians routinely used intravenous ephedrine for the prophylaxis and treatment of hypotension caused by spinal anesthesia particularly during caesarean section. In the past, ephedrine was used to treat Stokes–Adams attacks (complete heart block), and was also recommended as a treatment for narcolepsy. Over the years, ephedrine has been replaced by other, more effective agents, and the advent of highly selective β-agonists has mostly eliminated the need to use ephedrine in treating asthma.

European medical researchers have, for several years, used ephedrine to help promote weight loss, at least in the morbidly obese, and nutritional supplements containing naturally occurring ephedra alkaloids are sold in the United States for the same purpose. Clinical trials confirm that, taken as directed, use of these supplements does result in weight loss, though whether such losses are sustained has not been determined.

Prior to its banning by the Food and Drug Administration (FDA) in 2004, ephedra was found in many "*food supplements*," used by bodybuilders. Generally, it was compounded with other ingredients such as vitamins, minerals, and amino acids in products, which are said to increase muscle mass and enhance endurance. Performance improvement secondary to ephedrine ingestion has been established in a controlled clinical trials, and use of ephedrine has been prohibited by the International Olympic Committee.

Ephedra was also sold in combination with many other herbs in obscure combinations. Labels frequently listed 10 or 15 different herbs, but, analysis usually disclosed only the ephedra alkaloids and caffeine as present in sufficient quantities to be physiologically active. After several well-publicized accidental deaths, products clearly intended for abuse, such as "*herbal ecstasy*," and other "look-alike drugs" (products usually containing ephedrine or phenylpropanolamine designed to look like illicit methamphetamine, but in concentrations higher than recommended by industry or the FDA) were withdrawn from the market. Labels on these products were frequently misleading. For example, one might suppose that a product called "Ephedrine 60" contained 60 mg of ephedrine when, in fact, the actual ephedrine content was 25 mg.

Sources and Chemical Composition

Ephedra is a small perennial shrub with thin stems. It rarely grows to more than a foot in height, and at first glance, the plant looks very much like a small broom. Different, closely related species are found in Western Europe, southeastern Europe, Asia, and even the Americas. Some of the better known species include *Ephedra sinica* and *E. equisentina* from China (collectively known as ma haung), as well as *E. geriardiana*, *E. intermedia*, and *E. major*, which grow in India and Pakistan, and countless other members of the family Ephedraceae that grow in Europe and the United States (*E. distachya*, *E. vulgaris*).

Ephedra species vary widely in their ephedrine content. One of the most common Chinese cultivars, known as "China 3," contains 1.39% ephedrine, 0.361% pseudoephedrine, and 0.069% methyl-ephedrine. This mix is fairly typical for commercially grown ephedra plants. Noncommercial varieties of ephedra may contain no ephedrine at all, while others may contain more pseudoephedrine than ephedrine. Depending on the variety, trace amounts of phenylpropanolamine, (–) norephedrine, and methylephedrine may also be present, however (+) norephedrine does not occur naturally, and its presence is proof of adulteration.

Labels on herbal supplements listed total ephedra alkaloid content, usually 10 or 11 mg per serving. Depending on the raw materials used, different production runs of the same product contained ephedrine and pseudoephedrine in varying proportions. Occasionally, supplement makers were accused of adulterating their product by adding synthetic ephedrine or pseudoephedrine. Unlike with (+)-norephedrine, these compounds occur naturally and product adulteration should not have been alleged just because alkaloids other than ephedrine were detected in trace amounts, or because the ratio of ephedrine to pseudoephedrine was close to, or even greater than, 1:1. Of course, if one of the minor alkaloids, such as methylephedrine, were found to be present in concentrations approaching those of ephedrine, the ratio could only be explained by adulteration.

Products Available

Prior to its ban in 2004, no one government agency was tasked with tracking production of ephedrine-containing products. Nor were these products indexed by any industry or trade organization. Ephedrine-containing supplement products were mostly purchased at health food stores or over the Internet. Claims made by some of the Internet vendors were quite outrageous and totally unsupported by any scientific research.

The large supplement makers, of course, had web pages, many of which contained, or had links to, the most recent peer review studies. But in addition to the established names, hundreds of other, smaller manufacturers also advertised and sold over the Internet. These companies came into and went out of existence so rapidly that a detailed listing of their web sites would likely be outdated before the links were published. Even today, a simple search using the word "*ephedrine*," will disclosed numerous off-shore vendors, along with numbers of attorneys soliciting for ephedra-related class action legal cases.

In addition to selling their own proprietary mixture, many of these same web sites sold the same popular products as the herbal and general retail outlets, such as a previous Twin Labs best seller "Ripped Fuel," which contained ephedrine in the form of ma huang, combined with guarana, L-carnitine, and chromium picolinate. Metabolife 356TM contained guarana (40 mg caffeine), 12 mg ephedrine as ma huang, chromium picolinate 75 mg, and several other ingredients. Ever since ephedrine became the precursor of choice for making methamphetamine, federal regulators have severely restricted bulk sales of ephedrine, but these restrictions have been bypassed in some cases by illegally ordering from a foreign web site.

In most products, ephedrine content ranged anywhere from 12 to 80 mg per serving, with the majority of products falling into the lower range. Industry standards called for a total dose of ephedrine of less than 100 mg/day. The FDA, however, allowed a maximum daily dose of 150 mg/day of synthetic ephedrine. Unless fortified, the expected ephedrine content of ma huang capsules was generally less than 10%. Thus, a capsule said to contain 1000 g of ephedra would probably have contained no more than 80 mg of ephedrine.

In the United States, (+)-norpseudoephedrine, in its pure form, is considered a Schedule IV controlled substance. However, because of the small amounts of this alkaloid in ephedra plants or extracts, the Drug Enforcement Administration (DEA) had never stated or proposed that ephedra products were subject to the scheduling requirements of the Controlled Substances Act. Quite the contrary, DEA published a proposed rule in 1998 that stated DEA's intent to exempt legitimate ephedra products in finished form from regulation even as "chemical mixtures." Other regulatory sanctions and actions on ephedra rendered action on this regulation moot.

Pharmacological Effects

Studies have shown that resultant effects are similar, regardless of whether pure synthetic ephedrine or naturally occurring ephedra is ingested. There are, however, significant enantioselective differences between the enantomers in both pharmacokinetic and pharmacodynamic effects. All of the ephedra alkaloids have important effects on the cardiovascular and respiratory systems, but not to the same degree.

Ephedrine, the predominant alkaloid in ephedra, is both an α and β stimulant. It directly stimulates α_2 and β_1; receptors and, because it also causes the release of norepinephrine from nerve endings, it also acts as a β_2 stimulant. The resultant physiological changes are variable, depending on receptor distribution and receptor regulation. Tolerance to ephedrine's β agonist actions emerges rapidly, which is why ephedrine is no longer the preferred agent for treating asthma; receptor downregulation quickly occurs and the bronchodilator effects are lost.

Receptor distribution probably explains why ephedrine has no effect on diastolic pressure, and only minimal effect on systolic. β_2 Stimulation of vessels in peripheral muscles results in peripheral vasodilation and "*diastolic runoff*," which more than cancels ephedrine's other inotropic effects. The absence of any significant effect on blood pressure was firmly established during the late 1970s and early 1980s in dozens of double-blind, placebo-controlled studies performed to compare the effectiveness of ephedrine with that of newly synthesized adrenergic agents. The pharmacokinetic and toxicokinetic behavior of any isomer cannot be used to predict that of any other ephedrine isomer. The (+) isomer of methamphetamine, for example, is a potent CNS stimulant, but the (-)-isomer is merely a decongestant. There is a tendency in the literature to lump together all "*ephedrine alkaloids*" and use the term "*class effect*" to assume that all the different drugs in that class exert the same effects on the same biological targets. In fact, some of the drugs in the class will be similar in some regards and different in others.

The affinity of the various ephedrine isomers for human β-receptors has been measured and compared (as indicated by the amount of cyclic adenosine monophosphate produced compared to that of isoproterenol) in tissue culture. Activity of the different isomers is highly stereoselective, i.e., the different isomers had very different receptor-binding characteristics. For β_1-receptors, maximal response (relative to isoproterenol = 100%) was greatest for ephedrine (68% for 1R, 2S-ephedrine and 66% for the 1S, 2R-ephedrine isomer). Both of the pseudoephedrine isomers had much lower affinities (53%). When binding to 32-receptors was measured, the rank order of potency for 1R, 2S-ephedrine was 78%, followed by 1R, 2R-pseudoephedrine (50%), followed by 1S, 2S-pseudoephedrine (47%). The 1S, 2R-ephedrine isomer had only 22% of the activity exerted by isoproterenol, but was the only isomer that showed any significant agonist activity on human β_3-receptors (31%). Stimulation of β_3-receptors, which are thought to be located only in fat cells, may account for ephedrine's ability to cause weight loss.

Ephedrine is also an α agonist and, as such, is capable of stimulating bladder smooth muscle. At one time, it was used to promote urinary continence. In animal models, when compared to norepinephrine, ephedrine is a relatively weak α-adrenergic agonist, possessing less than one-third the activity of norepinephrine. Ephedrine's usefulness as a bronchodilator is limited by the number of β-receptors on the bronchi. The number of β-receptors located on human lymphocytes (which correlates with the number found in the lungs) decreases rapidly after the administration of ephedrine; the density of binding sites drops to 50% after 8 days of treatment and returns to normal 5 to 7 days after the drug has been withdrawn.

Clinical Studies

Bronchodilation

Banner et al. summarized studies where the effects of ephedrine and ephedra were compared to placebo in controlled studies in humans. None of the controlled trials disclosed any evidence of cardiovascular toxicity when ephedrine was given in doses as high as 1 mg/kg, even when it was administered to severe asthmatics with known cardiac arrhythmias. The trial reported by Banner et al. studied the respiratory and circulatory effects of orally administered ephedrine sulfate, 25 mg, aminophylline, 400 mg, terbutaline sulfate, 5 mg, and placebo in 20 patients with ventricular arrhythmia by a double-blind crossover method. The study was comprised of 20 patients, with an average age of 60 years and a preexisting history of both asthma and heart disease (as evidence by the presence of frequent premature ventricular contractions). The bronchodilator effect of terbutaline was similar to that of aminophylline over 4 hours but superior to ephedrine at hour 4. Both terbutaline and ephedrine exhibited chronotropic effects, with the effect of terbutaline greater than that of ephedrine at hour 4. The effect of aminophylline on heart rate (HR) did not differ from placebo. Only terbutaline was associated with an increase in ventricular ectopic beats. Ventricular tachycardia occurred in three patients treated with terbutaline and in one patient with ephedrine (which occurred before he was given ephedrine). There were no significant changes in blood pressure. Orally administered terbutaline should not be regarded as safer than orally administered ephedrine or aminophylline in patients with arrhythmias.

In 1992, Astrup studied the effects of ephedrine and caffeine in a group of obese patients. In a randomized, placebo-controlled, double-blind study, 180 obese patients were treated by diet (4.2 mJ/day) and either an ephedrine/caffeine combination (20 mg/200 mg), ephedrine (20 mg), caffeine (200 mg), or placebo three times a day for 24 weeks. Withdrawals were distributed equally in the four groups, and 141 patients completed the trial. Mean weight losses was significantly greater with the combination than with placebo from week 8 to week 24 (ephedrine/caffeine, 16.6 $\pm$ 6.8 kg vs placebo, 13.2 $\pm$ 6.6 kg [mean $\pm$ standard deviation {SD}], $P = 0.0015$). Weight loss in both the ephedrine and the caffeine groups was similar to that of the placebo group. Side effects (tremor, insomnia, and dizziness) were transient and after 8 weeks of treatment they had reached placebo levels. Systolic and diastolic blood pressure fell similarly in all four groups.

Weight Loss

The most recent of the studies examining weight control were designed to address concerns about long-term safety and efficacy for weight loss using a mixture containing 90 mg of ephedrine (from ephedra) and 192 mg of caffeine, derived from cola nuts. A 6-month randomized, double-blind, placebo-controlled trial was performed, in which a total of 167 subjects (body mass index 31.8 $\pm$ 4.1 kg/m^2) were randomized to receive either placebo ($n = 84$) or herbal treatment ($n = 83$). The primary outcome measurements were changes in blood pressure, heart function, and body weight. Secondary variables included body composition and metabolic changes. It was found that herbal vs placebo treatment decreased body weight (-5.3 $\pm$ 5.0 vs -2.6 $\pm$ 3.2 kg, $P < 0.001$), body fat (-4.3 $\pm$ 3.3 vs -2.7 $\pm$ 2.8 kg, $P = 0.020$), and low- density lipoprotein cholesterol (-8 $\pm$ 20 vs 0 $\pm$ 17 mg/dL, $P = 0.013$), and increased high-density lipoprotein cholesterol (+2.7 $\pm$ 5.7 vs -0.3 $\pm$ 6.7 mg/ dL, $P = 0.004$). Herbal treatment produced small changes in blood pressure variables (+3 to -5 mmHg, $P \sim 0.05$), and increased HR (4 $\pm$ 9 vs -3 $\pm$ 9 beats per minute, $P < 0.001$), but cardiac arrhythmias were not increased ($P > 0.05$). By self-report, dry mouth ($P < 0.01$), heartburn ($P < 0.05$), and insomnia ($P < 0.01$) were increased and diarrhea decreased ($P < 0.05$). Irritability, nausea, chest pain, and palpitations did not differ, nor did numbers of subjects who withdrew. *Conclusions*: In this 6-month placebo-controlled trial, herbal ephedra/caffeine (90/192 mg/day) promoted body-weight and body-fat reduction and improved blood lipids without significant adverse events.

Athletic Performance

In a series of studies, Bell et al. assessed the effects of ephedrine mixtures on performance, and found measurable improvement. One and one-half hours after ingesting a placebo (P), caffeine (C) (4 mg/kg), ephedrine (E) (0.8 mg/kg), or caffeine and ephedrine, 12 subjects performed a 10-km run while wearing a helmet and backpack weighing 11 kg. The trials were performed in a climatic suite at 12–13°C, on a treadmill where the speed was regulated by the subject. VO_2, VCO_2, V(E), HR, and rating of perceived exertion were measured during the run at 15 and 30 minutes, and again when the individual reached 9 km.

Blood was sampled at 15 and 30 minutes and again at the end of the run and assayed for lactate, glucose, and catecholamines. Run times (mean ± SD), in minutes, were for C (46.0 ± 2.8), E (45.5 ± 2.9), C + E (45.7 ± 3.3), and P (46.8 ± 3.2). The run times for the E trials (E and C + E) were significantly reduced compared with the non-E trials (C and P). Pace was increased for the E trials compared with the non-E trials over the last 5 km of the run. VO_2 was not affected by drug ingestion. HR was elevated for the ephedrine trials (E and C + E), but the respiratory exchange ratio (a measure of maximal exertion) remained similar for all trails. Caffeine increased the epinephrine and norepinephrine response associated with exercise and also increased blood lactate, glucose, and glycerol levels. Ephedrine reduced the epinephrine response but increased dopamine and free fatty acid levels. Bell concluded previously that the effects of caffeine, when taken with ephedrine, were not additive, and that all of the observed improvement could be accounted for by the presence of ephedrine.

PHARMACOKINETICS

Phenylpropanolamine is readily and completely absorbed, but pseudoephedrine, with a bioavailability of only approx 38%, is subject to gut wall metabolism, and absorption may be erratic. Pure ephedrine is well absorbed from the stomach, but absorption is much slower when it is given as a component of ma huang, rather than in its pure form. Ephedrine ingested in the form of ma huang has a t_{max} of nearly 4 hours, compared to only 2 hours when pure ephedrine is given. Like its enantiomers, ephedrine is eliminated in the urine largely as unchanged drug, with a half-life of approx 3–6 hours.

The rate at which any of the enantiomers is eliminated depends upon the urinary pH. At high pHs, excretion time is prolonged. At low pH ranges, excretion is accelerated. In controlled laboratory studies, where volunteer subjects were given either bicarbonate or ammonium chloride, the higher the urine pH, the more slowly the ephedrine and pseudoephedrine were excreted. Conversely, when the urine pH is low, excretion is accelerated. The importance of these observations is hard to assess, because without the addition of bicarbonate, urine pH values in the general population rarely approach 8.0. A study of pseudoephedrine pharmacokinetics in 33 volunteers who were not treated with drugs to alter urine pH found that these parameters could not be correlated to urine pH, mainly because there was little difference in pH between the different participants. Excretion patterns may be much more rapid in children, and a greater dosage may be required to achieve therapeutic effects. Patients with renal impairment are at special risk for toxicity.

Peak concentrations for the other enantiomers, specifically phenylpropanolamine and pseudoephedrine, occur earlier (0.5 and 2 hours, respectively) than for ephedrine, but all three drugs are extensively distributed into extravascular sites (apparent volume of distribution between 2.6 and 5.0 L/kg). No protein-binding data in humans are available. Peak ephedrine levels after ingestion of 400 mg of ma huang, containing 20 mg of ephedrine, resulted in blood concentrations of 81 ng/mL—essentially no different than the peak ephedrine levels observed after giving an equivalent amount of pure ephedrine. In another study, 50 mg of ephedrine given orally to six healthy, 21-year-old women produced mean peak plasma concentrations of 168 ng/mL, 127 min after ingestion, with a half-life of slightly more than 9 hours. The results are comparable to those obtained in studies done nearly 30 years earlier. Very high levels of methylephedrine have been observed in Japanese polydrug abusers taking a cough medication called BRON. Concentrations of methylephedrine less than 0.3 mg/L, the range generally observed in individuals taking BRON for therapeutic rather than recreational purposes, appear to be nontoxic and devoid of measurable effects. Methylephedrine is a minor component of most ephedra plants, but in Japan (where, unlike in the United States, methylephedrine is legally sold) it is produced synthetically, and is used in cough and cold remedies, especially BRON. In terms of catecholamine stimulation, methylephedrine appears comparable to ephedrine; however, it does not react with most standard urine screening tests for ephedrine. This can be a cause of some forensic confusion, because 10–15% of a given dose of methylephedrine is converted to ephedrine.

Although the issue has been raised in litigation, the amounts of methylephedrine and norephedrine contained in naturally occurring ephedra are so low as to be of no clinical consequence. For example, the study by Gurley et al. found that most of the commercial products tested had no methylephedrine whatsoever, but when it was present, it was usually in quantities of less than 1 mg per serving (range 0.2 to 2.2 mg). If the volume of distriution (Vd) of methylephedrine is assumed to be 3.5, approximately the same as ephedrine, then a 70-kg man ingesting a 2-mg serving of methylephedrine would produce a blood concentration of (dose = kg weight × blood concentration × Vd) 0–0.06 mg, undoubtedly below most laboratories' minimum level of detection, and a clinically insignificant finding. Similar considerations apply to the small amounts of norephedrine found in these products.

Adverse Effects and Toxicity

Two journal articles analyzing *adverse event reporters* (AERs) have been published in the peer-reviewed literature, and both reports have received wide publicity. The reports are, however, of limited use in assessing toxicity, because they are comprised of passively collected anecdotal data, which is often incomplete and unreliably reported. For example, one of the FDA ephedrine AERs "analyzed" in an article published in the *New England Journal of Medicine* described the sudden death of a teenage girl who had been born with a lethal cardiac malformation who died while playing volleyball. Postmortem blood and tissue tested negative for ephedrine, and the article failed to mention the existence of the cardiac malformation. In other AERs, massive doses of ephedrine were consumed (as with products intended for abuse, such as "*herbal ecstasy*," now withdrawn from the market). Toxicology testing was rarely performed in any of these cases, and it is not known with any certainty whether ephedrine was even taken. Even the authors of the two papers concede that anecdotal reports cannot be used to prove causality, stating that "Our report does not prove causation, nor does it provide quantitative information with regard to risk". There is little point in reviewing material that cannot be used to prove causality, and it is not included in the summaries that follow, which are comprised only of published, peer-review case reports, epidemiological surveys, and controlled clinical trials. An additional review of the utility of spontaneously reported adverse events involving supplements and, more specifically, ephedra was published by Kingston et al.. The review discussed the limitations of spontaneously reported data in assessing supplement safety and determining causality between exposure and adverse effects.

Despite conflicting data regarding the safety of ephedra from clinical studies and conclusions drawn from spontaneously reported adverse events, FDA banned the sale of ephedra-containing supplements in 2004.

Neurological Disorders

Many strokes attributed to ephedrine have actually been caused by the ingestion of ephedrine enantiomers, pseudoephedrine, phenylpropanolamine, and even methylephedrine. Two cases of ischemic stroke have been reported, but in neither case was their any toxicological testing to confirm the use of ephedrine. A decade-old report described the autopsy findings in three individuals with intracerebral hemorrhage and positive toxicology testing for ephedrine; however, one had hypertensive cerbrovasular disease and the other had a demonstrable ruptured aneurysm.

Intracerebral hemorrhage has also been described in suicide and attempted suicide victims who took overdoses of pseudoephedrine. There is also a report describing a patient who developed described arteritis following the intravenous administration of ephedrine during a surgical procedure. On the other hand, a large study to assess risk factors for stroke in young people (age 20–49) over a 1-year period was carried out in Poland, a country where ephedra-based products are widely used. Nearly one-half the cases of stroke were associated with preexisting hypertension, another 15% had hyperlipidemia, and 6% were diabetic. None of the individuals were ephedrine users.

Sometimes, especially in Japan and the Philippines, ephedrine is taken specifically as a psychostimulant. In Japan, BRON, the OTC cough medication containing methylephedrine, dihydrocodeine, caffeine, and chlorpheniramine, is very widely abused, and transient psychosis commonly results. Reports of ephedrine-related psychosis following prolonged, heavy use are fairly common. In general, psychosis is only seen in ephedrine users ingesting more than 1000 mg/day, and it resolves rapidly once the drug is withdrawn.

Ephedrine psychosis closely resembles psychosis induced by amphetamines: paranoia with delusions of persecution and auditory and visual hallucinations, even though consciousness remains unclouded. Typically, patients with ephedrine psychosis will have ingested more than 1000 mg/day. Recovery is rapid after the drug is withdrawn. The ephedrine content per serving of most food supplements is on the order of 10–20 mg, making it extremely unlikely that, in recommended doses, use of any of the products would lead to neurological symptoms.

Renal Disorders

Reports, particularly in the European literature, have described the occurrence of renal calculi in chronic ephedrine users. A review of cases from a large commercial laboratory specializing in the analysis of kidney stones found that 200 out of 166,466, or 0.064%, of stones analyzed by that laboratory, contained either ephedrine or pseudoephedrine. Unfortunately, the analytic technique used could not distinguish ephedrine from pseudoephedrine, and because pseudoephedrine is used so much more widely than ephedrine, it seems that the risk of renal calculus associated with ephedrine use must be quite small. There have been no new reports of ephedrine-related nephrolithiasis since 1999. Direct toxicity, with altered renal function and demonstrable kidney lesions related to ephedrine use, has never been demonstrated. Urinary retention, occurring as a consequence of drug overdose, was occasionally reported, but additional cases have not been described in more than a decade. The FDA and Commission E both warn against the possibility of urinary retention in patients with prostatic enlargement, but the theoretical basis for this concern is unclear, and, in any case, retention in patients with prostate disease has not been reported.

Small amounts of ephedrine are oxidized in to norephedrine and norpseudoephedrine in the liver. In patients with diminished renal function, these drugs may accumulate and have the potential to cause serious toxicity. None of the ephedrine enantiomers are easily removed by dialysis, and treatment of overdose remains supportive, using pharmacological antagonists to counter the α- and β-adrenergic effects of these drugs. Because excretion is pH-dependent, patients with renal tubular acidosis are also at risk. The FDA reports having received a number of accounts of hematuria after use of ephedra-based products, but no such cases have ever appeared in the peer-reviewed literature, and review of the reports published by the FDA shows that all of the affected individuals were taking multiple remedies, some capable of causing interstitial nephritis.

Cardiovascular Diseases

Ephedrine and pseudoephedrine share properties with cocaine and with the amphetamines because they: (1) stimulate β-receptors directly, and (2) also cause the increased release of norepinephrine. Chronic exposure to abnormally high levels of circulating catecholamines can damage the heart. This is certainly the case with cocaine and methamphetamine, but ephedrine-related cardiomyopathy is an extremely rare occurrence, occurring only in individuals who take massive amounts of drug for prolonged periods of time. Only two papers have ever been published on the subject. The two existing reports are uninterpretable, because histological findings were not described in either report, and angiography was not performed, thereby making it impossible to actually establish the diagnosis of cardiomyopathy.

Similar considerations apply to the relationship (if any) between myocardial infarction and ephedrine use. The report by Cockings and Brown described a 25-year-old drug abuser who injected himself with an unknown amount of cocaine intravenously. The only other published reports involved a woman in labor who was receiving other vasoactive drugs; and two pseudoephedrine users, one of whom was also taking bupropion, who developed coronary artery spasm.

Three cases of ephedra-related coronary spasm in anesthetized patients have also been reported, but multiple agents were administred in all three cases, and the normal innervation of the coronary arteries was disrupted in two of the cases where a high spinal anesthetic had been administered. One case of alleged ephedrine-related hypersensitivity myocarditis has been reported, but the patient was taking many other herbal supplements, and the responsible agent is not known with certainty. Although there are no reasons why ephedra alkaloids should not cause allergic reactions, the incidence appears to be extremely low.

Although clinical trials or epidemiological studies are lacking, it has been suggested that maternal use of OTC cold medication may result in fetal arrhythmias, but linkage between ephedrine and isomers and arrhythmia has never been demonstrated. The literature contains one case report describing arrhythmias occurring in a 14-year-old who overdosed on cold medications. The child had taken a total of 3300 mg of caffeine, 825 mg of phenylpropanolamine, and 412 mg of ephedrine. Clearly, large doses of ephedrine, and its enantiomers, are capable of exerting toxicity.

The paucity of peer-reviewed studies describing cardiovascular complication with ephedra alkaloids suggests that few such cases are occurring. This notion is support by the studies of Porta et al., who performed a follow-up study of more than 100,000 persons below age 65 years who filled a total of 243,286 prescriptions for pseudoephedrine. No hospitalizations could be attributed to the drug. There were no admissions within 15 days of filling a prescription for pseudoephedrine for cerebral hemorrhage, thrombotic stroke, or hypertensive crisis. There were a small number of hospitalizations for myocardial infarction, seizures, and neuropsychiatric disorders, but the rate of such admissions among the pseudoephedrine users was close to the expected rate in the population at large.

Workplace Drug Testing

Ephedra alkaloids, even when used in the recommended amounts, can cause positive urine screening tests for methamphetamine, sometimes yielding surprisingly high concentrations.

Postmortem Toxicology

Very few fatalities have ever been reported (or studied), but it appears that the therapeutic index for ephedrine is very great. A 1997 case report described a 28-year-old woman with two prior suicide attempts, who died after ingesting amitriptyline and ephedrine. The blood ephedrine concentration was 11,000 ng/mL, and the liver concentration was twice that value (kidney, 14 mg/kg; brain, 8.9 mg/kg). The amitriptyline concentration was 0.33 mg/kg in blood and 7.8 mg/kg in liver. Values in a second case report (where methylephedrine concentrations were nearly 6000 ng/mL) may or may not be relevant to the problem of ephedrine toxicity, as the individual in question took massive quantities of a calcium channel blocker, and it is not known whether methylephedrine exerts all the same effects as ephedrine. Baselt and Cravey mention the case of a young woman who died several hours after ingesting 2.1 g of ephedrine combined with 7.0 g of caffeine, but tissue findings were not described. Her blood ephedrine level was 5 mg/L, whereas the concentration in the liver was 15 mg/kg.

A report from the European literature describes the findings in a 19- year-old woman who committed suicide by taking 40 Letigen tablets (200 mg of caffeine and 20 mg of ephedrine) amounting to 10 g of caffeine and 1 g of ephedrine. She developed severe toxic manifestations from the heart, CNS, muscles, liver, and kidneys leading to several cardiac arrests, and died subsequently of cerebral edema and incarceration on the fourth day of hospitalization. Postmortem blood concentrations were not given.

Pseudoephedrine concentrations, but not measurements for ephedrine or any of the other enantiomers, have been published by the National Association of Medical Examiners in their Annual Registry report. In 15 children diagnosed with sudden infant death syndrome, the mean blood pseudoephedrine concentration was 3.55 mg/L, the median 2.3 mg/L, with a range of 0.07–13.0 mg/L (SD = 3.36 mg/L). The authors of the study take pains to point out that "The data do not allow definitive statements about the toxicity of pseudoephedrine at a given concentration".

In the only autopsy study yet published, all autopsies in the San Francisco Medical Examiner's jurisdiction from 1994 to 2001 where ephedrine or any its isomers (E+) were detected were reviewed. Cases where ephedrine or its isomers were detected were compared with those in a control group of drug-free trauma victims. Of 127 ephedrine-positive cases identified, 33 were the result of trauma. Decedents were mostly male (80.3%) and mostly Caucasian (59%). Blood ephedrine concentrations were less than 0.49 mg/L in 50% of the cases, with a range of 0.07–11.73 mg/L in trauma victims, and 0.02–12.35 mg/L in nontrauma cases. Norephedrine was present in the blood of only 22.8% (mean concentration of 1.81 mg/L, SD=3.14 mg/L) and in the urine of 36.2% of the urine specimens, with a mean concentration of 15.6 mg/L, SD=21.50 mg/L). Pseudoephedrine (PE) was detected in the blood of 6.3%. More than 88% of the decedents who tested positive for ephedrine or one of its isomers also tested positive for other drugs, the most common being cocaine (or its metabolites) and morphine. The most frequent pathological diagnoses were hepatic steatosis and nephrosclerosis. Left ventricular hypertrophy was common, and coronary artery disease was detected in nearly one-third of the cases. The most common findings in the ephedrine-positive deaths reviwed were those generally associated with chronic stimulant abuse. There were no cases of heat stroke and no cases of rhabdomyolysis.

Methamphetamine Manufacture

Either (–)-ephedrine or (+)-pseudoephedrine can be used to make meth- amphetamine by reductive dehalogenation using red phosphorus as a catalyst. If (–)-ephedrine is used as the starting material, the process will generate (+)-methamphetamine. If psuedoephedrine is used, the result will be dextromethamphetamine. As this synthetic route has become nearly universal, both state and federal governments have enacted laws limiting the amount of pure ephedrine or pseudoephedrine that can be purchased.

Drug Interactions

The ephedra alkaloids are all sympathomimetic amines, which means that a host of drug interactions are theoretically possible. In fact, only a handful of adverse drug interactions have been reported in the peer-reviewed literature. The most important of these involve the monoamine oxidase inhibitors (MAOI). Irreversible, nonselective MAOIs have been reported to adversely interact with indirectly acting sympathomimetic amines present in many cough and cold medicine. In controlled trials with individuals taking moclobemide, ephedrine's effects on pulse and blood pressure were potentiated, but only at higher doses than those currently provided in health supplements. Ephedrine-MAOI interaction may, on occasion, be severe enough to mimic pheo-chromocytoma. In addition, there is decreased metabolic clearance of pseudoephedrine when MAOIs are administered concurrently. At least one case report suggests that selective serotonin reuptake inhibitor antidepressants can react with pseudoephedrine, leading to the occurrence of "*serotonin syndrome*". Bromocriptine, the ergot-derived dopamine agonist can interact with pseudoephedrine, and would presumably interact with ephedrine as well. Surgical patients being treated with clonidine have an enhanced pressor response to ephedrine, apparently a result of clonidineinduced potentiation of α_1-adrenoceptor-mediated vasoconstriction. In some clinical trials, the coadministration of ephedrine with morphine has been shown to increase analgesia, but this approach to pain relief remains somewhat controversial.

Reproduction

Use of ephedra-containing products is likely unsafe during pregnancy because of reports of psychoses and cardiovascular effects.

Regulatory Status

In 2004, the FDA issued a final rule prohibiting the sale of dietary supplements containing ephedrine alkaloids (ephedra), citing concerns over safety and potential risk of illness or injury. The FDA reviewed evidence about ephedra' s pharmacology: peer-reviewed scientific literature on ephedra' s safety and effectiveness, adverse event reports, and a seminal report by the RAND Corporation, an independent scientific institute. Spontaneously reported adverse effects with high-profile sports figures and others raised public awareness and fueled the debate over safety. Subsequent to the ban, various trade groups and supplement companies have criticized the ban, and an appeal of the decision with temporary suspension of sanctions in some jurisdictions, pending further review, has occurred. Regardless of the regulatory outcome, reintroduction of OTC ephedra-containing supplements is not likely to occur. Although banned in the United States, use of ephedra in other countries is likely to continue.

12

Grapefruit and Other Citrus

The first report of grapefruit juice (GFJ) interacting with a drug, altering its bioavailability, was published in 1991. This accidental discovery was made in a study on ethanol–drug interactions—the bioavailability of felodipine was increased when subjects were consuming GFJ concomitantly with felodipine, associated with a lower dehydrofelodipine/felodipine area under the curve (AUC) ratio, decreased diastolic blood pressure, and an increased heart rate (1). Subsequent research in the area of fruit–drug interactions focused on grapefruit and grapefruit compounds of which several were found to affect the absorption or metabolism of certain drugs. GFJ was shown to alter the pharmacokinetics of several drugs such as statins, calcium channel blockers, antibiotics, and others. Other fruits, vegetables, and dietary supplements also have the potential to cause an adverse interaction with conventional drugs. Over 16% of all prescription drug users reported that they concurrently use at least one plant-based dietary supplement, including grapefruit and citrus products.

Many consumers have become more aware of the health benefits of antioxidants, phytochemical-rich fruits and vegetables, and products that contain these. In 1997, GFJ was purchased by 21% of all households as a popular antioxidant breakfast juice, predominantly preferred by the elderly. By-products from the citrus-processing industry, such as grapefruit seed extract, flavonoids, essential oils from the peel, and pectins may be added to other food products to improve taste, consistency, or overall quality. These by-products also may be used in the production of dietary supplements. Consequently, citrus compounds that have a potential for an interaction with drugs may find their way into other food products.

Hence, increased availability and consumption of drugs, dietary supplements, and phytochemical-containing antioxidant foods may increase the likelihood of an adverse interaction between foods and certain drugs. Absorption and metabolism of a drug may be adversely affected, shifting the administered dose outside of the therapeutic range, which may lead to a lower effectiveness of the drug or to an overdose associated with undesired or even dangerous side effects.

Based on our current knowledge, grapefruit compounds interact with drugs that are metabolized by cytochrome P450 3A4 (CYP3A4) and also have a low or variable oral bioavailability. The major mechanism leading to a grapefruit–drug interaction appears to be the reduction of the *"pre-systemic"* metabolism through the inhibition of intestinal CYP3A4. Some hydroxymethylglutaryl-coenzyme A (HMG-CoA) reductase inhibitors and calcium channel antagonists are among the affected drugs.

Other mechanisms of interaction have also been reported, such as altered activity of other enzymes within the CYP450 family. Moreover, GFJ may also inhibit the intestinal P-glycoprotein (P-gp)-mediated efflux transport of drugs such as cyclosporine to increase its oral bioavailability. GFJ and other fruit juices have recently been shown to be potent in vitro inhibitors of a number of organic anion-transporting polypeptides (OATPs).

Grapefruits and GFJ have a potential to interact with several oral medications when consumed in moderate amounts, such as one or two servings of GFJ. The concern that the concomitant administration of grapefruit products with certain drugs may lessen the effect of a drug or cause a toxic effect based on the increases in oral drug bioavailability has lead to the recommendation to avoid the consumption of grapefruit products in combination with these drugs of concern. The Food and Drug Administration (FDA) requires some drugs such as cyclosporine, sirolimus, simvastatin, lovastatin, and felodipine to carry a warning label regarding the possibility of an interaction. For example, Neoral, an immunosuppressant drug, carries a label stating"... Grapefruit and GFJ affect metabolism, increasing blood concentrations of cyclosporine, thus should be avoided". Procardia is labeled "... Co-administration of nifedipine with GFJ resulted in approximately a 2-fold increase in nifedipine AUC and C_{max} with no change in half-life. The increased plasma concentrations are most likely due to inhibition of CYP3A4 related first-pass metabolism. Co-administration of nifedipine with GFJ is to be avoided". In addition to GFJ, interactions with certain medications also have been shown for Seville orange juice, although Seville oranges are usually not processed to juice.

For most drugs in question, definitive recommendations regarding their concomitant administration with GFJ are not available, because conclusions regarding the clinical significance of the observed or predicted grapefruit–drug interactions are still limited and most of the data available are derived from in vitro experiments. Furthermore, the determination of the clinical relevance of observed interactions in human intervention trials is complicated by interindividual variability. Whereas the attention of patients and health care professionals is currently focused mainly on interactions of drugs with grapefruit products, other fruits and vegetables and dietary supplements, such as St. John's wort, green tea, and ginseng are an additional potential source for drug interactions. On the other hand, it should be noted that almost all drugs that show an interaction with grapefruit can be replaced by another drug within the same drug class that is without a known potential for an interaction.

Phenolic Compounds in Grapefruit and Citrus with Potential Drug Interactions

A major group of citrus compounds interacting with drugs are phenolics, which include hydroxycinnamic acids, flavonoids such as flavanones, flavones, and flavonols, and anthocyanins, as well as coumarins. Many of these phenolic compounds have been shown to have antioxidant and anticancer properties that may play an important role in cancer prevention, but also in prevention of other chronic diseases such as coronary heart disease, gout, and arthritis.

Interactions with numerous drugs have been demonstrated for furanocoumarins, especially for bergamottin (BG) and 6'7'-dihydroxybergamottin (DHBG). Overall, furanocoumarins appear to interact with susceptible drugs through CYP3A4 and P-gp, but also were shown to influence OATP. In in vitro studies, BG and DHBG have been demonstrated to inhibit the activity of CYP3A4 by reversible and irreversible mechanism-based inhibition. Several studies report an inhibitory effect of BG on the activity of CYP3A4, which leads to an increased C_{max} and AUC of diazepam in dogs, of nifedipine in rats, and of felodipine in humans. BG and DHBG inhibited CYP3A4 in human liver microsomes in vitro, leading to a decreased metabolism of saquinavir, whereas naringin and DHBG decreased the ratio of basolateral-to-apical to apical-to-basolateral (BA/AB) transport of saquinavir. BG, DHBG, bergaptol, and bergapten increased the steady-state uptake of [^{3}H]-vinblastine sulfate by Caco-2 cells, and BG and DHBG decreased the OATP-B–mediated uptake of estrone-3-sulfate into human embryonic kidney cells.

Several human intervention studies confirm the contribution of BG and DHBG to drug interactions with GFJ. After previous reports had been inconsistent regarding the potency of BG and DHBG, Paine

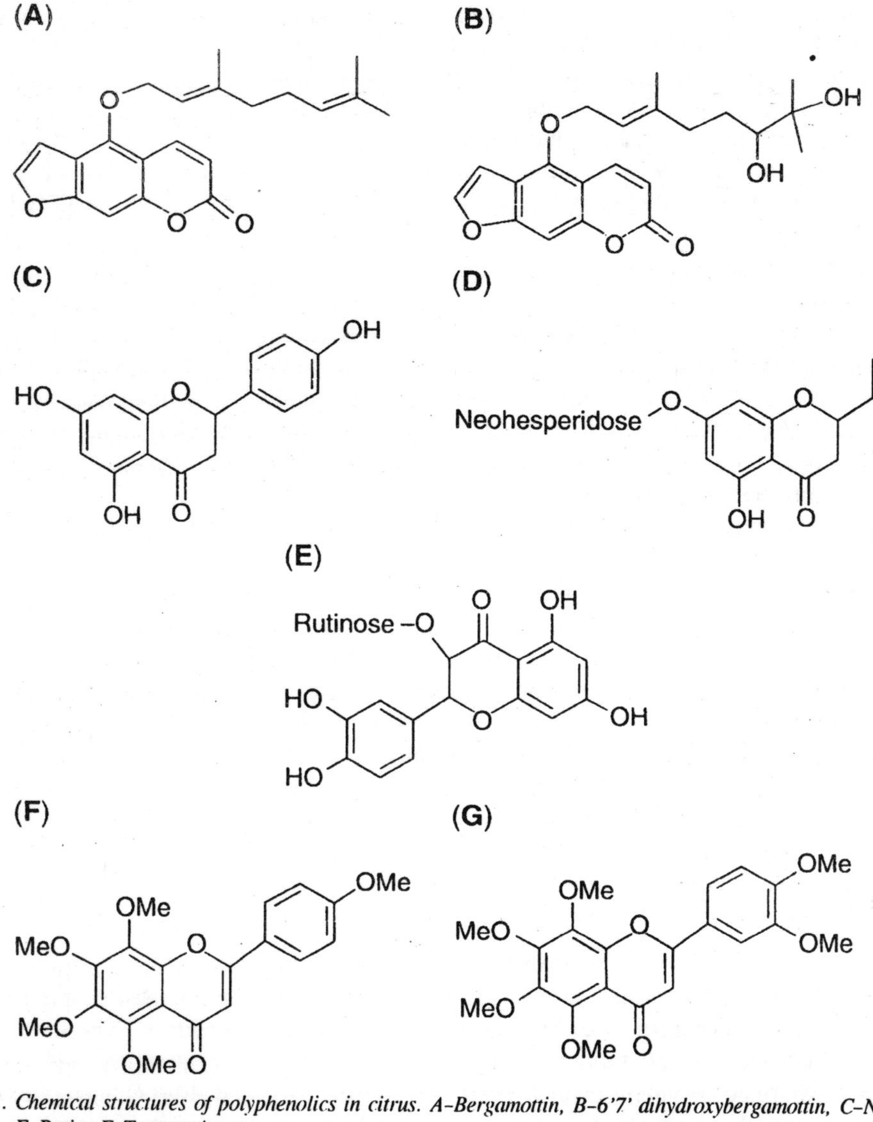

(A) (B)

(C) (D)

(E)

(F) (G)

Fig. 12.1. Chemical structures of polyphenolics in citrus. A–Bergamottin, B–6'7' dihydroxybergamottin, C–Naringenin, D–Naringin, E–Rutin, F–Tangeretin.

et al. investigated the kinetics of reversible and mechanism-based inhibition of CYP3A4 by BG and DHBG, using midazolam and testosterone as probes in human intestinal microsomes. In this study, it was found that the inhibition caused by DHBG was substrate-independent, reversible, and mechanism-based. BG was found to be a substrate-dependent reversible inhibitor, with an eightfold higher inhibition for midazolam than for testosterone.

Interactions with drugs have also been demonstrated for flavonoids from.citrus. Naringin and naringenin were shown to interact with simvastatin and saquinavir in in vitro experiments and caused alterations in the pharmacokinetics of quinine in rats. Moreover, naringenin and naringin were found to inhibit the OATP-B–mediated uptake of estrone-3-sulfate into human embryonic kidney cells.

Quercetin has been found to inhibit P-gp–mediated efflux of ritonavir in Caco-2 cells, to reduce the oxidation of acetaminophen in rat liver micro somes and HepG2 cells, and to inhibit the metabolism of midazolam and quinidine in human liver microsomes. It did not have an effect on CYP3A4-mediated metabolism and P-gp–mediated transport of saquinavir. Rutin was demonstrated to moderately increase the uptake of idarubicin in an isolated perfused rat lung model, and also the outflow recovery of the major metabolite idarubicinol, possibly by affecting P-gp. Nobelitin and tangeretin were shown to inhibit OATP-B–mediated uptake of estrone-3-sulfate into human embryonic kidney cells. For several phenolics from citrus, such as eriocitrin, poncirin, and sinapic acid and anthocyanins, which occur in red grapefruit varieties and blood oranges, no specific drug interactions and also no interactions with CYP3A4 and P-gp are reported.

The clinical relevance of data obtained from studies with single compounds is questionable, because most studies were performed in in vitro systems, limiting the predictability of the effects of the examined compounds in vivo. Moreover, some polyphenolics, such as quercetin, were shown to interact with the absorption or metabolism of drugs only at very high concentrations (50–100 μmol/L), which are likely to exceed the expected in vivo concentration after the consumption of a moderate amount of a grapefruit/citrus product. Also, flavonoids have been demonstrated to potentially induce apoptosis in cell lines at concentrations comparable to those used for some in vitro drug interaction studies. This potentially could have impaired the investigation of enzyme and transporter activities. In summary, studies with single compounds demonstrate that furanocoumarins and their dimers are primarily responsible for the interactions of GFJ and drugs.

In conclusion, grapefruit, sour orange (Seville), and also limes, which contain BG seem to have the highest potential among the citrus species for interacting with drugs, whereas other citrus varieties such as sweet orange seem to have an overall low potential for interfering with medications. However, in studies using juices rather than single compounds, orange juice has also been shown to interact with drugs in some studies. A more recent study in rats demonstrated that orange juice (and apple juice) decreased the oral exposure of fexofenadine, possibly through an inhibition of the influx transporter OATP. The interaction of orange juice and fexofenadine has also been demonstrated in HeLa cells, where orange juice at 5% strength inhibited the uptake of fexofenadine in a concentration- dependent manner by an array of human and rat OATPs. Also, orange juice inhibited the uptake of estrone-3-sulfate into human embryonic kidney 293 cells, probably mediated through OATP-B. Overall it can be concluded that orange juice has a minor potential for drug interactions.

Possible Mechanisms of Interaction

A food–drug interaction can be defined as the alteration of absorption, metabolism, or effects of a drug by food. The underlying mechanisms can be classified into two broad categories. The first category is pharmacokinetics, which includes alterations in absorption, distribution, metabolism, and excretion. The second category is pharmacodynamics, which describes alterations in the drug concentration–effect relationship. Changes in the pharmacokinetics of drugs are the more common consequences of citrus–drug interactions, which may shift the effect of the drug outside of its therapeutic window, possibly leading either to loss of effect or undesired side effects, or even toxicity. Originally, the liver was expected to be the major site of grapefruit–drug interactions. However, for felodipine, it was shown that the interaction only occurred when drugs are administered orally, but not intravenously, which indicated that the interaction may take place during the gastrointestinal absorption phase.

For several other drugs such as cyclosporine, midazolam, and nifedipine, the gastrointestinal mucosa has been demonstrated to be a major metabolic organ, where an inhibition of CYP3A4 caused an increase in oral bioavailability. Because most drugs exhibiting an interaction with GFJ are metabolized primarily by CYP3A4, it has been suggested that the effect of GFJ may be due to the inhibition of CYP3A4 activity. This effect may be particularly important for orally administered drugs, because CYP3A4 is located not only in the hepatocytes, but also in the epithelial cells of the intestine, where major interactions occur. In addition to CYP3A4, enterocyte efflux transport proteins such as P-gp, and enterocyte uptake proteins, such as OATPs also appear to be involved in grapefruit–drug interactions. The activity of P-gp has been reported to be altered by GFJ and orange juice in in vitro experiments, whereas the clinical relevance currently seems unclear, as demonstrated by inconsistent results from human clinical trials.

Regarding interactions between citrus and OATP, data from in vitro, animal and also human clinical studies are available. These studies demonstrated the inhibition of OATP-A in HeLa cells and of OATP-B–mediated uptake of estrone-3-sulfate in human embryonic kidney cells and also in the oral intake of fexofenadine in rats by fruit juices, including GFJ and orange juice. Human clinical trials suggest a potential role of OATP in grapefruit- drug interactions using fexofenadine as substrate.

Cytochrome P450 Family

The cytochrome P450 (CYP) enzyme family is the major catalyst of phase I drug biotransformation reactions. CYP enzymes are bound to membranes of the endoplasmatic reticulum and are predominantly expressed in the liver, although they are also present in extrahepatic tissues such as the gut mucosa. In humans, 16 gene families and 29 subfamilies have been identified to date. CYP3A4 is the most abundantly expressed isoform and represents approximately 30% to 40% of the total CYP protein in human adult liver. CYP3A4 is located mainly in the liver and in apical enterocytes of the small intestine. The high expression levels in the intestinal mucosa and the broad substrate specificity may contribute to the high susceptibility of CYP3A4 for citrus–drug interactions. Many drugs for which interactions with citrus have been demonstrated are metabolized by CYP3A4. Grapefruit inhibits the activity of intestinal CYP3A4, which can lead to an interaction with drugs during their first passage from the intestinal lumen into the systemic circulation.

The alteration of intestinal CYP3A4 by GFJ includes reversible and mechanism-based inhibition and also destruction of the CYP3A4 protein, whereas mRNA levels remain unaltered, indicating an accelerated degradation after mechanism-based inhibition. A moderate consumption does not appear to lead to an inhibition of hepatic CYP3A4 activity. Several drug classes such as dihydropyridine calcium antagonists and HMG-CoA-reductase inhibitors are affected by grapefruit-induced inhibition of CYP3A4. GFJ increased the AUC and maximal plasma concentration (C_{max}) for these calcium antagonists within an approximated range of 1.5- to 2.5-fold on average in a single-dose study design. For the HMG-CoA reductase inhibitor atorvastatin, double-strength GFJ increased the AUC 2.5-fold but not the C_{max}, when the GFJ was administered over three days and the drug was given on day three. In a very similar study design performed by the same group with simvastatin, double-strength GFJ increased the AUC 16-fold and C_{max} ninefold. The same group demonstrated that one glass of GFJ caused an increase of plasma triazolam concentrations, and the repeated consumption of GFJ induced a higher increase in triazolam concentrations and a prolonged half-life of triazolam.

The repeated consumption may cause an inhibition of hepatic CYP3A4. In a three day study, GFJ increased the AUC of simvastatin 3.6-fold and that of simvastatin acid 3.3-fold. C_{max} of simvastatin and simvastatin acid were increased 3.9-fold and 4.3-fold, respectively, when the GFJ was administered for three days and simvastatin on day three. In an in vitro study performed with several grapefruit compounds, it was shown that BG, DHBG, and the furanocoumarin dimers GF-I-1 and GF-I-4 inhibited CYP3A4-catalyzed nifedipine oxidation in a concentration-and time-dependent manner, which is consistent with the mechanism-based inhibition. DHBG was more potent than BG, while the dimers were more potent than the monomers. Not only CYP3A4 but also CYP2C9, CYP2C19, and CYP2D6 seem to be affected by citrus compounds. In the same study, the inhibitory effect of BG was stronger on CYP1A2, CYP2C9, CYP2C19, and CYP2D6 than on CYP3A4. In an intervention trial with healthy volunteers, GFJ (twice daily) decreased the activity of CYP1A2, as determined with caffeine as a probe. In another human study, GFJ and naringenin caused a minor reduction of the activity of CYP1A2. Overall, the inhibition of CYP enzymes other than CYP3A4 does not appear to be of great magnitude and may clinically be relevant only for drugs with a narrow therapeutic range.

Not much information is available regarding the reversible and mechanism-based inhibition kinetics for grapefruit compounds. In a study conducted with human intestinal microsomes by Paine and coworkers, DHB induced a substrate-independent reversible and mechanism-based inhibition on CYP3A4. In contrast, BG, being more lipophilic, was a substrate- dependent reversible inhibitor and a substrate-independent mechanism-based inhibitor. Similar trends resulted with cDNA-expressed CYP3A4. For BG, the inhibition for testosterone was more potent than for midazolam, possibly due to the higher affinity of BG for the testosterone-binding site than for the midazolam-binding site. As mechanism-based inhibitors, BG and DHBG are substrates for CYP3A4, but the binding sites are not known. The authors conclude that both furanocoumarins inactivate CYP3A4 by the binding of the furanoepoxide to the apoprotein, presumably at or near their respective substrate domains. The same group determined the onset time of inhibition by both compounds.

It was found that DHBG inhibited 85% of CYP3A4 activity independent of substrate within 30 minutes, whereas the onset for BG-induced inhibition was much later—a 70% inhibition was reached after three hours. The substrate-dependent inhibition caused by BG was more than 50% after 0.5 to 3 hours for testosterone 6-hydroxylation, while midazolam 1'-hydroxylation was unaffected, or activated, within one hour. Both furanocoumarins caused 40% to 50% reduction of CYP3A4 protein, probably due to intracellular degradation of the enzyme caused by mechanism-based inactivation. These data imply that, after the consumption of GFJ, DHBG causes the enzyme inhibition earlier than BG. Greenblatt et al. determined, in a human intervention trial, the time of recovery of intestinal CYP3A4 after the consumption of 300 mL of regular-strength GFJ and a single dose of midazolam at 2, 26, 50, or 74 hours after administering the juice. After two hours, the AUC was 1.65-fold increased and after 26, 50, and 74hours, the AUC was 1.29-, 1.21-, and 1.06-fold increased, respectively, in comparison to the control. The recovery half-life was estimated at 23 hours. These results indicate that a single dose of GFJ was able to impair the intestinal presystemic metabolism of midazolam when administered orally, which appeared to recover after 74 hours, consistent with a mechanism-based inhibition.

In summary, the presented in vitro studies confirm the inhibitory effects of grapefruit compounds on CYP-enzymes, with major effects on CYP3A4 with both, mechanism-based and reversible inhibition. Overall, DHBG appears to be more potent than BG; however, coumarin dimers seem to be more effective than monomers in the inhibition of CYP3A4. The human intervention trials examining the pharmacokinetic interaction of GFJ revealed a great interindividual variability, where subjects with the highest content of CYP3A4 showed the largest reduction of this enzyme after the consumption of grapefruit. Overall, GFJ seems to interact with orally administered drugs, not with intravenously administered drugs.

P-Glycoprotein

The interest in transporters as mediators of interactions between grapefruit and drugs is increasing. One of the most studied drug transporters is P-gp. P-gp is a 170 kDa plasma glycoprotein, which is encoded by the multidrug resistance (MDR) 1 gene and belongs to the family of ATP-binding cassette transporters. P-gp was first characterized in tumor cells, where it contributes to the MDR. P-gp is expressed constitutively at high levels on the apical surface of the small intestines, liver, pancreas, kidney, colon, and adrenal glands, but also can be found at the blood–brain barrier and blood–cerebrospinal fluid barriers. Striking overlaps of substrates and inhibitors between CYP3A4 and P-gp were reported by Wacher et al. Consequently, the inhibition of P-gp function may also play a role in the effects of GFJ.

Earlier studies demonstrated that GFJ did not influence the activity of P-gp; Lown and coworkers found that 8 oz of GFJ (three times per day for six days) did not alter P-gp concentrations in healthy volunteers. Eagling et al. confirmed these findings in their in vitro study in Caco-2 cells, where compounds from grapefruit were not found to modulate P-gp. In 1999, it was reported that GFJ increased P-gp–mediated transport in Madin-Darby canine kidney epithelial cells (MDCK)–MDR1 cells. However, this finding is controversial to the later findings and was attributed to an equipment-generated artifact by the authors. Takanaga et al. were the first group to demonstrate the inhibition of P-gp by GFJ in Caco-2 cells with vinblastine as probe. Vinblastine also is a substrate of CYP3A4, which limits the conclusions regarding the inhibition of P-gp. Therefore, the same group showed that GFJ and phenolic compounds from orange, such as tangeretin, nobiletin, and heptamethoxyflavone, which have been demonstrated not to alter CYP3A4 activity, increased the net influx of vincristine into adriamycin-resistant human myelogenous leukemia cells conclusively, through the inhibition of P-gp.

Clinical studies that compare the effects of orange juice with GFJ on drug bioavailability confirm the involvement of P-gp in grapefruit-induced alterations in drug absorption. A clinical intervention study conducted by Edwards et al. in the same year indicated that GFJ may interact with P-gp activity. In this study, AUC and peak concentrations of cyclosporine, a P-gp substrate, were increased by GFJ, whereas Seville orange juice did not have an influence on cyclosporine, while it reduced enterocyte concentrations of CYP3A4. DHBG did not inhibit P-gp in vitro. These data imply that the inhibition of P-gp activity by other compounds in GFJ may be responsible for the increased bioavailability of cyclosporine.

These results were confirmed by Malhotra et al. in a randomized three-way crossover intervention study in healthy volunteers who received felodipine with Seville orange juice, dilute GFJ (normalized to equivalent total concentration of BG and DHBG), and sweet orange juice. Seville orange juice and GFJ increase the AUC of felodipine. While Seville orange juice and GFJ probably interact with felodipine through inactivation of intestinal CYP3A4, the lack of interaction between Seville orange juice and cyclosporine indicates that grapefruit may cause interactions also through the inhibition of intestinal P-gp. Several in vitro studies with different probes (talinolol, digoxin, and vinblastine) also confirm the findings that GFJ inhibits the efflux of P-gp substrates. In addition to GFJ, orange juice and pomelo juice also have been shown to inhibit the activity of P-gp in vitro. Flavones from orange juice have been shown to be more potent than compounds from grapefruit in the inhibition of P-gp.

The pharmacokinetics of several drugs that are known P-gp substrates were not altered by GFJ in several clinical studies that investigated the effect of GFJ on the bioavailability of digoxin, amlodipine, and indinavir. Possible other unknown mechanisms and factors such as strength of the administered juices and length of consumption are relevant for the interactions of citrus with drugs.

Overall, it can be stated that grapefruit and other citrus may interact with several drugs through the combined inhibition of CYP3A4 and P-gp. The magnitude of interactions may strongly depend on variations in the polyphenolic profile of the GFJs and study design. The clinical significance of P-gp–related interactions between drugs and GFJ needs to be clarified in further clinical studies.

Organic Anion Transporting Polypeptides

The family of OATPs consists of membrane carriers that mediate the transport of anionic molecules, although not exclusively, and more recently, transport of nonanionic molecules has been observed. OATPs are located in the small intestines on luminal membranes of enterocytes, where they mediate the uptake of drugs. In the liver, OATPs facilitate the uptake of drugs into the hepatocytes. Whereas OATP-A is predominantly located in the brain, OATP-B has been found to be expressed on the membranes of intestinal epithelial cells. The inhibition of drug uptake mediated by OATP may alter the plasma concentration of OATP substrates.

In a more recent work, GFJ and orange juice have been reported to reduce the availability of fexofenadine and celiprolol. Both drugs are substrates for P-gp and OATP, but not CYP3A4. If P-gp had played a major role in the observed interactions, the bioavailability would have been increased instead of decreased. This led to the conclusion that a mechanism other than P-gp was involved. In theory, the inhibition of OATP could lead to a decreased absorption of OATP substrates into intestinal enterocytes. This hypothesis was tested by several in vitro studies.

Dresser et al. determined that GFJ and orange juice decreased OATP-A–mediated fexofenadine uptake into HeLa cells, and that this inhibition of OATP was more potent than the inhibition of P-gp. In a corresponding human trial, the same authors found that GFJ and orange juice decreased the bioavailability of fexofenadine in healthy volunteers. These results imply that citrus juices may be able to inhibit both forms of OATP, namely OATP-A, which occurs in the brain and was used in the in vitro experiments, and OATP-B, which occurs in the intestine and may be responsible for the reduction of the availability of fexofenadine in the human trial. The authors also considered that the apparent decreased bioavailability of fexofenadine in human subjects may have been caused indirectly by an increased drug intake when the drug was administered with water, due to the lower osmolarity of water. Therefore a nonpolar fraction of GFJ was tested in a clinical trial. This nonpolar fraction also significantly reduced the bioavailability of fexofenadine. In a study with human embryonic kidney 293 cells expressing OATP-B, different citrus juices were tested in their effect on the uptake of estrone-3-sulfate. GFJ, orange juice, BG, DHBG, quercetin, naringin, and naringenin significantly inhibited OATP-B–mediated uptake of estrone-3-sulfate.

The citrus compounds DHBG and tangeretin significantly inhibited OATP-B–mediated influx of the probe ibenclamide The effects of fruit juices on the oral availability of fexofenadine also have been tested in rats. In this study, orange juice decreased the oral bioavailability of the drug to a lesser extent than that observed in humans. The clinical relevance of this study for the situation in humans is not clear, because fexofenadine mainly is substrate for OATP-A, which in humans predominantly occurs in the brain, whereas OATP-B occurs in the intestines. Overall, it has to be considered that genetic differences in the OATPs between humans and other species may contribute to differences in the susceptibility to grapefruit-induced inhibition, which also is true for other transporters and enzymes.

Two reports of human clinical trials discuss the potential role of OATP in grapefruit-drug interactions using fexofenadine as substrate. These studies in human healthy volunteers revealed that consumption of GFJ reduced the rate and extent of absorption of fexofenadine, an OATP substrate. The C_{max} and AUC was decreased by a range of 30% to 60% compared to when the drug was taken with water. Both studies suggested an inhibition of the influx mediated drug transporter OATP by GFJ. A recent report evaluates the effect of GFJ on disposition of talinolol in healthy human volunteers. A single glass of GFJ decreased the talinolol area under the serum concentration-time curve (AUC), peak serum drug concentration (C_{max}) and urinary excretion values to around 55%, compared with water. In addition repeated ingestion of GFJ had a similar effect. Because both single and repeated ingestion of GFJ lowered rather than increased talinolol AUC, the findings suggest that constituents present in GFJ preferentially inhibit an intestinal uptake process such as OATP rather than P-glycoprotein. In summary, OATPs appear to play an important role in the influx of a number of drugs into enterocytes and hepatocytes. The inhibition of OATP activity has been demonstrated for orange and GFJ in in vitro experiments. The clinical relevance of this mechanism remains to be investigated in further human intervention trials.

Classes of Drugs Interacting with GFJ

The major drug classes for which grapefruit or other citrus interactions have been reported are described below and the interactions are summarized. Predicting the clinical significance of pharmacokinetic drug interactions is sometimes difficult especially for drugs where there are no robust methods to quantify effects or side effects. There has been recent effort in the United States by the FDA and the Pharmaceutical Research and Manufacturers of America (PhRMA) to establish some general guidelines to help drug companies, prescribers, and patients interpret the clinical significance of drug interactions.

These are based on the clinical experience gained from some well-known drug interactions, such as the inhibition of CYP3A4. For this interaction, it could be shown that the benzodiazepine midazolam is a reproducible probe that allows quantitative determination of the interaction potential of an enzyme inhibitor. The degree of interaction can be measured in the form of an increase in the AUC of the midazolam serum concentrations. It was recently proposed to classify changes of midazolam AUC being less than twofold as "weak," which is the case observed on midazolam coadministration with ranitidine, relatively small volumes of GFJ, roxithromycin, fentanyl, or azithromycin. AUC changes that range from two- to five-fold, which occur on midazolam coadministration with erythromycin, diltiazem, fluconazole, verapamil, relatively large volumes of GFJ, and cimetidine are classified as "moderate." Changes that exceed a fivefold increase in midazolam AUC are labeled as "strong." Examples of drugs demonstrating strong midazolam interactions include: ketoconazole, itraconazole, mibefradil, clarithromycin, and nefazodone.

Strong drug interactions are considered clinically significant and result in contraindications or strong warnings on the product label. The clinical significance of moderate inhibitors may include decisions about dose adjustments that should be based on the concentration–effect relationship. Where applicable, the clinical relevance of GFJ interactions in the present paper was assessed according to the PhRMA classification. If not otherwise stated studies are performed with human subjects.

Antiallergics

Interaction studies were performed for the antiallergic drugs desloratadine, fexofenadine, and terfenadine. When taken with GFJ, the mean exposure for desloratadine was not altered, whereas for fexofenadine it was decreased. In one of the terfenadine interaction studies, the concentrations of the control group could not be quantified. Three other studies demonstrated a significant increase in exposure to terfenadine. The maximum difference in exposure to terfenadine of 2.4-fold was shown by Clifford et al. No significant changes were observed in electrocardiogram (ECG) parameters for desloratadine and fexofenadine. A statistically significant increase in rate-corrected QT (QTc) intervals was reported for terfenadine when administered with GFJ and this drug was taken off the market. The mean effect of GFJ on desloratadine pharmacokinetics seems to be unlikely to be clinically relevant.

Antibiotics

The effects of GFJ were studied for clarithromycin and erythromycin. A decrease in the time to reach the maximal plasma concentration (T_{max}) was found for clarithromycin. The exposure of erythromycin was mildly increased when administered concomitantly with GFJ. It seems unlikely that the reported interactions with the above drugs would be relevant in a clinical setting.

Anticoagulants

Data derived from studies performed with coumarin are inconsistent. In one study, the percentage of 7-hydroxycoumarin excreted in urine was decreased; in a second study the appearance of the metabolite was delayed when 300 mL GFJ were given concomitantly, but the recovery in urine was unchanged, whereas four times 250 mL juice in 30-minute intervals increased the recovery by 100%. The delay in appearance of the metabolite could also be confirmed in a third study. A study performed with warfarin did not assess any pharmacokinetic parameters. However, when patients were pretreated with 8 oz GFJ three times a day for one week, no change in prothrombin time or International Normalized Ratio could be observed. More conclusive clinical studies are necessary to assess the overall effect of GFJ on these anticoagulants.

Antimalaria Drugs

GFJ increases the exposure of artemether, chloroquine in chicken and mice, and halofantrine. No signs of bradycardia or changes in the QTc interval were observed when artemether was administered with GFJ. No overt signs of toxicity of chloroquine were observed in the chicken study. Furthermore, it was reported that GFJ increased the QTc interval when administered concomitantly with halofantrine. No changes in pharmacokinetic parameters were observed when quinidine or quinine was coadministered with GFJ. However, one study assessing the interaction with quinidine reported a significant change in QTc interval prolongation, although this change was only seen at one hour after drug administration. According to the classification of Bjornsson et al. the interactions of quinidine and quinine would be considered weak and unlikely to be clinically relevant. However, a further evaluation of the pharmacodynamic parameters would be desirable. Artemether showed a moderate interaction and halofantrine exhibited a strong interaction. GFJ should be avoided with halofantrine and should be consumed only after a cautious risk and benefit assessment with artemether.

Antiparasitic Drugs

The interactions of GFJ with albendazole and praziquantel can be considered moderate and weak, respectively. GFJ increases the AUC of albendazole 3.1-fold and the C_{max} 3.2-fold. A concomitant consumption should only be considered after a cautious risk and benefit assessment. Regarding the pharmacokinetic parameters, an interaction of GFJ with praziquantel seems unlikely to be clinically relevant (AUC increases 1.62-fold and C_{max} increases 1.9-fold).

Anxiolytics

Alprazolam did not exhibit any changes in pharmacokinetic parameters when administered concomitantly with GFJ in a single dose experiment. When predosing existed, GFJ increased the exposure of alprazolam only in smokers. However, this interaction seems unlikely to be clinically relevant. In both parts of this study psychomotor function remained unchanged; however, there was a small decrease in cognitive speed in the single dose part at one and two hours after dosing. Midazolam and triazolam exposure increased when administered concomitantly with GFJ, but these increases fall into the category of weak interactions. However, changes in psychomotor function have been reported. On the other hand, one study showed a 2.06-fold increase in exposure in patients with liver cirrhosis. These results would be considered a moderate interaction and cannot easily be extrapolated to healthy patients. Diazepam plasma concentrations in dogs were increased by GFJ. Buspirone plasma concentrations also were increased, as was the overall subjective effect. No changes in psychomotor function were observed. Patients should not consume buspirone concomitantly with GFJ, because they may exhibit a strong pharmacokinetic interaction.

Calcium Channel Blockers

Amlodipine, diltiazem, nimodipine, nifedipine, pranidipine, and verapamil exhibit a weak interaction with GFJ with regard to their mean exposure. However, only for amlodipine, diltiazem, nimodipine, and verapamil, no changes in the pharmacodynamic parameters heart rate and blood pressure were reported. Blood pressure was increased after verapamil and GFJ administration only at eight hours after drug administration. No pharmacodynamic parameters were recorded in the studies performed with nifedipine and GFJ. Heart rate was increased after pranidipine administration with GFJ, however the blood pressure remained constant. Felodipine, nicardipine, nisoldipine, and nitrendipine exhibit a moderate interaction with GFJ. However, blood pressure decreased in one study with nisoldipine and no change in pharmacodynamic parameters were observed in the study examining nitrendipine. The heart rates after concomitant nicardipine and GFJ administration changed only at two hours. Changes in pharmacodynamic parameters were reported after GFJ was administered with felodipine. Felodipine-, nicardipine-, nisoldipine-, and nitrendipine-containing products should only be consumed with GFJ after a cautious risk and benefit assessment.

HIV Protease Inhibitors

Amprenavir and indinavir showed no changes in pharmacokinetic parameters when administered concomitantly with GFJ. Even 180 mL double-strength GFJ had no effect on indinavir pharmacokinetics. The AUC of saquinavir was increased after predosing with GFJ. The mean increase was 1.5-fold. In a study performed with one subject, a 5-fold increase was reported. The reported interactions can be considered weak and are unlikely to be clinically relevant.

HMG-CoA Reductase Inhibitors

Increases in exposure were reported for atorvastatin, lovastatin, and simvastatin. GFJ was shown not to have an effect on the pharmacokinetics of pravastatin. Atorvastatin exhibited a moderate interaction with GFJ regarding the overall exposure. Lovastatin and simvastatin exhibited a strong interaction. Pravastatin could be chosen as an alternative drug if patients want to ensure a lack of interaction.

Hormones

No effect of GFJ was observed on 17-beta estradiol or prednisone pharmacokinetics. AUCs were increased for ethinyl-estradiol and methylprednisolone. The increases in exposure can be considered weak and seem to be unlikely to be clinically relevant. It has to be mentioned that a decrease in morning cortisol plasma concentrations has been observed after administration of methylprednisolone with GFJ.

Immunosupressants

An increase in cyclosporine exposure was reported by 11 out of a total of 13 studies. Two studies reported no change in AUC induced by GFJ. Even administration of a large amount of GFJ was shown to increase the AUC only by 7%. Regarding the exposure to cyclosporine, the reported interactions seem unlikely to be clinically relevant.

Antitumor Drugs

GFJ was demonstrated to decrease the mean AUC of etoposide when administered concomitantly. However, there was no indication of whether the results were statistically significant. Furthermore, an increased uptake has been shown for vinblastine in Caco-2 cells. These results can only serve as an estimate. Further research will have to be conducted to develop recommendations for this drug class.

Over-the-Counter Drugs

Contradicting results were reported for GFJ when administered with caffeine. One study reported no changes in AUC, blood pressure, and heart rate. A second study reported increases in AUC and half-life. However, no assessment of pharmacodynamic parameters was performed. Furthermore, GFJ increased the fraction absorbed and the percentage of excreted dextromethorphan. The above-mentioned interactions can be considered weak regarding the overall exposure. Furthermore, dextromethorphan has a broad therapeutic window. The interactions of GFJ with caffeine and dextromethorphan do not seem to be of clinical relevance. Hence, no clinically significant interactions of an over-the-counter drug with GFJ have been reported.

Beta-Blockers

When celiprolol was administered concomitantly with GFJ, AUC and Cmax of celiprolol decreased by 95%; however, heart rate and blood pressure remained the same. The authors of this study conclude the observed interactions to be clinically relevant. In an animal study, talinolol exposure has been reported to increase after GFJ administration. Further human studies will have to be conducted to derive reliable conclusions.

Other Drugs

GFJ has been shown to increase the exposure of carbamazepine, cisapride, fluvoxamine, losartan, methadone, scopolamine, and sertraline. However, only the interaction of GFJ with carbamazepine and cisapride seems to be clinically relevant. No alteration in exposure was observed for clozapine, heophylline, haloperidol, and omeprazole. Reports of increased pharmacokinetic parameters of clozapine, theophylline, and haloperidol suggest that an interaction is unlikely to be clinically relevant. Contradicting results were reported for itraconazole, digoxin, and sildenafil. An increased effect on concomitant use of diclofenac and GFJ was observed in rats. Overall, the clinical relevance for this drug class appears to be low.

Citrus-based products, and grapefruit in particular, can interact with several orally administered medications. In most cases, the expected interactions will be minor and of little clinical relevance. However, the overall observation that a single serving of GFJ can induce a long-lasting increase in oral bioavailability of some drugs, which may lead to potential drug toxicity, does call for caution. In situations where toxicity can be expected, GFJ and other citrus products with a known interaction should be avoided during the whole period of drug treatment. For those patients who want to consume GFJ while they are being medicated, many of the drugs showing an interaction could be replaced by other drugs that have been shown to not interact with GFJ. It also should be considered that, for patients who regularly consumed GFJ before the dose of their medication was adjusted and continued with a constant consumption of GFJ during their medication, a potential toxicity appears relatively unlikely, because the dose for these patients would be lower than for those not consuming GFJ.

There are a few situations in which toxicity could potentially occur: (i) in patients taking unusually high doses of a susceptible drug, who then consume GFJ for the first time, where the GFJ may lead to a sudden decrease of intestinal CYP3A4 activity, (ii) in patients with severe liver disease, the exposure to the drug would be expected to be higher with the intestines being the major site of metabolism. Also in these patients a sudden decrease in CYP activity in the intestines may lead to an increase of drug concentration, and (iii) patients susceptible to toxic effects from drugs are likely to exhibit drug toxicity when the bioavailability is increased.

Additionally, it has to be considered that patients consuming GFJ are also prone to consume other fruits and vegetables and dietary supplements, which may cause an interaction. Sales of dietary supplements containing phytochemicals have been expanding, in part driven by increased health awareness. In particular, patients with chronic diseases have a high propensity to consume prescription drugs and concomitant dietary supplements. Between 1990 and 1995, the use of alternative medicines, including dietary supplements, increased from 34% to 42%, leading to an expenditure of $27 billion for patients. Worldwide, up to 75% of cancer patients use alternative medicine. In a survey conducted in 2002, 54.9% of all users of alternative medicines, including dietary supplements, thought that the natural remedy would help when consumed in combination with the conventional drug. Although GFJ can cause a drug interaction by itself, it should not be taken out of the context of the complete diet, which may also contribute to drug interactions.

The extent of the grapefruit–drug interaction in the case of CYP3A4 appears to be dependent on the patient intestinal enzyme activity. Subjects with a high activity of CYP3A4 appear to show a higher inhibition by GFJ, whereas the inhibition of CYP was lower for subjects with a low initial CYP activity. Theoretically, the concomitant administration of GFJ with a susceptible drug would cause the highest increase in AUC in subjects with a high intestinal CYP3A4 activity; however unexpectedly, these subjects tend to have a low AUC after intake of a standard dose of drugs without the administration of

GFJ. More studies will have to be performed, especially in the area of P-gp and OATP-mediated drug interactions, before sound recommendations for the concomitant intake of citrus with certain drugs can be given.

Future Directives

According to the current, still limited knowledge, responsible recommendations should be communicated effectively to health care personnel and patients. Here care must be taken to avoid careless prescription of susceptible drugs without unnecessary overreaction leading to complete avoidance of citrus products, because these contain significant amounts of antioxidant phytochemicals with significant health benefits. It also appears possible to develop grapefruit furanocoumarins as additives to certain drugs in order to improve their oral bioavailability and reduce the variability. On the other hand, citrus juice manufacturers could develop GFJs without furanocoumarins, which would reduce the potential for a drug interaction. Citrus fruits, other than grapefruit, and other food products, in general, will have to be investigated further in their potential to induce a drug interaction. Further clinical research will have to be conducted to conclusively determine the mechanisms and clinical relevance of these interactions.

GRAPEFRUIT JUICE INTERACTION

Communication to the public of medical product risks, such as drug–grapefruit juice interaction, is an important aspect of public health agency work. Health Canada, for example, advised the public in 2002 not to consume grapefruit products with medications used for certain medical conditions, such as anxiety, depression, and others. In the United States, the Food and Drug Administration (FDA) includes documented information on drug–grapefruit juice interaction in individual product labeling [also known as the package insert (PI)]. Information in the PI is typically based on the studies submitted by a drug's manufacturer. The FDA has also utilized spontaneous adverse event case reports as a tool in the evaluation of drug–grapefruit juice interaction labeling. Other sources of data that contribute to labeled information include studies or case reports from the medical literature.

Spontaneous Adverse Event Case Reports

To identify case reports containing information on drug and grapefruit juice interaction, the Adverse Event Reporting System (AERS) database was searched in April 2004 for any mention of a grapefruit-containing product (e.g., grapefruit, grapefruit juice, grapefruit seed) as either a "suspect" (the product that is suspected by the reporter to have caused the event) or a "*concomitant*" (other medical products that the patient was receiving at the same time) product. It must be understood that the FDA does not receive all reports of adverse events and product interactions that occur in medical practice. This is particularly true for a product such as grapefruit juice. First, there is no regulatory requirement to submit food (such as grapefruit juice)-related adverse events to the FDA. Further, grapefruit juice is not generally considered a "*medical product*," so it is possible that a reporter describing a "drug" adverse event report concerning such an interaction may not list "*grapefruit juice*" in either of the "*medical product*" blocks on the MedWatch form. Thus, all reports that mention grapefruit juice in the AERS database might not have been located. The grapefruit juice search described above identified 186 cases in the AERS database. At that point, we had a group of cases that could describe drug–grapefruit juice interaction. However, as with all searches in AERS for spontaneous case reports, each patient case must be scrutinized further, using either a case definition or a set of criteria to better focus on the safety concern of interest. The following criteria were chosen to better identify reports that were more likely to describe an interaction between a drug product and grapefruit juice:

1. The patient was documented to be stable on the drug product prior to receiving grapefruit juice, and,

2. The adverse event occurred after the initiation of grapefruit juice, is a known effect of the drug, and is usually dose-related. Allergic reaction events, lack of drug effect, and events describing gastrointestinal upset after grapefruit juice consumption were not included.

or,

1. The patient was documented to be consuming grapefruit juice each day prior to starting drug therapy, or started grapefruit juice and the drug product at the same time, and,
2. The adverse event occurred upon initiation of drug therapy, and,
3. The adverse event resolved upon retention of same dose of drug with discontinuation of grapefruit juice.

Among the original 186, forty case reports were identified that met the criteria. Thirty-three were from the United States, and seven were reported from foreign countries. Three case reports are presented in table 12.1.

Case 1

A 60-year-old male patient with hypertension, chronic lower extremity venous stasis/edema, renal insufficiency, non–insulin dependent diabetes mellitus, and a familial history of hyperlipidemia had been receiving lovastatin, 40mg tablet, twice a day for 5 to 10 years. Other therapy included gemfibrozil (600 mg b.i.d.), amlodipine, and an oral hypoglycemic agent. Early in the year, the patient's creatinine was 3.5 mg/dL. In October, the patient began drinking grapefruit juice in the morning for the first time in his life. During this time, the patient denied any strenuous exercise. Two weeks later, the patient was in so much pain that he went to an emergency room where his creatine phosphokinase was greater than 40,000 U/L. He was admitted to the hospital with a diagnosis of rhabdomyolysis. Therapy with lovastatin and gemfibrozil was discontinued. During that time, his creatinine increased to about 5.0 mg/dL. The patient was discharged from the hospital after approximately one week and was considered to be recovering. The physician felt that the patient's rhabdomyolysis was caused by the interaction of the grapefruit juice with lovastatin and gemfibrozil.

Case 2

A 92-year-old female with hypertension had been taking nifedipine 30mg daily for four years. While traveling in Florida, she took her nifedipine with grapefruit juice and experienced extreme fatigue, dizziness, vertigo, decreased appetite, and disorientation. She was hospitalized for three to four days, the grapefruit juice was stopped, and she recovered. After returning home from Florida, she took her nifedipine with grapefruit juice again and experienced a similar but milder reaction. Her pharmacist suspected an interaction between nifedipine and grapefruit juice.

Case 3

A 62-year-old female with a history of systemic lupus erythematosis, osteoporisis, angina pectoris, and renal failure began taking amlodipine. She had also been taking lisinopril and spironolactone for an unknown duration. A year later, she started eating grapefruit every day. She fell down after developing disturbance of consciousness for a few minutes. Four days later, she developed generalized fatigue and was admitted the next day to the hospital for shock symptoms consisting of decreased blood pressure, clouded consciousness, and vomiting. Her medications were discontinued, and she was treated for the shock symptoms. The patient recovered. Her physician suspected an interaction between amlodipine and grapefruit.

Case Reports Compared to Labeling

The next step in the assessment was to check the current status of product labeling for drug–grapefruit juice interaction. The Physicians' Desk Reference (PDR) was utilized as a tool for this

Table 12.1. Cases of possible drug–grapefruit interaction

Drug	*Events*
Amlodipine	Peripheral edema, asthenia, hypotension
Amlodipine	Dyspnea, anxiety, hypertension
Amlodipine	Dizziness
Amlodipine	Hypotension
Amlodipine	Head "fullness," strange sensations
Amlodipine	Hypotension, asthenia (feeling of tiredness)
Amlodipine	Dizziness, tachycardia, lightheadedness
Amlodipine	Near syncope
Amlodipine	Atrial fibrillation, ventricular tachycardia
Amlodipine	Increased LFTs and bilirubin
Amlodipine	Dizziness, asthenia
Amlodipine	Hypotension, syncope
Astemizole	Syncope, atrioventricular block
Atorvastatin	Anxiety
Atorvastatin	Epistaxis
Atorvastatin	Gingival bleeding
Atorvastatin	Myalgia, asthenia
Atorvastatin	Dizziness, nausea
Atorvastatin	Myalgia, myasthenia
Atorvastatin	Arm/shoulder pain
Atorvastatin	Paresthesia
Atorvastatin/azithromycin	Increased LFTs and bilirubin
Bupropion	Hypertensive crisis
Doxazosin	Near syncope
Estrogens, conjugated	Fluid retention
Gabapentin	Dizziness, syncope, slurred speech
Lisinopril	Headache, nausea, dizziness
Lisinopril	Dizziness, syncope, seizure
Lovastatin	Rhabdomyolysis
Nifedipine	Pedal edema
Nifedipine	Hypotension, gait abnormal
Nifedipine	Eye swelling, foggy feeling, headache
Nifedipine	Tiredness, asthenia
Nifedipine	Tachycardia
Nifedipine	Dizziness, asthenia, gingivitis
Nifedipine	Asthenia, dizziness, vertigo, disorientation
Risperidone/fluoxetine	Somnolence, fatigue
Sertraline	Sweating
Sibutramine	Insomnia
Verapamil	Palpitations, flushing, malaise, facial redness

purpose. An online query in April 2004 for "grapefruit" was performed, which identified relevant labeling. An important limitation of this strategy is that not all products are included in the PDR, so this search was not expected to identify 100% of drug product labeling containing information on grapefruit. There were 24 ingredients identified in the PDR online search, which described documented or theoretical effects of grapefruit consumption; these are listed in Table 2. The information was contained in a variety of labeling sections: "Clinical Pharmacology," "Warnings," "Precautions,"and "Dosage and Administration." For most drugs, placement of the information was in the Precautions/Drug Interactions section. The following three labeling examples were chosen to illustrate the range of information on drug-grapefruit interaction available.

Lovastatin (Mevacor)

Clinical pharmacology

Lovastatin is a substrate for cytochrome P450 isoform 3A4 (CYP3A4). Grapefruit juice contains one or more components that inhibit CYP3A4 and can increase the plasma concentrations of drugs metabolized by CYP3A4. In one study, 10 subjects consumed 200 mL of double-strength grapefruit juice (one can of frozen concentrate diluted with one rather than three cans of water) three times daily for two days and an additional 200 mL double-strength grapefruit juice together with and 30 and 90 minutes following a single dose of 80 mg lovastatin on the third day. This regimen of grapefruit juice resulted in a mean increase in the serum concentration of lovastatin and its (beta)-hydroxyacid metabolite (as measured by the area under the concentration–time curve) of 15-fold and 5-fold, respectively (as measured using a chemical assay—high performance liquid chromatography). In a second study, 15 subjects consumed one 8 oz glass of single-strength grapefruit juice (one can of frozen concentrate diluted with three cans of water) with breakfast for three consecutive days and a single dose of 40mg lovastatin in the evening of the third day. This regimen of grapefruit juice resulted in a mean increase in the plasma concentration (as measured by the area under the concentration–time curve) of active and total hydroxymethylglutaryl coenzyme A (HMG-CoA) reductase inhibitory activity [using an enzyme inhibition assay both before (for active inhibitors) and after (for total inhibitors) base hydrolysis] of 1.34-fold and 1.36-fold, respectively, and of lovastatin and its (beta)-hydroxyacid metabolite (measured using a chemical assay—liquid chromatography/tandem mass spectrometry—different from that used in the first study) of 1.94-fold and 1.57-fold, respectively. The effect of the difference in amounts of grapefruit juice in these two studies of lovastatin pharmacokinetics has not been studied.

Warnings

Myopathy caused by drug interactions

The incidence and severity of myopathy are increased by concomitant administration of HMG-CoA reductase inhibitors with drugs that can cause myopathy when given alone, such as gemfibrozil and other fibrates, and lipid-lowering doses (greater than or equal to 1 g/day) of niacin (nicotonic acid). In addition, the risk of myopathy may be increased by high levels of HMG-CoA reductase inhibitory activity in plasma. Lovastatin is metabolized by the CYP3A4. Potent inhibitors of this metabolic pathway can raise the plasma levels of HMG-CoA reductase inhibitory activity and may increase the risk of myopathy. These include cyclosporine; the azole antifungals itraconazole and ketoconazole; the macrolide antibiotics erythromycin and clarithromycin; HIV protease inhibitors; the antidepressant nefazodone; and large quantities of grapefruit juice (greater than 1 quart daily).

Precautions/Drug interactions

CYP3A4 interactions

Lovastatin has no CYP3A4 inhibitory activity; therefore, it is not expected to affect the plasma concentrations of other drugs metabolized by CYP3A4. However, lovastatin itself is a substrate for

CYP3A4. Potent inhibitors of CYP3A4 may increase the risk of myopathy by increasing the plasma concentration of HMG-CoA reductase inhibitory activity during lovastatin therapy. These inhibitors include cyclosporine, itraconazole, ketoconazole, erythromycin, clarithromycin, HIV protease inhibitors, nefazodone, and large quantities of grapefruit juice (greater than 1 quart daily).

Grapefruit juice contains one or more components that inhibit CYP3A4 and can increase the plasma concentrations of drugs metabolized by CYP3A4. Large quantities of grapefruit juice (greater than 1 quart daily) significantly increase the serum concentrations of lovastatin and its (beta)- hydroxyacid metabolite during lovastatin therapy and should be avoided.

Nifedipine (Procardia)

Clinical pharmacology

Coadministration of nifedipine and grapefruit juice resulted in an approximately twofold increase in nifedipine area under the curve (AUC) and Cmax with no change in half-life. The increased plasma concentrations are most likely due to the inhibition of CYP3A4-related first-pass metabolism.

Precautions/Other interactions

Grapefruit juice: Coadministration of nifedipine and grapefruit juice resulted in an approximately twofold increase in nifedipine AUC and C_{max} with no change in half-life. The increased plasma concentrations are most likely due to the inhibition of CYP3A4-related first-pass metabolism. Coadministration of nifedipine and grapefruit juice is to be avoided.

Dosage and administration

Coadministration of nifedipine and grapefruit juice is to be avoided.

Aripiprazole (Abilify)

Precautions/Drug interactions

Other strong inhibitors of CYP3A4 (itraconazole) would be expected to have similar effects and need similar dose reductions; weaker inhibitors (erythromycin and grapefruit juice) have not been studied.

Plausibility of Drug-Grapefruit Juice Interaction

When assessing cases identified in AERS, the plausibility of the drug–grapefruit juice interaction is also considered, based on the current knowledge of the mechanism of this interaction. Drugs that are likely to interact with grapefruit juice would have to be given orally, be a substrate for metabolism by CYP3A4, and have a relatively low bioavailability due to extensive presystemic extraction (first-pass metabolism) by CYP3A4.

The next step would be to compare the drugs identified in the AERS search with current labeling to check for potential drug–grapefruit juice interactions meriting further investigation and the possible need for labeling updates. One of the drugs is no longer marketed (astemizole). The following list categorizes the remaining drugs identified in the AERS cases in relation to product labeling.

1. Appropriate information on grapefruit juice interaction appears in product labeling: estrogens, lovastatin, nifedipine, and verapamil.
2. No information in product labeling, but low plausibility of grapefruit juice interaction: bupropion, doxazosin, gabapentin, lisinopril, risperidone, sertraline, and sibutramine.
3. Product labeling indicates lack of grapefruit juice interaction, but plausibility of interaction: amlodipine.
4. No information in product labeling, but plausibility of grapefruit interaction: atorvastatin.

This screening process indicated that the majority of drugs identified in the AERS cases contain appropriate information in their product labeling with the exception of the two drugs in categories 3 and 4. Other available resources, such as the literature and drug interaction studies submitted by manufacturers, will be sought to evaluate the need for labeling revisions for these two products. This process illustrates how the FDA continuously works with drug manufacturers to determine which drug products need revisions to their labeling to reflect accurate and clinically relevant information on drug–grapefruit juice interactions. This process has resulted in a number of changes to drug product labeling addressing grapefruit juice interaction in recent years.

Drugs that are already labeled for grapefruit juice interaction can also be reviewed to evaluate labeling for similar products, such as those that have a similar pharmacokinetic profile and are in the same drug class. For example, current labeling for tadalafil states that it is likely that grapefruit juice would increase tadalafil exposure. Two other drugs in this class, vardenafil and sildenafil, also appear to be sensitive substrates of CYP3A. This is suggested by high (10–49-fold) increases in AUC for these two drugs compared to a 2.2-fold increase with tadalafil after coadministration of strong CYP3A inhibitors such as ketoconazole and ritonavir. In addition, both have low oral bioavailability. There was one case of sildenafil–grapefruit juice interaction among the original 186 AERS cases that did not meet the specific criteria described above; however, the report was suggestive of an interaction. The FDA is evaluating the labeling for these two products to ensure that proper grapefruit interaction information is included.

A 52-year-old male with impotence experienced hypotension several hours after taking sildenafil 100 mg. The patient took sildenafil with grapefruit juice. He was also taking a mixture of dietary flavinoids and triamterene/hydrochlorothiazide. The reporting pharmacist did not state if the patient took sildenafil alone without problems before this episode. Spontaneous case reports have been an important addition to literature reports and manufacturer-submitted clinical studies to help the FDA continuously evaluate product labeling for drug–grapefruit interaction.

13

PHARMACEUTICAL PRODUCTS

Plants are amazing chemical factories! Provided only with the simplest and most inexpensive inputs of carbon dioxide, sunlight, water and minerals, plants produce thousands of sophisticated chemical molecules with different structures. A team of organic chemists starting with the same raw materials could never accomplish the same results in their lifetimes. Some of the chemicals that plants produce for their basic metabolic processes are the same ones found in all living organisms; amino acids, sugars, nucleic acids, and lipids are very similar or identical throughout the plant, animal, and microbial worlds. However, plants differ from other organisms in the vast diversity of additional products they produce. At least 50,000 different chemical structures have been characterized so far within the plant kingdom, and even the majority of plant species have not yet been closely analyzed! A single plant species produces only a fraction of this number of chemicals, but there are hundreds of thousands of different plant species, many of which produce chemicals that are unique to each one.

These so-called secondary metabolites apparently are not essential to the life of the cell, and are often produced only in certain cell types or at certain times. They can be classified into broad molecular types based on their distinguishing structural features. Why do plants produce these thousands of different chemical structures? For the vast majority of secondary products their specific function is unknown. In many cases, scientists have good evidence that they act as defense molecules or as deterrents against feeding by animals or microorganisms. Another possibility is that some compounds have no real function and represent the results of randomness in the evolutionary divergence of traits that are neutral for survival. There may be little or no negative impact on evolutionary fitness if a plant species uses some of its carbon and energy to produce additional structures.

Table 13.1. Diversity of chemical structures produced by plants

Molecular Class	*Number of Known Structures*	*Examples*
Isoprenoids	>25,000	Menthol, turpentine, rubber
Alkaloids	>12,000	Caffeine, nicotine
Phenolics	>8,000	Vanillin, anthocyanins
Fatty acids and derivatives	>500	Castor oil, urushiol (poison ivy toxin)
Miscellaneous	>5,000	Sorbitol, guar gum

In many cases, plants produce only small amounts of a secondary metabolite, but these metabolites can have extremely potent biological activities. Products that represent less than 1 % of the weight of a leaf or root can make the organ toxic to a potential grazer. In other cases, plants produce secondary metabolites that accumulate to very high levels often in specific cell types. The plant secretory glands found in specialized epidermal hairs of species such as mint represent a cell type able to produce large amounts of relatively pure organic compounds (such as menthol). Interestingly, epidermal hairs of other plant species also represent some amazing chemical factories. For example, hairs of some tomato species secrete sticky glucose fatty-acid esters that can represent 25% of the leaf dry weight, and cotton fibers represent specialized cellulose fiber factories that derive from epidermal hairs.

Harvesting Biochemical Diversity: Can (Green) Plants Replace (Chemical) Plants?

One of the major goals of 21st-century society will be to develop renewable and sustainable resources to replace limited petrochemical reserves. This goal is schematically illustrated and the following discussion outlines some factors that will affect the development of a more bio-based chemical industry.

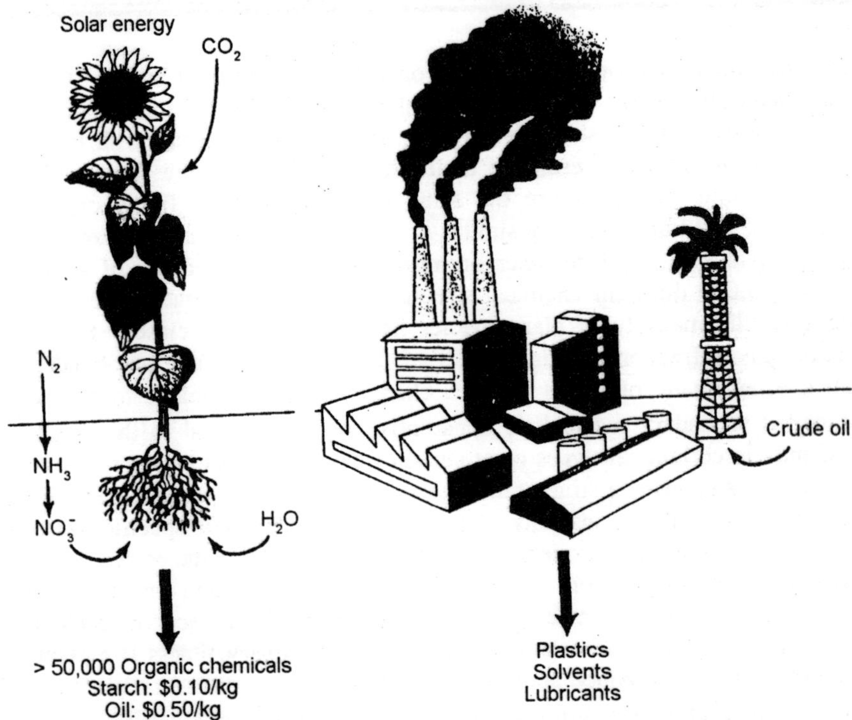

Fig. 13.1. Can plants replace plants? In green plants the inputs are carbon dioxide and solar energy, in chemical plants the input is petroleum.

Today people think of agriculture largely as a source of food for human consumption and feed for animals. There are of course many other uses for crops, such as the production of cotton, linen, or jojoba oil, but these represent only a small fraction of overall agricultural production. In the United States, nonfood and nonfeed uses of field crops account for only 10 to 20% of the total value of field crops. Thus, despite the amazing potential of plants, the agricultural products that humans now use are much more limited in chemical variety and in human applications. Seeds, harvested for their food value, constitute the economic value of most crops. Therefore, agricultural crop production largely takes the form of the starch, proteins, and oils that are the major constituents of field crop seeds.

For most of human civilization, plants provided a much wider range of products. Many ancient civilizations had highly evolved uses of plants to provide medicines, dyes, and other chemicals. One hundred years ago, a major proportion of human clothing, fuel, dyes, medicines, construction materials, and industrial chemicals was still derived from plants, and these products represented very substantial aspects of the agricultural economy. This situation changed dramatically over the past 50 to 100 years. Synthetic rubber and synthetic fibers for clothing (nylon, polyesters, and so on) are just two examples of petroleum derived products that were developed to replace products once obtained exclusively from agriculture. Over a 30-year period—between 1930 and 1960—the sources of industrial chemicals changed from coal and plants as the dominant raw materials to petroleum as the new feedstock. As of 2000, over 95% of organic chemicals that society uses are derived from petroleum, not from plants.

Why has the chemical industry switched to petroleum as the almost exclusive source of starting raw materials? Two major factors brought this about. First, the development of the automobile, which led to enormous demand for inexpensive liquid fuel, coincided with the discovery of vast petroleum reserves. Second, the chemical industry invented new processes that permitted the inexpensive conversion of oil into more valuable chemicals that replaced many products that society had long derived from agriculture. Furthermore, during the time the chemical industry was rapidly expanding, petroleum was a much less expensive raw material than agricultural products. For example, in the 1940s and 1950s, a kilogram of maize cost four to five times more than a kilo of crude oil. The development of the chemical industry, and in particular the plastics industry, is an amazing scientific and industrial success story that has led to the creation of an industry that produces US$ 110 billion in products per year in the United States, compared to US$ 90-100 billion for major U.S. field crops.

Technological advances and economic factors during the 20th century led people to replace many nonfood agricultural products with petroleum-derived products. Many believe this trend will be partially reversed in the 21st century by genetic engineering of plants and the continuing decline in the price of agricultural products. Over the past 50 years agricultural products have become less expensive because of the increased productivity of crops and decreased labor input, whereas the price of crude oil, even if one ignores the sharp peak in the 1980s, has been steadily rising. Relatively speaking, plants are a bargain. Additional factors that could bring about a switch from oil to plants may be the environmental, political, and social benefits that could accompany such a changeover. Petroleum is a limited commodity produced in a relatively small number of countries, and production at present levels cannot be sustained. Therefore, its cost is expected to continue to rise in the future and known supplies will likely be exhausted within the next 75-100 years, unless automobiles use a different source of energy. In contrast, over the past 50 years, agricultural production has continued to increase slightly faster than demand and crop prices have generally declined. Whether this trend continues, it seems almost certain that for the next 50 years or more agricultural raw materials will be less expensive than petroleum, a very different situation from when the chemical industry began its major expansion.

Table 13.2. Approximate cost of phyto- versus petrochemicals

Phytochemical	*US$/kg*	*Petrochemical*	*US$/kg*
Corn	0.10	Crude oil	0.12
Starch	0.11	Gasoline	0.2
Vegetable oil	0.55	Benzene	0.5
Glucose	0.22	Plastic	1.0
Ethanol	0.30		

If plant material is now less expensive than petroleum, why has the chemical industry not switched back to plants? A major reason is that the industry has over 50 years of experience using petroleum as its raw material and has invested huge sums in chemical plants that convert fossil hydrocarbons into plastics and other consumer goods. Additional comparisons of the advantages and disadvantages of fossil fuels versus plants as sources of raw material for the chemical industry are shown in Table 13.3.

Table 13.3 Industrial chemicals from fossil versus plant sources

Petroleum-based raw material	*Plant-based row material*
50-100 years of chemical experience in industry	Less chemical experience
Raw material is imported, and cost have increased	Raw material is domestic, and cost have declined
Petrochemical industry is major source of pollution	Less pollution
Chemical diversity is limited: Primarily highly reduced carbon	Chemical diversity is almost unlimited

Gene Manipulation will let Crop Plants Produce Specialty Chemicals and Permit Production of Biologically Based Plastics

The surplus capacity for agricultural production in industrialized countries and the innovation of plant genetic engineering are leading to new ideas about how people might use agriculture to benefit society. A recent coalition of government, industry, and academic groups suggested that the United States should attempt to use biological sources such as crops to supply 10% of industrial chemical needs by the year 2020 and to increase this to 50% by 2050. Can these ambitious goals be achieved? Ultimately, agriculture will produce almost all biological resources, and plant genetic engineering will be an essential tool for increasing use of renewable resources. An underlying concept of plant genetic engineering is that limitations to large-scale production and use of valuable plant chemicals can be overcome by transferring the genes for their biosynthesis from wild plants into crop species. In this way, much previously untapped chemical biodiversity can be harnessed for society. What are the prospects that biotechnology can create new uses for crops, and could this affect the cost of food?

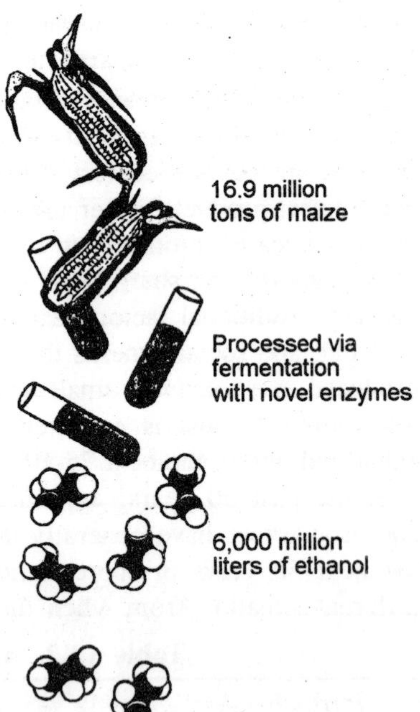

Fig. 13.2. Conversion of maize to ethanol by large-scale fermentation.

Currently, the major process for producing chemicals biologically is microbial fermentation. Worldwide, approximately 15 million tons of products such as ethanol, citric acid, lactic acid, and others are now produced by fungi or bacteria grown in huge 10,000-liter (or larger) stainless steel fermentors. The energy (carbon) source for growing the microorganisms in these fermentors is almost always derived from plant material, usually from maize starch. The microbes are always very highly selected or have been genetically engineered so that they efficiently convert feedstock (glucose) into end product (ethanol or other). Maize is the number one U.S. field crop, and the ability to produce maize starch at a price of less than 15 cents per kg has led to many uses of starch other than as food. Roughly 8% of the U.S. maize crop, or 14 million tons, is used in fermentation processes. Approximately 1 billion U.S gallons of ethanol are produced in this way for automobile fuel (compared to 130 billion gallons of gasoline used per year in the United States). By increasing demand for maize, production of fuel ethanol raises the price of maize (about 1 cent per kg of grain) and helps provide additional markets for excess U.S. agricultural capacity. Improvements over the past 10 years in ethanol yields and recovery have converted the overall ethanol production process from a net energy consumer to a positive energy balance. However, ethanol's ability to compete with low-cost petroleum still requires tax advantages to be practical.

The production of millions of tons of ethanol from maize has demonstrated the feasibility of large-scale conversion of crops into new products. In addition to ethanol, some new fermentation products are now becoming available; this will lead to higher-value products from agriculture that can fully compete with petroleum. Large-scale production facilities have recently been established for two precursor molecules that can be polymerized to make plastics. A plastic with the trade name NatureWorks is produced by CargillDow Polymers. The NatureWorks process is based on the fact that bacteria that are fed glucose derived from starch can produce very high yields of lactic acid. After recovery from the fermentor, technicians polymerize the lactic acid to polylactic acid (PLA), a biodegradable plastic with many attractive properties for products such as carpet fibers and thin films. By using a biologically produced feedstock (glucose), PLA uses 30 to 50% less fossil fuel than is required to produce conventional plastic resins. In 2001, a large-scale facility began production of 150 thousand tons per year. Analysts expect that up to 1 million tons of this plastic may be produced in 2004 at a cost comparable to most petroleum-derived plastics.

For another example, fermentation of glucose can produce 1-3 propane diol, which after combining with terephthalic acid can be polymerized to produce a plastic useful for fiber production. DuPont will market this bio-based plastic under the name Sorona. One key to economic production of both of these bio-based plastics has been genetic engineering of bacterial strains to achieve very high yield of the precursors for polymerization. In the case of PLA, almost 70% of the carbon from the plant glucose feedstock ends up in the final plastic.

An excellent application of PLA plastic films would be as a replacement for the polyethylene films now used to solarize soil for vegetable production. Covering the soil with plastic film reduces water evaporation, prevents weeds from growing, and, most importantly, kills harmful nematodes that accumulate when the same land is used year after year. This new method to prevent damage from nematodes could replace chemical fumigation of soils with methyl bromide, a potent pesticide that kills harmful and beneficial insects alike. If these films were made from a biodegradable plastic, the farmer could simply plow them under at the end of the growing season while preparing the new planting bed.

Chemical Production within a Plant may Offer Economic Advantages over Fermentation

Although fermentation can produce some chemicals at a relatively low cost, the expense of building and maintaining large fermentors increases the cost of the products. For this reason, most products sell for several-fold higher prices than starch or glucose. An ideal process would produce the desired chemical in the plant, where the major energy input is sunlight, and would require only minimal further processing after extraction from the plant. The production of the biodegradable plastic polyhydroxyalkanoate (PHA) in plants represents one effort to reach this goal. Many species of bacteria synthesize and accumulate biodegradable plastics called *polyhydroxyalkanoates* at levels up to 80% of their dry weight. One such PHA, called *polyhydroxybutyrate* (PHB), is produced commercially by fermentation with the bacterium *Alcaligenes eutrophus*. PHB combines biodegradability with water resistance, making it a suitable polymer for a wide variety of uses. The major drawback of bacterial PHB is its high production cost, making it substantially more expensive than synthetic plastics and thereby restricting its large-scale use. Could large-scale production of such plastics in plants rather than in bacteria lower their cost? In 1992, scientists at Michigan State University demonstrated that expression in the plant *Arabidopsis* of genes from the bacterium *Alcaligenes eutrophus* could lead to production of the plastic polyhydroxybutyrate (PHB) in plants. This landmark advance demonstrated that plants can be engineered to produce entirely new chemical products using genes from distant organisms. However, growth of the plants was stunted. In a 1994 study, three genes from this bacterium were engineered so that the proteins they encode would end up in the chloroplasts. The genetically engineered *Arabidopsis* plants produced PHB inclusions exclusively in their chloroplasts. The amount of PHB detected in fully expanded leaves of the plants was up to 10 mg/g fresh weight (about 14% of dry weight). No significant deleterious effect on growth or seed yield was detected in these plants. In response to these discoveries, industry has begun to explore production in plants of polyhydroxyalkanoate polymers that have physical properties superior to those of PHB.

Extraction, Purification, and Energy Costs can Greatly Influence the Success of Plant-produced Chemicals

Although at first it seemed that producing plastics directly in plants could offer the best approach toward a renewable, "green," chemical industry, many factors determine the balance of costs and benefits. To recover plastic from plants will require new extraction facilities that use organic solvents to extract the plastic. Calculations of the energy inputs needed to grow and harvest the crops, to operate the extraction facilities, and to recover the solvents suggest that the amount of fossil fuel consumed may actually be higher than if the same plastic were produced from petroleum. Such calculations require a number of assumptions that depend on future energy costs. If the energy used to run the extractions were derived from plant material, such as the crop residues after the plastic is extracted, the balance could again swing in favour of the plant-based production system. These considerations emphasize the complexities in judging the balance of costs and benefits for each chemical production system.

Plastic production via fermentation offers one useful comparison of how extraction and purification costs impact the cost of two biodegradable plastics produced biologically. PHA plastic (biopol) produced in bacteria at a level of 80% of their dry weight costs US$4- 5/kg. In contrast, polylactic acid (PLA) plastic is produced by chemical polymerization from lactic acid that is first produced by bacteria.

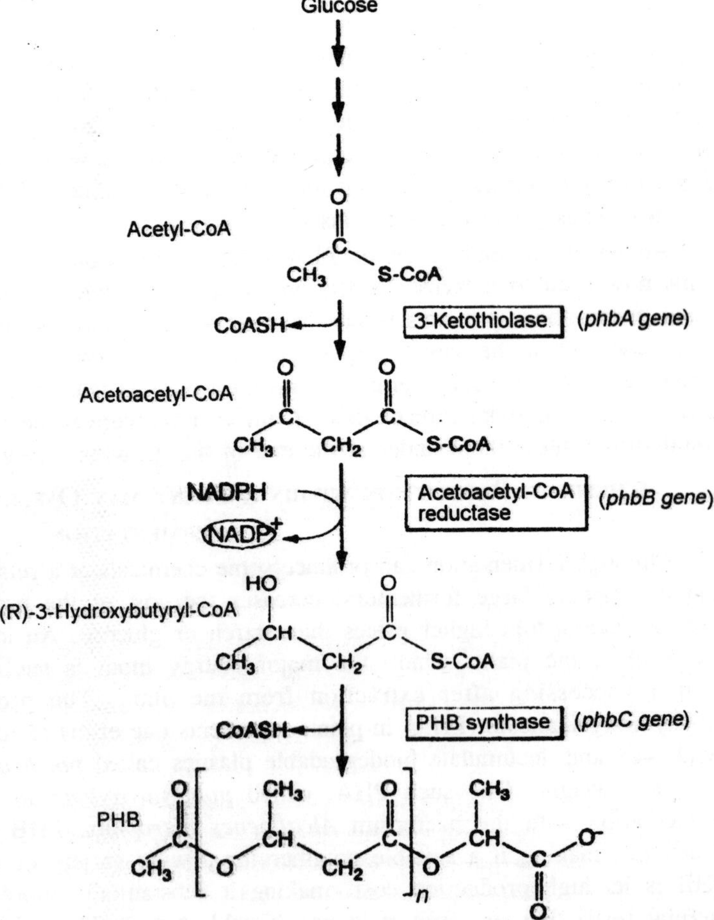

Fig. 13.3. Pathway of polyhydroxybutyrate (PHB) synthesis in chloroplasts of genetically modified plants.

PLA plastic can be produced for US$1-2/kg. This substantial price difference is in large part caused by the extraction and purification costs associated with recovering PHA polymer from bacterial cells, whereas the precursor of PLA is soluble and can be recovered from the fermentor at low cost. Thus, although all steps of PHA synthesis occur biologically, the extraction/recovery costs raise the final cost of the product above that of PLA, which requires two biological production steps (in plants and in bacteria), followed by a chemical polymerization. In many cases, producing a monomer or precursor in plants that is easily recovered, rather than producing a complex polymer, will lower overall costs. If the desired product can be recovered using existing plant-processing technology (such as oil extraction or wet milling), the economics of chemical production in plants may be even more favourable.

Plant Oils can be Engineered for New Industrial Uses

Different plant species use different polymers, usually oil or starch, to store energy needed for seedling growth. In oilseeds, up to 50% of the seed dry weight is in the form of oil. Some plants also produce large amounts of oil in other organs. For example, olive, avocado, and oil palm fruits contain high levels of oil. Here oil probably serves as an animal attractant, to aid in seed dispersal. Plant oils are triacylglycerols, also called *triglycerides* and consist of three fatty acids attached to glycerol that accumulate in discrete subcellular organelles called oil *bodies*. Oil bodies are droplets of vegetable oil surrounded by a monolayer of phospholipids.

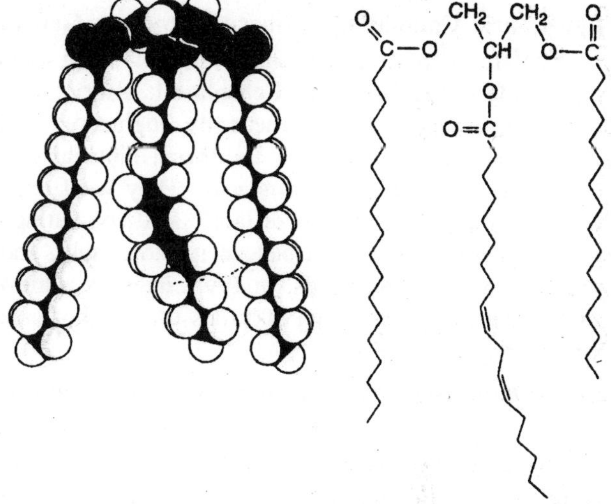

Fig. 13.4. Space-filling and conformational models of the structure of triacylglycerol, the major form of carbon and energy storage in oilseeds.

Vegetable oils are a major commodity, with a well-developed commercial infrastructure for production and use, supplying both food and industrial needs. World vegetable oil production, from soybean, palm, rapeseed (canola), and other crops is over 100 million metric tons per year and is valued at approximately US$50 billion in annual oil sales. The primary use for vegetable oils today is in the food industry and includes salad oils, margarine, and oils used in frying and baking. In most plants, the same five or six fatty acid structures that are also found in the phospholipids in the cell membranes occur in the triacylglycerols found in seeds. These 16- and 18-carbon fatty acids (primarily palmitic, oleic, and linoleic acids) are the major constituents of the vegetable oils that are consumed as foods.

In addition to food uses, about 30% of vegetable oils that are produced today are used by the oleo chemical industry for hundreds of products such as soaps, detergents, paints, lubricants, and polymers. In many cases, the fatty acid composition of the oils used for these applications differs from that found in edible oils, and these different structures lead to special applications. For example, the tropical oils from coconut and palm kernel are rich in lauric acid, which is a 12carbon saturated fatty acid. The properties of lauric acid lead to a balanced solubility in water and oil, which makes it ideal for production of soaps and detergents. As a result, the United States imports up to US$400 million of these tropical oils for use largely in soap and detergents. Thus, although these oils are edible, their major use is not for food.

Although most plant species store oils with the same five or six fatty acid structures in their seeds, analysis of seeds from more than 10,000 different plant species has revealed several hundred fatty acid structures that are considered "unusual." The chain lengths vary from 8 to 24 carbons and the number of unsaturated double bonds varies from zero to 4, or more. In addition, hydroxy, epoxy, acetylenic, cyclic, and conjugated unsaturated groups occur. As with lauric acid, the unique properties of these fatty acids have led to a number of industrial applications.

For over 100 years, chemists have explored plant fatty acid structures as interesting alternatives to petroleum. Thus, the chemical industry has a rich knowledge about the properties and chemical potential of plant oils. However, in most cases these unusual fatty acids are not available in large quantities and at a low enough cost to compete with petroleum alternatives in large-scale uses. The desirable fatty acid structures are generally produced only in wild plant species that produce low yields if planted in fields. Although efforts to domesticate some of these species are under way, this may take decades and may meet insurmountable barriers. Plant gene technology offers the potential to move genes that control production of these high-value fatty acids into high-yielding and well-developed oilseed crops such as canola.

High-Lauric Canola:Oil A Success Story in Genetic Engineering of Oils

The first commercial product to result from changing the composition of a plant seed via genetic engineering is high-lauric-acid canola oil. Lauric acid is found at a very high level in tropical oils, but could temperate crops be genetically engineered to produce the same high level of this short fatty acid? Scientists at Calgene, a California biotechnology company, discovered the biochemical pathway responsible for lauric acid synthesis. They used extracts of seeds from the California bay tree; these seeds accumulate high levels of lauric acid-containing oils. The scientists cloned the gene for the critical enzyme in the pathway, introduced it into canola, and dramatically changed the spectrum of fatty acids produced in the canola seeds. In 1995, industry achieved the first commercial production of a genetically engineered oil by extracting 500 tons of oil from canola seeds engineered to produce an oil with 40 to 50% lauric acid.

High-Oleic Soybean Oil has been Genetically Engineered with Improvements for both Food and Nonfood Uses

Soybeans are the largest source of vegetable oils in the world, and in the United States soybean oil accounts for about 70% of vegetable oil consumed. Most soybean varieties produce an oil rich in polyunsaturated fatty acids (about 50% linoleic acid or 18:2 and 10% linolenic acid or 18:3), and these fatty acids make the oil unstable and easily oxidized so that it becomes rancid. When heated, the oil develops objectionable flavors and odors. Thus unprocessed soybean oil is unsuitable for many applications, so it is chemically hydrogenated. This process adds to the cost of the oil and also introduces side reactions such as conversion of double bonds from the *cis* to *trans* configuration creating *trans*-fatty acids.

The biosynthesis of polyunsaturated fatty acids in plants is catalyzed by a series of enzymes; in the first step, an enzyme converts oleic acid (18:1) to linoleic acid (18:2). After the gene for this enzyme was isolated, molecular biologists succeeded in suppressing the expression of the gene in soybean. This decrease of the 18:1 fatty acid to 18:2 conversion step almost completely eliminated polyunsaturated fatty acids in the soybean oil.

The new transgenic soybean oil has 85% oleic acid, one of the highest oleic acid contents found in nature. The absence of polyunsaturated fatty acids eliminates the need for hydrogenation to stabilize the oil. Furthermore, an unanticipated benefit of the oleic increase was that the saturated fatty acid content of the oil fell from approximately 15% to less than 8%. The new soybean oil has a composition similar to olive and other high-oleic oils, which are considered to provide greater health benefits, compared to other plant and animal oils. The fatty acid trait was stable in field trials, and the oil yield of the crop was identical to the control lines. Thus, neither the transformation process nor the major change in fatty acid composition was detrimental to the high yield of the soybean line. This example is also instructive because it demonstrates how quickly some discoveries can be translated into new crops. With the resources of a major corporation, genetic engineers needed only five years from gene isolation to a field-tested transgenic soybean crop ready for commercialization of an industrial product. Marketing a food product would require additional safety tests mandated by U.S. regulatory agencies.

Vegetable oils have long been known to have useful properties as lubricants, and because they are biodegradable are ideal for applications where harm to the environment must be avoided. However, the tendency of the oils to break down or polymerize as a result of oxidation limits their use. With its very low polyunsaturated fatty acids, high-oleic soybean oil has an oxidative stability more than 10 times greater than most vegetable oils. As a result, it can substitute for mineral oil in many applications such as marine engines, chain saw lubricants, and other applications where oil spills are particularly damaging. A number of other industrial applications may also become possible, because chemical additions to the double bond can lead to polymers and other products that have desirable properties for certain plastics.

The Challenge of High-Level Production

In the two cases of high-lauric acid canola oil and high-oleic acid soybean oil, manipulation of only one gene produced a new commercial product in transgenic plants. However, these two cases may not be representative of the types of metabolic engineering that will be needed for most new products. A number of other attempts to engineer oilseed fatty acid composition have been less successful, because the amount of the desired product was too low. Although the genes were isolated from plants or other organisms that accumulate the unusual fatty acids at levels of 50 to 90% in their oils, when engineers transformed the genes into a crop plant, accumulation was much less. The activity of the introduced enzyme has generally not been limiting, so other factors probably limit product accumulation. In at least one case, the new fatty acid induces its own breakdown. Thus, it is clear that plant metabolic engineering will be technically challenging.

Table 13.4. The challenge of unusual fatty acid production in transgenic oilseeds. Specialty fatty acids are found at high levels in nature, but are produced at low levels in transgenic plants

Fatty Acid	*Level in Native plant*	*Level in Transgenic plant*
18: 1^{D6} (petroselinic)	85%	<10%
16:1^{D6}	85%	<10%
Cyclopropane	50%	<5%
Ricinoleic	90%	17%
Acetylenic	70%	25%
Epoxy	60%	15%
Conjugated	65%	17%
Lauric (+ 10:0)	65%(+25%)	50-60%

Furthermore, even if high-level production of a chemical is achieved, the costs of producing a new product in plants may be higher than production by existing methods. Such costs fall into several categories: First, the costs of crop breeding approximately doubles for each independently segregating gene that must be maintained in the breeding population. Second, any yield penalty associated with the transgene will add to the cost of the end product. Third, identity preservation of a GM crop adds to storage, transport, and processing costs. And fourth, perhaps most importantly, the cost of special extraction or processing of a new product can substantially increase the price of an end product. Plant metabolic engineering will be most successful if industry can recover the products at low cost and high yield.

Potential Impact of Large-scale Chemical Production in Temperate and Tropical Crops

If biotechnology can overcome the problems just discussed and many new chemicals are produced in plants, what will be the impact on agriculture? It could be huge, because of the area of land needed. Producing a chemical such as ricinoleic acid in a crop, rather than importing it, could be easily accommodated in the United States with little impact on land use or commodity prices. In contrast, producing the bulk of adipic acid for nylon manufacture could require 10 million hectares of canola. In comparison, only 2 million hectares of maize are needed to produce 1 billion gallons of ethanol. Large-scale production in plants of plastics or their monomers could greatly exceed the use of crops for ethanol production.

At present, genetic engineering of oilseeds is largely confined to temperate crops such as canola and soybean. The immediate goal is to expand the range of fatty acids available from crop species so that the commercial uses of plant fatty acids can be expanded. In the future, genetic engineering of perennial tropical species such as oil palm will increase in importance. Oil palm, which can grow year round, is capable of producing up to 100 barrels of oil per hectare per year. This production capacity is 3- to 10-fold higher than yields obtained from most annual temperate crops.

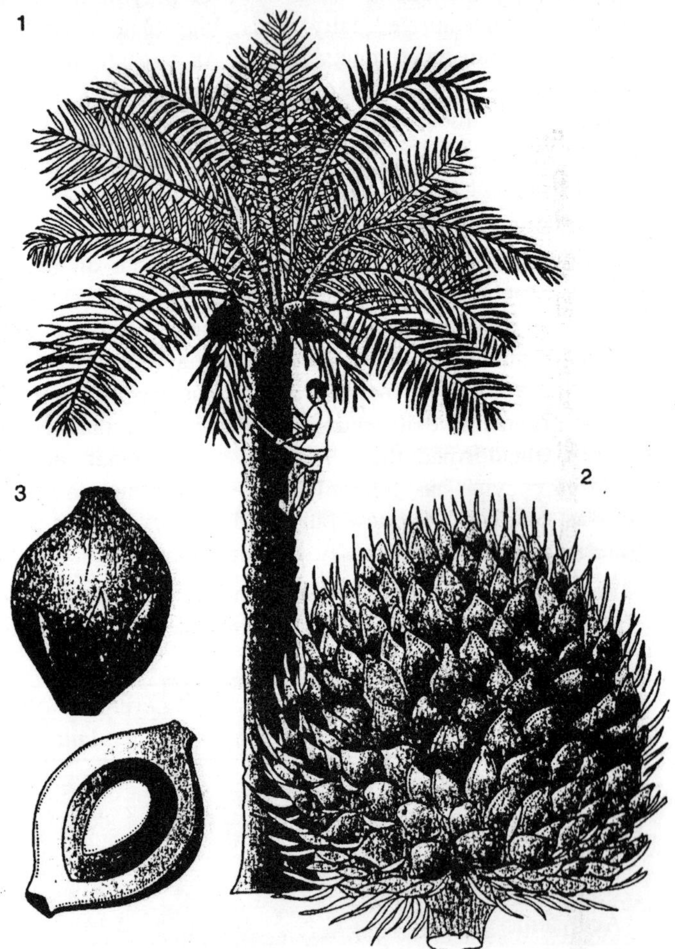

Fig. 13.5. Oil palm produces 3- to 10-fold more vegetable oil per hectare than oilseed crops grown in temperate climates.

Palm oil is currently the second largest type of vegetable oil worldwide, mostly coming from Malaysia, Indonesia, and the Philippines. However, in a number of tropical areas in Africa and South America oil palm production could be expanded. As genetic engineering of oil palm advances, this tree may produce a wide range of chemicals at a cost competitive with petroleum. Because oil palm can be grown in several less developed countries, such expansion could have important social benefits by bringing new agricultural income and exports to these areas. However, the governments will have to weigh the benefits of such cash crops against the need for food production.

WILL "PLANTS AS FACTORIES" BIOTECHNOLOGY HURT THE ECONOMIES OF DEVELOPING COUNTRIES?

Experts such as the Kenyan economist Calestous Juma warn that transferring production of high-value plant metabolites from farms in developing countries to biotech crops in developed countries will have a negative impact on the economies of the developing countries. They will lose an important source of foreign exchange, and many farmers will lose their livelihoods. In his book *The Gene Hunters: Biotechnology and the Scramble for Seeds* (1989, Princeton, N: Princeton University), Juma writes,

"The direction of this research is aimed at reducing dependence on imported raw materials. The impact on the countries exporting this product will be profound and irreversible. Over the years, large sections of the population have organized their lifestyles around the production of these crops. This is going to be changed by current developments in biotechnology. The impacts will not be only economic, but will have long-run political implications and the attempt by communities to reorganize themselves in response to the changed conditions. It is, therefore, in the interest of raw material exporters to closely monitor current trends in biotechnology and the use of genetic resources and modify their internal policies in anticipation of potential long-term effects."

One possible impact of biotechnology is illustrated by the present production of vanilla. Although the plant *Vanilla planifolia* is indigenous to Central and South America, commercial production of vanilla occurs largely in Madagascar, the Comoro Islands, and Reunion. Madagascar alone accounts for 75% of the world market (about US$80 million) and earns more than US$50 million in foreign exchange. Vanilla beans account for 10% of Madagascar's export earnings. The production of vanilla involves over 70,000 small landholders (owners of small farms) whose incomes depend on vanilla exports. The current price of vanilla is about US$70-75 per kg. By the two criteria mentioned in the previous section—price per kilogram and size of the market—vanilla is presently not threatened by genetically engineered crops. Yet research to try to change this is tempting, because the structure of vanilla is simple and may require modification of only one or two genes to produce vanilla in seeds.

Another example of the potential relocation of production from developing countries to developed countries is provided by the newly discovered sweet proteins such as thaumatin. Certain tropical fruits and leaves of tropical shrubs contain small proteins that are 1,500-3,000 times sweeter (based on weight) than sucrose. Some companies have established plantations in the tropics to produce these proteins. They can be used as sugar substitutes and advertised as "natural" sweeteners that are possibly safer than synthetic products such as saccharin and aspartame. Because they are proteins, each is encoded by a single gene, and such genes can easily be isolated and introduced into bacteria, yeast, or plants, which will then produce the sweet protein. Work on the genes of sweet proteins has attracted capital from some of the world's biggest companies, indicating the importance they attach to this emerging technology. Tate and Lyle, the British company that established plantations in Liberia, Ghana, and Malaysia to produce thaumatin, has also introduced the thaumatin gene into yeast by genetic engineering, allowing the company to manufacture the sweetener in large fermentors. The cheaper process will eventually displace the more expensive one.

It is important to note that the relocation of industries and production processes from one country to another has been going on for hundreds of years. The development of new chemical technology for producing synthetic rubber had an immense impact on the economics of rubber tree plantations in countries such as Malaysia. In some cases, the economy of a country can accommodate such changes, particularly if soil and climate conditions are favourable. In Malaysia, oil palm plantations replaced many of the rubber tree plantations and the agricultural economy remained strong. Many countries that were industrial leaders in the 20th century have seen their manufacturing industries move production facilities to locations with the lowest cost of labor and raw materials. Many products (such as electronics) once primarily produced in North America and Europe are now predominantly produced in (other) Pacific Rim countries. Thus the biotechnology industry is not unique or particularly to blame for long-standing trends where economic and political forces, rather than social considerations, drive relocation of production systems. Although in the short term the most developed countries may be the first to benefit from agricultural biotechnology, in the long term all countries with good climate, soil, and water will derive increased income and economic growth from the production of chemicals in plants.

Starch and Other Plant Carbohydrates have a Wide Range of Industrial Uses

Maize, wheat, rice, and other grains consist of 50 to 70% starch by dry weight. Potato and some other root crops are even richer in starch. Starch can be recovered at very low cost and in high purity by wet milling of grains. In this process maize is soaked 30 to 40 hours in water and then starch, gluten (protein), fiber, and the germ are physically separated. Very large-scale processing plants can recover starch at a price of less than 20 cents per kg. The other parts of the grain can be used as food or animal feed. This price for a complex organic polymer is substantially lower than for plastic polymers derived from petroleum (such as nylon or polystyrene), and the price comparison emphasizes the inherent efficiency of photosynthesis and large-scale agriculture. This low price has stimulated decades of research to find new uses for starch. For example, in the food industry starch is used as a thickening or gelling agent. The conversion of starch to high-fructose syrup is another important application. In the United States, more of this sweetener is now used as compared to sucrose (sugar beet or cane sugar), primarily because it is cheaper to produce. Starches are also used in hundreds of nonfood applications such as paper manufacture and adhesives and have been incorporated into plastic films to increase their breakdown in landfills. Chemical modifications of starch have led to dozens of other uses, such as absorbents. For example, Superslurper is a starch/acrylonitrile copolymer capable of absorbing 1,500 times its own weight in water and is used in diapers, bandages, and seed coatings. However, the major industrial use of starch has been for fermentation to produce ethanol and other chemicals described earlier.

Starch is a mixture of two glucose polymers: amylose, which is linear, and amylopectin, which is highly branched. The different properties of these two polymers and their relative proportions in starch determine the physical properties and uses of starch. A relatively small number of enzymes are involved in producing starch from glucose, and the relative activities of these enzymes determine the ratio between amylose and amylopectin. Genetic engineering of these enzymes might allow starches to be designed for specific end uses. However, almost all mutants in starch biosynthetic enzymes, although often providing useful starch properties, also lead to reductions in overall starch content and crop yields. Therefore, to date, genetic engineering of starch composition has not reached commercial success.

A number of bacterial and fungal carbohydrate polymers have important applications, and if plants could produce these in high yield the costs of production might be less than from fermentation. For example, xanthan gum—a glucose polymer produced by the bacterium *Xanthomonas campestris*—is used as a thickener and for other food and nonfood applications. Each year 20 million kg of xanthan gum are produced, with a price of US$6 to US$8/kg. This market size is sufficient to stimulate efforts to engineer plants to produce xanthan gum rather than starch.

Specialty Chemicals and Pharmaceuticals can be Produced in Plants

Most chemicals and products discussed so far can be considered "commodities." Such products are characterized by large market volume and low price. Both the agriculture and the petrochemical industries have excelled at producing low-cost commodity products and in some cases these two sectors compete with each other to capture markets. In such raw material markets, profit margins are small, and manufacturers often choose one supply source over another based on a price difference of 1 % or less. Specialty chemicals represent a different situation. These chemicals are more expensive to produce and almost always have a smaller volume of sales. Many thousands of chemicals fall into this category. For example, the Aldrich Chemical Company lists over 25,000 chemicals in its catalog and collectively these specialty chemicals have annual sales of many billions of dollars. Many flavors and fragrances, both synthetic and biologically produced, fall into this class.

What are the prospects that biotechnology will lead to alternative or improved plant sources for such compounds? Although scientists can envision schemes where plants are engineered to produce thousands of specialty chemicals, you have seen that metabolic engineering is not always as straightforward as expected. Despite much research in the past 10 years focused on new products, only a few processes have been commercialized. This is both because of complexities in engineering biochemical pathways and because of economic considerations. Economists estimate that research and development for a new product requires at least a la-year investment and that this investment can only be recovered if the product has an annual market of at least US$100 million. Only a small subset of new plant products falls into this category. For example, both jasmine and ajmalicine, components of perfume, are quite expensive (jasmine costs US$5,000 per kg and ajmalicine costs US$1,500 per kg), but the total market for jasmine is estimated to be only US$500,000 and for ajmalicine, US$5 million. Spearmint oil, in contrast, has a substantial market (US$100 million), but the relatively low price (US$30 per kg) means that new plant production systems must produce high yields. Currently, these considerations limit industry investments in research to products with very large market values. However, in the future, as more experience is gained in biotechnology, the process of rational plant metabolic engineering will become more cost effective and will be extended to a wider range of products.

Pharmaceuticals represent a type of specialty chemicals. As much as 25% of the pharmaceutical drugs that people use are either produced in plants or originated in plants. For example, morphine and codeine (and opium) are alkaloids derived from the opium poppy. Other legal and illegal plant-derived drugs consumed in large quantities include caffeine, nicotine, heroin, and tetrahydrocannabinol (from marijuana). Some medically important structures once isolated from plants are now produced synthetically, because synthesis provides a more reliable supply and sometimes at lower cost. However, some structures are so complex—taxol, for example—that chemical synthesis, although possible, is prohibitively expensive. However, in some cases, the complex structure makes chemical synthesis difficult. The anticancer drug, taxol, can be made from a precursor isolated from the European yew tree.

Can genetic engineering of plants result in increased yields of products such as taxol? In many cases, the biosynthetic pathways of these complex structures are still not completely understood. In addition, geneticists have not yet identified the genes that encode the enzymes that carry out the various synthetic steps. Over expression of the first enzyme in a pathway sometimes results in higher levels of the desired end product. However, even in some well-characterized pathways efforts to engineer higher flux (in both plants and microorganisms) have been disappointing. A promising strategy may be to discover global regulators, such as transcription factors that can regulate many steps in a biosynthetic pathway.

Fermentation and In-plant Production Systems are Complementary Technologies; both Depend on Agriculture

Microorganisms are often the preferred production system for producing small water-soluble molecules such as organic acids, because during fermentation the compounds are released into the growth medium, where they can be easily recovered. Citric acid, used in both food and nonfood applications is one example. In 1998, 200 million kg, with a value of US$500 million, were produced by fermentation in the United States. Although this is a large and attractive market, producing high concentrations of citric acid or other small molecules in a plant might be difficult, because if the compound accumulates to high levels in the cells the metabolic and osmotic consequences could be disastrous for the plant. Similarly, production of industrial enzymes now takes place largely by fungal or bacterial fermentation, with the enzymes being secreted into the medium. Tons of enzymes such as proteases, lipases, glucose isomerase, and pectinases can be produced at costs as low as US$3 to US$10 per kg. Subtilisin, a protease produced by bacteria, is added to laundry detergents, where it constitutes the largest single use of industrial enzymes. The detergent industry uses over 80 million kg of subtilisin worth US$500 million per year. At first glance it seems that large-scale production of these proteins in crops would be even cheaper than by fermentation. For example, soybean meal, which is 48% protein, sells for US$0.21 per kg and one hectare of soybeans can produce >1,000 kg of protein. If an industrial enzyme such as subtilisin could be produced at a level of 20% of the soy meal protein, then a single 100-hectare farm could produce 20,000 kg of enzyme. However, the crucial cost consideration in this example is purification of the product. Separating the desired enzyme from other proteins in the seed can be expensive, and therefore the fermentation process, although it requires more expensive inputs, often produces at a lower cost, a sufficiently pure enzyme, because the proteins are easily extracted from the medium.

A group of Canadian scientists has developed one ingenious solution that may help reduce the costs of purifying some proteins from plants. The gene of the desired protein is engineered by fusing it to the gene that encodes oleosin. Oleosin is an oil body surface protein with a long hydrophobic anchor and a cytoplasmic terminus found in the seeds of canola and other oil crops. The protein to which oleosin is fused will be firmly attached to the oil bodies. When oil is extracted from the seeds, the protein is recovered with the oil, where it is separated from the bulk of other proteins. The desired protein product can then be cleaved from its target sequence and recovered in relatively high purity.

There are other situations where producing industrial proteins in plants is likely to be cost effective. In some cases, the desired use of the enzyme does not require purification. For example, cell wall polymers cannot be digested by monogastric animals and therefore have no nutritive value. Adding enzymes that help break down cellulose or hemicellulose can free up more calories and increase feed efficiency. For example, a thermostable endoglucanase has been expressed in transgenic potato at levels up to 25% of leaf-soluble protein. A small amount of crop containing this enzyme could be mixed with other animal feed to partially digest cellulose, thereby enhancing feed use.

Plants are Ideal Production Stems for Diagnostic or Therapeutic Human Proteins Needed in Large Amounts

Mammalian genes can readily be expressed in plants, and plants are likely to become the production systems of choice for some mammalian proteins that are needed as diagnostics for detecting human diseases or as therapeutics for treating human and animal diseases. Research in "biopharming" is aimed at producing monoclonal antibodies, blood plasma proteins, peptide hormones, and cytokines in plants. The reasons that plants are emerging as the system of choice are purity and cost. Small amounts of human recombinant proteins can now be produced in cultured mammalian or insect cells, but many medical applications of monoclonal antibodies or plasma proteins require the production of as much as 1,000 kg of pure protein per year. Scaling up the production of antibodies by mammalian cells would require a capital investment of US$100 million, and the antibodies would cost thousands of dollars per gram. Extraction from crop plants, in contrast, can be done in a factory that costs US$10 million, and the antibodies will cost US$200 per gram. About 20 companies worldwide are now doing research to eventually produce recombinant human proteins in plants. Likely production systems are the leaves of tobacco plants, the endosperm of maize seeds, potato tubers, or the stems of sugar cane. The targets could include not only therapeutic proteins but also proteins such as collagen that are used in the cosmetics industry. Factories where human proteins made in plants will be purified have already been built.

Proteins could be produced after stable transformation of a plant or by transient expression after infecting the leaves with a virus that carries the gene of interest. One important aspect that scientists must consider is that if human proteins are made in seeds, the crops must be grown in relative isolation so that the human genes do not show up in maize used for human or animal consumption. Although the proteins might be harmless, their consumption could have unintended consequences. Genes could spread if, for example, pollen from maize plants that are producing the human proteins were to be blown to other fields. The fields therefore need to be isolated, so that there is no chance that pollen from the transgenic plants may cause the genes to spread to other fields. An advantage of using a vegetatively propagated crop such as sugar cane is that it eliminates the possibility of pollen spreading genes.

The problem of gene spread can also be circumvented with transient expression systems based on plant viruses that infect leaves. Transient expression means that the proteins are made for only a few days, but the goal is to have very high levels of expression during that short period. Viruses have small genomes consisting of only a few genes. It is sometimes possible to replace some of the viral genome with a segment of nucleic acid that carries the information to make the active fragment of an antibody molecule. When this is done, the virus may still be able to replicate and reproduce itself when it infects a plant. Whether or not this is possible depends on the type of virus. Thus, when a leaf of a plant is infected with a genetically engineered RNA virus that carries the active segment of a human gene, the virus will spread throughout the plant, making billions of copies of itself. If viral infection is done before the plant flowers, then the virus does not spread to the pollen. The plant cells will translate the messenger RNA and produce the antibody fragment. Engineering the virus and allowing the plants to make the protein can all be accomplished in two months. Such a rapid production system can be used to create antibodies that are customized for a particular patient—a requirement for treating non-Hodgkin's lymphoma, a type of cancer that attacks certain white blood cells.

The body uses antibodies to fight diseases (see later discussion), but antibodies also have many other uses in medicine. They can be used therapeutically, to deliver a killer drug specifically to cancer cells, for example, and also to diagnose diseases. Antibodies are directed at antigenic determinants (or epitopes) at the surfaces of proteins or other molecules. For example, an antigenic determinant could be a few amino acid side chains displayed on the surface of a protein. Antibodies directed at antigenic determinants displayed on the surface of cancer cells specifically bind to those cancer cells. Different antibodies will home in on different types of cancer cells. Such specific binding can be used to image the cancer, or to target the cancer cells for destruction. Antibodies produced in plants could also provide protection against intestinal pathogens such as hepatitis viruses, *Helicobacter pylori*, or toxic *E. coli*. They could target respiratory pathogens (rhinoviruses and influenza), sexually transmitted diseases, and dental caries, and could be used as contraceptives. Clinical trials with antibodies produced in plants are already under way, and these novel plant-derived pharmaceuticals may be in the marketplace by 2005.

Plants can be Used to Deliver Edible Vaccines for Serious Diseases of Humans and Domestic Animals

In humans and other warm-blooded animals, the primary task of the immune system is to defend against invading organisms—particularly against pathogenic bacteria and viruses. Most infections are initiated in the mucosal surfaces that line the digestive tract, the respiratory tract, and the urino-reproductive tract; these surfaces also contain the first line of defense the mucosal immune system. Killing the invaders by engulfment is the function of phagocytic cells such as the macro phages in the bloodstream. Engulfment can occur after the pathogens have been covered with antibodies—proteins produced by the immune system—that are directed at specific antigenic determinants on the surface of the invaders. Producing antibodies is the function of specialized cells of the immune system. The unique features of antibodies are that they recognize a single molecular structure (often a protein) and bind to it with very high affinity. A common type of antibody is the immunoglobulin G (IgG) protein, consisting of two heavy (large) polypeptide chains and two light (small) polypeptide chains.

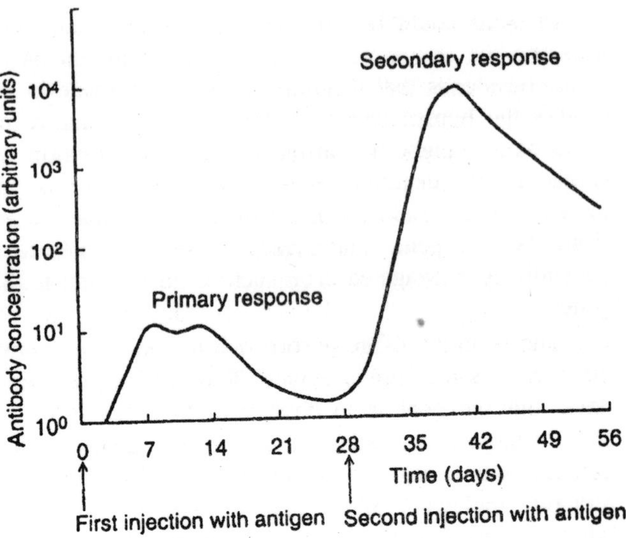

Fig. 13.6. Responses to initial and later injections of an antigen.

When a foreign invader comes calling the first time, the body's response in terms of antibody production has a considerable lag time and may not be strong enough to overpower the invader, so the animal may get very sick or even die. If the immune system has been exposed to the antigenic determinants on the invader's surface in the past, and now faces a second invasion, then the response is faster and more powerful. That is the basis of vaccination. Because detection of an invader can occur in the mucosal membranes or in the bloodstream, vaccines can be effective whether taken orally or injected. For example, there are effective oral and injected vaccines against the polio virus. Vaccination makes people immune because it triggers the immune system to produce antibodies by simulating that first invasion, not with virulent pathogens, but with a greatly attenuated strain of bacteria or viruses. Alternatively, immunity through vaccination can be achieved by injecting or ingesting only the antigenic determinants (often parts of proteins that are the antigens) that are displayed on the surface of the invading organisms.

Such vaccines are called subunit vaccines. The use of subunit vaccines increases vaccine safety by circumventing the need to use live bacteria and viruses. However, subunit vaccines are not as stable and need to be refrigerated, increasing their cost. Safe vaccines for many important diseases such as cholera and hepatitis are well beyond the reach of the health care system in many developing countries. Scientists realized that plant genetic engineering could solve the problem of providing low-cost vaccinations to people in developing countries, a realization that prompted the idea of expressing the antigens in plants and creating edible plant vaccines. Various pharmaceutical companies are targeting quite a few diseases for vaccine production in plants.

Table 13.5. Production of vaccines in plants

Potential Application	*Protein Expressed in Plants*
Dental caries	*Streptococcus mutans* surface protein
Cholera and *E. coli* diarrhea	*E. coli* heat-labile enterotoxin
Hepatitis B	Hepatitis B antigen
Hoof-and-mouth disease	Hoof-and-mouth viral antigen
Malaria	Malarial B-cell epitope
Influenza	Hemagglutinin
Rabies	Rabies virus glycoprotein
HIV	HIV epitopes gp41 and gp 120

Some of the antigenic determinants on the surface of viruses and bacteria are proteins made by the pathogen. The genes encoding these antigens can be isolated from the pathogens and expressed in plants using standard recombinant DNA technology. In addition, it is possible to incorporate an antigenic determinant (usually a short stretch of amino acids) into an existing plant protein or into a plant virus. If the plant makes the antigenic determinant, then eating the plant (under the right conditions) will amount to vaccination. If the antigenic determinant is incorporated into a plant virus and expressed at the surface of the viral particle, the plant virus can be used as a vaccine (plant viruses are quite harmless to humans and animals; when eating vegetables, people consume large numbers of various plant viruses). Researchers conducted the first human clinical trials with transgenic plant-derived vaccines in the late 1990s. They showed that such vaccines are protected from digestion in the human intestinal tract and stimulate both the mucosal and systemic immune responses. Vaccinating people against cholera by feeding them bananas, or vaccinating animals against foot- or hoof-and mouth disease by feeding them sugar beets could well be a reality by 2010.

14

Phytotoxicology

At the subcellular level, plants and animals display more similarities than differences. Consider the processes of replication of information, genetic recombination, development of subcellular structures, and respiration. Attention should be focused on the fact that, although plants and animals are different, they are evolving branches of a common origin. The study of phytotoxicology, then, is an examination of those products characteristic of one major part of the biota that produce or evoke a specific deleterious reaction when they interact with systems characteristic of the other. This view emphasizes the differences that have evolved between plants and animals despite their common beginnings.

In every plant poisoning case it should be possible to identify precisely the plant product and the animal system involved, the route by which they are brought together, and the specific subsequent events that occur until the animal either dies or recovers. However, according to a recent authoritative statement on this subject, "Probably no field of scientific endeavor exists in which it is more difficult to separate fact from fiction than in the study of poisonous plants. Examination of the pertinent literature will reveal considerable confusion tending to mask an even greater amount of ignorance.... Unbelievable chaos reigns in the area of plant identification and nomenclature as applied by the nonspecialist".

This chapter is directed toward three main topics: (i) the importance of poisonous plants to man and animals, (ii) an assessment of current knowledge of poisonous plants, and (iii) a discussion of the effects of toxic principles elaborated by plants.

Importance of Poisonous Plants to Man and Animals

Although incomplete, the best figures on incidence of human poisoning by plants are those collected from individual poison control centers and analyzed by the National Clearinghouse for Poison Control Centers. Ingestions are grouped by categories in the annual summaries. Plants as a category have consistently ranked in the top seven, accounting for about 4 percent of the reported ingestions until the 1970s. In 1970, plants (excluding mushrooms and toadstools) accounted for 4,059 reported ingestions, representing 4.8 percent of all reported ingestions for that year. This was greater than the number of disinfectant, tranquilizer, insecticide, hormone, acid and alkali, antiseptic, polish, or paint ingestions and was exceeded only by the incidence of ingestion of aspirin and soap-detergents-cleaners. Since the, however, plants as a category has come to the top of the list of "products" most frequently implicated in poisoning of children under 5 years of age. Now, about one out of every ten cases reported by poison control centers is related to plants. This conspicuous rise in importance of plants since 1965 is related to the equally dramatic decrease in cases of poisoning from aspirin. These have fallen from greater than 25 percent of reported incidents in 1965 to just 4.1 percent in 1976. Safety packaging, limited quantities per package, and increased public awareness of hazards—the result of governmental and private compaigns—have been the main contributing factors in reducing the importance of aspirin. Only the latter method of decreasing incidents is easily applicable to the unnecessarily high volume of ingestions involving plants now being experienced at poison control centers. According to reported figures there are approximately 75,000 human ingestions of poisonous plants in the United States each year. According to Canadian figures, ingestions of nonfood plants are more than ten times greater than the number of reported incidents involving venomous bites or stings.

The above figures need to be viewed with a certain amount of caution. First, the majority of the ingestions were of materials, or in amounts, that would not have proved capable of eliciting a toxic response, or else treatment, usually emesis, intervened before such a response occurred. These, strictly speaking, are better defined as ingestions than as poisonings, without implying whether a toxic reaction was or was not possible. Second, many incidents are treated by private physicians who do not report to a poison control center. In addition, some poison control centers fail to report to the National Clearinghouse. Thus, the precise number of toxic responses that occur from plants annually in the human population cannot be determined from these data.

Poisoning of pets and livestock is even more extensive than is poisoning of man. No reliable figures exist, but estimates as to loss of range livestock in western states consistently place the annual figure at more than one million dollars per state or region. Certainly, well-documented instances in which more than 1,500 animals have been killed at a single time, such as losses of sheep to *Halogeton glomeratus*, represent considerable economic impact and do not need to be multiplied by many such events or by many similar plants to assume major importance to the livestock industry. The resulting economic loss includes not only the value of the animals but also the diminution in real value of the range acreage following its infestation with a poisonous weed. Looses of pets and wild animals are so poorly documented that no useful generalizations are possible.

Current Knowledge of Poisonous Plants

Effective consideration of phytotoxicology requires a synthesis across wide academic boundaries and is thus difficult. This, and the diversity of natural phenomena dealt with, perhaps more than other factors, have inhibited a more rapid, rigorous, development of the subject. Comparison of the principal current references dealing with North American poisonous plants shows immediately that fundamental input and analysis are required from botany, physiology, and pathology. In all cases, full development of a particular topic also requires the attention of one or more academic or practical specialists, such as physician, veterinarian, toxicologist, clinician, organic chemist or biochemist, plant physiologist, pharmacologist, pharmacognosist, agronomist, horticulturist, geneticist, and animal husbandman. Consideration must be given to identification and description of a toxic reaction; practical understanding of the history, to aid in diagnosis and in formulating control; secure determination of the etiology; identification and perhaps isolation of a toxic principle; specific description of tis action; the seasonal, ecologic, and genetic control of production of the toxic principle by the plant; and the way in which man or animals were or may be exposed. The latter involves agricultural practices in the case of livestock. In man, a plethora of possibilities exist, from overuse of plant-derived drugs by adults to accidental or experimental ingestions of drugs or whole plant parts by children.

Approximately 700 species of North American plants are considered to be poisonous on the basis of case histories, experimental investigations, or other specific reasons. More will be discovered. This is only a small fraction of the perhaps 30,000 species of plants in the wild and cultivated floras of North America. Nevertheless, it is a large number, and no generalization emerges from a review of their botany, ecology, or management by man to allow systematizing them for easier comprehension. Poisonous species are scattered throughout the plant kingdom from algae, to ferns, to gymnosperms, to angiosperms, and in the latter large groups they appear almost randomly among the plant families.

One sometimes hears that certain groups such as the nightshade or potato family (Solanaceae) are particularly dangerous, but such statements are not entirely valid. Such families are the larger ones, and certain poisonous species more or less in relation to their size.

Toxicity usually exists at the level of the genus. If one species of a genus is toxic, some or all others in that genus usually display similar toxicity. This is why species names are sometimes omitted in general discussion of toxicity. This is a mistake. First, exceptions to the generalization are numerous and important. Second, even when similar species display similar toxicity, they may have other important differences. For example, most species of the genus *Asclepias* (milkweeds) are toxic, but the most toxic species (*A. labriformis*) is found only in a limited area of Utah. However, the most troublesome milkweeds (*A. subverticillata*, *A. eriocarpa*) differs in appearance from the former and have significantly less toxicity on a weight basis, but are much more widely distributed geographically. Important distinctions such as these are lost when species names are not used.

In some cases, groups of genera within a single family display similar toxicity. This is true, for example, of the laurel group of the heath family (Ericaceae). At the other extreme, however, are those instances in which closely related species differ in toxicity. *Eupatorium rugosum* is closely related to the numerous other species of *Eupatorium*, but only the former is known to be toxic. Occasionally the same toxic principle is found in plants of great botanical or habitat difference. For example, the only other plant known to contain the same poisonous principle as *Eupatorium rugosum* is *Aplopappus heterophyllus*. These two are in the same plant family (Compositae), but the former is found in woodlands of eastern North America while the latter is limited to the dry ranges of the Southwest. Nicotine, as another example, can be isolated from plants as botanically distant as tobacco (*Nicotiana tabacum*) of the nightshade or potato family of angiosperms, and from club moss (*Lycopodium* spp.) of the primitive, spore-bearing lycopods.

The content of a given poisonous principle, and even to some extent its molecular structure, can vary widely in some species with the environmental conditions under which the plant grew. This is particularly true of many glycosides. In other cases, certain alkaloids for instance, elaboration of the poisonous principle in a given species in under reasonably tight genetic control and varies little with growing conditions. Even when under strict genetic control, the content of a poisonous principle in a given plant may vary with stage of growth; it may concentrate in particular parts of the plant, or it may vary with the particular variety or strain of plant.

Exemplifying the last point, one of the nightshades exists in two distinct populations, which fortunately do not normally interbreed. The botanic distinctions between them are so small that the two populations were originally recognized as a single species (*Solanum nigrum*). One kind, however, is quite toxic. The other, now set off taxonomically as *Solanum intrusum*, is not known to be toxic. In fact, *S. intrusum* has entered trade as "garden huckleberry" or "wonderberry" and is sometimes recommended to home gardeners for its edible fruits. Poinsettia (*Euphorbia pulcherrima*) may represent a similar example. Its listing as toxic was found originally on a reported case of human mortality in Hawaii in 1919. Recent feeding experiments, prompted by the belief of horticulturists that the poinsettia is really not poisonous, showed no toxic effect in rodents fed large quantities of the red bracts. However, poison control centers continue to report instances of gastric distress in children after ingestion of poinsettia. A possible explanation for the apparent conflict in these several observations is the very great horticultural manipulation that has in recent years been devoted to breeding showier, longer-lasting, and differently colored poinsettias. It is reasonable to suppose that the ability to form a toxic principle may have been consciously selected against at the same time that desirable floral characteristics were sought. If this is the case, older varieties of poinsettia, or ones distinctly different from those experimentally fed to the rodents, may still be capable of eliciting a toxic reaction. These examples could be extended to demonstrate that toxicity can vary not only with species, sex, age, and even individuality of plants and animals but also with environmental factors. The examples quoted above should suffice to convey an idea of the complexity of phytotoxicology.

A major factor in much of the present confusion regarding poisonous plants is the problem of adequate identification of the plant material involved. A physician or veterinarian faced with a case of plant poisoning is, in a sense, like a chemist conducting experiments by guesswork from a stockroom of unlabeled reagents. Accurate identification of the plant is prerequisite to making use of existing pertinent toxicologic information.

The layman rarely appreciates the difficulties involved in the accurate identification of plant materials. Not only are the reagents unknown, but some of them may not be identifiable with the material and tests at hand. Botanic identification often requires a particular part of the plant—usually the mature reproductive parts (which evolve most slowly and hence show relationships best). This means the flower, and the flower is often a very transitory event in the life of a plant. Flowers often are not present when the berries of a plant attract children. Also, flowers do not generally survive well in the gut, so that examination of vomitus or of ruminal or stomach contents will rarely yield identifiable flowering stages, even if initially present.

Most actual cases of poisoning of man and livestock involve the identification of plant material by common name alone. This can be disastrous. Serious errors in reputable reports in the medical and veterinary literature can be traced to erroneous use of common names or failure to appreciate the lac of precision in such usage. All plant materials involved in toxicologic examinations of any kind should be identified by genus and species. If this requires the help of a taxonomic botanist, it should be obtained. If experimental work resulting in a published report is done with plant materials, voucher specimens should be prepared and deposited in a major herbarium, and the fact should be noted in the publication. One or more such herbaria exist in every state, usually at the department of botany in the state university, but sometimes at the state museum or at a major private university. Herbaria function, somewhat like libraries, as depositories of materials that can be recalled in good condition at a later date if it prove necessary to reassess the botanic materials involved in a particular investigation. Despite the obvious benefit of such a practice and its relative simplicity, it is rarely done.

The foregoing discussion presents reasons why current practices make it difficult to deal with poisonous plants. There are historic reasons as well. At the risk of oversimplification, they may be summarized as follows.

The Greeks and Romans were astute and recorded, in some detail and with commendable accuracy, their conclusions concerning the useful, medicinal, or toxic characteristics of the natural world. For most areas of knowledge, these collected observations went into eclipse during the Dark Ages, surviving largely as manuscripts copied from generation to generation without significant addition or modification. This was not the case with poisonous materials, for the practice of poisoning to obtain succession of royalty or ecclesiastic authority, inheritance of wealth, or defeat of armies became a highly developed art. Persons able to get results commanded high fees. In order to protect their hard-won trade secrets, these artisans compounded recipes with many esoteric ingredients, thereby obscuring the identity of the actual active principle. A perhaps overdrawn, but nonetheless illuminating example is the several-stanza list of ingredients that went into the witches' cauldron in Shakespeare's *Macbeth*; only two could be expected to do the job.

With the Renaissance, it became necessary to separate fact from fiction, something that was only imperfectly accomplished. Further confusion resulted from an innocent attempt to relate the reports of the classic authors, who dealt mainly with Mediterranean plants, to the flora of central and northern Europe. Keep in mind that the science of naming plants originated with Linneaus' *Species Plantarum* of 1753; before that time all names applied to plants had no more authority than the common names used today.

Herbals supplanted classic manuscripts around 1470; floras and tomes on materia medica supplanted herbals around 1670; all contained frequent reference to toxic capacities of plants. Monographs devoted solely to poisonous plants began to appear shortly before 1700. All of these works were European, and interested educated persons were expected to know the literature of their subject well enough so that citation of authority for particular statements was not deemed necessary. Hence, for the most part, the first books dealing with poisonous plants do not say where the information came from. Most later writers on poisonous plants have apparently been reluctant to eliminate anything that sounds reasonable, in spite of their inability to verify the information. Some patently unreasonable things are also perpetuated. Even today, books may appear in which medieval error is repeated unconsciously but with devastating effect if taken as scientifically accurate documents.

In 1814 M.J.B. Orfila published the first edition of his substantial work on toxicology, to which the use of an experimental approach in toxicology is generally traced. Orfila experimented with a number of plants, describing their effects and attempting to trace the distribution of the poisonous principle in the body. His chief experimental animal was the dog, an animal that vomits readily, and Orfila frequently found it necessary to excise and tie off the esophagus to ensure absorption when dealing with plants or plant materials administered orally. Although Orfila recognized the difficulties this caused in interpreting his results, it remains problematic to separate the effects of the toxic principle from the effects of this relatively drastic procedure.

The flora of North America east of the Mississippi is essentially similar to that of Europe, while west of that river it becomes increasingly foreign. Thus, the early settlers of the United States were able to bring with them not only their livestock but also their practical experience in dealing with the plants they found here. The few persons interested in pursuing investigations of poisonous plants could look up appropriate information in European compendia and apply it with some usefulness to local conditions. The dawn of scientific agriculture in North America can be traced to events a score of years apart: the founding of the Department of Agriculture and the passage of the Morrill Land Grant Act in support of agricultural experimentation and experiment stations in 1887.

These events meant, in practical terms, that as settler moved their livestock west into an increasingly foreign flora and began to put pressure on it, they could no longer turn to experience or to European knowledge to cope with the poisonings that occurred. On the other hand, as each state organized it took advantage of federal programs to establish a college of agriculture and an experiment station, and it was to these agencies that the problems were referred. Increasingly sophisticated experimentation ensured, and by 1900 all but three states west of the Mississippi had published work dealing in a practical way with poisonous plants. At the same time, reports of experimental work from eastern states remained almost nonexistent. Looking at the results of these influence at the present time, one sees that about a third of thee existing body of information derives from case histories in man and animals, about a third derives from experimental investigation, mostly from a veterinary point of view, and about one third entered our literature from European source. At the same time, the excellence of mush of the recent experimental work with poisonous plants, and the continuous long-term record of productivity of such laboratories as the federal poisonous plants investigational program at Logan, Utah, contrast sharply with the fact that apparently current information being used to treat human poisonings today may in fact be traced to experiments of Orfila and even to Dioscorides.

Given so many potentially harmful plants and the variety of syndromes they may provoke under particular circumstances, a common first reaction of those who must learn to deal practically with the problem of plant poisoning in man is to seek a list of the few most dangerous or troublesome ones.

The National Clearinghouse for Poison Control Centers published a detailed review of the collected reports of plant ingestions for 1965 that were treated as poisoning emergencies. If reporting were accurate, this list should contain the most troublesome plants of the United States. It is not difficult, however, to show that the clearinghouse list is actually of little value for this purpose. Assuming that the personnel of most poison control centers are competent, concerned, conscientious, and medically trained, this list represents a summation of their frustrations in obtaining useful histories, identifying plants, finding or interpreting appropriate literature, discovering that useful or recent experimental results do not exist for the plant in question, or finding that no tested treatments have been recommended for particular circumstances, as was suggested earlier in this chapter.

Taking the plants in order as named:

Standard botanical manuals for the United States list at least three different genera to which the name "pokeweed" or "pokeroot" is commonly applied. All three (*Phytolacca americana* [=*P. decandra*], *Veratrum viride*, and *Symplocarpus foetidus*) have histories of toxicity, but the syndromes differ greatly, as would appropriate treatments. All three are discussed later in this chapter. (*Symplocarpus foetidus*, also called skunk cabbage, is an aroid or member of the plant family Araceae.)

"Yew" is another common name that can cause trouble. It normally refers to a species of *Taxus*. One of these species, *T. canadensis*, is more commonly called "ground hemlock" in some areas. Thus, it is easy for the uninitiated to transfer inadvertently in the literature from "yew" to "hemlock," especially since the latter is a well-known name associated with toxic plants. However, "hemlock" is applied as a common name to at least four genera of plant, only one of which is not lethal. Again the syndromes and appropriate treatments vary.

"Philodendron" is both a scientific name and a common name. As the later, it is applied by the public, and nearly as loosely by many florists, to almost any viny, leafy, nondeciduous (not shedding) potted plant, with or without holes in the leaf blade. A survey of plants with these characteristics, made with the help of the staff at Cornell's Hortorium, resulted in a tally or several score species of plants in nine genera. Only a plant specialist could accurately identify the plant involved when a call comes to a poison control center that a child has eaten "philodendron," and then probably only by seeing the actual plant.

The name "bittersweet" is commonly applied to two entirely different plant genera, one of which has only an ancient European record of putative toxicity. The other is a species of *Solanum* (*dulcamara*), thus a "nightshade." "Nightshade" is perhaps the worst possible designation for a poisonous plant. Usually it refers to one of the multitude of species of *Solanum*. However, it can apply to more than one genus (for example, *Atropa belladonna* is commonly called "nightshade") and is also regularly used to designate the family that contains these plants, the Solanaceae. This is one of the larger plant families and contains such useful plants as potato (*Solanum tuberosum*) and tomato (*Lycopersicon esculentum*). Any member of the family can be called a "nightshade") in a general way, and many are poisonous ("deadly nightshade") under some circumstances including potato and tomato, despite their widespread daily use as food. The last plant named is also a nightshade (Jerusalem cherry = *Solanum pseudocapsicum*), but is at worst only mildly poisonous according to the available information. In conclusion, "nightshade" and "deadly nightshade" are words many parents know. They are likely to come up when a child has eaten a wild plant whose identity is not known by the distraught mother. Under these circumstances, the usefulness of "nightshade" for indicating the identity of a species of plant is virtually nil.

"Holly" is applied to a dozen species of the genus *Ilex*. Many are native American plants and have no record to toxicity. English holly (*I. opaca*) figures in a wealth of Middle Ages mythology. The only definite published reference to the toxicity of *Ilex* is in a second-hand French report from

1889, with authority not stated. Although this has been carried forward in texts to the present, and may even have some validity to it, the present-day physician would be unwise to base his treatment on this kind of information.

"Honeysuckle" refers to more than a score of species of plants in eight genera belonging to four different plant families. The plant commonly referred to in the East as "honeysuckle" is not known to be toxic.

"Pyracantha" is a scientific name and, although several common names are available for this shrub. Without a species name this particular plant is identified only partially, although probably better than with only a common name. The berries of *Pyracantha*, however, have been shown by experiments to be nontoxic in four species of laboratory animals. At present there is no reason to assume that *Pyracantha* berries are poisonous to man.

"Castor bean" is a satisfactory (i.e. unambiguous) common name designation for a single plant species. *Ricinus communis*. The seeds of this plant are well known to be lethal if ingested in small to moderate amount. However, the plant has been subjected to horticultural manipulation so that two main commercial selections now exist. One is grown for production of oil; the other, as a showy, ornamental hedge or garden plant. The unselected native type grows wild in Florida. Whether these three types have equivalent toxicity is not known.

Greater impetus for detailed experimentation with poisonous plants might develop if more investigators realized the important discoveries that have come from such research in the past. Some, such as ergot, digitalis, belladonna, and morphine, are classic. The discovery of dicoumarol as an anticoagulant and the development of warfarin as a rodenticide came directly from an experiment station investigation of why cattle fed moldy sweetclover hay (*Melilotus* spp.) bled to death. More recently investigation of the toxicity of cyads (*Cycas* spp.) has yielded important information about carcinogenesis of bracken fern (*Pteridium aquilinum*), about avitaminosis B_1 in nonruminants; of pokeweed (*Phytolacca americana*), about mitogenesis in leukocytes of hellebore (*Veratrum californicum*) and lupine (*Lupinus sericeus*), about teratogenesis; and of groundsels (*Senecio* spp.) about liver function.

Other sources of sophisticated information on the effects of plants on animals are the pharmaceutical industry and feed manufacturers, or academic laboratories functioning in these areas. The pertinence of investigations of natural drug products is obvious. Less obvious, perhaps, is the work that has been done to analyze and determine nutritional insufficiencies or minor toxicities associated with utilizing large amounts of particular crops for feedstuffs. A good example is the detailed analysis of gossypol, the toxic component of cottonseed (*Gossypium* spp.) that makes cottonseed meal potentially poisonous to livestock.

Effects of Toxic Principles Elaborated by Plants

Poisonous principles of plants range from single elements or simple salts accumulated by some species under certain circumstances (e.g., selenium in cereal crops or oxalates in *Halogeton glomeratus*) to the elaboration of complex molecules of high toxicity (the phytotoxin abrin, a protein, in *Abrus precatorius*, the infamous Rosary pea or Jequirity bean). The specific action of a toxic principle in an animal as presently understood may range from simple irritation of mucous tissues to disruption of an enzymatic process at the microsomal or mitochondrial level. Thus, it is difficult to organize the available information along lines of either molecular structure or fundamental physiology. This difficulty is compounded by the fact that precise knowledge of the toxicity of the 700 species of plants known to be toxic is lacking for more than half. Different authors use various more or less successful schemes based on physiology, pathology, chemistry, or some combination. Any attempt to categorize poisonous principles of plants on chemical grounds suffers not only from the fact that the exact chemistry is rarely known but also that the common categories employed for such purposes (alkaloids, glycosides, saponins, etc.) are not parallel and therefore not mutually exclusive.

Information on poisonous principles and actions is organized below according to observed responses to average toxicologic exposures, and this in turn is considered in sequence of major target organ or tissue as the poison passes through the body, assuming initial exposure by ingestion. Poisons and responses they elicit are so numerous and varied that what follows is more of a summary than a discussion. Only one example is given for each situation, although numerous examples may exist, and subsidiary effects or consequences in additional organs or tissues are ignored, although they are often clinically important. The discussion is also limited to effects of plants ingested as such, not to overdoses of drugs of plant origin, and does not include consideration of differential diagnosis, clinical signs, pathology, or treatment, except as specially important to the point singled out for attention.

External Structure and Mouth

The sap of some plants is acrid and irritating to skin and mucous membranes. The mown stubble of a field of spurge (*Euphorbia esula*) has caused inflammation and loss of hair on the legs of horses used to mow it. Some plants taken into the mouth cause intense stomatitis by direct irritation. The foliage or berries of the ornamental shrub daphne (*Daphne mezereum*), for example, cause corrosive lesions of the mouth, if chewed or eaten. Animals will taken such distasteful materials into the mouth out of curiosity, and most poisonings occur when prunings or clippings are thrown into a pasture or stall. Children exhibit similar curiosity and will swallow distasteful material as readily as they spit it out, just to get it out of the mouth. Either of these plants, and many others of course, will produce intense irritation of the esophagus and, if swallowed, the gut. The exact nature of the irritant is usually unknown. Many aroids (members of the plant family Araceae) cause a similar intense burning sensation in the mouth. Perhaps the most notorious is dumbcane (*Dieffenbachia* spp.) These plants contain needle-like crystals of calcium oxalate that may cause some mechanical as well as chemical irritation. The severity of the reaction has been traced, however, to the presence of a proteolytic enzyme in the plant that attacks the oral tissues. This reaction is often accompanied by swelling and glottic edema and has been fatal in man when the breathing passages have been blocked as a consequence.

Degree of mastication of seeds and fleshy or thick plant parts may determine the severity of the subsequent reaction, or whether it takes place at all. Seeds of precatory bean (*Abrus precatorius*), for example, are highly toxic if chewed, but will pass through the gut undigested if the seed coat is left intact.

Rumen, or Foregut

Ruminants may react to plant poisons quite differently from nonruminants due to differences in digestive structure and function between these two major groups of animals. Ability to vomit effectively is one difference. Some plants stimulate a strong vomit reflex. The vomiting disease of swine, for example, requires ingestion of only a very small amount of barley grain parasitized by a mold fungus (*Gibberella* sp.). Cardioactive glycosides such as those in foxglove (*Digitalis purpurea*) similarly provoke vomiting in most species of animals, even when administered parenterally. Many simple-stomached animals (e.g., man and dog) vomit easily. Ruminants can vomit, but the reflex is not as easily stimulated and the degree to which vomiting is effective in removing poisonous material from the gut is much less than in nonruminants. Vomiting in ruminants is not equivalent to normal eructation. Horses can vomit, but the structure of the oral cavity leads to complications if vomiting occurs. In horses, vomitus is directed into the trachea. Pneumonitis is a common result, and this can develop into pneumonia and result in death. In severe cases, death may result directly from asphyxiation after vomiting. Many glycosides, as they exist in plants, are not toxic to animals. Toxicity comes from breakdown of the glycoside to release a toxic component. Breakdown often occurs more readily or more rapidly in the rumen than in the digestive tract of monogastric animals. Also, small molecules can be absorbed at the rumen and thus enter the circulation rapidly. Breakdown of cyanogenic glycosides, such as amygdalin, from members of the rose family (Rosaceae) is an example. The ruminant is more likely to achieve toxic levels of cyanide in the blood in the balance between breakdown of the glycoside, absorption of cyanide, and its detoxification and excretion, than is the nonruminant. Occasionally, unexpected events occur. Sheep on dry range pasture may die of cyanide poisoning shortly after drinking water. The loss of life from ingestion of poison suckleya (*Suckleyea suckleyana*) is an example. It is hypothesized that the ruminal contents are too dry for effective breakdown of the glycoside until the animal drinks, whereupon the reaction is intense. Sometimes such instances are erroneously diagnosed as poisoning from the water itself. Many of the symptoms are similar to those of nitrate poisoning.

Ingestion of abnormally large amounts of various forages or grains, or sudden massive change in diet, can provoke unusual reactions in the rumen that have toxicologic consequences. These can range from a simple pH shift to the elaboration of specific highly toxic molecules. For example, it has been suggested that, under unusual circumstances, silage of high nitrate content can undergo a reaction in the rumen that yields nitrogen oxide gases. Taken into the lungs, small amounts of nitrogen oxides cause severe, irreversible pulmonary emphysema. The same thing happens when nitrogen oxides are formed in silage made from forage (usually corn, *Zea mays*) of high nitrate concentration. Being heavier than air, the nitrogen oxides accumulate around the base of the silo. Breathed by man, they cause a similar syndrome, which is called "silo-filler's disease."

A syndrome of cattle, associated clinically with reduced levels of magnesium in the blood, is characterized by staggering that develops shortly after they are placed on lush pasturage. It has been postulated that, under these pasturage conditions, sufficient ammonia is formed in the rumen to react with, and tie up enough, magnesium (as hydroxide) to produce dietary insufficiency.

Some plants cause ruminal stasis. When this happens, a low-grade toxemia commences that becomes more severe with time. Also, signs of starvation appear. Mesquite bean (*Prosopis juliflora*) poisoning, recognized on southwestern ranges in cattle, is an example. Stasis is complete or nearly so. Mesquite beans have been found in the rumen on postmortem examination of animals that had not had access to this plant for as long as nine months. Under these conditions, the seeds have sometimes sprouted and begun to grow in the rumen.

Ruminants are more susceptible to bloat (the foamy entrapment of gases) than are monogastric animals. Bloat, if unrelieved, can have lethal consequences. Some plants promote bloat. These include some common leguminous forage crops and certain wild plants. Among the latter is the wild larkspur (*Delphinium* spp.) of western ranges. Under range conditions bloat may not be observed in time to be treated effectively.

Just as the rumen can promote the release of a toxic compound from an innocuous precursor, so can it sometimes aid in the detoxification of an initially poisonous compound. Sheep fed high-calcium alfalfa hay, for example, are protected to some degree against the toxic effects of halogeton (*Halogeton glomeratus*), which contains soluble oxalates. It is postulated that calcium is precipitated by the oxalate ions in the rumen, thus making the oxalate unavailable for absorption from the gut.

Another situation in which the ruminant has the advantage over the nonruminant occurs in the case of ingestion of plants containing a thiaminase, such as field horsetail (*Equisetum arvense*). In the horse, continued exposure to such plants causes destruction of the thiamine in the diet and the development of a definite and eventually lethal B_1 deficiency with classic signs of polyneuritis. This is one of the few instances where the ultimate biochemical lesion, disruption of carboxylation in the Krebs cycle, is known. In a ruminant, the microflora of the rumen manufacture copious quantities of thiamine, which apparently is carried intact in the bacterial cell to a point in the gut beyond which the thiaminase is inactivated or destroyed. There it is released by digestion of the bacterial cell and absorbed by the ruminant. In any event, ruminants do not suffer from thiamine deficiency despite having a level of thiaminase in the diet that would kill a horse.

The initial distribution of ingested materials in the ruminant is determined partly by density. Seeds tend to pass quickly through the rumen and to concentrate in the abomasum. Here irritant substances may be released from seeds in concentrated form and promote irritation and hemorrhage of the abomasal wall seeds of members of the mustard family (Cruciferae) are examples.

Lower Gut

Two important actions, irritation and absorption, may occur in the stomach and intestines. Many plants cause irritation, varying in severity from mild to ulcerative, and the principal signs or symptoms in many cases of poisoning are simply those of gastroenteritis. Pokeroot (*Phytolacca americana*), for example, experimentally produces hemorrhagic gastritis, and ulcers are found in postmortem examination at locations where pieces of root lie against the mucosa of the gut. Pokeweed poisoning in cattle results in copious, almost explosive diarrhea that may contain signs of hemorrhage. In contrast, although blood-tinged feces are characteristic of oak (*Quercus* spp.) poisoning of cattle, the gastroenteritis is usually accompanied by constipation.

Absorption of toxins into the bloodstream normally takes place in the lower gut. Some toxic principles are large molecules, not readily absorbed. Saponins, such as those of the cockles (*Saponaria* spp.), are examples. As irritants, however, they promote their own absorption. Cardioactive glycosides are saponic in physical properties. It has been shown that the nature of the sugar portion of the intact glycoside is important in determining the solubility of these molecules and, therefore, their physiologic availability.

Liver

Poisonous principles absorbed into the circulatory system pass first to the liver. Here many different things can happen. A number of plant substances are severely poisonous to hepatic tissue, causing rapid destruction and necrosis where contact is made. A common finding on necropsy is a pathologic liver, but the exact pathology varies considerably. In many cases, the assault is somewhat chronic, and the appearance and function of the liver represent whatever current balance exists between destructive and regenerative changes.

A great deal of recent work has gone into elucidation of the exact hepatotoxic effects of pyrrolizidine alkaloids. Alkaloids of this configuration are found in a number of genera of higher plants (such as *Senecio* spp.). The primary lesion is a characteristic megalocytosis accompanied by venous obstruction or occlusion. Some plants. (e.g., *Trifolium subterraneum*, subterranean clover) accumulate copper from soils of high copper content. Copper then accumulates in and produces degenerative changes of the liver. In akee (*Blighia sapida*) poisoning of man, a hypoglycemic poisonous principle in the plant reduces the glycogen content of the liver to nearly zero.

One of the types of liver dysfunction that commonly occurs in plant poisonings is the reduced ability to eliminate certain pigmented molecules in the bile. These, instead, enter general circulation. When they reach the capillaries of the skin, they react with light and cause the capillaries to leak serum. This reaction and its consequences constitute the syndrome of photosensitization. If edematous swelling is severe, the involved tissues die and are sloughed off. Thus, in range sheep, the disease known as bighead is characterized initially by erythema, then by edematous swelling, and ultimately by necrosis of portions of the ears, cheeks, and lips, if severe Animals that have lost the lips are unable to forage effectively and die of starvation. The identity of all pigments involved in photosensitization is not yet known, but several have been established. One or more are breakdown products of chlorophyll and are normally present in the hepatic portal circulation. In other cases, the photosensitizing pigment is contained in the plant itself and passes unchanged through digestion, absorption, and the normal liver. Photosensitization caused by the first type of pigment, always accompanied by signs of liver damage, and usually by icterus, is termed secondary. Photosensitization involving a normal liver and lacking any signs of liver dysfunction is termed primary. A range plant commonly provoking secondary photosensitization (bighead) is horsebrush (*Tetradymia* spp.). Primary photosensitization is less common, but can be caused by ingestion of St. Johnswort (*Hypericum perforatum*).

Circulatory System

The discussion of photosensitization has taken us from the liver to the general circulation. A number of other disease syndromes may occur in the circulatory system or involve the blood itself. The toxic principle of sweet pea (*Lathyrus odoratus*), β-aminopropionitrile, causes dissecting aneurysm of the aorta in small animals. A number of plants contain toxins that provoke lysis of red blood cells and consequent hemolytic anemia. Among them are cultivated onion (*Allium cepa*) and rape forage (*Brassica napus*). Saponins and phytotoxins cause lysis of red blood cells *in vitro*. Their role *in vivo* is less clear.

Coumarin, contained in sweetclover hay (*Melilotus* spp.), is converted to dicoumarin by molds under certain conditions. This compound interferes with prothrombin synthesis and results in a hemorrhagic disease when molded sweetclover hay is ingested over a period of time. Animals bleed to death from minor injuries or, in advanced cases, bleed to death internally. Large subdermal hemorrhages are usually found in these cases. Soluble oxalates in the diet can result in the precipitation of calcium ions in the circulating blood. The resulting ionic imbalance has neurologic consequences. Hypomagnesemia and its consequences have already been mentioned.

Kidney

Once a toxin is in the blood, all organs are exposed to its effect unless a membrane barrier intervenes. Many plant toxins are destructive to parenchymatous organs in general. Digenerative effects are seen primarily in liver and kidneys on postmortem examination. In some cases described above, the liver is the primary target organ. In a few others, the kidney shows the major pathology. The latter includes the effects of tannins in oak (*Quercus* spp.) poisoning, and the crystallization of oxalates in kidney tubules in oxalate poisoning as by halogeton (*Halogeton glomeratus*).

Heart

In many lethal poisonings, the immediate cause of death is heart failure. Heart dysfunction can be brought about by malfunction of innervation or of the heart's conducting tissues, or it may be a result of a more direct effect on the heart musculature. Ingestion of foxglove (*Digitalis purpurea*) is similar to an overdose of the drug digitalis, which acts by stimulating the vagus center. At toxic levels, cardioactive glycosides produce cardiac irregularities and heart block. The alkaloid of yew (*Taxus cuspidata*) depresses the conducting tissue of the heart and stops it, often very quickly, in diastole. Hypotensive alkaloids, such as in false hellebore (*Veratrum viride*), cause a marked slowing of the heart rate.

Bone

Bracken fern (*Pteridium aquilinum*) contains two toxins. One is a thiaminase. The other, more slowly acting, has its target the bone marrow. Thus, ruminants exposed to a steady diet of bracken for many weeks develop a blood dyscrasia, characterized particularly by diminished counts of leukocytes and platelets, which may be traced in origin to severe destruction of bone marrow. Clinical signs of this disease are unusual for a poisoning and consist chiefly of hemorrhaging throughout the body due to thrombocytopenia, and invasion of the body by ordinarily nonpathogenic bacteria due to leukocytopenia, and the concomitant development of an elevated temperature. This disease is basically indistinguishable from radiation poisoning.

Pokeweed (*Phytolacca americana*), in contrast, contains an active principle that promotes the division of white blood cells. The pokeweed mitogen is also associated with the stimulation of production of interferon. Some unusual plant toxins act on the skeletal structure itself. Sweet pea (*Lathyrus odoratus*) poisoning in laboratory animals consists primarily of disturbance of the normal deposition and resorption of bone in such a way that cartilage proliferates. The gross effect includes a twisting of the vertebral column. The poisonous principle is the same as that causing dissecting aneurysms in other animals. "Crooked-calf disease" is associated with lupines (*Lupinus sericeus*).

Lung

Under some circumstances, ingestion of rape (*Brassica napus*) forage results in lesions of the lung. Edematous swelling and emphysema are followed by rupture of the alveoli and passage of air from the lungs to collect subdermally on either side of the backbone, where it may be palpated. Somewhat similar lesions are produced by nitrogen gases, as described above.

Thyroid

Many members of the cabbage family (Cruciferac) contain glycosides that release goitrogenic factors (thiocyanates and thiooxazolidone) when digested. The thyroid responds to these compounds by enlarging, and other signs of goiter appear. These goitrogens are iodine responsive.

Eye

The eyes are sometimes involved in signs of poisoning. Severe mydriasis that may provoke visual disturbance is a common sign of poisoning by alkaloids of the tropane configuration, such as atropine found in jimsonweed (*Datura stramonium*). Blindness accompanies a number of poisonings in which other signs may be primary. Often the eye appears normal in structure and function, and blindness may be ascribed to malfunction in the central nervous system. An example of a plant that causes functional blindness is tansy mustard (*Descurainia pinnata*). In man, ingestion of poppy (isoquinoline) alkaloids as from prickly poppy (*Argemone mexicana*) often causes generalized edema and has been held responsible for producing glaucoma.

Nervous System

Nervous signs are perhaps second in frequency only to gastroenteric signs in case of plant poisoning. Nervous involvement and the consequent specific signs are extremely varied and may depend as much on species of animal as on the poisonous principle itself. When nervous tissue is the principal target, lesions are usually absent. There are exceptions. Cerebral demyelination has been reported in lambs born to ewes fed heavily on seaweeds. Liquefactive necrosis of cerebellar areas has been associated with molded forage and with ingestion of sensitive fern (*Onoclea sensibilis*) by horses. More recently, focal necrosis in the anterior globus pallidus of the cerebrum and substantia nigra of the mesencephalon has been described as the primary lesion in "chewing disease" of horses fed on yellow star thistle (*Centaurea solstitialis*). Pigs born to sows that have fed heavily on white clover (*Trifolium repens*) may under some circumstances display demyelination of the spinal cord and be unable to suckle . The exact signs associated with lesions of cranial and spinal nervous tissue vary widely and depend on exactly what areas are involved. An example of a type of poisoning in which obvious lesions of the nervous tissue do not occur is the convulsive syndrome characteristic of water hemlock (*Cicuta maculata*) poisoning.

In many cases a nervous syndrome may be characterized grossly as excitatory or depressive: An example of the former is the hyperexcitable condition produced in animals after ingestion of Dallis grass (*Paspalum dilatatum*) parasitized by an ergot (*Claviceps paspali*). In contrast, weakness and paralysis are characteristic of the effect of poisons that interfere with the normal action of the voluntary musculature through its innervation. For example, guajillo (*Acacia berlandieri*) provokes a classic syndrome of ascending posterior paralysis. In this case, the identity of the poisonous principle, N-methyl-β-phenylalanine, has been worked out. Coniine, the alkaloid of poison hemlock (*Conium maculatum*), the first alkaloid to be synthesized, has an action that is essentially similar in gross effect. Sleepy grass (*Stipa robusta*), found in a limited area of the American Southwest, provokes drowsiness in horses in small amounts and deep sleep in larger doses. The active principle is unknown.

Additional nervous symptoms can be described in man. Paresthesias and hallucinations are associated with particular types of poisoning. The alkaloid aconitine (e.g., from monkshood, *Aconitum napellus*) causes the former, and jimsonweed (*Datura stramonium*) in moderate amount gives rise to the latter.

In gangrenous ergotism, chronic ingestion of ergot (usually *Claviceps purpurea* on rye—*Secale cereale*) causes constriction of the musculature of arterioles. This results in a predisposition to thromboses or other occlusions of the circulation, particularly in the extremities where blood pressure is minimal. The tissue distal to the occlusion dies, becomes septic, and eventually is lost.

Respiratory System

Respiration can be inhibited or prevented at any level, from the mouth to the respiring cells themselves. Cases involving the oral cavity, the lungs, and the cardiovascular system have already been mentioned. In addition to these, respiration can be blocked by the inability of blood to carry oxygen, as in nitrate poisoning, or of the cells to use it, as in cyanide poisoning. Many plants can take up dangerously high levels of nitrogen under conditions of heavy fertilization; a few (such as corn—*Zea mays*) are predisposed to do so. Treatment of crops with 2,4-D can upset nitrogen metabolism with similar results. Nitrate in plants is largely converted to nitrite in the gut. This reacts with hemoglobin to form methemoglobin. The ability of the blood to carry oxygen is impaired in proportion to conversion. Cyanide, released from cyanogenetic glycosides of various kinds elaborated in a large number of plants (e.g., wild cherries—*Prunus* spp.), blocks the action of cytochrome oxidase and thereby interferes with the uptake of oxygen into cellular respiration. In both cases the general signs are those associated with asphyxiation.

Reproductive System

Some plants can elaborate estrogenic factors under certain conditions. Certain forage legumes are known to do so. Corn (*Zea mays*) molded by an unknown fungus has provoked a similar syndrome. Poisoning is characterized by vaginal swelling and prolapse in female animals. Some effects may also be observed in male animals.

Some plants, ingested by pregnant animals, have a pronounced teratogenic effect on the developing embryo. Perhaps the best example is the cycloptic lambs born to ewes that have been exposed to ingestion of hellebore (*Veratrum californicum*) on the fourteenth day of gestation. At that time the embryo is sensitive in development of facial structure. A massive exposure to the plant results in severely inhibited facial development. While the lower jaw remains more or less normal, the upper jaw disappears nearly entirely, and both eyes occupy a single orbit in the center of the forehead, or there may be but one central eye.

Some poisonous principles (fortunately few) are readily excreted in milk. These include especially those of high fat solubility, which may be concentrated in the butterfat. While excretion of toxic principles is beneficial to the animal doing so, the consumption of milk from poisoned animals can induce poisoning in man or nursing animals. The secondary poisoning can be more severe than that in the lactating animal because of the concentration of the poison and the lesser ability of the nonlactating consumer to eliminate it. Perhaps the best example is given by white snakeroot (*Eupatorium rugosum*). The poisonous principle (tremetol) it contains provokes a disease in cattle called trembles. Ingestion of milk from poisoned cattle causes a serious debilitating disease, "milksickness," in man. Farmers commonly associate abortion in livestock with ingestion of weeds. The actual causes of abortion are many. Some plants in toxic amounts (e.g., broomweed—*Gutierrezia microcephala*) undoubtedly can lead to abortion in pregnant animals, but the mechanism is undetermined.

Hair and Skin

The effect of photosensitization on the skin has been mentioned above. Thickening of the skin (hyperkeratosis), usually accompanied by loss of hair, can result from a variety of causes but is often related to a deficiency of vitamin A. It has been shown that some isolates of the imperfect fungus *Aspergillus* can induce the formation of a factor in the substrate that produces well-developed cases of hyperkeratosis in experimental animals. Some forage plants, particularly legumes, grown on high molybdenum soils accumulate enough molybdenum to become toxic to grazing animals. One of the chief signs of poisoning is depigmentation of the hair. The incidence of this disease in cattle at pasture can be mapped by airplane. Depigmentation, which develops slowly, is ascribed to the inability of tyrosinase to mediate the formation of melanin in the absence of copper, the availability of which in the body is reciprocally related to the bodily concentration of molybdenum. Loss of long hair and hoof deformity accompany one type of selenium poisoning. On certain seleniferous soils, grain crops may develop concentrations of selenium, in organic combination, of 5 ppm or more. At this level, continued ingestion of grain or forage produces the syndrome known as "alkali disease." It appears that selenium substitutes for sulfur in amino acids. Hoof deformity and sloughing will become severe enough in time that affected animals will graze from a kneeling position and eventually starve to death. The amino acid mimosine (as in koa haole, *Leucaena glauca*) is held responsible for causing a similar syndrome when ingested.

Poisonous plants are numerous, ubiquitous, troublesome, and poorly studied. No general botanic, ecologic, or geographic relationship exists among them to bring order to an understanding of the diversity of toxic principles contained in plants, viewed either chemically or physiologically, and in the syndromes provoked by them in man and animals. Syndromes in which a plant poison is the prime etiologic factor may involve the integument and hair, the mouth and any portion of the digestive system (with a major distinction possible between ruminant and simple-stomached animals), the parenchymatous organs (especially the liver and kidney), the circulatory system (including the heart, vascularization, and blood itself), the skeletal system, the lungs and respiratory system (including cellular respiration), various glands (especially the thyroid), the central and autonomic nervous systems, the eye, and the reproductive system. Full elucidation of a poisonous plant syndrome usually requires the efforts of a team of research specialists. Some past investigations have yielded results of great medical or economic importance. Results are needed even more urgently so that reaction to poisoning by plants, its treatment, and prevention can be more intelligently founded than at present.

15

VALERIAN

Valerian is a perennial herb comprised of grooved hollow stems and saw-toothed green leaves. White, pale pink, or reddish flowers appear from June to August. Valerian grows to heights of 3–5 feet in the temperate climates of North America, western Asia, and Europe, often in moist soil along riverbanks. The vertical rhizome and attached roots of valerian are parts used medicinally, and are best harvested in the autumn of the second year. Although the fresh drug has no distinctive odor, over time hydrolysis of compounds present in the volatile oil produces isovaleric acid, which has an offensive, somewhat putrid odor. Fortunately, the smell can be removed from the skin and utensils by washing with sodium bicarbonate. Even though valerian has a disagreeable odor, people in the 16th century considered it a fragrant perfume. Traditional uses include treatment of insomnia, migraine headache, anxiety, fatigue, and seizures. It has also been applied externally on cuts, sores, and acne. Traditional Chinese uses include treatment of headache, numbness caused by rheumatic conditions, colds, menstrual difficulties, and bruises. The pharmacological effects of valerian have been attributed to the constituents of volatile oils, monoterpenes, valepotriates, and sesquiterpenes (valerenic acid). Some of these constituents have been shown to have a direct action on the brain, and valerenic acid inhibits enzyme-induced breakdown of γ-amino butyric acid (GABA) in the brain resulting in sedation.

CURRENT PROMOTED USES

Valerian is promoted in the United States primarily as a sedative-hypnotic for treatment of insomnia, and as an anxiolytic for restlessness and sleeping disorders associated with anxiety.

SOURCES AND CHEMICAL COMPOSITION

Also referred to as: *Valeriana officinalis* (L.), *Valeriana wallichii* DC. (Indian valerian), *Valeriana alliariifolia Vahl*, *Valeriana sambucifolia Mik*, *Radix valerianae*, red valerian (*Centranthus ruber* [L.] DC), valerian root, *Valerianae radix*, garden heliotrope, all heal, amantilla, and setwall.

PRODUCTS AVAILABLE

Crude valerian root, rhizome, or stolon is dried and used either "as is" or to prepare an extract. Valerian is available as a capsule, tablet, oral solution, or tea. Valerian is also administered externally as a bath additive.

PHARMACOLOGICAL/TOXICOLOGICAL EFFECTS

Insomnia

Several studies have examined the effects of valerian on sleep. Donath and colleagues performed a randomized, double-blind, placebo- controlled, cross-over study assessing the short-term (single dose) and long-term (14-day multiple dosage) effects of valerian extract on sleep structure and sleep quality. There were significant differences between valerian and placebo for parameters describing *slow-wave sleep* (SWS) and shorter sleep latency, with very low adverse events. Leathwood and colleagues demonstrated valerian's effect on sleep quality. A freeze-dried aqueous extract of valerian root (*Rhizomȧ valeriana officinalis* [L.]) 400 mg was compared to two Hova (valerian 60 mg and hop flower extract 30 mg per tablet) tablets and placebo (finely ground brown sugar) in this crossover study involving 128 volunteers. Study participants took the study medication 1 hour before retiring, and filled out a questionnaire the following morning. This was repeated on nonconsecutive nights, such that each of the three treatments, identified only by a code number, was administered in random order three times to each patient. Valerian caused a significant improvement in subjectively evaluated sleep quality and a significant decrease in perceived sleep latency. The self-reported improvement in sleep quality was especially notable in smokers, those patients who considered themselves poor or irregular sleepers, and those who reported having difficulty falling asleep on a prestudy questionnaire. Hova did not demonstrate any beneficial effect, but it was reported to cause a "*hangover effect*" the next morning. Because subjective sleep questionnaires may not correlate with sleep *electroencephalogram* (EEG) results, a parallel EEG sleep study was performed comparing valerian to placebo in 10 young men. There was not a statistically significant difference between valerian and placebo in this small study. The authors hypothesized that the results of this experiment might have differed from the questionnaire-assessed study because of small sample size and differences in study populations. The larger study involved young and older individuals, men and women, and good and poor sleepers, whereas the EEG study involved young men with no reported sleep abnormalities. Rather than place more credence on the objective study, the investigators concluded that the questionnaire provides a more sensitive means of detecting mild sedative effects.

A double-blind, placebo-controlled study was performed in eight volunteers recruited from among the research staff at Nestle Products and their families who reported that they "usually have problems getting to sleep." Sleep latency was measured using an activity monitor and questionnaire. The investigators documented a small (7 minute) but statistically significant decrease in sleep latency with 450 mg of an extract of valerian (*V. officinalis* [L.]). No further improvement was demonstrated with a 900-mg valerian dose; however, patients receiving the higher dose were more likely to feel sleepy the next morning. Sleep quality, sleep latency, and sleep depth also improved according to a nine-point subjective rating scale. However, the appropriateness of the statistical analysis used to interpret the results of the subjective portion of the study is unclear.

A more objective double-blind, placebo-controlled trial evaluated the effect of 450- and 900-mg doses of an aqueous valerian extract (*V. officinalis* [L.]) on two groups of healthy, young (21–44 years of age) volunteers at home and in a laboratory setting. The effect of valerian on sleep was measured using a questionnaire and night-time motor activity recordings in both settings. The effects of valerian on the volunteers in the sleep laboratory were also measured using polysomnography and spectral analysis of the sleep EEG. Both groups demonstrated the mild hypnotic effects of valerian; however, the benefits of valerian were statistically significant only under home conditions.

Another double-blind, placebo-controlled crossover study evaluated Valerina Natt, a preparation equivalent to 400 mg of valerian root composed mainly of sesquiterpenes from *V. officinalis* [L.], on subjective sleep quality assessed using a three-point rating scale. Study subjects were 27 consecutive patients seen in a medical clinic for evaluation of sleep difficulty and fatigue who were willing to participate in the investigation. Statistically significant improvement in sleep quality was noted with the valerian preparation. Valerian was rated as better than placebo by 21 subjects, two rated the preparations equally, and four preferred placebo. No adverse effects were reported. Although some study subjects had experienced nightmares when using conventional hypnotics, nightmares were not reported in the study. The effects of repeated doses (three tablets three times daily) for 8 days of Valdispert Forte (135 mg of dried extract of *V. officinalis* [L.]) in 14 elderly women with sleeping difficulties was assessed using polysomnography in a particularly well-designed study. Inclusion criteria were well defined: sleep latency longer than 30 minutes, more than three nocturnal awakenings per night with inability to go back to sleep within 5 minutes, and total sleep time less than 5 hours. Subjects could not have medical, psychological, or weight-related causes of sleep difficulty, and had to have normal health status for their age. Sedatives, hypnotics, and other *central nervous system* (CNS)-active drugs were discontinued 2 weeks prior to the study, and drug screening for morphine, benzodiazepines, barbiturates, and amphetamine was done prior to study commencement. Results showed an increase in SWS, and a decrease in sleep stage 1. There was no effect on *rapid eye movement* (REM) sleep, sleep latency, time awake after sleep onset, or self-rated sleep quality.

In aggregate, the results of these clinical studies suggest that at doses of approx 450 mg of the aqueous extract, valerian has mild hypnotic effects, possibly by affecting non-REM sleep in patients with reduced SWS. Unlike benzodiazepines, valerian appears not to adversely affect SWS or REM sleep, and does not appear to cause nightmares or hangover. Further well-designed studies are needed to objectively evaluate valerian. Results of animal studies reflect the clinical data. Sedative properties of Valdispert (dried aqueous extract of *V. officinalis* [L.]) in mice were documented based on reduced spontaneous movement and an increase in thiopental-induced sleep time; however, these effects were slightly less than those of diazepam and chlorpromazine. No significant anticonvulsant effect was observed.

Hendriks and colleagues tested several components of the volatile oil, obtained by steam distillation of *V. officinalis* [L.], on mice. The essential oil, its hydrocarbon fraction, its oxygen fraction, valeranone, valerenal, valerenic acid, and isoeugenyl-isovalerate were injected intraperitoneally at various doses ranging from 50 to 1600 mg/kg, with three mice receiving each dose. The mice were observed between 15 and 30 minutes post-injection for various symptoms suggestive of CNS stimulation or depression, analgesia, sympathomimetic or sympatholytic activity, vasodilation, or vasoconstriction. It was concluded that components of the essential oil, particularly valerenic acid and valerenal, which are present in the oxygen fraction, have a sedative and/or muscle relaxant effect. The authors tested the effect of intraperitoneal valerenic acid compared to diazepam, chlorpromazine, and pentobarbital on ability to walk on a rotating rod and grip strength in mice. The effects of valerinic acid on spontaneous motor activity and on pentobarbital-induced sleeping time were also assessed. Diazepam, a muscle relaxant, affected the grip test but not the rotarod test, whereas chlorpromazine, a neuroleptic, affected the rotarod test but not the grip test. Valerenic acid, like pentobarbital, decreased performance in both the rotarod and grip tests. The authors concluded that valerenic acid, like pentobarbital, has general CNS depressant activity. Valerenic acid also decreased spontaneous motor activity and prolonged pentobarbital-induced sleeping time. Dose–response effects of valerenic acid were also observed by the investigators. At a dose of 50 mg/kg, a decrease in spontaneous motor activity occurred. At 100 mg/kg, mice exhibited ataxia, then remained motionless. Muscle spasms occurred at 150–200 mg/kg and convulsions at 400 mg/kg, followed by death in six of seven mice within 24 hours.

Sedation is mediated predominantly through the inhibitory neurotransmitter GAB A. Although the mechanism of action of valerian as a sleep aid is not fully understood, it may involve inhibition of the enzyme that breaks down GABA. Dihydrovaltrate, hydroxyvalerenic acid, a hydroalcoholic extract containing 0.8% valerenic acid; a lipid extract; an aqueous extract of the hydroalcoholic extract, and another aqueous extract of *V. officinalis* (L.) were assessed for in vitro binding to rat GABA, benzodiazepine, and barbiturate receptors. The results indicated that an interaction of some component of the hydroalcoholic extract, the aqueous extract derived from the hydroalcoholic extract, and the other aqueous extract had affinity for the $GABA_A$ receptor. Because hydroxyvalerenic acid (a volatile oil sesquiterpene) and dihydrovaltrate (a valepotriate) did not show any notable activity, the investigators could not identify the specific constituents responsible for this activity. The lipophilic extract derived from the hydroalcoholic extract, as well as dihydrovaltrate, showed affinity for barbiturate receptors, and some affinity for peripheral benzodiazepine receptors.

Other in vitro studies have also yielded results that suggest GABA-mediated activity; however, the active constituent was unidentified. Cavadas and colleagues verified that valerenic acid (0.1 mmol/L) was not able to displace [^{3}H] muscimol from the $GABA_A$ receptor, although both an aqueous and a hydroalcoholic extract were able to do so. The investigators then attempted to identify other compounds in the extracts capable of displacing [^{3}H] muscinol. Both glutamate and glutamine, amino acids present in the aqueous extract, had little inhibitory effect on [^{3}H] muscinol binding. However, glutamine can cross the blood-brain barrier (BBB) and can be taken up by nerve terminals and converted to GABA inside GABA-nergic neurons. Thus, glutamine could be responsible for the sedative effect of the aqueous extract, but not the hydroalcoholic extract, in which it is not present. GABA is found in both extracts, but GABA itself cannot explain the sedative effects of valerian because it is unlikely to cross the BBB in amounts significant enough to cause sedation. However, the amount of GABA present in the aqueous extract is sufficient to have effects on peripheral GABA receptors, perhaps resulting in muscle relaxation. Another study suggests a different mechanism of action involving inhibition of neuronal GABA uptake and stimulation of GABA release from synaptosomes. These investigators did not attempt to elucidate which constituent of the aqueous extract was responsible for these effects.

The CNS-depressant component of valerian is still unknown. Thus far, three major constituents of valerian have been identified: the volatile or essential oil, containing sesquiterpenes and monoterpenes, nonglycosidic iridoid esters (valepotriates), and a small number of alkaloids. Valepotriates are unstable compounds and are easily hydrolyzed by heat and moisture. In addition, valepotriates are not water soluble, and aqueous extracts contain small amounts. For example, the aqueous extract used in the study by Balderer and Borbely, described previously was analyzed using thin-layer chromatography, and no valepotriates were detectable. Furthermore, valepotriates are not well absorbed orally. Therefore, the likelihood that valepotriates are a major contributor to valerian's effects is questionable. Because of the low amount of alkaloid present in preparations, their contribution is also questionable. It is postulated that a combination of volatile oils, valepotriates, and possibly certain water-soluble constituents that have not yet been identified are responsible for valerian's sedative effects.

Antidepressant effects of valerian were identified by Oshima and associates using a methanol extract of *V. fauriei* roots. They found a strong antidepressant activity in mice as measured by the forced swimming test. One active component isolated was α-kessyl alcohol, a volatile oil component. At 30 mg/kg intraperitoneally, α-kessyl alcohol exhibited an effect similar to imipramine, a commonly used antidepressant. Kessanol and cyclokessyl acetate, guaiane-type sesquiterpenoids, also exhibited antidepressant activity. Kanokonol, kessyl glycol, and kessyl glycol diacetate, valerane-type sesquiterpenoids, did not exhibit an effect. A 30% ethanol extract of the Japanese valerian root extract (4.1 g/kg and 5.7 g/kg) and imipramine (20 mg/kg) also demonstrated statistically significant antidepressant effects compared to placebo as measured by the forced swimming test in rats. As in the Oshima study, kessyl glycol diacetate exhibited no antidepressant activity in the forced swimming test. Because the forced swimming test can be affected by stimulants, anticholinergics, and antihistamines as well as antidepressants, the effect of the valerian extract on reserpine-induced hypothermia, a test for antidepressant activity and inhibition of neuronal reuptake of monoamines, was measured. Both valerian (11.2 g/kg) and imipramine (20 mg/kg) reversed reserpine-induced hypothermia, suggesting that the antidepressant effect of valerian is caused by reuptake of monoamine neurotransmitters, as with conventional antidepressants. More evidence is needed to evaluate the use of valerian in children. One study using a combination product of valerian root extract and lemon balm leaf extract found that symptoms of dyssomnia or pathological restlessness might decrease in children under age 12.

Anxiety

A few studies have examined the effects of valerian on anxiety. Cropley and colleagues investigated whether kava or valerian could moderate physiological stress induced under laboratory conditions in healthy volunteers. Subject (n = 18-kava, and n = 18-valerian) and comparison group (n = 36) volunteers performed a standardized mental stress task 1 week apart. Cases had their blood pressure, heart rate, and subjective ratings of pressure assessed at rest and during the mental stress task (time 1 = *T1*). The valerian subjects took a standard dose for 7 days (time 2 = *T2*). In the valerian group, heart rate reaction to mental stress was found to decline, systolic blood pressure decreased significantly, and subjects reported less pressure during mental stress test tasks at *T2* relative to *T1*. Behavioral performance on the standardized mental stress test task did not change between the groups over the two time points. There were no significant differences in blood pressure, heart rate, or subjective reports of pressure between *T1* and *T2* in the control group. Kohnen and Oswald conducted a study on the effects of valerian, propranolol, and combinations on activation, performance, and mood of healthy volunteers under social stressor conditions. The results of this study were equivocal and published over 15 years ago; however, it is mentioned here for historical reference.

Andreatini and colleagues examined the effect of valerian extract (valepotriates) using a randomized, parallel, double-blind placebo-controlled pilot study design in patients with generalized anxiety disorder (GAD). After a 2-week wash-out period, 36 patients with GAD as defined by the *Diagnostic and Statistical Manual of Mental Disorders,* were randomized to one of the following three treatment groups for 4 weeks: valepotriates, mean daily dose of 81.3 mg; diazepam, mean daily dose of 6.5 mg; or placebo. There was a significant reduction in the psychic factor of the Hamilton anxiety scale in the valepotriates group; however, the principal study analysis using between group comparisons on total Hamilton anxiety scale scores found negative results. The conclusion of this study suggests that there may be a potential anxiolytic effect of valepotriates on the psychic symptoms of anxiety, but the total number of subjects per group (n = 12) was very small and results must be viewed as preliminary.

Musculoskeletal Relaxation

Isovaltrate and valtrate (valepotriates) and valeronone, an essential oil component, isolated from *V. edulis* ssp. procera Meyer (Valeriana "mexicana") caused suppression of rhythmic contractions in guinea pig ileum in vivo at a dose of 20 mg/kg administered intravenously via the jugular vein. The investigators also demonstrated that the same compounds as well as dihydrovaltrate isolated from the same valerian species produced relaxation of carbachol-stimulated guinea pig ileum preparations in vitro. They concluded that these compounds have a musculotropic action in concentrations from 10^{-5} to 10^{-4} *M*.

Pharmacokinetics

One study has evaluated the pharmacokinetics of valerian following administration to humans. Following administration of a single 600-mg dose of valerian, the pharmacokinetics of valerenic acid were measured. The T_{max} occurred between 1 and 2 hours and the C_{max} was between 0.9 and 2.3 ng/mL. Concentrations of valerenic acid were measurable for at least 5 hours following the dose. The elimination half-life was approx 1 hour. The authors suggest that based on the expected use of valerian (sedative effects), dosing 30 minutes to 2 hours prior to bedtime would be appropriate based on the previously mentioned pharmacokinetics.

Adverse Effects and Toxicity

Reproductive System

There has been a theoretical concern with regard to pregnant women taking valerian because of possible effects on uterine contractions, but no problems were noted in three cases of intentional overdose with 2–5 g of valerian during weeks 3–10 of pregnancy. A mentally retarded child was born to a woman who overdosed on valerian 3 g, phenobarbital, glutethamide, amobarbital, and promethazine at 20 weeks of gestation, but this same woman delivered a mentally retarded child 2 years later after an overdose attempt with glutethamide, amobarbital, and promethazine.

V. officinalis (L.) was tested on rats and their offspring. A mixture, containing three valepotriates (80% dihydrovaltrate, 15% valtrate, and 5% acevaltrate), was orally administered to female rats for 30 days at 6-, 12-, and 24-mg/kg doses. Each dose was given to 10 rats, and placebo was given to another 10. No changes were noted in the average length of the estrus cycle, or the number of estrus phases during the 30-day observation period. The valepotriate mixture or placebo were also administered to 40 pregnant rats in the manner described previously from the day 1 through day 19 of pregnancy. Valerian did not increase the risk of fetotoxicity or external malformation. However, internal examination revealed a significant increase in the number of fetuses with retarded ossification with the 12- and 24-mg/kg doses. No developmental changes were detected in the offspring after treatment during pregnancy.

Cardiovascular System

Pharmacological investigations using a particular valepotriate fraction called Vpt2 extracted from the roots of *V. officinalis* (L.) have shown antiarrhythmic activity and ability to dilate coronary arteries in experimental animals. Moderate positive inotropic and a negative chronotropic effect were also observed. Vpt2 contains valtratum (50%), valeridine (25%), and valechlorin (3%), with trace amounts of acevaltrate, dihydrovaltratum, and epi-7-desacetyl-isovaltrate.

Alcoholic extracts of *V. officinalis* (L.) root (labeled V103 and V115) demonstrated hypotensive effects in rats, cats, and dogs. The V115 fraction showed greater potency and was extracted by a countercurrent distribution to yield three fractions. The first two fractions demonstrated hypotensive effects in rats, with the first fraction showing a hypotensive effect at 30 mg/kg. The third fraction produced hypertensive effects at a dose of 200 mg/kg. The authors noted that, apparently, with each succeeding extraction, less of the hypotensive principle was extracted. The hypotensive effect of the V103 fraction in rats was demonstrated at a dose of 500 mg/kg, and was hypothesized to act via a parasympathomimetic effect, blockade of the carotid sinus reflex, and CNS depression.

Cytotoxicity

The valepotriates valtrate/isovaltrate and dihydrovaltrate were isolated from *V. mexicana* and *V. wallichii*, respectively. The valepotriates tested were cytotoxic to granulocyte/macrophage colony-forming units (GM-CFCUs), lymphocytes, and erythrocyte colony-forming units (E-CFCUs). Valtrate was found to be a more potent inhibitor of GM-CFCUs (ID50 $\sim 3.7 \times 10^{-6}$ *M* vs $\sim 1.7 \times 10^{-5}$ *M*) and T-lymphocytes (ID50 $\sim 2.8 \times 10^{-6}$ *M* vs $\sim 3 \times 10^{-5}$ *M*) than dihydrovaltrate. Valtrate and dihydrovaltrate were similar in their activity against E-CFCUs (ID50 $\sim 2.3 \times 10^{-8}$ *M* vs $\sim 4.2 \times 10^{-8}$ *M*). Because pharmaceutical products containing valepotriates are orally administered, their cytotoxicity to gastrointestinal mucosal cells is of concern.

The effects of valtrate, dihydrovaltrate, and deoxido-dihydrovaltrate, valepotriates extracted from *V. wallichii* (DC.), on cultured rat hepatoma cells have been studied. Valtrate killed 50% of the cell population at a concentration of 5 μM, Deoxido-dihydrovaltrate and dihydrovaltae demonstrated this same toxicity at double the dose. Valtrate was also the most potent inhibitor of DNA and protein synthesis. These results suggest a mechanism by which valerian may cause hepatotoxicity.

Case Reports of Toxicity

Four cases of women who sustained liver damage after taking valerian-containing herbal medicines to relieve stress have been described. In addition, valerian was used by a patient who exhibited hepatotoxicity attributed to Chaparral.

Hospitals admitted 23 patients for treatment of intentional overdose with Sleep-Qik (75 mg of valerian dry extract, 0.25 mg of hyoscine hydrobromide 2 mg of cyproheptadine hydrochloride) between 1988 and 1991. Of these 23, 9 were men and 14 were women, with a mean age of 23.8 years (range 15–37 years). They were previously healthy, except for two patients with histories of psychiatric illness. The mean number of Sleep-Qik tablets taken per patient history was 33 (range 6–166), for an average of 2.5 g (range 0.5–12 g) of valerian. Four patients were asymptomatic. The other 19 patients reported drowsiness (n = 11), dilated pupils (n = 11), tachycardia (n = 6), nausea (n = 4), confusion (n = 3), urinary retention (n = 3), visual hallucination (n = 2), flushing (n = 2), dry mouth (n = 1), and dizziness (n = 1). Coingestants were alcohol (n = 2), a pesticide (n = 1), and Pansedan (n = 1) (*Passiflora* extract, *Viscum album* extract, *Uncariarhyncophylla* extract, and *Humulus lupulus*). One patient who was drowsy had also taken Panseden, and one who was confused had ingested alcohol.

Most patients received gastric lavage (n = 14), and one received syrup of ipecac. The patient who took 60 tablets of Sleep-Qik required ventilatory support. Liver function tests were performed on 12 patients approx 6–12 hours after ingestion with normal results. Drowsiness and confusion resolved within 24 hours. All patients recovered completely and were discharged after an average of 1.7 days (range 1–6 days). At an average of 43 months (range 27–65 months) after presentation, 10 patients were contacted by telephone. They had all remained well after discharge and none continued taking Sleep-Qik. Delayed onset of severe liver damage was ruled out via telephone interview, but subclinical disease could not be ruled out.

Subsequently, Chan reported on 24 cases of overdose of a product containing valerian dry extract· 75 mg, hyoscine hydrobromide 0.25 mg, and cyproheptadine hydrochloride 2 mg. Six patients developed vomiting, and 15 underwent gastric lavage. Co-ingestants included alcohol (n = 10), cold products (n = 3), hypnotics (n = 2), unknown drugs (n = 2), and gasoline (n = 1). Symptoms were mainly CNS depression and anticholinergic symptoms. One patient required ventilatory support. Liver function tests were performed in 17 cases, and all were normal. Over the next 22–48 months postingestion, none of the patients returned to the hospital or clinic for any reason, suggesting that serious hepatotoxicity did not occur. The author points out that gastric lavage and spontaneous vomiting may have limited the amount of valerian absorbed in these patients, thus decreasing the risk of any delayed adverse effects. Other adverse effects attributed to overdose or chronic use of valerian include headaches, excitability, restlessness, uneasiness, blurred vision, and cardiac disturbances.

In another reported suicide attempt, an 18-year-old female ingested between 40 and 50, 470-mg capsules (18.8–23.5 g valerian) of 100% powered valerian root. The patient complained of fatigue, crampy abdominal pain, chest tightness, tremor of the hands and feet, and lightheadedness 30 minutes after ingestion. She presented to the emergency room 3 hours postingestion. Her vital signs were: blood pressure 111/64 mmHg, pulse 72 beats/minute, respiratory rate 14 breaths/minute, and temperature 37.6°C. Physical exam was unremark-able except for mydriasis (6 mm bilaterally). Electrocardiograph, complete blood count, and chemistry profile including liver function tests were normal. Toxicology screen was positive for marijuana, which she admitted using 2 weeks previously. She denied ingesting anything else. After two doses of activated charcoal, her symptoms resolved within 24 hours.

A withdrawal syndrome was described after abrupt discontinuation of valerian root extract in a 58-year-old man who had taken 530–2000 mg/dose five times daily as an anxiolytic and hypnotic for many years. Withdrawal symptoms included sinus tachycardia of up to 150 beats/minute, tremulousness, and delirium after recovery from general anesthesia (propofol, nitrous oxide, isoflurane, and thiopental) for open biopsy of a lung nodule. Medical history included coronary artery disease, hypertension, and congestive heart failure with an ejection fraction of 30–35%. Medications included isosorbide dinitrate, digoxin, furosemide, benazepril, aspirin, lovastatin, ibuprofen, potassium, zinc supplement, and vitamins.

The biopsy was complicated by multiple episodes of oxygen desaturation, and after extubation, the patient experienced tacycardia, oliguria, and increasing oxygen requirement. Despite naloxone administration, symptoms worsened. Swan-Ganz catheterization revealed high-output heart failure. At this time, interview with family members revealed the patient's long-standing valerian use. Because valerian withdrawal was suspected, midazolam 1 mg/hour (total dose 11 mg in 17 hours) was administered. Signs and symptoms improved, and stabilized by the third postoperative day. He was switched to lorazepam 1 mg/hour as needed (total dose 5 mg in 24 hours), and then to a tapering dose of clonazepam. He was discharged on postoperative day 7, and was stable at 5-month follow-up. Other causes of high-output heart failure were ruled out, but because of the patient's multiple medical problems, postsurgical status, and medications administered, the cause of the patient's symptoms is unclear. The authors of this case report note that valerian has been reported to attenuate benzodiazepine withdrawal in rats.

Interactions

Two alcoholic valerian extracts were found to potentiate pentobarbital sleeping time in mice, and Valdispert, an aqueous extract prepared from *V. officinalis* (L.), increased the thiopental sleeping time in a dose-dependent manner in rats. Based on these animal studies, in vitro studies of valerian's effect on GABAnergic transmission, as well as the case series reported by Chan and colleagues, valerian would be expected to have at least an additive effect with barbiturates, alcohol, benzodiazepines, and other CNS depressants. Valerian may have the potential to increase the level of drugs metabolized by the cytochrome P-450 3A4 (CYP3A4) enzyme. In vitro studies have found that valerian may have an inhibitory effect on CYP3A4. A clinical research study suggested that low to moderate doses of valerian did not significantly inhibit CYP3A4, although taking valerian extract 1000 mg/day increased alprazolam levels by 19%. Therefore, it may be wise to use valerian cautiously in patients taking medications that are CYP3A4 substrates such as lovastatin, ketoconazole, itraconazole, fexofenadine, alprazolam, triazolam, and various chemotherapeutic agents.

Reproduction

No information is available concerning any potential effects of valerian on female reproductive function. However, Mkrtchyan and colleagues reported that valerian had no effect on human male sterility.

Regulatory Status

Valerian was included as an official drug in the US Pharmacopeia until 1936 and in the National Formulary until 1946. Currently, the USP advisory panel does not recommend valerian's use owing to lack of adequate scientific evidence and conflicting study results. They encourage further research. Valerian is generally recognized as safe as a food and beverage flavoring by the FDA. The German Commission Monograph E has approved valerian as a sleep-promoting and calmative agent to be used in the treatment of unrest and sleep disturbances caused by anxiety. In Australia, valerian is acceptable as an active ingredient in the "listed products" category of the Therapeutic Goods Administration. In Belgium, subterranean parts, powder extract, and tincture are allowed for use as traditional tranquilizers. The Health Protection Branch of Health Canada allows products containing valerian as a single agent in the form of crude dried root in tablets, capsules, powders, extracts, tinctures, drops, or tea bags intended for use as sleeping aids and sedatives.

16

VITEX AGNUS-CASTUS

Vitex agnus-castus is a botanical plant that has the following National Oceanographic Data Center Taxonomic Code: Kingdom, Plantae; Phylum, Tracheobionta; Class, Magnoliopsida; Order, Lamiales; Family, Verbenaceae; Genus, *Vitex* L.; Species, *Vitex agnus-castus* L. The genus name *Vitex* is a Latin derivation for plaiting or weaving. The species name *agnus-castus* combines two Latin word origins: "*agnus*," which means lamb, and "*castitas*," which means chastity. *V. agnus-castus* is a large deciduous shrub, native to Mediterranean countries and central Asia, and is also used in America as an ornamental plant. *V. agnus-castus* has long, finger-shaped leaves and displays fragrant blue-violet flowers in midsummer. Its fruit is a very dark-purple berry that is yellowish inside, resembles a peppercorn, and has an aromatic odor. Upon ripening, the berry is picked and allowed to dry. The twigs of this shrub are very flexible and were used for furniture in ancient times.

References to *V. agnus-castus* go back more than 2000 years, describing it as a healing herb. Ancient Egyptians, Greeks, and Romans used it for a variety of health problems. In 400 BCE, Hippocrates recommended chaste tree for injuries and inflammation. Four centuries later, Greek botanist Dioscorides recommended *V. agnus-castus* specifically for inflammation of the womb and lactation. Use of *V. agnus-castus* continued into the Middle Ages, where folklore persists that medieval monks chewed *V. agnus-castus* tree parts to maintain their celibacy, used the dried berries in their food, or placed the berries in the pockets of their robes in order to reduce sexual desire; thus, the synonym of Monk's pepper. Use of *V. agnus-castus* has persisted to modern times. Though its use was initially concentrated in the Mediterranean area, its popularity has increased in England and America since the mid-1900s.

Traditional medicinal uses of *V. agnus-castus* lie predominantly around the oral ingestion of the shrub's fruit; however, other plant parts such as leaves and flowers have been used in some preparations. The dry or liquid extract of, or oils from, the berry have been used for a variety of symptoms, most commonly related to the female reproductive system. Other uses include the treatment of hangovers, flatulence, fevers, benign prostatic hyperplasia, nervousness, dementia, rheumatic conditions, colds, dyspepsia, spleen disorders, constipation, and promoting urination. Traditional topical medicinal uses of *V. agnus-castus* include acne, body inflammation, and insect bites and stings. Use of *V. agnus-castus* is not commonly employed in traditional Chinese medicine or traditional Indian medicine (Ayurveda); however, other *Vitex* species (*negundo*, *trifoliata*) are used in these therapies.

CURRENT PROMOTED USES

Current promoted uses of *V. agnus-castus* relate to treatment of disorders of the female reproductive system such as short menstrual cycles, *premenstrual syndrome* (PMS), and breast swelling and pain (mastodynia/ mastalgia). The Commission E has approved the use of *V. agnus-castus* for irregularities of the menstrual cycle, premenstrual complaints, and mastalgia. Recent randomized, placebo-controlled studies have been conducted and found *V. agnus-castus* to be effective and well-tolerated for the relief of PMS symptoms, especially the physical symptoms of breast tenderness/fullness, edema, and headache. *V. agnus-castus* is not considered effective for PMS-related symptoms of abdominal bloating, craving sweets, sweating, palpitations, or dizziness.

V. agnus-castus is not used in foods and is not recommended for use in children, adolescents, pregnant women, or women who are breast- feeding. *V. agnus-castus* should be avoided in patients receiving exogenous sex hormones, including oral contraceptives, as *V. agnus-castus* may counteract the effectiveness of birth control pills by its effect on prolactin.

Sources and Chemical Composition

V. agnus-castus (L.), Agnolyt, arbre chaste, chaste berry, chasteberry, chaste tree, chaste tree fruit, chastetree, chastetree berry, Cloister Pepper, *Fructus Agni Casti*, Fruit de Gattilier, Gattilier, Hemp Tree, Keuschlamm, Mönchspfeffer, Monk's Pepper, *V. agnus castus*, *V. agnus castus fructus*, Vitex. The major constituents of *V. agnus-castus* include the following. Flavonoids: flavonol (kaempferol, quercetagetin) derivatives, the major constituent being casticin. Additional flavonoids found include penduletin, orientin, chrysophanol D, and apigenin. Water-soluble flavones: vitexin and isovitexin. Alkaloids: viticin. Diterpenes: rotundifuran (labdane-type); vitexilactone; 6-β,7-β-diacetoxy-13-hydroxy-labda-8,14-diene; 8,13-dihydroxy-14-labden; X-hydroxy-y-keto-15,16-epoxy-13, 14-labdadien;X-acetonxy-13-hydroxylabda-y,14-dien; cleroda-x,14-dien-13-ol; cleroda-x,y, 14-trien- 13-ol. Iridoid glycosides: In the leaf: 0.3% aucubin, 0.6% agnuside (the *p*-hydroxybenzoyl derivative of aucubin), and 0.07% unidentified glycosides. In the flowering stem (6'-O-foliamenthoyl-mussaenosidic acid [agnucastoside A], 6'O (6,7- dihydrofoliamenthoyl) mussaenosidic acid [agnucastoside B], and 7-O-trans-*p*-coumaroyl-6'-O-trans-caffeoyl-8-epiloganic acid [agnucastoside C], aucubin, agnuside, mussaenosidic acid, 62-O-*p*-hydroxybenzoylmussaenosidic acid, and phenylbutanone glucoside [myzodendrone]. Essential oil of leaves and flowers: monoterpenes (major chemicals found: limonene, cineole, sabinene, and α-terpineol, linalool, citronellol, camphene, myrcene) and sesquiterpenes (major chemicals found: β-caryophyllene, β-gurjunene, cuparene, and globulol). Depending on the maturity of the fruits used and the distillation processes, the components of the essential oil can vary greatly. Other constituents: fatty acids (including stearic, oleic, linoleic, and palmitic acids), amino acids (glycine, alanine, valine, leucine), castine (a bitter principle), vitamin C and carotene, and trace amounts of hormones from leaves and flowers (progesterone and 17 α-hydroxyprogesterone). Main components of the volatile oil 0.5%: mixtures of monoterpenes and sesquiterpenes, cineol, and pinene.

Products Available

V. agnus-castus is available as bulk berries, bulk powder, crushed fresh or dried berry, tea (loose or in tea bags), extract, tonic, elixir, or tincture. Chasteberry products may consist of the herb alone or in combination with other herbs and vitamins. Topically, it is used primarily as the essential oil, mixed in combination with other products in cream form.

A proprietary preparation (Agnolyt) containing an alcoholic extract of *V. agnus-castus* (0.2% w/w) has been available in Germany since the 1950s. Other products using *V. agnus-castus* synonyms in their nomenclature are available by a myriad of manufacturers. Some single-entity brand names include Agnofem, Agno-Sabona, Agnucaston, Agnufemil, Agnuside, Agnumens, Agnurell, Antimast N, Antimast T, Gynocastus, Mastodynon, and Vitex Extract. Many combination products also contain *V. agnus-castus* and include, but are not limited to, the following: Herbal Premens, Herbal Support For Women Over 45, Dong Quai Complex, Emoton, Feminine Herbal Complex (FM), Menosan, Mulimen,

Phytoestrin, Virilis-Gastreu SR41, Femisana, Lifesystem Herbal Formula 4 Women's Formula (FM), PMT Complex (FM), Presselin Dysmen Olin 3 N (FM), Women's Formula Herbal Formula 3 (FM), and others. The amount of *V. agnus-castus* contained in oral tablets or capsules varies depending on whether the product contains crushed fruit or extract of the berry. For example, tablet or capsule formulations have included the following: Chaste Berry 450 mg/Chase Berry Extract 50 mg, Chastetree fruit 500 mg, Chaste Berry Dried Extract 1.6–3.0 mg corresponding to 20 mg *Vitex*, Chaste Tree Berry Extract (0.5% agnuside) 225 mg. Commercial extract forms of chasteberry are usually standardized to contain 6% agnuside constituent. Chasteberry liquid extract may or may not contain alcohol.

Dosage

Generally, the Expanded Commission E Monographs reports the following total daily dosages:

- 30 to 40 mg of dry or fluid extracts of crushed fruit
- 0.03 to 0.04 mL of fluid extract 1:1 (g/mL), 50–70% alcohol (v/v)
- 0.15 to 0.2 mL of tincture 1:5 (g/mL), 50–70% alcohol (v/v)
- 2.6 to 4.2 mg of dry native extract (9.5–11.5:1 (w/w)

Other references have reported total daily dosages ranging from 20 to 1800 mg/day of crude *V. agnus-castus* extracts. *V. agnus-castus* is considered safe when used orally and appropriately.

Pharmacological/Toxicological Effects

Prolactin Secretion

Evidence of varying levels of discrimination exists that demonstrate *V. agnus-castus* inhibits the secretion of prolactin by the pituitary gland. In a randomized, placebo-controlled, double-blind study, Milewicz et al. examined whether *V. agnus-castus* affected elevated pituitary prolactin reserve. Participants were 52 women with luteal phase defects caused by latent hyperprolactinemia. Intervention was *V. agnus-castus* 20 mg daily or placebo, for 3 months. Only 37 women (20 = placebo, 17 = *V. agnus-castus*) completed the study. Outcome measures were pre- and posthormonal analysis (blood draws taken on days 5–8 and day 20 of menstrual cycle) 1 month prior to treatment and after 3 months of treatment and latent hyperprolactinemia analysis (monitoring prolactin release 15 and 30 minutes after intravenous injection of 200 μg *thyrotropin-releasing hormone* (TRH). Results from this study showed that compared to preintervention, the *V. agnu-scastus* group had statistically significant reduced prolactin release after 3 months, whereas the control group did not. The study's information came from an English abstract of a German publication. Information about inclusion/exclusion criteria, study specifics, study funding, or author disclosures were not available.

In an open and intraindividual comparison study, Merz et al. conducted a clinical study of tolerance and prolactin secretion of *V. agnus-castus* using 20 healthy male subjects between the ages of 18 and 40 years. Placebo and three doses of *V. agnus-castus* (total daily dosages of 120, 240, and 480 mg were divided into 8-hour administration times) were given in an increasing sequence. Prolactin concentration profile after TRH stimulation was assessed by determining the maximum concentrations (C_{max}) and the area under the curve over a period of one hour ($AUC_{0\text{-}1h}$). These procedures were identical in all four study phases. Results for the $AUC_{0\text{-}24h}$ showed that as daily doses increased, prolactin levels decreased ([$AUC_{0\text{-}24h}$ {μIU · hour}/μΛ ± standard deviation]; placebo: 6182 ± 1827; 120 mg: 6874 ± 1790; 240 mg: 5750 ± 1594; 480 mg: 5998 ± 1664), with statistically significant findings for the 120-mg dosage only.

Wuttke reported in a 1996 abstract the results of experiments demonstrating that 3 months of *V. agnus-castus* therapy (double-blind clinical study vs placebo) significantly reduced basal prolactin levels in patients. However, details of the experiments and study subjects were not outlined or referenced.

Follicle-Stimulating Hormone, Luteinizing Hormone

There are a limited number of human studies regarding how *V. agnus-castus* directly effects *luteinizing hormone* (LH) or follicle-stimulating hormone (FSH). In a 1994 case report by Cahill et al., a 32-year-old woman undergoing unstimulated in vitro fertilization (IVF) treatment took *V. agnus-castus* for one cycle without consulting her physician. During this cycle, she had symptoms of mild ovarian hyperstimulation in the luteal phase. Her FSH and LH levels prior to day 13, the predicted day of LH surge in the IVF cycle, were reviewed and found to be much higher than normal. Reviewing five other cycles of this patient and finding normal pituitary gonadotrophin file and normal follicular ovarian responses, the authors suggest that *V. agnus-castus* was the causative agent.

In the 1996 study by Merz et al. that primarily examined prolactin secretion in male subjects, initial hormone levels of FSH and LH were measured on days 1 and 13 (beginning and near-end of placebo phase) and from blood samples taken during the prolactin secretion profiling. The authors state that *V. agnus-castus* had no effect on FSH or LH levels, but no other details were provided.

Progesterone/Testosterone Synthesis

In a randomized, placebo-controlled, double-blind study, Milewicz et al. examined the effect of *V. agnus-castus* on prolactin reserve and luteal phase progesterone synthesis in 52 women. The intervention was *V. agnus-astus* 20 mg daily or placebo, for 3 months. Results from the 37 women (20 = placebo, 17 = *agnus-castus*) who completed the study showed that compared to preintervention, the *V. agnus-castus group* had statistically significant increases in luteal phase progesterone synthesis. The study information came from a German publication with an English abstract. Information about inclusion/exclusion criteria, study specifics, study funding, or author disclosures were not available.

In the 1996 study by Merz et al. that primarily examined prolactin secretion in male subjects, initial hormone level of testosterone was measured on days 1 and 13 (beginning and near-end of placebo phase), and from blood samples taken during the prolactin secretion profiling. The authors state that *V. agnus-castus* had no effect on testosterone levels, but no other details were provided.

Infertility

Gerhard et al. studied the influence of a commercially available preparation of *V. agnus-castus* on infertility. Using a randomized, placebo-controlled, double-blind design, 96 women with fertility disorders (31 with luteal insufficiency; 38 with secondary amenorrhea; 27 with idiopathic infertility) received either *V. agnus-castus* or placebo twice a day for 3 months. The dose of *V. agnus-castus* was 30 drops of Mastodynon twice a day (*agnus-castus* or casticin-standardization not mentioned). The outcome measures were: (i) pregnancy or spontaneous menstruation for women with secondary amenorrhea, and (ii) pregnancy or improved luteal hormone levels in women with luteal insufficiency or idiopathic infertility. A total of 66 women were suitable for evaluation. No differences were noted between the placebo and *V. agnus-castus* groups with respect to effect.

PMS and Menopausal Symptoms

In a 1997 multicenter, randomized, double-blind, controlled trial, Lauritzen et al. examined the efficacy and tolerability of a commercially available capsule formulation of *V. agnus-castus* (Agnolyt) compared with pyridoxine in women with PMS. Inclusion criteria were females aged 18 to 45 years, PMS symptoms in luteal phase of menstrual cycle, PMS symptoms with each cycle, PMS symptoms affecting quality of life, and no drug therapy for PMS in 3 months preceding the study. Of 175 participants, 85 were in the *V. agnus-castus* group (took one capsule twice a day, with one capsule containing 3.5 to 4.2 mg of *V. agnus-castus*, the second capsule containing placebo), and 90 were in the pyridoxine group (days 1–15, took one capsule twice a day, each capsule containing placebo; days 16–35, took one capsule twice a day, each capsule containing 100 mg of pyridoxine). Women in both treatment groups had equal reductions in PMS scores (*V. agnus-castus*: 15.2 to 5.1; pyridoxine: 11.9 to 5.1; $p = 0.37$), suggesting no differences in effect.

In 2000, Loch et al. conducted an open label, uncontrolled study examining the efficacy and safety of a new oral *V. agnus-castus* treatment for PMS complaints. Suffering from PMS was the only inclusion criterion and pregnancy was the only exclusion criterion. A questionnaire on mental and somatic PMS symptoms was completed by 857 gynecologists after interviewing 1634 females at the start of Femicur therapy (20 mg daily), and after a period of three menstrual cycles under therapy. Physicians reported that 42% of women reported that they had no more PMS symptoms, 51% showed a decrease in symptoms ($p < 0.001$), and 1% had an increase in number of symptoms. After 3 months of treatment, both psychic and somatic complaints were dramatically lowered. Although 30% of the women still complained about mastodynia after *V. agnus-castus* treatment, most reported complaints of lower intensity. Physicians described the patients' tolerance of this *V. agnus-castus* product as good or very good in 94% of women. Although one of the authors works for the pharmaceutical company that makes Femicur, the article did not contain funding disclosure statements.

In 2000, Berger et al., using a prospective, multicenter trial design, examined the efficacy of an oral, casticin-standardized *V. agnus-castus* therapy on 43 women diagnosed with PMS. Treatment phases consisted of baseline (two cycles, pretreatment), treatment (three cycles), posttreatment (three cycles, no treatment). The dose was 20 mg, but no placebo control was included. At the end of the trial, Moos' menstrual distress questionnaire (MMDQ) scores were reduced by 43% compared with start (statistically significant, $p < 0.001$), but that improvement decreased gradually in the post-treatment phase. At the end of the posttreatment phase, patients had improved compared to the start of therapy ($p < 0.001$) and for up to three cycles thereafter. A group of 20 women had baseline MMDQ scores that were reduced by at least 50% at the end of treatment phase. *Visual analogue scale* (VAS) and global efficacy scales showed similar findings ($p < 0.001$ and global efficacy rated excellent by 38 women). Areas of improvement included symptoms related to pain, behavior, negativity, and fluid retention. The most frequent adverse events were acne, headaches, and menstrual spotting.

In 2001, using a prospective, randomized, double-blind, placebo-controlled, parallel-group comparison design, Schellenberg studied the efficacy and tolerability of *V. agnus-castus* extract on PMS. Participants were female outpatients, 18 years of age or older, of six general medicine clinics and had a PMS diagnosis according to the *Diagnostic and Statistical Manual of Mental Disorders*. Dose of *V. agnus-castus* was 20 mg daily for 3 months. Results showed that the group receiving *V. agnus-castus* had significant improvements ($p < 0.001$) in all symptoms except bloating compared to the placebo group. Sensitivity analyses removing women taking contraceptives did not alter results. Tolerability was good with acne, itching, and mid-cycle bleeding as the adverse events noted.

An uncontrolled study in 2002 by Lucks examined the effects of 3-month dermal application of *V. agnus-castus* essential oil (oil distilled at some point in the shrubs' development of the fruit but while some leaves were still on the plant) on menopausal and perimenopausal symptoms. A 1.5% solution of the essential oil was incorporated in a bland cream or lotion and applied once a day, 5 to 7 days/week, for 3 months. Descriptive outcome measures were self-report via a survey of symptomatic relief (major, moderate, mild, none, worse) and side effects. A total of 33% of women reported major improvement in symptoms, with the most often area of improvement being hot flashes/night sweats. Both improvement and worsening occurred in the areas of emotions and menstruation flow. Subjects who were also on progesterone supplementation reported breakthrough bleeding.

Mastodynia

In a 1987, Kubista et al. reported results from a placebo-controlled study comparing the effects of lynestrenol, *V. agnus-castus* (Mastodynon), and placebo therapy in women with severe mastopathy with cyclic mastalgia. More women in the lynestrenol and *V. agnus-castus* groups than placebo group reported good relief of PMS symptoms (82, 54, 37%, respectively).

In 1998, Halaska and colleagues examined the tolerability and efficacy of *V. agnus-castus* extract on mastodynia/mastalgia (breast pain, breast tenderness). The study was a double-blind, placebo-controlled, parallel-group (50 women each) design. Length of treatment (*V. agnus-castus* [60 drops daily dose] or placebo) was 3 months. Efficacy was determined using a VAS. Results of study showed that the intensity of mastodynia diminished more quickly in the *V. agnus-castus* group with low incidence of side effects.

Luteal Phase Length

In the 1993 randomized, placebo-controlled, double-blind study where Milewicz et al. examined the effect of *V. agnus-castus* on pituitary prolactin reserve, they also examined luteal phase length. Of the 37 women (20 = placebo, 17 = *V. agnus-castus*) who completed the study (1 month prior to, and 3 months treatment), the shortened luteal phases of the *agnus-castus* group became normal. The study information came from a German publication with an English abstract. Information about inclusion/exclusion criteria and study specifics were not available.

Premenstrual Dysphoric Disorder

Premenstrual dysphoric disorder (PMDD) is characterized by markedly depressed mood, anxiety, affective lability, and decreased interest in daily activities during the last week of luteal phase in menstrual cycles of the last year. In 2002, Atmaca et al. conducted an 8-week, randomized, single-blind, rater-blinded, prospective- and parallel-group, flexible-dosing trial to compare the efficacy of fluoxetine with *V. agnus-castus* for the treatment of PMDD in 42 females. Both fluoxetine and *V. agnus-castus* had dose ranges from 20 to 40 mg. Outcome measures included the Penn daily symptom report, Hamilton depression rating scale, clinical global impression (CGI)-severity of illness scale, and CGI-improvement scale. Both drugs were well tolerated. No statistically significant differences between groups were found. The authors concluded that fluoxetine was more effective (a decrease of more than 50% in rating symptoms) for psychological symptoms (depression, irritability, insomnia, nervousness), whereas *V. agnus-castus* helped with physical symptoms (irritability, breast tenderness, swelling, cramps). Lack of placebo-control and short duration of treatment were significant limitations.

Toxicological Effects

No systematic toxicological studies have been conducted, according to the Expanded Commission E Mongraphs.

Adverse Effects and Toxicity

Throughout years of use, *V. agnus-castus* has shown only mild adverse effects. Pruritus, rash (unspecified), urticaria, increased menstrual blood flow, persistent headaches, and gastrointestinal discomfort have been reported. Few adverse events related to chasteberry have been reported to the Food and Drug Administration.

Case Reports of Toxicity Caused by Commercially Available Products

An extensive search of all standard references, as well as reports of studies in humans, shows that there have been no case reports of toxic exposure to *V. agnus-castus* use. However, one case of nocturnal seizures, possibly attributed to *V. agnus-castus,* has been reported. The patient was taking concomitantly black cohosh root (*Cimicifuga racemosa*), *V. agnus-castus*, and evening primrose as well. Thus, attribution of effect to a specific agent was not possible. In animals, an adverse influence on nursing (lactation) performance has been observed; *V. agnus-castus* could potentially interfere with proper lactation.

Interactions

According to the German Commission E Monographs, drug interactions with *V. agnus-castus* are unknown. However, with animal experiments showing evidence of a "*dopaminergic effect*," it is generally recommended that the effect of *V. agnus-castus* can be diminished in cases when there is concurrent ingestion of dopamine-receptor antagonists (e.g., haloperidol). Similarly, because *V. agnus-castus* inhibits the secretion of prolactin via a dopamine-agonist action, drug interactions may occur with the D_2 family of dopamine-receptor agonists (bromocriptine, pergolide, pramipexole, ropinirole, cabergoline).

Some liquid formulations contain large percentages of alcohol (≥50% vol); health risks from ethanol may exist, and in certain populations, use of the dried extract formulations instead would be advisable.

Reproduction

Although there are no known case reports of toxicity in human reproduction, because of possible endocrine effects, *V. agnus-castus* could disrupt fetal development or proper gestation. A case of ovarian hyperstimulation syndrome and multiple follicular development resulting in no pregnancy occurred in a woman who took *V. agnus-castus* prior to one of her IVF protocol cycles. Tests showed that her serum gonadotropin and hormone evels were out of the desired range.

Regulatory Status

V. agnus-castus is available for use without a prescription in all Member States of the European Union and the United States. In the United States, *V. agnus-castus* is categorized as a dietary supplement. Throughout the world, *V. agnus-castus* is available through pharmacies, health-food shops, mail order companies, supermarkets, and department stores.

17

Medicinal Plants

Medicinal plants were known to the early civilization. As a matter of fact the history of the durg plants is as old as the hisotory of these civilizations. The Chines have used drug plants quite earlier in 5,000 to 4,000 B.C. The Assyrians, Babylonians, Herbrews and Egyptians knew many drug plants in about 1600 B.C. The works of Greeks, viz., Aristotle (384-322 B.C.), Hippocrates (460-370 B.C.), Pythagoras and Theophrastus (370-287 B.C.) have numerous references of many of the present day drugs. In 77 B.C., a Roma physician Dioscordies wrote 'De Materia Medicia' which described the nature and properties of all the 500 medicinal plants known at that time. His book was was considered to be the most authentic work on medicinal plants for the next 16 centuries or so. There was no advancement in the knowledge of durg plants during the Dark Ages.

After the introduction of printing in Europe in the fifteenth century many persons published 'herbalts' which contained many true and and false informations. Some people advanced the 'Doctrine of Signs of Signatures' which meant that the appearance of a plant or its organs indicated its utility, the sign being placed there by the Creator. Superstition about one such plant Mandrake (*Mandragora officinarum*) to be useful in

Fig. 17.1. Glycosides ingested by the caterpillar of this butterfly while feeding on milkweed are stored in the body and after metamorphosis, appear in the adult monarch's body.

treatment of human disease as due to its human body like appearance. In the present times medicinal science has paid great attention to the study of drug plants. The branch of medical science which deals with the drug plant is called *pharmoacognosy*, whereas the study of the action of drugs is called *pharmacology*. Usually the morphology of drug yielding plant is the main basis of drugs' classification.

Drugs Obtained from Roots

Rauwolfia

The drug Rauwolfia is obtained from the roots of *Rauwolfia serpentina* of *Apocynaceae*. It is native of India. The genus is found growing in the tropical regions of Asia, Africa and America. The Asian countries are India, Bangla Desh, Sri Lanka, Burma, Malaysia, Thailand and Indonesia. Five species of *Rauwolfia* grow in India. *R. Serpentina*, the most important of the five species is found in the sub-Himalayan tract from Punjab through Nepal, Sikkim, Bhutan to Assam, eastern and western ghats, Central India and the Andamans. The plant is grown commercially in Uttar Pradesh, Bihar, Orissa, West Bengal, Assam, Andhra Pradesh, Tamil Nadu, Karnataka, Keralan and Maharashtra. The plant is an erect perennial shrub. It is of a height of ½ ft. to 1½ ft. and may sometimes attain a height of 3 ft. The plant is evergreen. Leaves are in whorls of three. Flowers are small white or pink and are borne in cyme in large number. The fruit is a drupe.

Hot and humid climates of the tropics is best suited for its growth. It prefers shady habitats of the forests. It grows well in areas of very high rainfall (250 cm to 500 cm) and temperature range of 10 to 38°C. It can grow in different kinds of soils ranging from sandy alluvial loam to red lateritic loam. Well drained humus rich clayey soil is best for its growth. The plant is usually propagated by means of seeds. Rauwolfia drug is obtained from the root of the plant. The bark of the roots yields several alkaloids (about 80), the most important of which is *reserpine*.

Reserpine is extracted from another two species—*R. vomitoria* of Africa and *R. tetraphylla* of America. Reserpine is of great medicinal use in treatment of violent kinds of insanity and high blood pressure. It acts as a sedative and depressant in hypertension and chronic psychoses. It is also used for treatment of insect bites, fevers and dysentery. Reserpine is now widely used by research workers to study physiological systems. Other important alkaloids are *deserpidine*, *rescinnamine*, *reserpinine*, *serpentine*, *serpentinine*, *ajmaline*, etc.

Fig. 17.2. The snakelike root of Rauwalfia contains an alkaloid used in the treatment of hypertension and schizophrenia.

Aconite

The drug is obtained from the tuberous roots of *Aconitum napellus* of the *Ranunculaceae*. The plant is a perennial herb. The plant is also known as monk's hood or wolfbane. It is indigenous to the mountains of Europe (Alps, Pyrenees, etc.and western Asia.) It is also cultivated as an ornamental plant is most of the countries. Several alkaloids are obtained from the roots. Aconite, the most important alkaloid, is poisonous. Aconite is usually used externally for the treatment of neuralgia and rheumatism. Internally it is taken to relieve pain and fever.

Drugs obtained from Barks

Quinine

Quinine is obtained from the bark of several species of *Cinchona* of the *Rubiaceae*. The plant is native to Andean highlands of tropical America. It still grows in Peru and Bolivia at altitudes of 3,000 feet to 9,000 feet. The medicinal properties of the bark of the tree were discovered in 1638 when it cured the malarial fever of the wife of the Viceroy of Peru, the Countess of Cinchon. She carried it to Spain in 1639. The plant was named *Cinchona* by Linnaeus in the eighteenth century after her name.

Fig. 17.3. A flowering and fruiting branch of Cinchona officinalis, of which yields quinine, a remedy for malaria.

The bark of the plant called Jesuit's bark or Peruvian bark and it has got the attention of the people throughout the world. From mid-seventeenth to the mid-nineteenth century the wild trees in south America were ruthlessly cut down to get the bark. The British and the Dutch secured the seeds from South America and started large plantations in India and Java respectively. The Britishers had collected seeds from *Cinchona succirubra*. The Dutch plantations of Java were established from the seeds of *Cinchona ledgeriana* named after a British resident of Bolivia, Charles Ledger, who had collected the seeds. About 90 percent of the world export trade in quinine is carried on by Java. Other species of Cinchona used for the drug are *C. officinalis* and *C. calisaya*. The tea plantation has replaced the plant in Sri Lanka because the drug has quite uneconomic price.

The other *Cinchona*-growing regions are Burma and Tanganyika (Tanzania). *Cincona* plant is a fairly large tree and may attain a height of fifty feet or more. Depending upon the species of plant, the bark covering the stem can be light coloured, pale, brown or dark brown. The leaves are opposite. Yellow or pink coloured flowers are borne in terminal panicles. They are produed from the third or fourth year of growth. Cross pollination takes place. The plants are limited in distribution. They grow in the topics between 10° north and 20° south of the Equator at altitudes more than 1,000 feet. They grow well in regions of very high temperature and high rainfall (60 inches or more). Good porous soils are good for their growth.

Cinchona is propagated by seeds. Mixing of characters and consequent variation in plant is attained as a result of natural hybridization. It is now usually propagated by means of graftings and cuttings. The trees are usually felled down after they are 10 years old since the bark develops maximum alkaloid content by that time. The bark is removed from the stems, branches and roots and is slowly dried at not very high temperature. The dried bark is sent to the factory where the alkaloids are extracted by solvent extraction.

Cinchona bark has four important alkaloids (totaquine) which are quinine, quinidine, cinchonine and cinchonidine. Quinine is the most important of the alkaloides. *C. ledgeriana* has the maximum percentage of quinine in relation to the other alkaloids. Quinine is a antimalarial drug and is very bitter and white granular in appearance. It is also useful as a tonic and anti-septic. Some synthetic antimalarial drugs are also manufactured these days.

Drugs obtained from Stems

Ephedrine

The drug ephedrine is an alkaloid. It is obtained from several species of *Ephedra* of the *Gnetaceae*. Two important species are *E. sinica* and *E. equisetina*. The plant is native of Asia. In China it has been in use for over 5,000 years. The plant is leafless shrub with green stems. The plant is woody and xerophytic and grows in arid regions of the world. The plant is dioecious. Entire plant is the source of the alkoloid. However, its importance as a source of ephedrine has been reduced due to synthetic manufacture of the drug. The drug is used in the treatment of nasal and bronchial congestion, colds, asthma, hay fever and other ailments. It also acts as a stimulant.

Drugs obtained from Leaves

Eucalyptus

Eucalyptus oil is used in medicine. It is obtained from the leaves of several species of *Eucalyptus*. The plant is native to Australia. It is also grown in the Mediterranean region. U. S. A. and many other places. The plant is a tall tree which attains a height of 200 to 300 feet. The leaves are scythe-shaped in *E. globulus* and of different shapes in other species. Eucalyptus oil is used in the treatment of cold, malaria, nose and throat troubles and fevers.

Cocaine

Cocaine obtained from the leaves of a South American plant, *Erythroxylon coca* of the *Erythroxyllaceae*. The plant is indigenous to the Andean regions of Peru and Bolivia. The plant is now widely cultivated in the tropical regions of South America, Java, Sri Lanka and Taiwan (Formosa). The plant grows are higher altitudes. The plant is a small, much branched shrub or tree. Small, sessile, elliptical leaves are alternately arranged. The flowers are borne is small axillary groups. The leaves are harvested by hand when the trees are two to three years old. The leaves are picked three or four times in a year. Dried leaves are immediately exported in boxes. The alkaloid cocaine ($C_{17}H_{21}O_4N$) is ususlly extracted in the importing countries by solvent process. The drug is chiefly used as a local anesthetic. It is also used as a tonic for the digestive and nervous symptoms. The leaves are chewed by the natives of South America for stimulating physical and mental activities. The coca is 'de-alkaoidized' in U. S. A. to be used as 'cola' flavourings.

Fig. 17.4. The autumn crocus yields colchicine used to treat gout and cell malignancies.

Digitalis

Digitalis is obtained from the dried leaves of foxglove (*Digitalis purpurea*) of the *Scrophulairceae*. The plant is a native of Southern and Central Europe, which continues to be its chief grower. However it is grown as beautiful ornamental plant in other parts of world. The plant is a pubescent biennial or pernnial herb. The plant stem can attain a height of 5 feet but leaves are crowded in lower portion of stem. The stem bears purplish flowers in a terminal spike. The leaves are dried in shaded conditions. The active glycosides are extracted fron the dried leaves by solvent (alcohol) process. The glycosides are digitoxin, digitaline and digitalein and digiton. They are very important as stimulants of heart and as diuretics. Digitalis tone up the circulatory system by inducing the heart to make powerful and complete contraction.

Fig. 17.5. A branch and flower of a coca plant, which yields the alkaloid cocoine.

Belladona

The drug is obtained mainly form the dried leaves of the deadly nightshade, *Atropa belladona* of the *Solanaceae*. The plant is indigenous to Europe and Asia Minor. It is widely grown in Europe, U.S.A. and India. The plant is a perennial herb with a creeping root stock. It bears alternately arranged ovate leaves. The stem is hollow. The flowers are purplish. The plant bears brownish or blackish berry. The leaves are harvested at the flowering time (May-June). The leaves are then dried at low temperature for a period of 2 to 15 weeks. The alkaloids present in the plant are extracted with the help of solvents.

The plant yields several alkaloids-atropine ($C_{17}H_{22}O_3N$) hyoscyamine, scopolamine, apoatropine, belladonine, norhyocyamine noratropine, hyoscine, tropacocaine and meteloidine out of which the first three are of greater importance. They are used as a stimulant to the sympathetic nervous system, as diuretics, in dilating the pupil of the eye and in the treatment of palsy. Externally it is used in ointment to relieve pain. Dilation of eye, to check the excessive perspiration, stimulation of circulation, local pain releaf and counteracting the muscle sperm are the various purposes for which Atropine or hyoscyamine is used. Scopolamine is used as a narcotic anti-insomniac and anesthesia.

Drugs Obtained from Flowers, Fruits and Seeds

Opium

Immature capsules of opium poppu (*Papaver somniforvm*) of the *Papaveraceae* are source of opium. The plant is indigenous to Asia Minor and India. It is also grown in China, West Asia and the Mediterranean region. The plant is an annual herb of a height of 2 to 4 feet.

Plant leaves are alternate and white, pink or reddish flowers are borne. The fruit is a globular pale green capsule. A number of fine parallel slits are made into the unripe capsules in the evening and in dry weather. The exuding latex is collected in the moring. It is rolled into balls which are covered by dried petals. Whereas the 'alba' varietes are grown in India and China for getting commercial opium, the 'glabra' varieties are used for exraction of the alkaloids to be used in medicines. There are about 30 alkaloids out of which the important ones are morphine, codeine, narcotine and papaverine. Morphine and codeine are widely used as sedatives to relieve pain and cause sleep. They are usually taken orally, rectally or by injections to cause local insensitiveness. Morphine is also used in the treatment of cough.

While opium is a very valuable, it is also a dangerous drug. Opium has narotic effects. It is eaten and smoked by millions of people for deriving pleasant feelings of exhilaration and peace. However, physical and mental degradation debility, delirium and even death can be caused as a result of addition of opium, morphine codeine, heroine (artificial derivative of morphine) etc. The drug should be used in very limited quantities and under strict supervision of physician.

Strychnine

The drug is obtained from the seeds of *Strychnos Nuxvomica* of the *Loganiaceae*. The plant is indigenous to Australia, India, Sri Lanka and Cochin China. The plant also grows widely in South America and Europe. It was introduced in Europe in the 16th century. The plant is a woody vine of the tropical forests. The stem bears ovate leaves in opposite manner. The plant bears small flowers and large indehiscent fruits. A fruit of *Strychnos* contains 3 to 5 hard and greyish seeds.

The alkaloids present in the seeds are strychnine ($C_{21}H_{22}N_2O_2$) and brucine ($C_{23}H_{26}$ N_2O_4). They are expressed from the seed with boiling sulphuric acid. The alkaloids are first precipitated and are then purified. *Strychnine* is a very poisonous substance. It is of medicinal uses in the treatment of nervous disorders and paralysis It stimulates the central nervous system and acts as a tonic. It is famous as arrow poison (curare).

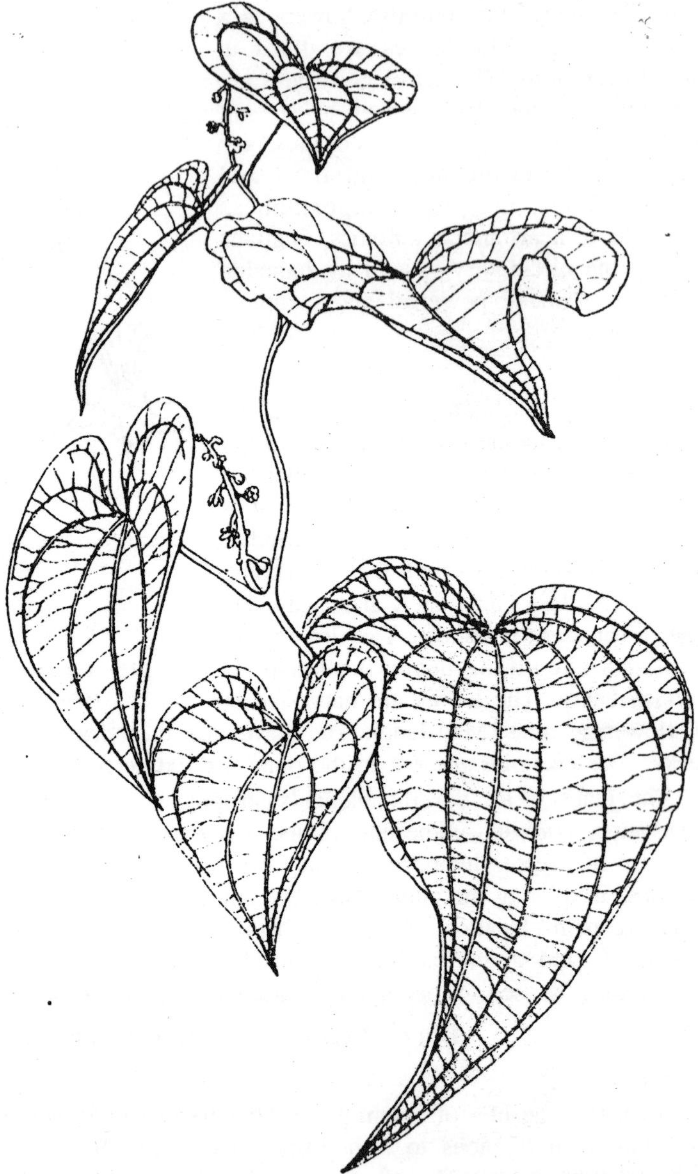

Fig. 17.6. The roots of yam vines are the principal sources of steroid precursors used to produce active compounds in oral contraceptives, and to treat hormone imbalances and heart.

Drugs Obtained from Various Plants

Hordeum vulgare

Hordeum vulgare is grass that may be either a winter or a spring annual of the Poaceae (Graminae) family. It forms a rosette type of growth in fall and winter, developing elongated stems and flower heads in early summer. Winter varieties form branched stems or tillers at the base, so several stems rise from a single plant. The stems of both winter and spring varieties may vary in length from 30 to

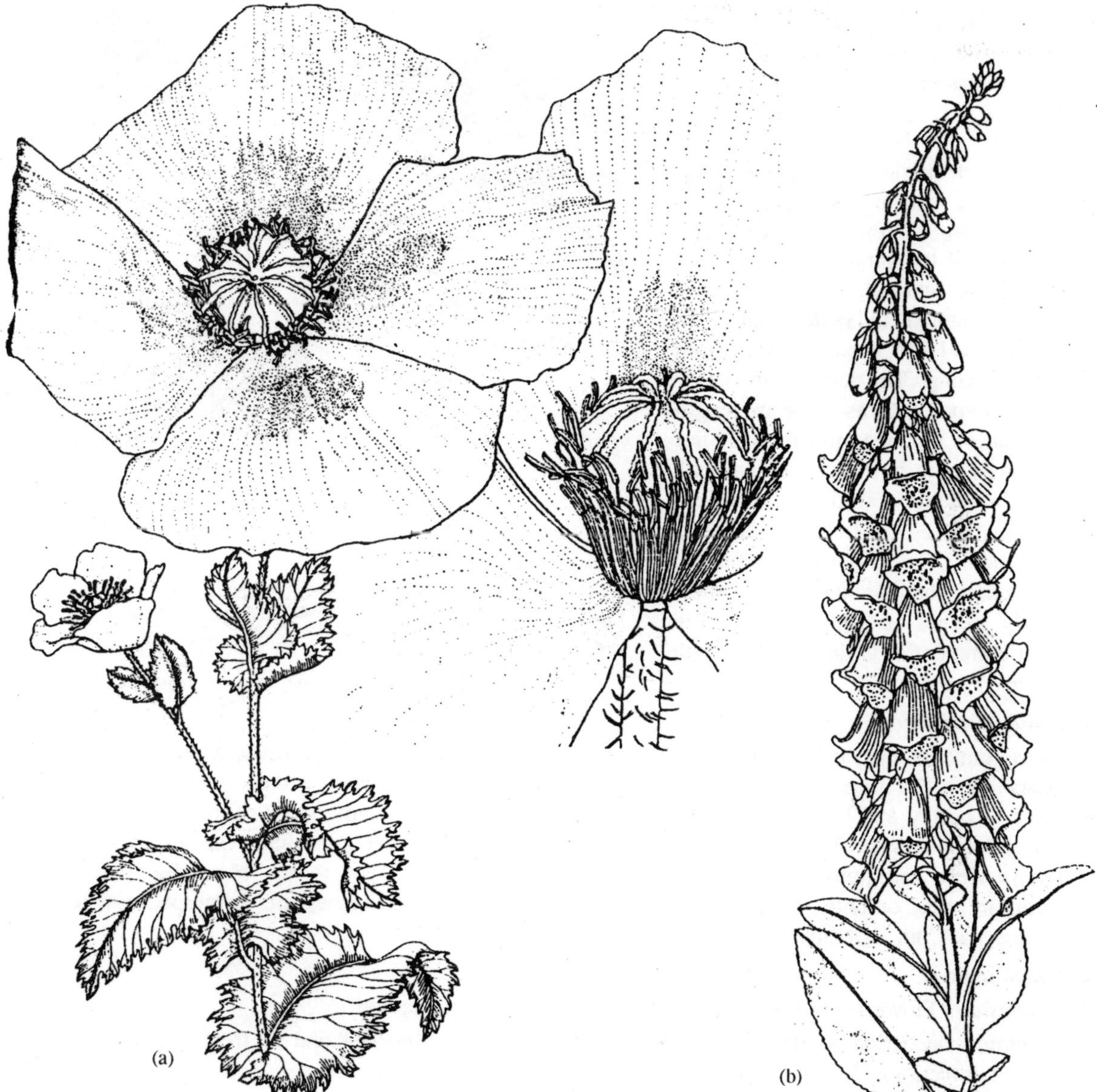

Fig. 17.7. (a) The opium poppy is the source of morphine, papaverine, and codeine. (b) The leaves of the garden ornamental foxglove which yields digitoxin, a steroidal glycoside effective in stabilizing the heart's action.

120 cm, depending on variety and growing conditions. Stems are round, hollow between nodes, and develop five to seven nodes below the head. At each node, a clasping leaf develops. In most varieties, the leaves are coated with a waxy chalk-like deposit. Shape and size of leaves vary with variety, growing conditions, and position on the plant. The spike contains the flowers and consists of spikelets attached to the central stem or rachis. Stem intervals between spikelets are 2 mm or less in dense-headed varieties and up to 4–5 mm in lax or open-headed kinds. Three spikelets develop at each node on the rachis. *Hordeum vulgare* is six-row variety, where all three of the spikelets at each node develop a seed. Each spikelet has two linear to lanceolate glumes rising from near the base and flat and terminates

in an awn. The glumes, minus the awn, are approximately half the length of the kernel in most varieties, but this varies from less than half to equal to the kernel in length. Glumes may be covered with hairs, weakly haired, or hairless. The awns on the glumes may be shorter than the gume, equal in length, or longer. The barley kernel consists of the caryopsis, or internal seed, the lemma, and palea. In most barley varieties, the lemma and palea adhere to the caryopsis and are a part of the grain following threshing. The lemmas in barley are usually awned. Awns vary in length from very short up to as much as 12 in. Edges of awns may be rough or "barbed" (bearded) or nearly smooth. Awnless varieties are also known. In six-row barley, awns are usually more developed on the central spikelets than on the lateral ones. The barley kernel is generally spindle shaped. In commercial varieties, the length ranges from 7 to 12 mm.

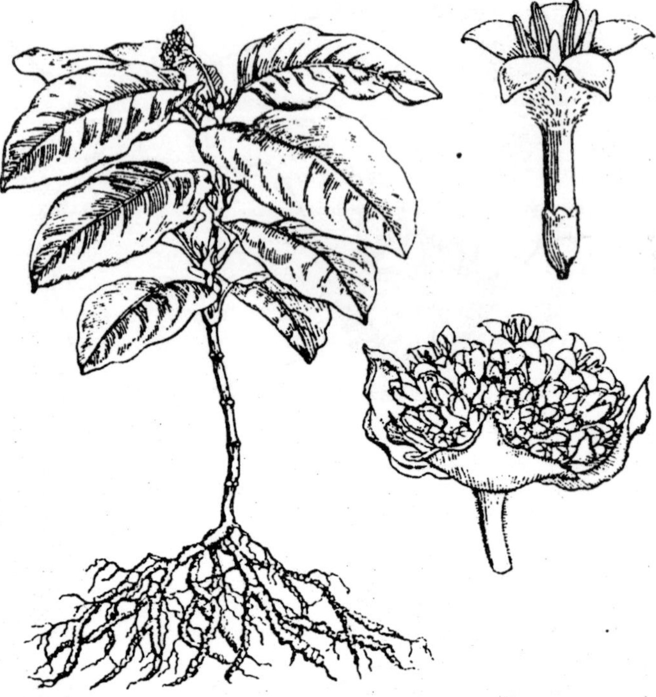

Fig. 17.8. Cephaelis, from which ipecac is obtained.

Origin and distribution

Grains found in pits and pyramids in Egypt indicated that barley was cultivated there more than 5000 years ago. The most ancient glyph or pictograph found for barley is dated approx 3000 BC. References to barley and beer are found in the earliest Egyptian and Sumerian writings. The origin of barley is still not known. There are differing views among researchers regarding whether the original wild forms were indigenous to Eastern Asia, particularly Tibet, or to the Near East, Eastern Mediterranean area, or both. Varieties are constantly changing as new ones are developed and tested while others pass out of cultivation.

Traditional uses

Afghanistan. Flowers are taken orally by females for contraception.

Argentina. Decoction of the dried fruit is taken orally for diarrhea and to treat respiratory and urinary tract infections.

Egypt. Dried fruits are smoked as a treatment for schistosomiasis. The fruit is used intravaginally as a contraceptive before and after coitus. Fifty-three percent of 1200 puerperal women interviewed practiced this method, of whom 47% depended on indigenous method and/or prolonged lactation.

Guatemala. Hot water extract of the dried seed is taken orally for renal inflammation and kidney disease. Hot water extract of the dried seed is used externally for dermatitis, inflammations, erysipelas, and skin eruptions.

India. Powdered flowers of *Calotropis procera*, fruits of *Piper nigrum*, seed ash of *Hordeum vulgare*, and rose water are taken orally for cholera.

Iran. Flour is used as a food. A decoction of the dried seed is used externally as an emollient and applied on hemorrhoids and infected ulcers. A decoction of the dried seed is taken orally as a diuretic and antipyretic and used for hepatitis, diarrhea, scorbutism, nephritis, bladder inflammation, gout, enema, and its tonic effect. Decoction of the dried seed is applied to the nose to reduce internasal inflammation.

Italy. Seeds are eaten as a urinary antiseptic. Compresses of boiled seeds are used to soothe rheumatic and joint pains. Infusion of the dried seed is used as a galactogogue.

Korea. Hot water extract of the dried entire plant is taken orally for beriberi, coughs, influenza, measles, syphilis, nephritis, jaundice, dysentery, and ancylostomiasis; for thrush in infants; and as a diuretic. Extract of the dried entire plant is used externally for prickly heat.

Peru. Hot water extract of dried fruits is used externally for measles and as an emollient and taken orally as a diuretic.

South Korea. Hot water extracts of the fruit and dried seeds are taken orally by pregnant women to induce abortion. Hot water extracts of the fruits taken orally by females as a contraceptive.

Turkey. Decoction of the fruit is taken orally for common colds.

United States. Infusion of the dried seed is taken orally for dysentery, diarrhea, and colic and for digestive and gastrointestinal disorders.

Fig. 17.9. Hordeum vulgare. A–Flowering plant, B–Flower, C–Spikelet.

Medicinal uses

Acquired immune deficiency syndrome therapeutic effect

Hot water extract of the dried fruit, administered orally to patients with acquired immuodeficiency syndrome (AIDS) four to five doses/week for a year for the purpose of in clearing heat and detoxifying the blood, produced an improvement in the patient's health.

Allergenic activity

Extract of the dried seed, administered externally to male adults at a concentration of 10%, was active. Allergens, administered by ingestion or inhalation to 40 children aged 3–6 months who suffered from diarrhea, vomiting, eczema, or weight loss after the introduction of the cereal in the diet and to 18 food-allergic adults and eight patients with Baker's asthma, produced strong effect in children. Protein Z(4), administered to four patients with beer allergy, provoked weak positive response to skin testing in two of the patients and was recognized by the four individual sera tested. Lipid transfer protein 1 showed reactivity with three of four individual sera and induced strong positive skin prick test responses in all four of the patients tested. Two cases of severe systemic reactions resulted from beer ingestion: one case of anaphylaxis requiring emergency care and one of generalized urticaria and angioedema were reported. Barley was recognized as the specific ingredient responsible for the observed allergic reaction. Beer and malt allergens were found in three patients with urticaria. Urticaria from beer was an immunoglobulin E (IgE)-mediated hypersensitivity reaction induced by a protein component of approx 10 kDa deriving from barley. A 50-year-old man, who developed bronchial asthma after exposure to barley flour, was confirmed by skin prick test and serum-specific IgE. Bronchial challenge test with every allergen showed no response, except for an immediate response to barley flour. The most relevant clinical feature was an immediate asthmatic response developed after oral provocation with either barley-made beer or barley flour itself that indicated IgE-mediated food-induced bronchial asthma. A 32-year-old storeman developed occupational asthma resulting from barley grain dust in the packaging of flour, barley, and peanuts. He developed immediate symptoms of sneezing, cough, and dyspnea on exposure to barley only. Bronchial provocation test to the barley confirmed the diagnosis.

Anti-atherogenic activity

β-Glucan in barley cellulose, administered to Syrian golden F(1)B hamsters at doses of 2, 4, or 8 g/100 g in a semipurified hyper-cholesterolemic diet of 0.15 g/g cholesterol, 20 g/100 g of hydrogenated coconut oil and 15 g/100 g of cellulose, produced cholesterol-lowering effect. Compared with control hamsters, dose-dependent decreases that were similar in magnitude in plasma total and low-density lipoprotein cholesterol concentrations were observed in hamsters fed the β-glucan diet at weeks 3,6, and 9. Liver cholesterol concentrations were also reduced significantly in hamsters consuming 8 g/100 g β-glucan.

Antibacterial activity

Decoction of the dried fruit, on agar plate, was inactive on *Pseudomonas aeruginosa*. Ethanol (95%) and water extracts of the dried fruit, on agar plate at a concentration of 50 μL/plate, were inactive on *Staphylococcus aureus*. Water extract of the dried fruit, on agar plate, at a concentration of 1 mg/mL, was inactive on *Salmonella typhi*. Hot water extract of the dried fruit, on agar plate at a concentration of 62.5 mg/mL, was inactive on *Escherichia coli* and *Staphylococcus aureus*. Tincture of the dried seed, on agar plate at a concentration of 30 μL/disc, was inactive on *Escherichia coli*, *Pseudomonas aeruginosa*, and *Staphylococcus aureus*. Extract of 10 g plant material in 100 mL ethanol was used.

Anticoagulation activity

Serpin BSZx (an inhibitor of trypsin and chemotrypsin) inhibited thrombin, plasma kallikrein, factor VIIa/tissue factor, and factor Xa at heparin-independent association rates. Only factor Xa turned a significant fraction of BSZx over as substrate. Activated protein C and leukocyte elastase were slowly inhibited by BSZx, whereas factor XIIa, urokinase and tissue type plasminogen activator, plasmin and pancreas kallikrein, and elastase were not or only weakly affected. Trypsin from *Fusarium* was not inhibited, while interaction with subtilisin Carlsberg and Novo was rapid, but most BSZx was cleaved as a substrate.

Antidiabetic activity

The plant, administered to diabetic rats, produced a decrease of blood glucose concentration, water consumption, and weight loss. No differences were found in healthy animals.

Antidiarrheal activity

Extract of the germinating seeds, administered in the ration of male rats, was active vs cecocolectomy-induced diarrhea. Germinated barley and scutellum fraction of germinated barley, administered to rats, prevented diarrhea caused by cecocolectomy and increased the protein content and sucrose activity of small intestinal mucosa. The aleurone and scutellum fractions of barley grains before and after germination, administered to rats with diarrhea, were active. The addition of fractions of germinated barley and not barley collected before germination increased the fecal output and jejunal mucosal protein content. The effect of malted barley was similar to that of germinated barley food-stuff.

Antifungal activity

Dried stem, on agar plate, was active on *Sphacelia segetum*. Hot water extract of the dried fruit, on agar plate at a concentration of 62.5 mg/mL, was inactive on *Aspergillus niger*. Water extract of the seed, on agar plate at a concentration of 5 mg/mL, was inactive on *Helicobacter pylori*. Protein fraction of the seed without seed coat, on agar plate at a concentration of 2 μg/disc, was active on *Neurospora crassa* and *Trichoderma* sp. Protein fraction of the seed without seed coat, on agar plate at a concentration of 2 μg/disc, was active on *Neurospora crassa*.

Antihepatotoxic activity

Methanol extract of the dried fruit, administered by gastric intubation to rabbits at a dose of 0.5 g/kg, was active vs CCl_4-induced hepatotoxicity. A mixture of *Machilus* sp., *Alisma* sp., *Amomum xanthioides*, *Bulboschoenus maritimus*, *Artemisia iwaymogis*, *Atractylodes japonica*, *Crataegus cuneata*, *Hordeum vulgare*, *Citrus sinensis*, *Polyporus umbellatus*, *Agastache rugosa*, *Raphanus sativus*, *Poncirus trifoliatus*, *Curcuma zeodaria*, *Citrus aurantium*, *Saussurea lappa*, *Glycyrrhiza glabra*, and *Zingiber officinale* was used.

Antihypercholesterolemic activity

Dried bran, administered in ration of male rats, was active. Methanol extract of the dried fruit, administered by gastric intubation to rabbits at a dose of 500 mg/kg, was active vs CCl_4-induced hepatotoxicity. A mixture of *Machilus* sp., *Alisma* sp., *Amomum xanthioides*, *Bulboschoenus maritimus*, *Artemisia iwaymogis*, *Atractylodes japonica*, *C. cuneata*, *Hordeum vulgare*, *Citrus sinensis*, *Polyporus umbellatus*, *Agastache rugosa*, *Raphanus sativus*, *Poncirus trifoliatus*, *Curcuma zeodaria*, *Citrus aurantium*, *Saussurea lappa*, *Glycyrrhiza glabra*, and *Zingiber officinale* was used. Results were significant at $p < 0.01$ level. Gum, administered orally to male rats for 4 weeks, was active. Biological activity reported had been patented. Chromatographic fraction of the green leaf juice, administered to rats at a dose of 1% of diet, was active vs cholesterol-loaded animals. The results were significant at $p < 0.005$ level. Seeds, administered to 20 men with hypercholesterolemia aged 41 ± 5 years, resulted in significant fall in serum total cholesterol, LDL cholesterol, and phospholipids, and LDL and very low-density lipoprotein (VLDL). A dose of 50/50 w/w mix with rice, administered to seven women with mild hypercholesterolemia aged 56.± 7 years twice daily for 2–4 weeks, produced a significant improvement of serum lipid profiles. In the normolipemic subjects, serum lipids were unaffected. Bran flour and oil extract, administered to 79 patients with hypercholesterolemia, aged 48.2 years at a dose of 3 g oil extract or 30 g flour for 30 days, significantly decreased total serum cholesterol. LDL cholesterol was decreased 6.5% with addition of bran flour and 9.2% with oil. High-density lipoprotein (HDL) cholesterol decreased significantly in the bran flour group but not in the oil group. Fiber (nonstarch polysaccharides), administered to 21 men with mild hypercholesterolemia aged 30–59 years for 4 weeks, produced a significant fall in plasma total cholesterol and LDL cholesterol. The triglyceride and glucose concentrations did not change significantly.

Antihyperglycemic activity

Dried seeds, administered orally to six patients with noninsulin-dependent diabetes mellitus at a dose of 50 g/person, was active. A single dose resulted in a glycemic index of 53.4. Water extract of the dried fruit, administered intragastrically to rats at a dose of 150 mg/kg, produced weak activity on blood vs streptozotocin-induced hyperglycemia. Dried seeds, administered orally to eight adults with normal glucose tolerance at a dose of 50 g/person, were active. Flour, administered in the ration of male rats, was active vs streptozotocin- induced hyperglycemia. Barley gum, administered to 4-week-old male Sprague–Dawley rats as a 2% dietary supplement for 14 days, lowered serum cholesterol concentration and suppressed the elevation of serum and liver triglyceride concentrations. Thick and thin rolled oat made from raw or preheated kernel, administered to healthy subjects, produced high glucose, insulin, and metabolic responses. Barley flour naturally high in β-glucan and β-glucan-enriched flour, administered to 11 healthy men, resulted in a decrease of the insulin response. Plasma glucose and insulin concentrations increased significantly. Cholesterol concentration dropped below the fasting concentration 4 hours after the meal and was significantly lower than after low-fiber meal. The cholecystokinin remained elevated for a long time after the barley-containing meals. Boiled intact and milled kernels with different amylase-amylopectin ratios, administered to healthy subjects, produced lower metabolic responses and higher satiety scores when compared to white wheat bread. The boiled flours produced higher glucose and insulin responses than did the corresponding boiled kernels. The impact of amylase to amylopectin on the metabolic responses was marginal. The intact kernels, administered to healthy subjects at concentrations of 40 and 80% (SCB-40 and SCB-80), produced the glycemic and insulinemic indices 39 and 33 for SCB-80, compared to pumpernickel bread 69 and 61, respectively. The glycemic index for SCB-40 was 66.

Anti-inflammatory activity

Germinated barley foodstuff, administered to mice with DSS-induced colitis, prevented disease activity and loss of body weight after induction of colitis. Serum interleukin (IL)-6 level, mucosal STAT3 expression, necrosis factor-κB activity, and mucosal damages were decreased and cecal butyrate content increased. The germinated barley foodstuff-fed mice had lower bile acid concentration than the control group. Green barley extract, in LPS-activated human monocytes cell line culture (THP-1), was active.

Antioxidant activity

Water extract of the roasted seed, at a concentration of 1 mg/mL, produced strong activity vs a liposome model system. A concentration of 25 mg/mL was active vs 2,2-diphenyl-1-picryl-hydrazyl-hydrate-induced radical. A concentration of 5 mg/mL was inactive vs linoleic acid system. Ethanol (80%) extract of the freeze- dried leaf, at a concentration of 60 μg/mL, was active vs oxidation of ethyl linoleate by Fenton's reagent. Young leaf extract, administered orally to 36 patients with type 2 diabetes at a dose of 15 g daily for 4 weeks, enhanced the scavenging of oxygen free radicals, saved the LDL-vitamin E content, and inhibited LDL oxidation. Purified green barley extract, in human mononuclear culture of cells isolated from perithelial blood and synovial fluid of patients with rheumatoid arthritis, was active. Leaf essence, administered to atherosclerotic New Zealand White male rabbits at a dose of 1% of diet, produced a decrease of plasma total cholesterol, triacylglycerol, lucigenin-chemiluminescence, and luminal-chemiluminescence levels. The value of T_{50} of red blood cell hemolysis and the lag phase of LDL oxidation increased in barley- treated group compared with the control. Ninety percent of the intimal surface of the thoracic aorta was covered with atherosclerotic lesions in the control group, but only 60% of the surface was covered in the barley group. This inhibition was associated with a decrease in plasma lipids and an increase in antioxidative abilities.

Anti-tumor activity

Commercial barley bran (13% dietary fiber) from the aleurone/ subaleurone layer; outer-layer barley bran, including the germ (25.5% dietary fiber); and spent barley grain bran (product of the brewery including the hull) (47.7% dietary fiber) were administered to male Sprague–Dawley rats as a 5% dietary supplement for 7 months. Commercial barley bran was most effective in reducing tumor incidence and burden. Tumor burden and tumor mass index were reduced significantly by outer-layer barley bran and spent barley grain bran. Commercial barley bran and spent barley grain bran, administered to rats with 1,2-dimethylhydrazine-induced intestinal tumor, produced a higher incidence and burden of tumor. Fiber was administered to 4-week-old male Sprague–Dawley rats with dimethylhydrazine-induced tumors at a dose of 5% of diet. The insoluble fiber-rich fraction (spent barley grain) was significantly more effective at preventing induced tumors than the soluble commercial barley bran. The incidence of rats affected, tumor mass index, and plasma cholesterol concentration were reduced by spent barley grain. Outer-layer barley bran was moderately effective in cancer prevention. The crude and partially purified lunasin, in stably *ras*-transfected mouse fibroblast cell culture, suppressed colony formation induced with isopropylthiogalactoside. This fraction also inhibited histone acetylation in mouse fibroblast (NIH 3T3) and human breast (MCF-7) cells in the presence of the histone deacetylase inhibitor sodium butyrate.

Anti-ulcer activity

Water extract of the green leaf juice, administered by gastric intubation to rats at a dose of 500 mg/kg, was active vs stress-induced (restraint) ulcers. The results were significant at $p < 0.001$ level. Water extract was active vs acetic acid-induced and aspirin-induced ulcers. The results were significant at $p < 0.01$ and $p < 0.005$ levels, respectively. Water extract was inactive vs pylorus ligation-induced ulcers. Extract of the dried seedling, administered orally to adults at a dose of 30 g/person, was active. Germinated barley foodstuff, administered orally to male Sprague–Dawley rats on day 6 after initiation of colitis, was active vs dextran sodium sulfate-induced colitis. Germinated barley foodstuff treatment reduced colonic inflammation with an increase in cecal butyrate levels. Fiber and protein factions of germinated barley foodstuff, administered to dextran sodium sulfate induced colitis Sprague–Dawley rats, significantly attenuated the clinical signs of colitis and decreased serum α1-acid glycoprotein levels, with an increase in cecal butyrate production, whereas germinated barley foodstuff- protein did not. Germinated barley foodstuff with or without salazosulfapyridine, administered to rats after the onset of colitis, accelerated colonic epithelial repair and improved clinical signs. Germinated barley foodstuff, administered orally to patients with mild to moderate active ulcerative colitis at a dose of 30 g/person daily for 4 weeks, produced a significant clinical and endoscopic improvement independent of disease extent.

The improvement was associated with an increase in stool butyrate concentrations and in luminal *Bifidobacterium* and *Eubacterium* levels. After the end of treatment, the patients had an exacerbation of the disease. Germinated barley foodstuff with or without *Clostridium butyricum*, administered to 3% dextran sodium sulfate-induced colitis in Sprague–Dawley rats for 8 days, prevented bloody diarrhea and mucosal damage and increased the fecal short-chain fatty acid levels. Germinated barley foodstuff, administered to Sprague–Dawley rats for 5 days, prevented bloody diarrhea and mucosal damage, elevated fecal acetic acid and *N*-butyric acid levels, and tended to increase the number of *Eubacteria* and *Bifidobacteria*. The number of *Enterobacteriaceae*, the total number of aerobes and *Bacteroidaceae*, were lowered by germinated barley foodstuff treatment. Germinated barley foodstuff, administered to HLA-B27 transgenic rats for 13 weeks, produced an increase of bacterial butyrate production and the decrease of cecal occult blood, colonic mucosal hyperplasia, colonic mucosal necrosis factor-κB-DNA binding activity, and the production of IL-8. Butyrate from germinated barley foodstuff, administered orally or intracecally to Sprague–Dawley rats, produced reduction of mucosal damage only by intrathecal administration. Bacterial butyrate production and reduction of mucosal damage depended on the dose of germinated barley foodstuff in the diet.

Germinated barley foodstuff and scutellum fraction of germinated barley, administered to Sprague–Dawley rats with colitis induced by 3% dextran sodium sulfate, prevented bloody diarrhea and mucosal damage in colitis. The germinated samples did not produce a protective effect. Germinated barley foodstuff increased mucosal protein and RNA content in the colitis model. Germinated barley foodstuff, administered to 18 patients with mildly to moderately active ulcerative colitis at a dose of 20–30 g germinated barley foodstuff daily for 4 weeks, produced a significant decrease in clinical activity index scores compared to the control group. No side effects related to germinated barley foodstuff were observed. Germinated barley foodstuff therapy increased fecal concentrations of *Bifidobacterium* and *Eubacterium limosum*. Germinated barley foodstuff, administered to patients with mild to moderate active ulcerative colitis, irresponsible to or intolerant of standard treatment at a dose of 20–30 g germinated barley foodstuff daily for 4 weeks, resulted in a significant clinical and endoscopic improvement associated with an increase in stool butyrate concentrations.

Cardiovascular activity

β-glucan, administered to 18 men with mild hypercholesterolemia with a mean body weight index of 27.4 ± 4.6 at a dose of 8.1–11.9 g β-glucan per day, produced no significant change in total, LDL and HDL cholesterol, triacylglycerol, fasting glucose, and postprandial glucose.

Cholesterol biosynthesis inhibition

The inhibitor I from oily nonpolar fraction of flour, administered to chicken at a dose of 2.5–20 ppm, produced a significant decrease in hepatic cholesterogenesis and serum total and LDL cholesterol and an increase in lipogenic activity.

Cholesterol-7-α-hydroxylase inhibition

Petroleum ether extract of the fresh fruit, administered to pigs at a concentration of 3.5 g/kg of diet for 29 days, produced 40% inhibition of the hepatic enzyme activity.

Cytotoxic activity

Water extract of the dried fruit, in cell culture at a concentration of 500 μg/mL, produced weak activity on CA-mammary-microalveolar. Ethanol (50%) extract of the seed, in cell culture, was inactive on CA-9KB, ED_{50} greater than 20 μg/mL. Methanol extract of the dried seed, in cell culture, was inactive on SNU-1 human cells, IC_{50} greater than 0.3 mg/mL, and on SNU-C4 human cells, IC_{50} greater than 0.3 mg/mL. Protein fraction of the seed without seed coat, in cell culture at a concentration of 2 μg/disc, was active on CA-Ehrlich ascites. Methanol extract of the aerial parts, in cell culture at a concentration of 50 mg/mL, was equivocal on CA-9KB.

Gastrointestinal activity

Water extract of the green leaf juice, administered by gastric intubation to rats at a dose of 500 mg/kg, was inactive vs pylorus ligation-induced ulcers. Fiber, administered orally to young male Wistar rats at a dose of 500 g extrudates/kg of diet for 6 weeks, produced a higher concentration of neutral sterols in the intestinal content of the barley-fed group than in the control group ($p < 0.005$) and affected indirectly the amount of formed secondary bile acids. Fiber, administered orally to young male Wistar rats at a dose of 50 g/100 g extrudates or mixtures for 6 weeks, produced greater food intake in the last 2 weeks and increased ceca and colon masses, cecal and colon contents, concentration of resistant starch in cecal, most of colon contents, and β-glucan level in the small intestine, cecum, and colon. The numbers of coliforms and *Bacteroides* were lower and those of *Lactobacillus* were higher than in the control group. The dose increased weight gain in the sixth week. Short-chain fatty acids were higher in the cecal, colon, and feces content of the test group. The proportion of secondary bile acids was lower and the amount of neutral sterol was higher in feces of fiber-treated animals. The concentrations of excreted bile acids increased up to 30% during the feeding period. Germinated barley food-stuff, administered to Sprague–Dawley rats fed on various diets with the same protein and dietary fiber levels, produced an increase fecal output compared with commercial water-soluble and insoluble dietary fibers.

The dietary fiber from germinated barley foodstuff increased the fecal output and mucosal protein content. The protein fraction of germinated barley foodstuff degraded to the peptide form did not increase the fecal output or mucosal protein content[HV152]. Germinated barley foodstuff from aleurone and scutellum fractions of germinated barley, administrated to healthy volunteers at a dose of 9 g daily for 14 days, significantly increased fecal butyrate content and fecal *Bifidobacterium* and *Eubacterium*. Ten anaerobic microorganisms selected from intestinal microflora were cultured in vitro in germinated barley foodstuff medium. After 3 days of incubation, seven strains (*Bifidobacterium breve*, *Bifidobacterium longum*, *Lactobacillus acidophilus*, *Lactobacillus casei* ssp. *casei*, *Bacteroides ovatus*, *Clostridium butyricum*, and *Eubacterium limosum*) lowered the medium pH producing short-chain fatty acid. Germinated barley foodstuff changed the intestinal microflora and increased probiotics such as *Bifidobacterium*. Butyrate was produced by the mutual action of *Eubacterium* and *Bifidobacterium*. Germinated barley foodstuff from aleurone layer, scutellum, and germ, administered to 10 healthy volunteers at a dose of 30 g/day/person for 28 days, produced an increased fecal butyrate content, fecal weight, and water content.

There were no significant changes in body weight and major abnormalities in hematologic and urinary analysis. Fiber, administered to nine patients with ileostomies at a dose of 35 g/day, increased the ileal excretion of starch. Flaked and finely milled barley, eaten by patients with ileostomies, showed that only 2 ± 1% of starch remained undigested after the consumption of finely milled barley and 17 ± 1% resisted digestion, partly as oligosaccharides but largely as intact unpitted starch granules bound by intact cell walls. The energy excretion from the stoma was three times higher after flaked that after milled barley. Nonstarch polysaccharide, starch, and fat made almost equal contributions to the higher energy excretion. Bran flour, administered to 44 volunteers at a dose of 30 g/day, decreased the transit time by 8.02 hours from baseline and increased daily fecal weight by 48.6. Groats were administered to volunteers at a dose of 1 g carbohydrate/kg body weight, three times or at three different doses of 0.75, 1, and 1.5 g carbohydrate/kg body weight. After consumption of 1 g carbohydrate/kg body weight, produced a mean mouth to cecum transit time of 8.4 ± 0.4 hour. After consumption of the high dose, a mean mouth-to-cecum transit time of 9.0 ± 0.5 hours was produced. Particle size did not significantly affect the mouth-to-caecum transit time.

Germinated barley food-stuff, administered to Sprague–Dawley rats, prevented diarrhea and mucosal damages; increased mucosal protein, DNA, and RNA content; and depressed bacterial translocation and elevation of myeloperoxidase activity induced by methotrexate. β-glucan-rich barley fraction, administered to ileostomy subjects at a dose of 13.0 g β-glucan/day for 2 days, increased the cholesterol excretion higher than with the oat bran with β-glucanase and wheat flour diets. Bile acid excretion was 755 (133–1187) mg/day. Carbohydrates, administered to healthy subjects at a dose of 90 g for dinner in random order 1 week apart, significantly increased the breath hydrogen and improved glucose tolerance. No difference in the rates of glucose disappearance or gut glucose absorption was observed. Serum-free fatty acid concentrations were significantly reduced the morning after the barley meal. Germinated barley foodstuff, administered to Sprague–Dawley rats with constipation induced by loperamide, produced an increase of bowel movements, fecal water content, and concentration of short-chain fatty acids in cecal content, especially butyrate.

Glucose tolerance effect

Fiber, administered orally to type 2 diabetic Goto– Kakizaki male rats for 9 months, improved the area under the plasma glucose concentration time curves, lowered the fasting plasma glucose and glycosylated hemoglobin levels, and decreased plasma total cholesterol, triglycerides, and free fatty acid levels. Fiber, administered orally to 8- week-old male Goto–Kakizaki strain rats, at a dose of 1.79 g/day/rat for 3 months, improved glucose tolerance and lowered the plasma cholesterol and triglyceride levels. The fasting plasma glucose level was significantly lower in comparison to rice and corn starch-fed rats. High barley (high-fiber diet), administered to 10 women (20.4 ± 1.3 year-old,

19.2 ± 2 kg/m^2) for 4 weeks with a 1-month interval, resulted in lowering plasma total and LDL cholesterol concentrations and reduced plasma triacylglycerol concentration. The barley diet increased stool volume. There was no significant difference in glucose tolerance between diet regimens. Barley bread containing lactic acid and reference barley bread, administered in the morning to 10 healthy men and women, produced a significant lowering of the incremental glycemic area and of the glucose response at 95 minutes after ingestion of the bread with lactic acid. At 45 minutes after the meal, the insulin level was significantly lower after the lactic acid bread, compared with reference barley bread.

Glucose-6-phosphate dehydrogenase inhibition

Petroleum ether extract of the fresh fruit, administered to pigs at a concentration of 3.5 g/kg of diet for 29 days, was active on hepatic enzymes.

Glutamate-oxaloacetate-transaminase inhibition

Methanol extract of the dried fruit, administered by gastric intubation to rabbits at a dose of 500 mg/kg, was active vs CCl_4-induced hepatotoxicity. A mixture of *Machilus* sp., *Alisma* sp., *Amomum xanthioides*, *Bulboschoenus maritimus*, *Artemisia iwaymogis*, *Atractylodes japonica*, *Crataegus cuneata*, *Hordeum vulgare*, *Citrus sinensis*, *Polyporus umbellatus*, *Agastache rugosa*, *Raphanus sativus*, *Poncirus trifoliatus*, *Curcuma zeodaria*, *Citrus aurantium*, *Saussurea lappa*, *Glycyrrhiza glabra*, and *Zingiber officinale* was used. Results were significant at $p < 0.01$ level.

Glutamate-pyruvate-transaminase inhibition

Methanol extract of the dried fruit, administered by gastric intubation to rabbits at a dose of 500 mg/kg, was active vs carbon tetrachloride (CCl_4)-induced hepatotoxicity. A mixture of *Machilus* sp., *Alisma* sp., *Amomum xanthioides*, *Bulboschoenus maritimus*, *Artemisia iwaymogis*, *Atractylodes japonica*, *Crataegus cuneata*, *Hordeum vulgare*, *Citrus sinensis*, *Polyporus umbellatus*, *Agastache rugosa*, *Raphanus sativus*, *Poncirus trifoliatus*, *Curcuma zeodaria*, *Citrus aurantium*, *Saussurea lappa*, *Glycyrrhiza glabra*, and *Zingiber officinale* was used. Result was significant at $p < 0.01$ level.

Hypocholesterolemic activity

Fixed oil of the bran, administered orally to adults of both sexes at a dose of 30 mg/day, was active. Flour bran, administered orally to adults at a dose of 3 g/day, was active. Dried bran, administered in the ration of male rats, was active. Petroleum ether extract of the fresh fruit, administered to pigs at a concentration of 3.5 g/kg of diet, produced a decrease of serum total cholesterol, LDL cholesterol, and HDL cholesterol after 29 days of feeding. Flour, administered orally to adults with hypercholesterolemia at a dose of 44 g/day, produced a decrease of total and LDL cholesterol levels.

Hypoglycemic activity

Water extract of the fermented root, administered intravenously to rabbits, was active. The dried seed, administered orally to eight healthy volunteers at a dose of 50 g/person, was active. A single dose resulted in a glycemic index of 68.7 and an insulinemic index of 71.1.

Hypolipemic activity

Fiber, administered orally to nine adults with ileostomies at a dose of 13 g/day, increased the excretion of cholesterol. Petroleum ether extract of the fresh fruit, administered to pigs at a concentration of 3.5 g/kg of diet, was inactive. Purified green barley extract, in human mononuclear culture of cells isolated from perithelial blood and synovial fluid of patients with rheumatoid arthritis, was active. Leaf essence, administered to atherosclerotic New Zealand White male rabbits at a dose of 1% of diet, produced a decrease of plasma total cholesterol, triacylglycerol, lucigenin-chemiluminescence, and luminal-chemiluminescence levels. The value of T_{50} of red blood cell hemolysis and the lag phase of LDL oxidation increased in barley-treated group compared with the control. Ninety percent of the intimal surface of the thoracic aorta was covered with atherosclerotic lesions in the control group, but only 60% of the surface was covered in the barley group. This inhibition was associated with a decrease in plasma lipids and an increase in antioxidative abilities.

Hypotriglyceridemic activity

Fixed oil of the bran, administered orally to adults of both sexes at a dose of 30 mg/day, was active. Flour bran, administered orally to adults at a dose of 3 g/day, was inactive.

Laxative effect

Powdered dried bran, administered orally to 44 adults at a dose of 30 g/person, was active on gastrointestinal motility. Transit time decreased by 8 hours, and fecal mass increased by 48.6 g/day.

Lipid metabolism

Fiber, administered orally to male type 2 diabetic Goto– Kakizaki rats for 9 months, improved the area under the plasma glucose concentration time curves, lowered the fasting plasma glucose and glycosylated hemoglobin levels, and decreased plasma total cholesterol, triglycerides and free fatty acid levels.

Lipolytic effect

Ethanol (95%) extract of the dried entire plant in combination with *Rhizoma zingiberis*, *Ligustrum chuanxiong*, *Lilium brownii*, *Nephelium longa*, and *Polygonum multiflorum*, administered in drinking water to C57BL/6J obese mice at a concentration of 5%, was active.

Lung function

Exposure of six men to barley dust for 2 days decreased ventilatory capacity. Five volunteers not previously exposed to barley dust, when exposed to the dust for 2 hours, decreased the ventilatory capacity ranging from 200 mL to 800 mL, with recovery taking up to 72 hours. All of the subjects had decreases in flow at 50% vital capacity but little or no change in flow at 75% vital capacity. In three subjects, there was a drop in specific conductance that lasted for less than 24 hours. Sixty-ine of 80 dockworkers handling grains reported evening feverish episodes/symptoms not related to smoking or atopic status. No gross deficits in lung function were detected.

Malic enzyme inhibition

Petroleum ether extract of the fresh fruit, administered to pigs at a concentration of 3.5 g/kg of diet for 29 days, was active on hepatic enzymes.

Mineral utilization

Germinated barley foodstuff, administered orally to 5-week-old Sprague–Dawley rats for 14 days, promoted the absorption of calcium (Ca) and magnesium (Mg) by the gastrointestinal tract. The absorption of iron and potassium was not attenuated and mineral absorption was not inhibited. Barley husk, administered to 5- and 9-week-old rats at different doses, produced a lowering of zinc (Zn) and Ca absorption already at dose 20 g dietary fiber/kg dry matter and had a small negative effect on potassium absorption. Phytate did not appear as a major factor affecting mineral absorption in barley husk. All of the diets containing barley husk had very low molar ratios (phytate:Zn was 4). Processed or unprocessed barley was administered to healthy subjects in two single meals containing porridge or breakfast (60 g) cereals for 2 months. Zn absorption from hydrothermally-treated barley porridge was significantly higher than from the control porridge; Ca absorption did not differ. Zn absorption from breakfast cereals of malted barley with phytase activity was significantly higher than from flakes of barley without phytase activity; Ca absorption was not significantly different. Standard barley and β-glucan-enriched barley dehulled grains was administered to 10 healthy hydrogen-producing adults at a dose of 35 g. The percentage of the ^{13}C dose oxidized was greater after standard barley than after enriched barley consumption. The area under the curve for H_2 was greater after enriched barley intake. There was no difference in CO_2 production. Hull fiber extract, in Caco-2 cell culture, produced no effect on the rate of transepithelial ^{45}Ca transport across Caco-2 cell monolayers and the uptake of ^{45}Ca into Caco-2 cells. A low-phytate barley-fiber concentrate was administered to young women at a dose of 15 g barley fiber (high-fiber, high-protein diet) and 15 g barley fiber (high-fiber, low-protein diet). The mean daily intake of the cations was 25.4 and 22.9 mmol Ca, 10.1 and 10 mmol Mg, 166.8 and 119.3 μmol Zn, and 186.2 and 154 μmol Fe, respectively. Mean balances were 1.9 and -0.8 mmol Ca, -0.2 and -0.5 mmol Mg, -4.6, and -18.4 imol Zn, respectively. The mean apparent iron absorption was 5.4 and -23.2 μmol.

Monocytic differentiation

Prodelphinidin B-3, T1, T2, and T3 from bran polyphenol extract, in HL60 human myeloid leukemia cell culture, induced 26–40% nitro blue tetrazolium positive cells and 22–32% α-naphthyl-butyrate esterase-positive cells. Proanthocyanidins potentiated all-*trans*-retinoic acid-induced granulocytic and sodium butyrate-induced monocytic differentiation in HL60 cells.

Mutagenic activity

Ethanol (70%) extract of the dried seed, on agar plate at a concentration of 50 mg/mL, was inactive on *Escherichia coli* PQ 37. The water and chloroform extracts of the ethanol (70%) extract were inactive. Metabolic activation had no effect on the results.

Oxidative effect

Ethanol (95%) extract of the dried entire plant, administered in drinking water to C57BL/6J obese mice at a concentration of 5%, increased glucose oxidation in epididymal fat pads. Extract of mixture of following plants: *Hordeum vulgare*, *Rhizoma zingiberis*, *Ligustrum chuanxiong*, *Lilium brownii*, *Nephelium longa*, and *Polygonum multiflorum* was used.

Pepsin inhibition

Water extract of the green leaf juice, administered by gastric intubation to rats at a dose of 500 mg/kg, was inactive vs pylorus ligation-induced ulcers.

Phosphogluconate dehydrogenase inhibition

Petroleum ether extract of the fresh fruit, administered to pigs at a concentration of 3.5 g/kg of diet for 29 days, was active on hepatic enzymes.

Proteinemic effect

Methanol extract of the dried fruit, administered by gastric intubation to rabbits at a dose of 500 mg/kg, produced an increase in serum albumin and protein content vs CCl_4-induced hepatotoxicity. A mixture of *Machilus* sp., *Alisma* sp., *Amomum xanthioides*, *Bulboschoenus maritimus*, *Artemisia iwaymogis*, *Atractylodes japonica*, *Crataegus cuneata*, *Hordeum vulgare*, *Citrus sinensis*, *Polyporus umbellatus*, *Agastache rugosa*, *Raphanus sativus*, *Poncirus trifoliatus*, *Curcuma zeodaria*, *Citrus aurantium*, *Saussurea lappa*, *Glycyrrhiza glabra*, and *Zingiber officinale* was used. Results were significant at $p < 0.01$ level.

Respiratory effect

Barley ear inhaled by a 2.5-year-old child produced fever, dyspnea, right paracardiac infiltrate with pleural reaction on X-rays, and normal bronchoscopy after 8 days. On day 11, extensive right pneumothorax, and on day 20, right axillary inflammatory lesion were observed. On day 28, the ear of barley was expulsed and there was complete recovery. Barley spike, inhaled into the tracheobronchial tree of 18 children under the age of 5 years, produced coughing and choking in 14 of the children. The spikes were removed by laryngoscopy in 12 patients and by rigid bronchoscopy in two. Four patients with history of cough, dyspnea, fever, and serious respiratory diseases, such as pneumothorax, lobar pneumonia, and pleural empyema, required surgical intervention. All of the children made satisfactory recoveries. Dust extract of barley, in cell culture on nonsensitized guinea pig tracheal smooth muscle pretreated with drugs, produced constrictor effect that was significantly inhibited by atropine indicating an interaction of the extracts with parasympathetic nerves. Inhibition of contraction of other mediators was less effective and varied with the dust extract.

Toxic effect

β-glucan-enriched soluble barley fiber, administered orally to Wistar rats at concentrations of 0.7, 3.5, and 7.0% β-glucan for 28 days, increased the number of circulating lymphocytes in males. The increase was not dose-dependent and was not observed in females. A dose-dependent increase in full and empty cecum weight was observed. There were no adverse effects on general condition and behavior, growth, feed and water consumption, feed conversion efficiency, red blood cell and clotting potential parameters, clinical chemistry values, and organ weight. Necropsy and histopathology findings revealed no treatment-related changes in any organ evaluated. β-glucan (64%) preparation (barley β-fiber), administered to CD-1 mice at concentrations of 1,5, or 10% of diet (0.7, 3.5, and 7% β-glucan) for 28 days, produced no adverse effect in hematological or clinical chemistry measurements, in organ weights and immunopathology in either sex after treatment or after the recovery period. Azidoalanine and the azide treated extracts, in Chinese hamster and normal human skin fibroblast cell cultures, significantly increased the frequency of sister chromatid exchanges observed in both cultures. This increase was approximately twofold, as compared with the control.

Toxicity assessment

Ethanol (50%) extract of the seed, administered intraperitoneally to mice, produced a maximum tolerated dose of 1 g/kg. Hexane extract of the green leaf juice, administered in ration of rats, produced lethal dose$_{50}$ greater than 10 g/kg.

Weight gain inhibition

Ethanol (95%) extract of the dried entire plant, administered in drinking water to C57BL/6J obese mice at a concentration of 5%, was active. The extract also contained *Rhizoma zingiberis*, *Ligustrum chuangxiong*, *Lilium brownii*, *Nephelium longa*, and *Polygonum multiflorum*.

Camellia sinensis

Camellia sinensis is an evergreen tree or shrub of the *Theaceae* family that grows to 10–15 m high in the wild, and 0.6–1.5 m under cultivation. The leaves are short- stalked, light green, coriaceous, alternate, elliptic-obovate or lanceolate, with serrate margin, glabrous, or sometimes pubescent beneath, varying in length from 5 to 30 cm, and about 4 cm wide. Young leaves are pubescent. Mature leaves are bright green in color, leathery, and smooth.

Flowers are white, fragrant, 2.5–4 cm in diameter, solitary or in clusters of two to four. They have numerous stamens with yellow anthers and produces brownish-red, one- to four-lobed capsules. Each lobe contains one to three spherical or flattened brown seeds. There are numerous varieties and races of tea. There are three main groups of the cultivated forms: China, Assam, and hybrid tea, differing in form. *Camellia sinensis assamica*, the source of much of the commercial tea crop of Ceylon is a tree that, unpruned, may attain a height of 15 m and has proportionally longer, thinner leaves than typical species.

Origin and distribution

The cultivation and enjoyment of tea are recorded in Chinese literature of 2700 BC and in Japan about 1100. Through the Arabs, tea reached Europe about 1550. Native to Assam, Burma, and the

Chinese province of Yunnan, it is highly regarded in southern Asia and planted in India, southern Russia, East Africa, Java, Ceylon, Sumatra, Argentina, and Turkey. China, India, Indonesia, and Japan produced about a half of the total world production.

Traditional uses

India. Decoctions of the dried and fresh buds and leaves are taken orally for headache and fever. Powder or decoction of the dried leaf is applied to teeth to prevent tooth decay. Fresh leaf juice is taken orally for abortion, and as a contraceptive and hemostatic.

Mexico. Hot water extract of the leaf is taken orally by nursing mothers to increase milk production.

Turkey. Leaves are taken orally to treat diarrhea.

China. Hot water extract of the dried leaf is taken orally as a sedative, an antihypertensive, and anti-inflammatory.

Guatemala. Hot water extract of the dried leaf is used as eyewash for conjunctivitis.

Kenya. Water extract of the dried leaf is applied ophthalmically to treat corneal opacities. The infusion is used for chalzion and conjunctivitis.

Fig. 17.10. Camellia sinensis – flowering branch.

Thailand. Hot water extract of the dried leaf is taken orally as a cardiotonic and neurotonic. Hot water extract of the dried seed is taken orally as an antifungal.

Medicinal values

Antibacterial activity

Alcohol extract of black tea, assayed on *Salmonella typhi* and *Salmonella paratyphi* A, was active on all strains of Salmonella paratyphi A, and only 42.19% of Salmonella typhi strains were inhibited by the extract. Hot water extract of the dried entire plant and the tannin fraction, on agar plate, were active on *Escherichia coli*, *Pseudomonas aeruginosa*, and *Staphylococcus aureus*.

Anticancer activity

Catechin, administered to pheochromocytoma cells in cell culture, was active. The cells were incubated with different concentrations of catechin at short-term (2 days) and long-term (7 days) in Dulbecco's modified Eagle medium. The activity of superoxide dismutase was measured and its mRNA assayed by Northern blotting. After incubation for 2 days, catechin significantly increased the activity of copper/zinc superoxide dismutase. However, it did not produce significant effect at 7 days. The magnesium superoxide dismutase activity produced significant changes in both short- and long-term treatment groups. The amount of mRNA also showed similar changes.

Anticarcinogenic activity

The anti-carcinogenic activity of tea phenols has been demonstrated in rats and mice transplantable tumors, carcinogen-induced tumors in digestive organs, mammary glands, hepatocarcinomas, lung cancers, skin tumors, leukemia, tumor promotion, and metastasis. The mechanisms of this effect indicated that the inhibition of tumors may be the result of both extracellular and intracellular mechanisms indicating the modulation of metabolism, blocking or suppression, modulation of DNA replication and repair effects, promotion, inhibition of invasion and metastasis, and induction of novel mechanisms. The association of green tea and cancer has been investigated in 8552 Japanese women 40 years of age. After 9 years of follow-up study, 384 cases of cancer were identified. There was a negative association between cancer incidence and green tea consumption, especially among females consuming more than 10 cups of tea a day. A slow down in increases of cancer incidence with age was observed among females who consumed more than 10 cups daily. Tea, taken by lung cancer patients at a dose of two or more cups per day, reduced the risk by 95%. The protected effect was more evident among Kreyberg I tumors (squamous cell and small cells) and among light smokers. The green tea polyphenols, epigallocatechin-*3*-gallate, applied topically to human skin, prevented penetration of ultraviolet (UV) radiation. This was demonstrated by the absence of immunostaining for cyclobutane pyrimidine dimers in the reticular dermis. Topical administration to the skin of mice inhibited UVB-induced infiltration of $CDIIb^+$ cells. The treatment also results in reduction of the UVB-induced immuno-regulatory cytokine interleukin (IL)-10 in the skin and draining lymph nodes, and an elevated amount of IL- 12 in draining lymph nodes. Green tea extract, in human umbilical vein endothelial cells, did not affect cell viability but significantly reduced cell proliferation dose-dependently and produced a dose-dependent accumulation of cells in the gastrointestinal phase. The decrease of the expression of vascular endothelial growth factor receptors fms-like tyrosine kinase and fetal liver kinase-l/kinase insert domain containing receptor in the cell culture by the extract was detected with immunohistochemical and Western blotting methods. Green and black tea, administered orally to hairless mice in the absence of any chemical initiators or promoters, resulted in significantly fewer skin papillomas and tumors induced by UVA and UVB light. Black tea however, provided better protection against UVB-induced tumors than green tea. Black tea consumption was associated with a reduction in the number of sunburn cells in the epidermis of mice 24 hours after irradiation, although there was no effect of green tea. Other indices of early damage such as necrotic cells or mitotic figures were not affected. Neutrophil infiltration as a measure of skin redness was slightly lowered by tea consumption in the UVB group. Epigallocatechin-3-gallate, in cell culture, activated proMMP-2 in U-87 glioblastoma cells in the presence of concanavalin A or cytochalasin D, two potent activators of MT1-MMP, resulted in proMMP-2 activation that was correlated with the cell surface proteolytic processing of Mt1-MMP to it's inactive 43 kDa form. Addition of epigallocatechin-3-gallate strongly inhibited the MT1-MMP-driven migration in the cells. The treatment of cells with non-cytotoxic doses of epigallocatechin-3-gallate significantly reduced the amount of secreted pro MMP-2, and led to a concomitant increase in intracellular levels of that protein. The effect was similar to that observed using well-characterized secretion inhibitors such as brefeldin A and manumycin, indicative that epigallocatechin could also potentially act on intracellular secretory pathways. Green tea polyphenols, at a dose of 30 mg/mL, inhibited the photolabeling of P-glycoprotein (P-gp) by 75% and increased the accumulation of rhodamine-123 in the multidrug-resistant cell line CH(R)C5. This result indicated that green tea polyphenols interact with P-gp and inhibited its transport activity. The modulation of P-gp was a reversible process. Epigallocatechin-3-gallate potentiates the cytotoxicity of vinblastine in CH(R)C5 cells. The inhibitory effect on P-gp was also observed in human Caco-2 cells.

Anticataract activity

Tea, administered in culture to enucleated rat lens, reduced the incidence of selenite cataract in vivo. The rat lenses were randomly divided into normal, control and treated groups and incubated for 24 hours at 37°C. Oxidative stress was induced by sodium selenite in the culture medium of the two groups (except the normal group). The medium of the treated group was additionally supplemented with tea extract. After incubation, lenses were subjected to glutathione and malondialdehyde estimation. Enzyme activity of superoxide dismutase, catalase, and glutathione peroxidase were also measured in different sets of the experiment. In vivo cataract was induced in 9-day-old rat pups of both control and treated groups by a single subcutaneous injection of sodium selenite. The treated pups were injected with tea extract intraperitoneally prior to selenite challenge and continued for 2 consecutive days thereafter. Cataract incidence was evaluated on 16 postnatal days by slit lamp examination. There was positive modulation of biochemical parameters in the organ culture study. The results indicated that tea act primarily by preserving the antioxidant defense system.

Antifungal activity

Ethanol (50%) extract of the entire plant, in broth culture at a concentration of 1 mg/mL, was inactive on *Aspergillus fumigatus* and *Trichophyton mentagrophytes*. Hot water extract of the leaf on agar plate at a concentration of 1.0% was active on *Alternaria tenuis*, *Pythium aphanidermatum*, and *Rhizopus stolonifer*. Saponin fraction of the leaf on agar plate was active on *Microsporum audonini*, minimum inhibitory concentration (MIC) 10 mg/mL; *Epidermophyton floccosum* and *Trichophyton mentagrophytes*, MICs 25 μg/ mL.

Antihypercholesterolemic activity

Tea supplemented with vitamin E, administered to male Syrian hamsters, reduced plasma low-density lipoprotein (LDL) cholesterol concentrations, LDL oxidation, and early atherosclerosis compared to the consumption of tea alone by the hamsters. The antioxidant action of vitamin E is through the incorporation of vitamin E into the LDL molecule. The hamsters were fed a semi- purified hypercholesterolemic diet containing 12% coconut oil, 3% sunflower oil, and 0.2% cholesterol (control), control and 0.625% tea, control and 1.25% tea or control and 0.044% tocopherol acetate for 10 weeks. The hamsters fed the vitamin E diet compared to the different concentrations of tea significantly lower plasma LDL cholesterol concentrations, -18% ($p < 0.007$), -17% ($p < 0.02$), and -24% ($p < 0.0001$), respectively. Aortic fatty streak areas were reduced in the vitamin E diet group compared to the control, -36% ($p < 0.04$) and low tea -45% ($p < 0.01$) diets. Lag phase of conjugated diene production was greater in the vitamin E diet compared to the control, low tea, and high tea diets, 41% ($p < 0.0004$), 40% ($p < 0.0004$), and 39% ($p < 0.0008$), respectively. Rate of conjugated diene production was reduced in the vitamin E diet compared to the control, low tea, and high tea diets, -63% ($p < 0.002$), -57% ($p < 0.005$), and -59% ($p < 0.02$), respectively. Infusion of black tea leaves was taken by 31 men (ages 47 ± 14) and 34 females (ages 35 ± 13) in a 4-week study. Six mugs of tea were taken daily vs placebo (water, caffeine, milk, and sugar) and blood lipids, bowel habit, and blood pressure measured during a run-in period and at the end weeks 2, 3, and 4 of the test period.

Compliance was established by adding a known amount of *p*-aminobenzoic acid to selected tea bags and then measure it excretion in the urine. Mean serum cholesterol values during run-in, placebo and on tea drinking were 5.67 ± 1.05, 5.76 ± 1.11, and 5.69 ± 1.09 mmol/L ($p = 0.16$). There were also no significant changes in diet, LDL-cholesterol, high-density lipoprotein (HDL) cholesterol, triacylglycerols, and blood pressure in the tea intervention period compared with placebo. Stool consistency was softened with tea compared with the placebo, and no other differences were observed in bowel habit. The results were unchanged within 15 "non-compliers" whose *p*-aminobenzoic acid excretion indicated that fewer than six tea bags had been used, were excluded from the analysis, and when differenced between run-in and tea periods were considered separately for those who were given tea first or second.

Anti-inflammatory effect

Epigallocatechin-3-gallate was shown to mimic its anti- inflammatory effects in modulating the IL-I β-induced activation of mitogen activated protein kinase in human chondrocytes. It inhibited the IL-I β-induced phosphorylation of c-Jun N-terminal kinase (JNK) isoforms, accumulation of phosphoc-Jun and DNA-binding activity of AP-1 in osteoarthritis chondrocytes, IL-I β but not epigallocatechin-3-gallate, and induced the expression of JNK p46 without modulating the expression of JNK p54 in osteoarthritis chondrocytes. In immune complex kinase assays, epigallocatechin-3-gallate completely blocked the substrate phosphorylating activity of JNK but not p38-mitogen activated protein kinase (MAPK). Epigallocatechin-3-gallate had no inhibitory effect on the activation of extracellular signal-regulated kinase p44/p42 (ERKp44/p42) or p38-MAPK in chondrocytes. Epigallocatechin-3-gallate did not alter the total nonphosphorylated levels of either p38- MAPK or ERKp44/p42 in osteoarthritis chondrocytes. Epigallocatechin-3-gallate administered to primary human osteoarthritis chondrocytes at a concentration of 100 μM in cell Culture, inhibited the IL-I β-induced production of nitric oxide by interfering with the activation of nuclear factor (NF)κB. Tea, in culture with bovine nasal and meta-carpophalangeal cartilage and human nondiseased osteoarthritis and rheumatoid cartilage with and without reagents known to accelerate cartilage matrix breakdown, produced chondroprotective effect that may be beneficial for the arthritis patient by reducing inflammation and the slowing of cartilage breakdown. Individual catechins were added to the cultures and the amount of released proteoglycan and type II collagen were measured by metachromatic assay and inhibition enzyme-linked immunosorbent assay (ELISA), respectively. Possible nonspecific or toxic effects of the catechins were assessed by lactate output and proteoglycan synthesis. Catechins, particularly those containing a gallate ester, were effective at micromolar concentrations at inhibiting proteoglycan and type II collagen breakdown.

Antimutagenic activity

The anticarcinogenic activity of tea phenols has been demonstrated in rats and mice, transplantable tumors, carcinogen-induced tumors in digestive organs, mammary glands, hepatocarcinomas, lung cancers, skin tumors, leukemia, tumor promotion, and metastasis. The mechanisms of this effect indicated that the inhibition of tumors maybe the result of both extracellular and intracellular mechanisms indicting the modulation of metabolism, blocking or suppression, modulation of DNA replication and repair effects, promotion, inhibition of invasion and Metastasis, and induction of novel mechanisms. Green and black teas, administered orally to human adults, were effective. Between 60 and 180 minutes after the teas were administered, the antimutagenic active compounds were recovered from the jejunal compartment by means of dialysis. The dialysate appeared to inhibit the mutagenicity of the food mutagen 2-amino-3,8-dimethylimidazo[4,5-f]quinoxaline on *Salmonella typhimurium*. The maximum inhibition was measured at 2 hours after administration and was comparable for black and green teas. The maximum inhibition observed with black tea was reduced by 22, 42, and 78% in the presence of whole milk, semi-skimmed milk, and skimmed milk, respectively. Whole milk and skimmed milk abolished the antimutagenic activity of green tea by more than 90% and semi- skimmed milk by more than 60%.

When a homogenized breakfast was taken with black tea, the antimutagenic activity was eliminated. When tea and mutagen 2-amino-3,8-dimethylimidazo[4,5-f]quinoxaline were added to the system, 2-amino-3,8-dimethylimidazo[4,5-f]quinoxaline mutagenicity was efficiently inhibited, with green tea showing a slightly stronger antimutagenic activity than black tea. The addition of milk had only a small inhibiting effect on the antimuta-genicity. The antimutagenic activity corresponded with reduction in antioxidant capacity and with a decrease of concentration of catechin, epigallocatechin gallate, and epigallocatechin. Chinese white tea, tested on rat liver S9 in assay for methoxyresorufin *O*-demethylase, inhibited methoxyresorufin *O*-demethylase activity and attenuated the mutagenic activity of 3 - methylimidazo [4,5-f] quinoline (IQ) in absence of S9. Nine of the major constituents found in green and white teas were mixed to produce artificial teas according to their relative levels in white and green teas.

The complete tea exhibited higher antimutagenic potency compared with the corresponding artificial tea. Green and black tea polyphenols, applied to the surfaces of ground beef before cooking, inhibited the formation of the mutagens in a dose-related fashion. Green or black tea polyphenols sharply decreased the mutagenicity of a number of aryl- and heterocyclic amines, of aflatoxin B_1, benzo[a]pyrene, 1, 2-dibromoethane, and more selectively of 2-nitropropane, all involving an induced rat liver S9 fraction. Good inhibition was found with two nitrosamines that required a hamster S9 fraction for biochemical activation. No effect was found with 1-nitropyrene and with the direct-acting (no S9) 2-chloro-4-methyl-thiobutanoic acid. Hot water extract on the leaf was evaluated in cell cultures on various systems vs decaffeinated and caffeinated teas. On mouse mammary gland vs decaffeinated and caffeinated teas, ICs_{50} were 10 mg/mL and 10 μg/mL on CA-A427, IC_{50} 27 mg/mL and 31 μg/mL, and on epithelial cells, IC_{50} 0.01 ng/mL and 0.3 ng/mL. Hot water extract of the leaf, on agar plate at a concentration of 1 mg/plate, was active on Salmonella typhimurium TA98 vs 2-amino-3-methylimidazo [4,5-f]quinoline-induced mutagenesis and produced weak activity vs benzo[a]pyrene-induced mutagenesis. Infusion of the leaf, on agar plate at a concentration of 0.7 mg/plate, was active on *Salmonella typhimurium* TA98 and TA100 vs 2-amino-3-methylimidazo [4,5-f]quinoline-; 3-amino-1,4-dimethyl-5H-pyrid[4,3-b]indole(Trp-1); aflatoxin B1-; 2-amino-6-methyl-dipyrido[1,2-A:3,2-d] imidazole-, and benzo[a]pyrene-induced carcinogenesis. Infusion of the leaf, on agar plate at a concentration of 50 mg/plate, was active on *Salmonella typhimurium* TA98 vs 2-amino-3-methylimidazo[4,5-f]quinoline-; 2-amino-3,4-dimethyl-imidazo[4,5-f]quinoline-;2-amino-3,8-dimethylimidazo[4,5 - f]quinoxaline-; 2-amino-1-methyl-6-phenylimidazo [4,5-b]-pyridine-;2-amino-3,7,8-trimethylimidazo[4,5-f]quinoxaline-; 2-amino-3,4,7,8-tetramethyl-3H-imidazo-[4,5-f]quinoxaline-inoxaline-; 3-amino-1,4- dimethyl-5 H-pyrid[4,3-b] indole (Trp-P-I)- and 3-amino-1-methyl-5H-pyrido [4,3-b] indole-induced mutagenesis. Metabolic activation was required for positive results.

Anti-neoplastic effect

Green tea, administered orally at a dose of 6 g per day in six doses to 42 patients who were asymptomatic and had manifested, progressive prostate specific antigen elevation with hormone therapy, produced limited antineoplastic activity. Continued use of lute inizing hormone-releasing hormone agonist was permitted. However, patients were ineligible if they had received other treatments for their disease in the preceding 4 weeks or if they had received a long-acting antiandrogen therapy in the preceding 6 weeks.

The patients were monitored monthly for response and toxicity. Tumor response, defined as a decline of 50% or greater in the baseline prostate-specific antigen (PSA) value, occurred in a single patient, or 2% of The cohort (95% confidence interval [CI], 1–14%). This one response was not sustained beyond 2 months. At the end of the first month, the median change in the PSA value from baseline for the cohort increased by 43%. Infusion of the leaf, administered in the drinking of female mice at a concentration of 1.25%, was active vs UV radiation-induced papillomas and tumors[CS172]. Leaves in the drinking water of female mice at a dose of 0.6% reduced lung tumor multiplicity and volume in 4-(methyl-nitrosamine)-1-(3-pyridyl)-1-butanone (NNK) treated mice.

Antioxidative effect

Tea, administered orally to rats, decreased the thiobarbituric acid reactive substances (TBARS) contents in urine and lowered the esterified and total cholesterol contents in plasma as compared with a control group. TBARS contents in liver, plasma, and cholesterol levels in the liver were not affected. The lower plasma cholesterol concentration could not be explained by increased fecal excretion of cholesterol or bile acids. On the other hand, a relationship between decreased plasma cholesterol and significantly higher acetate concentrations in the cecum, colon, and portal blood of rats was assumed. Copper absorption was significantly increased while iron absorption was not affected. Epigallocatechin gallate, tea polyphenols, and tea extract were added to human plasma and lipid peroxidation induced by the water-soluble radical generator 2,2'-azobis (2-amidinopropane) dihydrochloride. Following a lag phase, lipid peroxidation was initiated and it occurred at a rate that was lower in a dose that was lowered in a dose-dependent manner by the polyphenols. Similarly, epigallocatechin gallate and the extract added to plasma strongly inhibited 2,2′-azobis(2-amidinopropane) dihydrochloride-induced lipid peroxidation.

The lag phase preceding detectable lipid peroxidation was the result of the antioxidant activity of endogenous ascorbate, which was more effective at inhibiting lipid peroxidation than the tea polyphenols and was not spared by these compounds. When eight volunteers consumed the equivalent of six cups of tea, the resistance of their plasma to lipid peroxidation did not increase over a period of 3 hours. Black tea leaves, administered to human red blood cells, was effective against damage by oxidative stress induced by inducers such as phenylhydrazine, Cu^{2+}-ascorbic acid, and xanthine/xanthine oxidase systems. Lipid peroxidation of pure erythrocyte membrane and of whole red blood cell was completely prevented by black tea extract. Similarly, the tea provided total protection against degradation of membrane proteins. Membrane fluidity studies as monitored by the fluorescent probe 1,6-diphenyl-hexa-1,3,5-triene showed considerable disorganization of its architecture that could be restored back to normal on addition of black tea or free catechins.

The tea extract in comparison to free catechin seemed to be a better protecting agent against various types of oxidative stress. Ethanol/water (7:3) extract of green tea, tested on 2,2-azino-di-3-ethylbenzthiazoline sulphonate, produced antioxidant activity compared with that of ascorbic acid (10 mmol/L). The Nonpolyphenolic fraction of residual green tea (after hot water extraction) produced a significant suppression against hydroperoxide generation from oxidized linoleic acid in a dose-dependent manner. Using silica gel TLC plate, chlorophylls a and b, pheophytins a and b, β-carotene, and lutein were isolated. All of these constituents exhibited significant antioxidant activites, the ranks of suppressive activity against hydroperoxide generation were chlorophyll a > lutein > pheophytin a > chlorophyll b > b-carotene > pheophytin b.

Antiproliferative activity

Green tea fractions, tested on human stomach cancer (MK-1) cells, indicated six active flavan-3-ols, epicatechin, epigallocatechin, epigallocatechin gallate, gallocatechin, epicatechin gallate, and gallocatechin gallate. Among the six active flavan-3-ols, epigallocatechin gallate and gallocatechin gallate produced the highest activity. Epigallocatechin, gallocatechin, and epicatechin gallate followed next, and the activity of epicatechin was lowest. This suggests that the presence of the three adjacent hydroxyl groups (pyrogallol or galloyl group) in the molecule would be a key factor for enhancing the activity.

Antiviral activity

Epigallocatechin-3-gallate, administered to Hep2 cells in culture, produced a therapeutic index of 22 and an IC_{50} of 25 μM. The agent was the most effective when added to the cells during the transition from the early to the late phase of viral infection suggesting that the polyphenol inhibits one or more late steps in virus infection. Ethanol (50%) extract of the entire plant, in broth culture at a concentration of 50 μg/mL, was inactive on Raniket and Vaccinia viruses. Hot water extract of the leaf in cell culture was active on Coxsackie A9, B1, B2, B3, B4, and B6 viruses, Echo type 9 virus, herpes simplex virus, poliovirus III, vaccinia virus, and REO type 1 virus.

Anti-yeast activity

Ethanol (50%) extract of the entire plant, in broth culture at a concentration of 1 mg/mL, was inactive on *Candida albicans*, *Cryptococcus neoformans*, and *Sporotrichum schenckii*. Ethanol extract of the leaf on agar plate produced MIC 9.3 mg/mL on *Candida albicans*.

Cytochrome P50 expression

Fresh leaves of green, black, and decaffeinated black tea enhanced lauric acid hydroxylation. The decaffeinated black tea produced no significant effect. Green tea and black tea but not decaffeinated black tea, stimulated the *O*-dealkylations of methoxy-, ethoxy-, and pentoxy-resorufin indicating upregulation of cytochrome P50 (CYP) 1A and CYP2B. Immunoblot analysis revealed that green and black tea, but not decaffeinated black tea, elevated the hepatic CYP1A2 apoprotein levels. Hepatic microsomes from green and black tea-treated rats, but not those from the decaffeinated black tea-treated rats, were more effective than controls in converting IQ into mutagenic species in the Ames test.

Dental enamel erosion

Herbal tea and conventional black tea, tested on teeth, resulted in erosion of dental enamel. After exposure to tea, sequential profilometric tracings of the specimens were taken, superimposed, and the degree of enamel loss calculated as the area of disparity between the tracings before and after exposure. Tooth surface loss resulted from herbal tea (mean 0.05 mm^2) was significantly greater than that which resulted from exposure to conventional black tea (0.01 mm^2), and water (0.00 mm^2). Tannin, catechin, caffeine, and tocopherol, tested in vitro on tooth enamel, demonstrated that these components possess the property of increasing the acid resistance of tooth enamel. The effects increased dramatically when the components were used in combination with fluoride. A mixture of tannic acid and fluoride showed the highest inhibitory effect (98%) on calcium release to an acid solution. Tannin in combination with fluoride inhibited the formation of artificial enamel lesions in comparison with acidulated phosphate fluoride (APF) as determined by electron probe microanalysis, polarized-light microscopy, and Vickers microhardness measurement.

DNA effect

Green tea extract, in cell culture at a dose of 10 mg/L corresponding to 15 mmol/L EGCg for 24 hours, did not protect Jurkat cells against H_2O_2-induced DNA damage. The DNA damage, evaluated by the Comet assay, was dose-dependent. However, it reached plateau at 75 mmol/L of H_2O_2 without any protective effect exerted by the extract. The DNA repair process, completed within 2 hours, was unaffected by supplementation.

Fluoride retention

Tea, used as a mouth rinse, demonstrated strong avidity of enamel for tea and salivary pellicle components. Thirty-four percent of the fluoride was retained in the oral cavity. Differences in retention at the tooth surface in the presence and absence of an acquired pellicle were not statistically significant at incisor or molar sites. Fluoride from tea showed strong binding to enamel particles, which was only partially dissociated by solutions of ionic strength considerably greater than that of saliva.

Gastrointestinal effect

Green tea, administered to rats fasted for 3 days, reverted to normal the mucosal and villous atrophy induced by fasting. Black tea ingestion had no effect. Ingestion of black tea, green tea, and vitamin E before fasting protected the intestinal mucosa against atrophy. Characterization of melanin extracted from tea leaves proved similarity of the original compound to standard melanin. The Langmuir adsorption isotherms for gadolinium (Gd) binding were obtained using melanin. Melanin–Gd preparation demonstrated low acute toxicity. LD_{50} for the preparation was in a range of 1.25–1.50 g/kg in mice.

Magnetic resonance imaging (MRI) properties of melanin itself and melanin-Gd complexes have been estimated. Gadolinium-free melanin fractions possess slighter relaxivity compared with its complexes. The relaxivity of lower molecular weight fraction was 2 times higher than relaxivity of Gd(DTPA) standard. Postcontrast images demonstrated that oral administration of melanin complexes in concentration of 0.1 m*M* provides essential enhancement to longitudinal relaxation times (T[1])-weighted spin echo image. The required contrast and delineation of the stomach wall demonstrated uniform enhancement of MRI with proposed melanin complex.

Hypocholesterolemic effect

Green tea, in human HepG2 cell culture, increased both LDL receptor-binding activity and protein. The ethyl acetate extract, containing 70% (w/w) catechins, also increased LDL receptor-binding activity, protein, and mRNA, indicating that the effect was at the receptor level of gene transcription and that the catechins were the active constituents. The mechanism by which green tea upregulated the LDL receptor was investigated. Green tea decreased the cell cholesterol concentration (–30%) and increased the conversion of the sterol-regulated element binding protein (SREBP-1) from the inactive precursor form to the active transcription-factor form. Consistent with this, the mRNA of 3-hydroxy-3-methylglutaryl coenzyme-A reductase, the rate limiting enzyme in cholesterol synthesis, was also increased by green tea.

Immunomodulatory effect

To determine the effects of tea on transplant-related immune function in vitro lymphocyte proliferation tests using phytohemagglutinin, mixed lymphocytes culture assay, IL-2, and IL-10 production from mixed lymphocyte proliferation were performed. Tea had immunosuppressive effects and decreased alloresponsiveness in the culture. The immuno-suppressive effect of tea was mediated through a decrease in IL-2 production. Tea, assayed in cell culture, enhanced neopterin production in unstimulated peripheral mononuclear cells, whereas an effective reduction of neopterin formation in cells stimulated with concanavalin A, phytohemagglutinin or interferon (IFN)-γ was observed. Theaflavins potently suppressed IL-2 secretion, IL-2 gene expression, and the activation of NF-κB in murine spleens enriched for CD4(+) T-cells. Theaflavins also inhibited the induction of IFN-γ mRNA. However, the expression of the T(H2) cytokines IL-4 and IL-5, which lack functional NF-κB sites within their promoters was unexpectedly suppressed by theaflavins as well.

Insulin-enhancing effect

Tea, as normally consumed, was shown to increase insulin activity more than 15-fold in vitro in an epididymal fat cell assay. The majority of the insulin-potentiating activity for green and oolong teas was owing to epigallocatechin gallate. For black tea, the activity was present in addition to epigallocatechin gallate, tannins, theaflavins, and other undefined compounds. Several known compounds found in tea were shown to enhance insulin with the greatest activity due to epigallocatechin gallate followed by epicatechin gallate, tannins, and theaflavins. Caffeine, catechin, and epicatechin displayed insignificant insulin-enhancing activities. Addition of lemon to the tea did not affect the insulin-potentiating activity. Addition of 5 g of 2% milk per cup decreased the insulin-potentiating activity one-third, and addition of 50 g of milk per cup decreased the insulin-potentiating activity approx 90%. Non-dairy creamers and soymilk also decreased the insulin-potentiating activity.

Iron absorption

Tea, administered by gastric intubation to rats, did not affect iron absorption when tea was consumed for 3 days but when delivered in tea the absorption was decreased. Rats maintained on a commercial diet were fasted overnight with free access to water and then gavaged with 1 mL of ^{59}Fe labeled FeCl3 (0.1 m*M* or 1 m*M*) and lactulose (0.5 *M*) in water or black tea. Iron absorption was estimated from Fe retention. Intestinal permeability was evaluated by lactulose excretion in the urine. Iron absorption was lower with given with tea at both iron concentrations but tea did not affect lactulose excretion.

Lipid peroxidation activity

Solubilized green tea, administered orally to rats for 5 weeks, reduced lipid peroxidation products. The treatment produced increased activity of glutathione (GSH) peroxidase and GSH reductase, increased content of reduced GSH, a marked decrease in lipid hydroperoxides and malondialdehyde in the liver, an increase in the concentration of vitamin A by about 40%. A minor change in the measured parameters was observed in the blood serum. GSH content increased slightly, whereas the index of the total antioxidant status increased significantly. In contrast, the lipid peroxidation products, particularly malondialdehyde, was significantly diminished. In the central nervous tissue, the activity of superoxide dismutase and glutathione peroxidase decreased, whereas the activity of GSH reductase and catalase increased after drinking green tea. Moreover, the level of lipid hydroperoxides, 4-hydroksynonenal, and malondialdehyde decreased significantly[CS036].

Neuromuscular-blocking action

Thearubigin fraction of black tea was investigated for neuro-muscular-blocking action of botulinum neurotoxin types A, B, and E in the mouse phrenic nerve-diaphragm preparations. On binding, A (1.5 n*M*), B (6 n*M*), and E (5 n*M*) abolished indirect twitches within 50, 90, and 90 minutes, respectively. Thearubigin fraction mixed with each toxin protected against the neuromuscular-blocking action of botulinum neurotoxin types A, B, and E by binding with the toxins.

Oral submucousal fibrosis effect

Tea, administered orally to 39 patients with oral submucous fibrosis, indicated that the treatment was effective for patients with abnormal hemorheology. The patients were divided into control and experimental groups. The control group included 22 oral submucous fibrosis patients who were treated by oral administration of vitamins A and D, vitamin B complex, and vitamin E. The experimental group included 17 patients who were treated with vitamins and tea pigment after their examination of hemorheology. The results showed that 7 of 12 patients in the experimental group with abnormal hemorheology had average 7.9 mm improvement on the open degree (58.3%), and the open degree of the other five patients whose hemorheology was normal only increased 2 mm (20%). The therapeutical results of the experimental group (58.3%) were significantly better than that of the control group (13.6%) ($p < 0.005$).

P-glycoprotein activity

Green tea polyphenols (30 μg/mL) inhibited the photo-labeling of P-gp by 75% and increased the accumulation of rhodamine-123 threefold in a multidrug-resistant cell line CH(R)C5, indicating that the polyphenols interact with P-gp and inhibit its transport activity. The modulation of P-gp transport by polyphenols was a reversible process.

Photoprotection effect

Tea extracts, administered topically, produced a dose- dependent inhibition of the erythema response evoked by UV radiation. The (–)- epigallocatechin-3-gallate and (–)-epicatechin-3 -gallate polyphenolic fractions were most efficient at inhibiting erythema, whereas (–)-epigallocatechin and (–)-epicatechin had little effect. On histological examination, skin treated with the extracts reduced the number of sunburn cells and protected epidermal Langerhans cells from UV damage. The extract also reduced damage that formed after UV radiation. Green tea polyphenols, applied topically to the human skin, prevented UVB-induced cyclobutane pyrimidine dimers, which are considered to be mediators of UVB-induced immune suppression and skin cancer induction. The treatment, prior to exposure to UVB, protected against UVB-induced local as well as systemic immune suppression in laboratory animals. Additionally, treatment of mouse skin inhibited UVB-induced infiltration of CD11b cells. CD11b is a cell-surface marker for activated macrophages and neutrophils, which are associated with induction of UVB-induced suppression of contact hypersensitivity responses. The treatment also resulted in reduction of the UVB-induced immunoregulatory cytokine IL-10 in skin as well as in draining lymph nodes, and an elevated amount of IL-12 in draining lymph nodes.

Protease inhibition

Epigallocatechin-3-gallate, in cell culture at a concentration of 100 μM, reduced virus yield by 2 orders of magnitude producing an IC_{50} of 25 μM and a therapeutic index of 22 in Hep2 cells. The agent was the most effective when added to the cells during the transition from the early to the late phase of viral infection, suggesting that it inhibited one or more late steps in virus infection. One of these steps appears to be virus assembly, because the titer of infectious virus and the production of physical particles were much more affected than the synthesis of virus proteins. Another step might be the maturation cleavages carried out by adenain. When tested on adenain, epigallocatechin-3-gallate produced an IC_{50} of 109 μM.

Radical scavenging activity

Green tea, evaluated using the 1,1-diphenyl-2-picrylhydrazyl radical, indicated that the galloyl moiety showed more potent activity. The contribution of the pyrogallol moiety in the B-ring to the scavenging activity seemed to be less than that of the galloyl moiety.

Tetanus toxin protection

Thearubigin fraction of black tea was investigated for neuromuscular-blocking action on tetanus toxin in the mouse phrenic nerve-diaphragm preparations and on binding of this toxin to the synaptosomal membrane preparations of rat cerebral cortices. Tetanus toxin (4 μg/mL) abolished indirect twitches in the mouse phrenic nerve–diaphragm preparations within 150 minutes. Thearubigin fraction mixed with tetanus toxin blocked the inhibitory effect of the toxin.

Toxicity

Green tea, administered orally at a dose of 6 g per day in six doses to 42 patients who were asymptomatic and had manifested, progressive prostate specific antigen elevation with hormone therapy, produced grade 1 or 2 toxicity in 69% of the patients and included nausea, emesis, insomia, fatigue, diarrhea, abdominal pain, and confusion. However, six episodes of grade 3 toxicity and one episode of grade 4 toxicity also occurred, with the latter manifesting as severe confusion.

Toxicity assessment

Ethanol (50%) extract of the entire plant, administered intraperitoneally to mice produced lethal dose $(LD)_{50}$ 316 mg/kg. Ethanol (95%) extract of the leaf, administered by gastric intubation to mice, produced LD_{50} 10 g/kg. Intraperitoneal administration produced CD_{90} 0.7 g/kg.

Cocos nucifera

Cocos nucifera is an unbranched monoecious plant of the *Palmae* family. It grows to 30 m tall, with a crown of 25–35 paripinnate leaves, producing 12–16 new leaves per year. There is a central bud, which if cut off, leads to the death of tree. The trunk is straight or gently curved, with marked foliar scars, 30–50 cm in diameter, rises from a thickened base, and increases in height at a decreasing rate with age. The leaves are horizontal or somewhat hanging, 4–8 m in length and divided. Segments of leaves are numerous, linear-lanceolate, 0.5–1 m long and tapering. In the axil of each leaf is a spathe enclosing a long, stout, straw, or an orange-colored spadix. The spadix is composed of up to

40 branches, each bearing up to 300 small, fragrant, male flowers, and a few female, 2 cm long, globose flowers. The male and female flowers are produced separately in the leaf axils, usually on a long stalk. Approximately one-third of the female flowers develop into four to eight ripe fruits in 12–13 months, per inflorescence. The fruit is ovoid, three-angled drupe, up to 30 cm long, usually with thick, fibrous mesocarp (husk) and a hard, green-brown endocarp (shell) enclosing one seed. The seeds consist of 10 to 20 mm-thick white, fleshy endosperm (meat), covered by thin brown testa, surrounding a cavity party filled with a watery, sweet fluid (coconut water or milk). The latter is found mainly in immature fruits.

Fig. 17.11. Cocos nucifera. A-Plant a fruiting stage; B-Fruit (entire); C-Fruit (L.S.).

Origin and distribution

Evidence has been found that the place of origin is "submerged land to the north west of New Guinea." The major coconut areas lie between 20°N and 20°S of the equator. Although it is found beyond this region, 27° N and 27° S, cultivation has not been successful and the palm does not fruit. Many varieties are found in Melanesian region. It is most widely cultivated in the tropics: India, Ceylon, Malaysia, Indonesia, Philippines, South Sea Islands in the Pacific, East Africa, and Central and South Americas, up to 800 m above sea level, on humus-rich and porous soil or pure sand in coastal regions.

Traditional uses

Admiralty islands. The young root or leaves of the coconut plant are chewed for diarrhea.

Cook islands. Water extract of grated endosperm and *Citrus aurantium* juice is used to soak affected part in fractures and sprains. Endosperm is taken orally for asthenia. Oil, mixed with crushed *Phyllanthus virgatus*, is rubbed around the ear for ear infections. Extract of different dried parts of the palm are taken orally for filariasis. A water solution of the crushed dried bark or husk and grated bark of *Hibiscus tiliaceus* is used externally to soak fractures and sprains. Crushed aerial root tips of *Ficus prolixa* are fried with coconut cream made of fresh endosperm, and the resulting oil is taken orally as a laxative in treating serious diseases.

Fiji. Oil is used externally to prevent hair loss. The oil is warmed with crushed onion and garlic and applied aurally for earache. Water of the unripe fruit is taken orally for kidney problems.

Ghana. Coconut milk is taken orally for diarrhea.

Guatemala. Hot water extract of the dried fruit is taken orally as a febrifuge and sudorific and for renal inflammation and scrofula. Hot water extract of the dried fruit is applied externally on wounds, ulcers, bruises, sores, skin infections, mucosa, dermatitis, inflammations, abscesses, and furuncles.

Haiti. Decoction of the dried pericarp is taken orally for amenorrhea. Fresh essential oil is applied externally on burns.

India. Infusion of the inflorescence is taken orally every morning for 3 days, coinciding with the menstrual cycle for leukorrhea and problems associated with the menstrual cycle. A dose of 50 g daily of a mixture of *Cocos nucifera* fruit and *Ficus-benghalensis* latex is taken for 3 months to increase sexual potency in men. Fruit is taken orally as a remedy for tapeworms.

Indonesia. Coconut oil is applied externally to treat wounds and injuries by the ethnic group of Ngada. Shell is used as incense. Hot water extract of the root is taken orally for fever, bloody diarrhea, and dysentery. Milk is taken orally by adults for poisoning. Fruit milk is taken orally by females as a contraceptive. Fruit juice is believed to diminish libido or fertility. Seed oil with lemon juice and various tree roots is taken orally as an abortifacient. Fruit ointment is applied externally for swollen legs. Fresh flowers, chewed with *Borassus flbellifer*, are used for gonorrhea.

Jamaica. Hot water extract of the dried shell is taken orally for diabetes. Hot water extract of the root, with seven other plants, wine, and rum, is used as a tonic. Extract of different parts are taken orally for diabetes.

Marquesas islands. Fruit juice is mixed with *Cordia subcordata* and other plants and used for general menstrual disorders.

Mexico. A plaster made of fresh milk mixed with egg white is applied externally to prevent miscarriage.

Mozambique. Fruit is eaten by males as an aphrodisiac and used for relief of tumors.

Papua New Guinea. The fresh root of a young coconut is dug out and washed, then chewed and swallowed to relieve stomachache. Fruit milk of *Cocos nucifera* is taken orally together with leaves of *Cleroden* sp., *Pouzolzia microhylla*, or *Macaranga tanarius* by pregnant women as an abortifacient agent.

Peru. Hot water extract of the fresh fruit is taken orally for blennorrhagia and asthma and as a diuretic, tenifuge, and galactagogue.

Thailand. Hot water extract of the fresh fruit juice is taken orally as a cardiotonic and neurotonic.

Tonga. Infusion of fresh kernel of coconut and Euodia hortensis are taken orally to treat retention of blood clots in the uterus after childbirth (locally called "toka-ala"). A mixture of *Cocos nucifera*, *Glochidion concolor*, *Vigna marina*, *Morinda citrifolia*, *Euodia hortensis*, and *Premna taitensis* with lemon juice is taken orally by pregnant women to treat severe bleeding during early pregnancy. Infusion of fresh kernel, taken twice daily with "tongan oil" prepared from *Aleurites moluccana* is used for dysuria caused by "kahi," a locally described syndrome affecting the gastrointestinal and genitourinary systems.

Trinidad. Hot water extract of the root is taken orally for amenorrhea.

Vanuatu. Hot water extract of the fruit, drunk in large quantities when very hot, is said to induce abortion. Juice of the crushed root mixed with water is drank after delivery to restore strengh.

West Indies. Hot water extract of the root is taken orally for amenorrhea. Hot water extract of the mesocarp is taken orally by females for dysmenorrhea, amenorrhea, and menorrhagia.

Medicinal uses

Acquired immunodeficiency syndrome activity

Coconut oil and "monolaurin"—a coconut oil byproduct—were administered to 12 women and 3 men who were in the early stage of human immunodeficiency virus infection. Ten patients took different doses of monolaurin, and five patients took coconut oil. It was believed that the treatment would lead to higher CD4 counts and a lower viral load. The trial was abandoned because it received only lukewarm approval from the government.

Allergenic activity

Dried fruit juice, administered intraperitoneally to guinea pigs at a dose of 10 mL/animal, was active. Dried fruit juice, administered intravenously to guinea pigs at a dose of 5 mL/animal, was active[CN036]. A patient with coconut anaphylaxis was confirmed by skin prick test. In vitro serum specific immunoglobulin E (IgE) was present. Two patients with allergy manifested by life-threatening systemic reaction after consumption of coconut were investigated. Sera IgE from both patients indicated reduced coconut allergens with molecular weight of 35 and 36.5 kDa. IgE from 1 patient also bound a 55-kDa antigen. Preabsorption of sera with nut extracts suppressed IgE binding to coconut proteins. Preabsorption of sera with coconut produced a disappearance of IgE binding to protein bands at 35 and 36 kDa on a reduced immunoblot of walnut protein extract in one patient and suppression of IgE binding to a protein at 36 kDa in another patient. Three cases of individuals allergic to cocamide diethanolamide were investigated. In two of the cases, multiple other cutaneous allergies were present. In both instances, cocamide diethanolamide was present in several personal care products used by the patients. In the third case, occupational exposure was suspected.

Coconut diethanolamide, administered to six patients with occupational contact dermatitis caused by coconut diethanolamide, produced sensitization from a barrier cream in two patients, hand-washing liquid in three, and one had been exposed to a hand-washing liquid and to a metalworking fluid containing coconut diethanolamide. Leave-on products (hand-protection foams) produced sensitization more rapidly (2–3 months) than rinse-off products (5–7 years). There was no contact allergy to another coconut-oil-derived sensitizer (cocamidopropyl betaine). Coconut diethanolamide, administered to a dentist, produced an occupational allergic contact dermatitis for hand-washing liquids. Pollen extract, administered to 24 patients with allergy, asthma, and rhinitis produced positive skin prick test in all cases, and 19 of them were phadezym radioallergosorbent test-positive. Bronchial provocation test was positive in seven out of eight patients, and no late response or nonspecific reactions were observed. An 8-month-old baby fed from birth with maternal milk was investigated. The first milk induced a severe gastrointestinal disorder, which disappeared when second milk was used. The third milk caused a relapse.

The allergen was coconut, which was physicochemically modified in the second milk. It was confirmed by positive reintroduction test, positive skin test, and positive specific IgE test. Pollen extract immunotherapy was studied in 96 allergic patients for 6–12 months. The clinical status measured by the symptom-medication scores demonstrated that the patients had significant clinical improvement after pollen extract immunotherapy. Serological study indicated a significant reduction of specific IgE and elevation of specific IgG in posttherapeutic patients' sera. There was no correlation between symptom-mediation scores and changes in specific serum IgE or IgG levels. Proteins from fresh coconut and commercial extracts of coconut, administered to a patient with coconut anaphylaxis, produced a positive skin prick test. In vitro serum-specific IgE was present for coconut, hazelnut, Brazil nut, and cashew. Immunoblots demonstrated IgE binding to 35- and 50-kDa protein bands in the coconut and hazelnut extracts. Inhibition assays using coconut demonstrated complete inhibition of hazelnut specific IgE, but inhibition assays using hazelnut showed only partial inhibition of coconut-specific IgE. A total of 100 patients (59 females and 41 males, aged 10–59 years, mean age 27.9 years) with allergic rhinitis underwent a skin prick test with 30 aeroallergens.

Coconut produced 12% of positive results. Oil, administered to 3-lactoglobulin-treated brown Norway rats at a dose of 10% of diet supplemented with 0.5% curcumin for 3 weeks, lowered the circulatory release of rat chymase II in response to antigen. The triethanolamine (TEA) salt of the condensation product of coconut fatty acids with a complex of polypeptides and amino acids derived from collagen (TEA-Coco-hydrolyzed protein), administered to a 21-year-old woman, produced a severe dermatitis of the face after using a proprietary skin cleanser. Patch testing revealed delayed hypersensitivity to TEA-Coco-hydrolyzed protein but not to other ingredients of the cleanser and positive results with other condensates of fatty acids and protein hydrolysates.

Aminopeptidase activity influence

Oil, administered to mice testis using acrylamides as substrates, produced no difference in soluble glutamyl-aminopeptidase angiotensin among the groups tested. Soluble aspartyl-AP and soluble pyroglutamyl-AP progressively decreased with the degree of saturation of the fatty acids used in the diet. Membrane-bound glutamyl-AP progressively increased with the degree of saturation of the fatty acids used in the diet. For membrane-bound aspartyl-AP activity, mice that were fed diets containing fish oil showed significantly higher levels than those fed sunflower oil, olive oil, and lard, but not those fed coconut oil.

Antibacterial activity

Ethanol (50%) extracts of the leaf, on agar plate at concentrations above 25 μg/mL, were inactive on *Bacillus subtilis*, *Escherichia coli*, *Salmonella typhosa*, and *Staphylococcus aureus*. Tincture of the dried fruit (10 g plant material in 100 mL ethanol), on agar plate at a concentration of 30 μL/disc, was inactive on *Pseudomonas aeruginosa* and *Staphylococcus aureus* and produced weak activity on *Escherichia coli*. Water extract of the husk fiber and fractions from adsorption chromatography were active on *Staphylococcus aureus*. Hydrogenated oil, administered orally to Balb/c mice at a dose of 20% by weight for 4 wk and then infected with *Listeria monocytogenes* or treated with *N*-acetyl-L-cysteine (25 mg/mL intraperitoneally), produced no effect on survival. *N*-acetyl-L-cysteine reduced the recovery of *Listeria monocytogenes* from the spleen of the mice fed coconut oil diet. There was a reduction in lymphocyte proliferation. An important increase in the production of reactive oxygen species was found after 12 h of the incubation with *Listeria monocytogenes*. Hydrogenated oil, administered to *Listeria monocytogenes*-infected mice, produced a significant increase in peritoneal cells of coconut oil-fed mice and a reduction of bacterial recovery from the spleen. Oil, administered to mice injected with a nonlethal dose of *Escherichia coli* at a dose of 20% by weight for 5 weeks, produced a decrease of peak plasma tumor necrosis factor (TNF)-α, interleukin (IL)-1β, and IL-6 concentrations. Peak plasma IL-10 concentrations were higher in the coconut-fed group than in those fed the other diets, coconut oil diminished production of proinflammatory cytokines in vivo.

Anticarcinogenic effect

Coconut cake, administered to rat with 1,2-dimethylhydrazine-induced colon cancer at a dose of 25% of diet for 30 weeks, produced a significant decrease of the incidence and number of tumors, and β-glucuronidase and mucinase activities. Hydrogenated oil, administered to Wistar female rats at doses of 8% and 24%, with or without a 24 mg/day/rat phytosterol supplement, produced no significant influence. Colonic glands were found in area of lymphoid follicles in all the groups but were more frequently in rats on high-fat diet.

Antifungal activity

Ethanol (50%) extract of the leaf, on agar plate at a concentration of 25 μg/mL or more, was inactive on *Microsporum canis* and *Trichophyton mentagrophytes*. Oil, on agar plate at a concentration of 05 mL/plate, was active on *Absidia corymbifera*, *Aspergillus flavus*, *Aspergillus niger*, and *Penicillium nigricans*. Ethanol (95%) extract of the dried shell, on agar plate at a concentration of 100 μg/mL, was active on *Microsporum audouini*, *Microsporum canis*, *Microsporum gypseum*, *Trichophyton mentagrophytes*, *Trichophyton rubrum*, *Trichophyton tonsurans*, *Trichophyton violaceum*, and *Epidermophyton floccosum*. All fractions of the husk fiber, from adsorption chromatography, were inactive on *Candida albicans*, *Fonsecaea pedrosoi*, and *Cryptococcus neoformans*.

Antihypercholesterolemic activity

Oil was administered orally to male Wistar rats at doses of 12 or 24% of diet *ad libitum* for 4 weeks. Absorption of oleic acid in rats fed 24% oil was significantly greater than in controls during 0–8 hours but was not significantly different during 0–24 hours. There were no differences among groups in the distribution of cholesterol and oleic acid either in the lymph lipoproteins or in the lipid classes.

Oil was administered to 61 healthy males and 22 females aged 20–34 years, group I received a coconut–palm–coconut dietary sequence; group II, coconut–corn–coconut; group III, coconut oil during all three 5-week dietary periods. Compared with entry-level values, coconut oil raised the serum total cholesterol concentration greater than 10% in all three groups. The entry level of the ratio of low-density lipoprotein (LDL) to high-density lipoprotein (HDL) was not altered by coconut oil. Oil was administered to 6- month-old ovariectomized rats with or without taurine for 28 days. Body mass gain, food intake, liver weight, and plasma apoliprotein (apo) A-I, apo B, LDL, and very low-density lipoprotein (VLDL) concentrations were not affected by the diet. Taurine lowered the plasma total cholesterol and increased liver total lipid and triglyceride in rats that were fed corn oil but not in those fed coconut oil. Taurine increased the 3 -hydroxy-3-methylglutaryl coenzyme A (HMG-CoA) reductase messenger (m)RNA level in the liver of coconut-fed rats, but not in those fed corn oil. Structured lipids (10%) synthesized from oil triglycerides, coconut oil, and coconut oil–safflower oil blends (1:0.7 w/w) were administered to rats for 60 days. The structured lipids lowered serum and liver cholesterol levels. Most of the decrease observed in serum was found in LDL fraction. Oil, administered orally to male Wistar rats at a dose of 11% w/w for 6 months, produced a significant increase of serum total cholesterol (twofold), serum triglycerides (92.6%), LDL cholesterol (92.3%), and body weight gain (2.8-fold).

Antilipidemic activity

Triglycerides structured lipids from coconut oil, administered to rats at a dose of 10% of diet for 60 days, produced a 15% decrease in total cholesterol and a 23% decrease in LDL cholesterol levels in the serum compared to coconut oil-fed rats. Total and free cholesterol levels in the liver of structured lipid-fed rats were lowered by 31 and 36%, respectively. The triglycerides in the serum and liver were decreased by 14 and 30%, respectively.

Anti-nociceptive activity

Aqueous extract of the husk fiber, administered orally to mice at doses of 200 or 400 mg/kg, produced an inhibition of the acetic acid- induced writhing response.

Antioxidant activity

Aqueous extract of the husk fiber, in cell culture, was active vs 2,2-diphenyl-1 -picryl-hydrazyl-hydrate radicals. Juice, in cell culture, was active vs 1,1 -diphenyl-2-picrylhydrazyl, 2,2'-azino-*bis*(3-ethylbenz-thiazoline-6-sulfonic acid) and superoxide radicals but promoted the production of hydroxyl radicals and increased lipid peroxidation. The activity was most significant for fresh samples and diminished significantly when heated or treated with acid, alkali, or dialysis. Maturity of coconut drastically decreased the scavenging activity. Juice protected hemoglobin from nitrite-induced oxidation when added before the autocatalytic stage of the oxidation. Acid, alkali, or heat-treated or dialyzed juice showed a decreased ability in protecting hemoglobin from oxidation.

Antiparasitic activity

Polyphenolic-rich extract of the husk fiber, in cell culture at a concentration of 10 μg/mL, produced a reduction approx 44% of the association index between peritoneal mouse macrophages and *Leishmania amazonensis* promastigotes. It inhibited the growth of promastigote and amastigote developmental stages after 60 minutes. There was a concomitant increase of 182% in nitric oxide production by the infected macrophage in comparison to nontreated macrophages.

Antiproteinemic effect

Hydrogenated oil was administered orally to rats fed a protein- deficient diet at a dose of 200 g casein and 50 g coconut oil or 20 g casein and 50 g coconut oil for 28 days. The treatments produced a low concentration of protein and triacylglycerol (in serum VLDL, LDL–HDL 1 and 2–3), of cholesterol (in LDL–HDL1), and of phospholipids (in VLDL) in the protein-deficient groups. Relative amounts of linoleic and arachidonic acids in phospholipids of VLDL and HDL 2–3 were also lowered in the 20 g casein and 50 g coconut oil group. Coconut fat, administered to rats, produced an increase in plasma esterase-1 activity. Oil, administered to male Wistar rats at doses of 20% casein and 5% coconut oil or 2% casein and 5% coconut oil for 28 days, produced low concentration of protein, triacylglycerol, and VLDL in the plasma of 2% casein and 5% coconut oil group. Malondialdehyde content of the 2% casein and 5% coconut oil group was significantly higher than of control group. The lowest level of malondialdehyde was observed in the coconut oil group.

Anti-yeast activity

Tincture of the dried fruit, on agar plate at a concentration of 30 μL/disc, was inactive on *Candida albicans*. Extract of 10 g plant material in 100 mL ethanol was used. Ethanol (50%) extract of the leaf, at a concentration of 25 μg/mL or more, was inactive on *Candida albicans* and *Cryptococcus neoformans*. Seed oil, on agar plate at a concentration of 0.05 mL, was active on *Candida albicans*.

Apo B synthesis

Oil, administered to 1-month-old calves fed on a conventional milk replacer containing coconut oil, produced a twofold lower concentrations of total (^{35}S) proteins, (^{35}S) albumin, and (^{35}S) apo B in liver cells than in beef tallow-fed calves. The total amount of proteins secreted (including albumin) was similar in both groups. The amount of VLDL-(^{35}S) apo secreted was twofold lower in coconut oil-fed group. Oil, administered to rabbits at a dose of 14% of the diet for 4 weeks, produced an increase in HDL-cholesterol (C) level from 170 to 250% over chow-fed control, with peak differences occurring at 1 week. Plasma apo A-I levels were also increased from 160 to 180%. After 4 weeks, there was no difference in plasma VLDL-C or LDL-C levels in the both groups. Hepatic level of apo A-I mRNA was increased in the coconut-fed groups. Treatment of cultured rabbit liver cells and sera with various saturated fatty acids did not alter apo A-I mRNA levels as observed in vivo.

Atherosclerotic influence

Oil was administered orally to rabbits fed a semipurified, cholesterol-free atherogenic diet at a dose of 14% fat: coconut oil (CNO), corn oil (CO), palm kernel oil (PO), and cocoa butter (CB). Serum cholesterol levels (mg/dL) at 9 months were CO, 64; PO, 436; CB, 220; and CNO, 474. HDL cholesterol (%) was CO, 37; PO, 8.6; CB, 25; and CNO, 7. average artherosclerosis (arch + thoracic/2) was CO, 0.15; PO, 1.28; CB, 0.53; and CNO, 1.60. Oil, administered to British Halflop rabbits at a dose of 3% of the diet, produced an increase of cholesterol, triacylglycerol, apo B, apo C-III, and apo E as compared to chow-fed rabbits.

The apo A level did not differ from the chow-fed rabbits. There was a decrease in the low fractional catabolic rate of LDL-C. Oil, administered with cholesterol (30 g/kg) to male golden Syrian hamsters at a dose of 150 g/kg of diet, developed lipid-rich lesions in the ascending aorta and aortic arch after 4 weeks. The lesions continued to progress throughout the next 8 weeks. Removal of cholesterol from the diet halted this progression. Oil, administered without supplemental cholesterol in the same dose for 16 weeks, doubled the size of lesions in the ascending aorta and decreased linearly the lesion size in the aortic arch.

Blood pressure effect

Fruit juice, administered intravenously by infusion to dogs at a dose of 3 mL/minute for 100 minutes, was active. Initial effect was a decrease in blood pressure. Oil, administered to male weanling rats at a dose of 10% of diet for 5 weeks, produced significantly higher blood pressure than other groups. Systolic blood pressure was found related to the dietary intakes of saturated and unsaturated fatty acids. Prenatal exposure of the rats to a maternal low-protein diet abolished the hypertensive effect of the coconut oil diet.

Butyryl cholinesterase activity

Oil was administered to rats at different doses with or without clofibrate for 15 days. The hypolipidemic action of clofibrate was not influenced by the amount of fat. Clofibrate did not affect lower cholesterol concentration in rats fed the low fat diet, but it counteracted the rise in liver cholesterol seen in rats fed the high-fat diet. The high-fat diet produced slightly higher levels of butyryl cholinesterase in the small intestine but markedly raised intestinal esterase-1 activity.

Carcinogenic activity

Coconut oil acid diethanolamine condensate in ethanol was administered externally to 10 male and 10 female F344/N rats at doses of 25, 50, 100, 200, or 400 mg/kg body weight five times per week for 14 weeks. All of the rats survived the study. Final mean body weights and body weight gains of the 200 and 400 mg/kg males and females were significantly less than those of the controls. Clinical findings included irritation of the skin at the site of application in the 100, 200, and 400 mg/kg males and females. Cholesterol concentrations were significantly decreased in 200 and 400 mg/kg males. Histopathological lesions of the skin at the site of application included epidermal hyperplasia, sebaceous gland hyperplasia, chronic active inflammation, parakeratosis, and ulcer.

The incidences and severities of skin lesions generally increased with increasing dose in males and females. The incidences of renal tube regeneration in the 100, 200, and 400 mg/kg females were significantly greater than the vehicle control incidence, and the severities in the 200 and 400 mg/kg females were increased. Coconut oil acid diethanolamine condensate in ethanol was administered externally to 10 male and 10 female B6C3F1 mice at doses of the 50, 100, 200, 400, or 800 mg/kg body weight five times per week for 14 weeks. All mice survived until the end of the study. Final mean body weights and body weight gains of dosed males and females were similar to those of the vehicle controls. The only treatment-related clinical finding was irritation of the skin at the site of application in males and females administered the 800 mg/kg dose. Weights of the liver and kidney of 800 mg/kg males and females, the liver of the 400 mg/kg females, and the lung of the 800 mg/kg females were significantly increased compared to the controls. Epididymal spermatozoal concentration was significantly increased in the 800-mg/kg males. Histopathological lesions of the skin at the site of application included epidermal hyperplasia, sebaceous gland hyperplasia, chronic active inflammation, parakeratosis, and ulcer.

The incidences and severities of these skin lesions generally increased with increasing dose in males and females. Coconut oil acid diethanolamine condensate in ethanol was administered externally to 50 male and 50 female F344/N rats at doses of 50 or 100 mg/kg body weight five times a week for 104 weeks. The survival rates of treated male and female rats were similar to those of the vehicle controls. The mean body weights of dosed males and females were similar to those of the vehicle controls throughout most of the study. The only chemical-related clinical finding was irritation of the skin at the site of application in the 100-mg/kg females. There were marginal increases in the incidences of renal tubule adenoma or carcinoma (combined) in the 50-mg/kg females. The severity of nephropathy increased with increasing dose in the female rats. Nonneoplastic lesions of the skin at the site of application included epidermal hyperplasia, sebaceous gland hyperplasia, parakeratosis, and hyperkeratosis, and the incidences and severities of these lesions increased with increasing dose. The incidence of chronic active inflammation, epithelial hyperplasia, and epithelial ulcer of the forestomach increased with dose in female rats and the increases were significant in the 100-mg/kg group. Coconut oil acid diethanolamine condensate in ethanol was administered externally to 50 male and 50 female B6C3F1 mice at doses of 100 or 200 mg/kg body weight five times a week for 104 to 105 weeks. Survival of the mice was generally similar to that of the controls. Mean body weights of 100-mg/kg females from week 93 and 200-mg/kg females from week 77 were less than in the controls. The only clinical finding attributed to treatment was irritation of the skin at the site of application in males administered 200 mg/kg. The incidences of hepatic neoplasms (hepatocellular adenoma, hepatocellular carcinoma, and hepatoblastoma) were significantly increased in both sexes. Most of the incidences exceeded the historical control ranges. The incidences of eosinophilic foci in dosed groups of male mice were increased relative to that in controls. The incidences of renal tubule adenoma and renal tubule adenoma or carcinoma (combined) were significantly increased in the 200-mg/kg males. Several nonneoplastic lesions of the skin at the site of application were considered treatment related. Incidence of epidermal hyperplasia, sebaceous gland hyperplasia, and hyperkeratosis were greater in all dosed groups than in the controls. The incidence of ulcer in 200-mg/kg males and inflammation and parakeratosis in 200-mg/ kg females were greater than in the controls. The incidence of thyroid gland follicular cell hyperplasia in all of the dosed groups was significantly greater than those in the control groups.

Cardiovascular effect

Coconut and coconut oil, administered to 32 coronary heart disease patients in 16 age- and sex-matched healthy controls with no difference in the fat, saturated fat, and cholesterol consumption, produced no effect. Hydrogenated oil, administered to young male Wistar rats, at a dose of 10% of diet for 10 weeks, produced an increase in the risk of ventricular arrhythmias under conditions of both ischemia and reperfusion. The incidence of ventricular fibrillation was 67% in the oil fed group. The time until the first occurrence of extrasystole, the incidence of ventricular tachycardia, and the incidence of reperfusion-induced ventricular fibrillation were influenced in a similar manner between groups. The fatty acid composition of myocardial tissue, the ratio of *n*-3 to *n*-6 fatty acids, and the double-bond index were significantly affected by the various diets. Oil was administered by gastric intubation to male Sprague–Dawley rats at a dose of 86:14 w/w coconut oil/safflower oil for 10 days, containing either 0.8 mg Zn/kg or 111 mg Zn/kg. Zinc-deficient rats that were fed coconut oil had higher concentrations of triglycerides and total fatty acids in the heart than the control rats. The concentrations of phospholipids and total cholesterol were not different between zinc-deficient and control rats. Concentrations of lauric acid, myristic acid, palmitic acid, palmitoleic acid, and oleic acid were 65 to 192% higher in the hearts of zinc-deficient rats fed coconut diet than in the control rats. The level of arachidonic acid in phospholipids, which may represent desaturation activity, was not different in the zinc-deficient rats and control rats. Oil was administered orally to rats fed a copper-deficient diet at doses of 10 g/100 g coconut oil or 10 g/100 g coconut oil with 1 g/100 g cholesterol. Rats fed the diet with coconut and cholesterol had left ventricular chamber volumes that were twofold larger than those of rats fed diet with coconut oil only.

Copper deficiency reduced left ventricular chamber volume only in rats fed coconut oil and cholesterol. The results indicated that preload and contractility in the hearts of coconut oil-fed rats were greater than cardiac response to cholesterol addition to the coconut oil diet. Hearts in copper-deficient rats fed coconut oil and cholesterol exhibited eccentric hypertrophy and ventricular dysfunction. Oil, administered orally to rats for 4 weeks, produced a significant decrease in 5'-nucleotidase, phosphodiesterase I and *p*-nitrophenylphosphatase activity of cardiac sarcolemma. Sarcolemma from coconut-fed rats contained a significantly lower concentration of total polyunsaturated fatty acids (PUFAs) and a higher concentration of total monounsaturated fatty acids than that from safflower-fed rats. The fatty acid composition of the phosphatidylcholine exhibited the largest alterations because of coconut oil feeding. No dietary effect was observed in the sarcolemma content of cholesterol and phospholipids.

Cholestatic liver disease

Medium-chain fat from the oil, administered to bile duct-ligated rats, resulted in a decrease of consumption of the fat source compared to control rats; however, carbohydrate and protein intakes were not affected. Body weight gain was significantly greater in coconut-fed rats than in rats fed long-chain fat (Crisco vegetable shortening). Mortality was 44% in bile duct-ligated animals fed the long-chain fat and 0% in those fed medium-chain fat.

Cholesterol metabolism

Hydrogenated oil, administered orally to hamsters at a dose of 20% of diet for 4 weeks, induced hypercholesterolemia. Oil feeding had no effect on cholesterol synthesis but markedly inhibited cholesterol esterification in both the liver and the intestine. The diet-induced hypercholesterolemia was strongly correlated with an increase in acyl-CoA/cholesterol acyltransferase activity. The hypercholesterolemia increased aortic uptake of cholesterol and hence acyl-CoA/cholesterol acyltransferase activity. Coconut fat, administered orally to rabbits with partial ileal bypass, produced a significant increase of serum total cholesterol and phospholipids concentrations. The effect on serum lipids of the type of fat was similar in control and partial ileal bypass rabbits.

Coconut—a main source of energy for two Polynesian groups, Tokelauans and Pukapukans—was investigated. Tokelauans obtained a much higher percentage of energy from coconut than the Pukapukans, 63% compared with 34%, so their intake of saturated fat is higher. The serum cholesterol levels are 35–40 mg higher in Tokelauans than in Pukapukans. Analysis of a variety of food samples and human fat biopsies showed a high lauric (12:0) and myristic (14:0) acids content. Vascular disease is uncommon in both populations, and there is no evidence of the high saturated fat intake having a harmful effect in these populations. Oil, administered orally to growing male Sprague–Dawley rats fed a diet containing unoxidized or oxidized cholesterol (5 g/kg) with coconut or salmon oil (100 g/kg) for 5 weeks, produced an effect independent of the dietary fat. Hydrogenated coconut oil was administered to F(1)B Golden Syrian hamsters at a dose of 20 g/100 g for 10 weeks.

The hamsters were ranked according to their plasma VLDL and LDL cholesterol as a low or medium group. Low-coconut oil group had significantly higher aortic total and esterified cholesterol concentrations than the low-cholesterol-fed group. Hamsters in the low-cholesterolfed group had significantly higher aortic TNF-α concentrations than hamsters in the low-coconut oil group. Hamsters in the med-coconut oil group had significantly higher aortic IL-1-β concentrations than hamsters in the med-cholesterol group. Hamsters in both cholesterol fed groups had significantly lower plasma total cholesterol concentrations than hamsters in the low-coconut oil group.

Cholesterolemic effect

Oil, administered to phospholipids transfer protein knockout (PLTP0)-deficient mice, produced an increase of phospholipids and free cholesterol in the VLDL-LDL region of PLTP0 mice. Accumulation of phospholipids and free cholesterol was dramatically increased in PLTP0/HL0 mice compared to PLTP0 mice. Turnover studies indicated that coconut oil was associated with delayed catabolism of phospholipids and phospholipids/free cholesterol-rich particles. Incubation of these particles with hepatocytes of coconut-fed mice produced a reduced removal of phospholipids and free cholesterol by scavenger receptor BI, even though scavenger receptor BI protein expression levels were unchanged.

Coagglutination activity

Oil was administered to females at doses of a high-saturated fatty acids diet (HSAFA diet; 38.4% of energy from fat, polyunsaturated/saturated [P/S] fatty acid ratio 0.14) or low-saturated fatty acids diet (LSAFA; 19.7% of energy from fat, P/S ratio 0.17). The postprandial plasma concentration of tissue plasminogen activator (t-PA) antigen was decreased in the HSAFA-fed group. Plasma t-PA antigen was correlated with plasminogen activator inhibitor type 1 (PAI-1) activity when the participants consumed the HSAFA and LSAFA diet, although the diets did not affect the PAI-1 levels. There were no significant differences in postprandial variations in t-PA activity, factor VII coagulant activity, or fibrinogen levels as a result of the diets. Serum-fasting Lp(a) levels were lower in the HSAFA group and were lower in the LSAFA group. Serum Lp(a) concentrations did not differ when the women consumed the HSAFA and LSAFA diets.

Cytotoxic activity

Oil, in cell culture at a concentration of 300 μg/mL, was active on Ca-colon-HT29. Coir fiber, administered intratracheally to guinea pigs, resolved the granulomas. Coir fiber and ash, in cell culture, produced a hemolytic activity and macrophage cytotoxicity more marked with ash compared with the fiber alone. Extract of the husk fiber, in cell culture at a concentration of 10 μg/mL, inhibited the growth of promastigote and amastigote developmental stages of *Leishmania amazonensis* after 60 minutes. Extract of the fiber husk (rich in catechins), in erythroleukemia cell line (K562) and normal human peripheral blood lymphocytes activated by phytohemagglutinin or phorbol ester, produced a dose-dependent effect.

For phytohemagglutinin, this effect was irreversible, being already established on the first hours of culture. Oil, administered to C57B16 mice at a dose of 21% fat by weight, produced a significant decrease in macrophage-mediated death of P815 cells by nitric oxide and did not alter the death of L929 cell by macrophages. Liposaccharide-stimulated TNF-α production by macrophages decreased with increasing unsaturated fatty acid content of the diet: fish oil < safflower oil < olive oil < coconut oil < low-fat diet.

Desensitization effect

Saline extract of the dried pollen, administered subcutaneously to 96 allergic adults at variable doses, produced a clinical improvement and decreased IgE levels.

Dietary macronutrient distribution influence

Oil, administered to diet-induced overweight rats fed an energy-restricted diet with 60% of coconut oil, produced no difference between control and fat-fed group in weight loss and serum parameters. The coconut-fed group produced a greater reduction in the subcutaneous fat depot and of total body fat. Hepatic glycogen and glycogenic amino acid were altered in coconut-fed rats.

Diuretic activity

Decoction of the dried fruit, administered nasogastrically to rats at a dose 1 g/kg, was active. Fruit juice, administered intravenously by infusion to dogs at a dose of 3 mL/minute for 100 minutes, produced weak activity. Ethanol extract of the leaf, administered intraperitoneally to saline-loaded male rats at a dose of 0.185 mg/kg, was active. Urine was collected for 4 hours after treatment.

Erythrocytic effect

Hydrogenated coconut oil, administered orally to healthy rats for 10 weeks, produced a significant effect on five of the six classes of erythrocytes identified. The proportion of cells in each class was dependent on the diet. There was no significant effect of diet on erythrocyte filterability index and no statistical correlation between erythrocyte filterability index and morphology.

Exocrine pancreatic secretion

Oil, administered to piglets at a dose of 10 g/100 g of diet, did not affect the output of carboxylester hydrolase. Protein, chymotrypsin, carboxypeptidase A, elastase, and amylase outputs were not different among the dietary treatment groups. The outputs of trypsin and colipase were higher in the coconut-fed group. Oil, administered to three barrows at a dose of 15 g fat/100 g diet, produced a significant increase of chymotrypsin secretion.

Fatty acid composition influence

Oil was administered orally to mice bearing the L1210 murine leukemia cells at a dose of 16% of diet. A microsome-rich fraction prepared from the L1210 cells produced more monoenoic fatty acids (37% vs 12%) compared to mice fed the sunflower oil. Oil, administered to chicken at concentrations of 10 and 20% of diet, produced changes in fatty acid composition of free fatty acid and triglyceride fractions of chick plasma parallel to that of the experimental diet. Plasma phospholipids incorporated low levels of 12:0 and 14:0 acids, whereas 18:0, the main saturated fatty acid of this fraction, increased after coconut oil feeding.

The percentage of 18:2 acid significantly increased after coconut oil feeding. Oil, administered to rats at a dose of 15% of diet for 4 weeks, produced an increase of total cholesterol in the liver and the decrease of total liver phospholipids. Coconut/soy oil, administered by infusion into the duodenum of rats, produced the proportion of capric and lauric acids in the lymphatic triacylglycerol, reflecting the fatty acid composition in the diet. Results indicated that more than 50% of the capric and lauric acids could have been absorbed from the intestine as sn-monoacylglycerols. Hydrogenated coconut oil was administered to rats at a dose of 7% of the diet (essential fatty acids deficient in both *n*-6 and *n*-3) for 28 weeks (1 week gestation, 3 weeks lactation, and 24 weeks thereafter). The fatty acid compositional changes indicative of an essential fatty acid deficiency, such as the decreases in the levels of 18:2 *n*-6 along with an accumulation of 20:3 *n*-9 were observed in all the salivary glands. In the submandibular glands, the proportions of 16:1, 18:1 *n*-9 and 18:1 *n*-7 were higher in hydrogenated coconut oil-fed group than in the other groups. Some differences in the fatty acid composition of the three glands were found.

Fatty acid metabolism

Oil was administered to preruminant Holstein Friesian male calves fed a conventional milk diet containing coconut oil for 19 days. Fatty acid oxidation was determined by measuring the production of CO_2 (total oxidation) and acid-soluble products (partial oxidation). Production of CO_2 was 1.7–3.6-fold lower and acid soluble products tend to be lower in liver slices of coconut oil-fed than beef-tallow-fed calves. Fatty acid esterification as neutral lipids was 2.6- to 3.1-fold higher in liver slices of coconut oil-fed group. The increase in neutral lipid production did not stimulate VLDL secretion by the hepatocytes, leading to a triacylglycerol accumulation in the cytosol of the calves fed coconut oil. Oil, administered to preruminant calves fed a milk replacer containing coconut oil for 19 days, produced no significant difference in weight of the total body and tissue between the treated groups. Plasma glucose and insulin concentrations were lower in the coconut oil-fed group. Feeding on the coconut oil diet induced an 18-fold increase in the hepatic concentration of triacylglycerol.

The perixomal oxidation rate of oleate was 1.5-fold higher in the hearts of calves fed coconut oil. The cytochrome C oxidase/citrate synthase activity ratio was lower in the liver of the coconut oil-fed animals. Oil, administered as a ratio 7:1 w/w coconut oil/safflower oil by gastric intubation to zinc-deficient rats, reduced δ-desaturase activity in liver microsomes of rats fed coconut oil. Zinc-deficient rats on the coconut oil diet had unchanged δ-6-desaturase activity with linoleic acid as substrate and lowered activity with α-linolenic acid as substrate.

Genotoxic activity

Coconut oil acid diethanolamine condensate with or without S9 activation enzyme, in cell culture, produced no effect on *Salmonella typhimurium*. The treatment did not produce an increase in mutant L5178Y mouse lymphoma cell colonies, and no increased in the frequencies of sister chromatid exchanges or chromosomal aberrations in Chinese hamster ovary cells. In a peripheral blood micronucleus test in male and female mice from the 14-week test, positive results were obtained.

Genotype and diet effect

Coconut oil was administered to female Zucker rats throughout mating and lactation. Homozygous lean male and female rats, obese male, and lean heterozygous female rats were bred. Additional male rats were maintained on the same diet as their mothers until 11–12 days of age. Obese sucking rats had higher body weights than lean pups. Inguinal fat pad weights and pad-to-body weight ratios followed the pattern of obese greater than lean (FA/fa genes) pups that were greater than lean (FA/FA) pups. A similar relationship was found for adipose tissue lipogenic enzyme activities. At 11–12 weeks of age, measurements followed the general pattern of obese rats having greater value than lean rats (i.e., FA/fa = FA/FA). Coconut oil-fed fa/fa rats had lower hepatic lipogenic enzyme activities and lower fat cell numbers than safflower-fed fa/fa rats. High-fat diet did not result in a heterozygous effect in young adult lean male rats. Hydrogenated coconut oil, corn oil, or menhaden oil were administered to diabetes-prone BHE/ cdb and normal Sprague–Dawley rats. Both fat source and strain affected the temperature dependence of succinate-supported respiration. The transition temperature was greater in BHE/cdb rats than in the Sprague–Dawley rats. The efficiency of adenosine triphosphatase synthesis as reflected by the adenosine diphosphatase/O ratio was decreased in the BHE/cdb rats compared to Sprague–Dawley rats.

Glucose metabolism

Hydrogenated oil, administered to prediabetic weanling BHE rats fed a 6% fat and 64% sucrose diet, produced an increase of the fractional irreversible glucose turnover rates, fractional glucose carbon recycling, hepatic fatty acid synthesis rates, adipose fatty acid synthesis rate, lower muscle glycogen, and lower rates of incorporation of glucose into muscle glycogen than corn oil-fed rats. The diet had no effect on glucose mass and space, hepatic glycogen, or blood glucose levels. Oil, administered to male Wistar rats at a dose of 25% by weight for 26 days, produced a significant increase in serum total cholesterol, LDLs, liver cholesterol, and liver weight. There was a decrease in serum HDLs, triacylglycerol levels, and abdominal fat weight; an increase in 3-hydroxy-3- methylglutaryl-CoA reductase activity in the liver; reduction in plasma lecithin-cholesterol acyltransferase activity; lower level of serum insulin; and liver glycogen in the coconut oil-fed rats. Glucose use was altered because lower glucose-6-phosphatase and increased glucokinase activities in the liver of coconut oil-fed rats were found. Coconut palm wine was administered to rats at a dose of 24.5 mL/kg body weight/d for 15 days before conception and throughout gestation. On the 19th day of gestation, hypoglycemia was observed in both wine- and ethanol-treated groups. Synthesis of glycogen was elevated on exposure to ethanol/wine, but its degradation was enhanced only in ethanol-exposed rats. Key enzymes of the citric acid cycle and gluconeogenesis were inhibited on administration in both groups. The activities of glycolytic enzymes were increased. Hydrogenated oil (6%), administered to BHE rats at a dose of 5% of diet, produced no influence on glucose. Hepatocytes isolated from rats fed coconut oil had a significantly lower affinity for insulin than menhaden oil-treated rats.

Glycemic index

The glycemic index of different commonly consumed products supplemented with increasing levels of coconut flour was determined in 10 normal and 10 diabetic subjects. The test food with 200–250 g coconut flour/kg had significantly low, gastrointestinal (GI) (< 60); with 150 g coconut flour/kg had GI ranging from 61.3 to 71.4. A very strong negative correlation (r –0.85, $p < 0.005$) was observed between the GI and dietary fiber content of food. Oil, administered to lean Zucker rats at a dose of 5% of diet (87% saturated fatty acids) for 2 weeks, produced no differences in food intake, body weight, tissue level of glucagons-like peptide-1, plasma insulin, and glucagons levels in coconut oil- and olive oil-fed groups.

Hair damage prevention

Coconut oil application has a strong effect on hair as compared to sunflower and mineral oils, Among the three oils investigated, coconut oil was the only one that reduced the protein loss remarkably for both damaged and undamaged hair when used as a prewash and postwash grooming product.

Hemostatic effect

Coconut water, in citrated plasma of eight healthy volunteers, was observed. Replacement of up to 50% of diluted plasma by water did not influence initiation of coagulatión. Replacing 50% of citrated plasma by coconut water reduced maximum amplitude of thrombelastography recording dose by 39%.

Hepatic activity

Hydrogenated oil, administered orally to male weanling rats at a dose of 10% of diet for 9 weeks, produced no significant alteration of hepatic 5-, 6-, and 9- desaturase activity. Addition of oil to the diet, simultaneously with lowering of the carbohydrate level, diminished the stimulatory effect of dietary sucrose vs glucose on 9-desaturase activity.

Hepatic mitochondrial effect

Oil, administered to chicken at a dose of 20% of the diet, produced a clear damage to the hepatic mitochondria accompanied by an accumulation of glycogen and lipid droplets in the hepatocyte cytoplasm. Pharmaceutical coconut oil induced a high percentage of cellular death when administered for 14 days. Fatty acid profiles in liver and hepatic mitochondria changed during 24 hours after pharmaceutical and cooking oils supplementation to the diet. The accumulation of shorter chain fatty acids (12:0) and (14:0) was higher after pharmaceutical than after cooking oil diet feeding. Mitochondrial ratios of saturated/unsaturated and saturated fatty acids (SUFAs)/PUFAs rapidly changed in parallel to these ratios in both diets. Most of the mitochondrial parameters measured recuperate to the control values when diets were supplied for 5–14 days. The maintenance of these ratios after 14 days pharmacological oil diet feeding was significantly higher than those in control.

Hyperalphalipoproteinemic activity

Oil, administered orally to rabbits fed commercial chow or chow plus 14% w/w coconut oil, resulted in double increase the plasma levels of HDL cholesterol, phospholipids, and protein for up to 4 months without affecting HDL lipid and apoprotein composition. After 3 months also increased VLDL (107%) and LDL cholesterol (40%) levels, but the absolute increases in each of these lipoprotein fractions was less than half of that of HDL. Isotope kinetic studies of 125I-HDL protein indicated a double rate of production of HDL and no change in the efficiency of removal of HDL from plasma.

Hypercholesterolemic activity

Seed oil, administered by gastric intubation to dogs, was active. Seed oil, administered orally to adults, was active. Oil, administered orally to rats at a dose of 79 g/day for 3 weeks, produced a significant increase of the final plasma cholesterol in groups that consumed yeast with coconut oil (91 mg/dL), in comparison to group consuming soybean protein with coconut oil (36 mg/dL). Coconut fat, administered to hyporesponsive and hyperresponsive inbred strains of rabbits with high or low response of plasma cholesterol to dietary SUFAs vs PUFAs, produced no influence on the efficiency of cholesterol absorption. Fat, administered orally to young normolipidemic males at a dose of 30% of the diet with a polyunsaturated/saturated fat ratio of 4 or 0.25 for 8 weeks, produced a significant increase of total plasma cholesterol level. Oil, administered to Mongolian gerbils, produced a hypercholesterolemia associated with elevations in VLDL, LDL, and HDL.

The type of dietary fat did not influence lipoprotein composition and size. Fresh and thermally oxidized oil, administered to rats at a dose of 20% of diet, produced an increase of total cholesterol, LDL and VLDL cholesterol, and triacylglycerol and phospholipids levels and a decrease in HDL cholesterol. Oil, administered to 25 women at doses of 38.4% of energy from fat (HSAFA; PUFA/SUFA P/S ratio = 0.14) or 19.7% energy from fat (LSAFA; P/S ratio 0.17) for 3-week periods, produced no difference in serum total cholesterol, LDL cholesterol, and apo B concentrations between the diet periods. HDL cholesterol and apo A-I were 15% and 11%, respectively, higher during HSAFA diet period than during LSAFA diet period. The LDL/HDL-C and apo B/apo A-I ratios were higher during LSAFA diet period.

Hyperlipidemic activity

Palm wine was administered orally to female albino rats at a dose of 24.5 mL/kg body weight/day for 15 days before conception and during pregnancy. On days 13 and 19 of gestation, liver function and hyperlipidemia were seen in the fetuses. Hyperlipidemia was caused by increased biosynthesis since the incorporation of ^{14}C acetate into lipids and activity of HMG-CoA reductase and lipogenic enzymes were elevated. Oil, administered to miniature pigs fed a pig chow supplemented with 17.1% of coconut oil for 30 d, produced a significant increase of cholesterol, triglyceride, HDL cholesterol and subfractions, LDL cholesterol and subfractions, and lipoprotein lipase activity in both genders. For cholesterol, triglyceride, HDL cholesterol (HDL-C and HDL[2]-C), LDL cholesterol (LDL-C, LDL[1 and 2]-C), and hepatic lipase, the female response to the diet was exaggerated compared to the male response.

Hyperthermic effect

Coconut oil, administered orally to rats with human recombinant TNF-α at a dose of 190 g/kg for 12 weeks, produced an antihypothermic effect and changes in serum albumin and Cu content 8 hours after treatment and in muscle and liver protein after 24 hours. The results indicated that changes in ecosanoid metabolism may be involved in the modulatory effect of the coconut oil-enriched diet.

Hypertriglyceridemic activity

Coconut oil, administered orally to rats at a dose of 14%:0.5% oil/cholesterol diet, produced an increase of lipids and apo B in the VLDL and intermediate-density lipoprotein fractions. The particle diameters of lipoproteins were similar in coconut oil- and olive oil- fed groups. The rates of triglyceride hydrolysis of both groups' VLDL by postheparin lipoprotein lipase in vitro were the same. The average fractional removal of apo B did not differ between diet groups. Oil, administered to rabbits at a dose of 14%: 0.5% coconut oil/cholesterol of diet, produced an increase of plasma triglycerides 15 times higher than basal level. Postprandial triglyceride responses after the first high-fat/ cholesterol meal were more prolonged in coconut-fed rabbits than in olive-fed group. Postprandial triglyceride responses after chronic coconut oil/cholesterol feeding were significantly greater compared to the olive oil-fed group. One coconut oil/cholesterol meal was associated with a 40% increase of postheparin plasma lipoprotein lipase activity and changed little in chronically fed coconut oil/cholesterol group.

Hypocholesterolemic activity

Kernel protein, administered orally to rats on a coconut oil diet, lowered the levels of cholesterol, phospholipids, and triglycerides in the serum and most tissues when compared to casein-fed animals. The increase of hepatic degradation of cholesterol to bile acids and hepatic cholesterol biosynthesis and the decrease of esterification of free cholesterol were noted. In the intestine, cholesterogenesis was decreased. The kernel proteins also decreased lipogenesis in the liver and intestine.

Hypoglycemic activity

Ethanol extract of the leaf, administered orally to rats at a dose of 250 mg/kg, produced less than a 30% drop in blood sugar level. Fruit juice, administered intravenously by infusion to dogs at a dose of 3 mL/min for 100 min, was active. Hot water extract of the dried shell, administered by gastric intubation to dogs at a dose of 200 mL/animal (20 g of air-dried plant material), produced weak activity. The neutral detergent fiber from kernel, administered orally to rats at doses of 5%, 15%, and 30% of diet, resulted in significant decrease in the level of blood glucose and serum insulin, with increasing in the intake of fiber. The increase of fecal excretion of Cu, Cr, Mn, Mg, Zn, and Ca was present. Neutral detergent fiber from coconut kernel, administered to rats at doses of 5, 15, and 30% of diet, produced an increase of fecal excretion of Cu, Cr, Mn, Mg, Zn, and Ca.

Hypolipidemic activity

Protein, administered orally to hypercholesterolemic rats, reduced total, LDL, and VLDL cholesterol; triglycerides; and phospholipids levels in the serum and increased the level of serum HDL cholesterol. The concentration of total cholesterol, triglycerides, and phospholipids in the tissues was lower than in the control group. There was increased activity of superoxide desmutase and catalase. An increase of hepatic cholesterogenesis, conversion of cholesterol to bile acids and fecal excretion of bile acids, and excretion of urinary nitrate and an decrease of malonaldehyde level in the heart were observed. The neutral detergent fiber of kernel digested with cellulase and hemicellulase was administered orally to rats. Hemicellulose-rich fiber showed decreased concentration of total cholesterol and LDL and VLDL cholesterol and increased HDL cholesterol. Cellulose-rich fiber showed no significant alteration. There was increased HMG-CoA reductase activity and increased incorporation of labeled acetate into free cholesterol. Rats fed hemicellulose-rich fiber produced lower concentration of triglycerides and phospholipids and a lower release of lipoproteins into circulation. There were an increased concentration of hepatic bile acids and increased excretion of fecal sterols and bile acids.

Ileal oleic acid uptake

Hydrogenated oil, administered to rats at a dose of 5 g/100 g of diet, produced saturable kinetics in ileal brush border membrane vesicles: $V_{max} = 0.23 \pm 03$ μmol/mg protein/5 minutes and $K_m = 196 \pm 50.3$ nmol for controls, and $V_{max} = 04 \pm 01$ μmol/mg protein/5 minutes and $K_m = 206 \pm 85.3$ nmol for coconut oil-fed group.

Immune function

Coconut oil, administered to rats fed a fat-rich diet (corn oil) or a diet poor in linoleate (coconut oil) at high and low concentrations, completely abolished the responses to *Escherichia coli* endotoxin.

Intestinal brush border membrane

Oil, administered orally to rats at a dose of 10% for 5 weeks, produced an increase in level of saturated fatty acids in the brush border membrane from coconut oil-fed animals. Membrane fluidity was as follows: coconut oil less than commercial pellet diet less than corn oil less than fish oil. The membrane hexose content was high in the coconut-fed rats. Hexamines were elevated in coconut- treated rat brush borders. The activities of alkaline phosphatase, sucrase, and lactase were increased.

Intestinal esterase activity

Oil was administered to rats at different doses with or without clofibrate for 15 days. The hypolipidemic action of clofibrate was not influenced by the amount of fat. Clofibrate did not affect lower cholesterol concentration in rats fed the low-fat diet, but it counteracted the rise in liver cholesterol seen in rats fed the high-fat diet. The high-fat diet produced slightly higher levels of butyryl cholinesterase in the small intestine but markedly raised intestinal esterase-1 activity.

Intestinal neoplasia

Oil was administered to 8-week-old male Fischer rats divided into two groups of 60 each (sedentary and exposed to moderate exercise), at a dose of 21% of diet for 38 weeks. The exercising and sedentary rats fed coconut oil were significantly heavier than rats fed corn oil. In the rats fed coconut oil diet, nine carcinomas were recorded in the sedentary groups and five in the exercised rats, which developed significantly fewer neoplasms than corn oil-fed group.

Intestinal transport

Coconut water in different stages of maturation, administered to rats, produced a jejunal water absorption (17 $\pm$ 0.45 μL/minutes/cm), sodium excretion (–1694 $\pm$ 296 μEq/minutes/cm), and glucose absorption (5212.70 $\pm$ 2098.47 μg%/minutes/cm) in all the stages studied.

Intravenous hydration

The use of coconut water as a short-term intravenous hydration fluid for Solomon Island residents was investigated. Fresh young coconut water, administered to eight healthy male volunteers in three doses in separate trials representing 50, 40, and 30% of the 120% fluid loss at 30 and 60 minutes of the 2-hour rehydration period. The percent of body weight loss than was regained (used as index of percent rehydration) was 75 $\pm$ 5%. The rehydration index, which provided an indication of how much of what was actually ingested and used for body weight restoration, was 1.56 $\pm$ 0.14. There was no difference at any time in serum Na^+ and Cl^-, serum osmolality, and net fluid balance among the trials. Coconut water was significantly sweeter, caused less nausea and more fullness and no stomach upset, and was easier to consume in a larger amount compared to carbohydrate-electrolyte beverage and plain water. Water, administered to children with diarrhea, was inactive. The results indicated that coconut water composition, sodium and glucose concentrations, and osmolality values vary during maturation of the fruit. In no instance did the coconut water contain sodium and glucose concentrations of value as an oral rehydration solution.

Iron bioavailability

Oil, administered to suckling rats dosed with ^{59}Fe-labeled diet, produced a higher percentage of ^{59}Fe in the blood than those fed other fat sources. Administration to weanling rats produced a significantly higher percentage of ^{59}Fe retention than rats fed a formula-blend fat diet.

Jejunal oleic acid uptake

Hydrogenated oil, administered to rats at a dose of 5 g/100 g of diet, produced saturable kinetics in jejunal brush border membrane vesicles: V_{max} = 0.15 $\pm$ 01 μmol/mg protein/5 minutes, and K_m = 136 $\pm$ 29.1 nmol for controls, and V_{max} = 03 $\pm$ 01 μmol/mg protein/5 minutes and K_m = 124.5 $\pm$ 72.6 nmol for the coconut oil-fed group.

Lauric acid incorporation

Lauric acid (50%) from the oil, administered to rats for 6 weeks, produced no significant difference between the experimental distribution of triacylglycerol types and the random distribution, calculated from the total fatty acid composition.

Lipid metabolism

Oil, administered orally to female C57BL/6 mice weaned at 21 d of age at a dose of 15% w/w for 6 weeks, increased the total lipids, triglycerides, LDL and VLDL cholesterol, and thiobarbituric acid-reactive substances (TBARS) and reduced glutathione concentrations, without changes in phospholipids or total cholesterol concentrations compared to controls. The concentrations of total cholesterol, free and esterified cholesterol, triglycerides, and TBARS were increased in the macrophages of coconut-fed mice, whereas the content of total phospholipids did not change. The phospholipids composition showed an increase of phosphatidylcholine and a decrease of phosphatidyl-ethanolamine. Incorporation of [^{3}H]-cholesterol into the macrophages and into the cholesterol ester fraction was increased. The coconut oil diet did not affect [^{3}H]-AA uptake, induced an increase in [^{3}H]-AA release, and enhanced AA mobilization induced by lipopolysaccharide. Oil, administered to 28 persons with moderately elevated cholesterol level, decreased total cholesterol and LDL cholesterol (6.4 ± 0.8 and 4.2 ± 0.7 mmol/L), respectively, compared to butter diet (6.8 ± 0.9 and 4.5 ± 0.8 mmol/L). Apos A-1 and B were significantly higher on coconut oil and on butter than on safflower oil. In the group as a whole, HDL did not differ significantly in the three diets, whereas levels in women fed coconut oil were significantly higher than in the safflower oil group. Triacylglycerol level was lower in coconut oil group, but results were significant statistically only in women.

Lipid peroxide formation stimulation

Seed oil, administered to rats at a dose of 15% of diet for 6 weeks, was inactive on rat liver microsomes. Malondialdehyde concentration was unchanged among animals given different oils. Vitamin E level decreased among those fed soybean oil.

Lipogenetic effect

Kernel protein, administered orally to rats on a coconut oil diet, decreased lipogenesis in the liver and intestine. The kernel proteins also lowered the levels of cholesterol, phospholipids, and triglycerides in the serum and most tissues when compared to casein-fed animals. There was an increase in hepatic degradation of cholesterol to bile acids and hepatic cholesterol biosynthesis and a decrease in esterification of free cholesterol. In the intestine, cholesterogenesis was decreased.

Lipoprotein composition

Oil, administered to newborn chicken at a dose of 20% for 2 weeks, increased cholesterol concentration in all the lipoprotein fractions, whereas 10% coconut oil only increased cholesterol in LDL and HDL, an increase that was significant after 1 week of treatment. Similar results were obtained for triacylglycerol concentration after 2 weeks of treatment. Changes in phospholipids and total protein levels were less profound. Coconut oil decreased LDL and fluidity. Oil, administered with or without 0.5% cholesterol to 36 young male Syrian hamsters for 6 weeks, produced higher plasma total triglyceride and total cholesterol in coconut oil without cholesterol supplementation-fed group than in the fish oil-fed group. With cholesterol supplementation, there was no significant difference in plasma total triglyceride level among the three dietary groups. The hepatic cholesteryl ester content was higher, and there was lower liver microsomal acyl-CoA/cholesterol acyltransferase activity in the cholesterol-supplemented coconut oil group compared to other groups. There was no significant difference in the excretion of fecal neutral and acidic sterols among the three dietary groups.

Lipoprotein lipase activity

Oil, administered to preruminant calves, produced no effect on palmitate oxidation rate by whole homogenates and induced higher palmitate oxidation by intermyofibrillar mitochondria. Carnitine palmitoyltransferase I activity did not significantly differ between the groups. Heart and longissimus thoracis muscle of calves fed coconut oil had higher lipoprotein lipase activity but produced no differences in fatty acid-binding protein content or activity of oxidative enzymes.

Liver function effect

Palm wine was administered orally to female albino rats at a dose of 24.5 mL/kg body weight/day) for 15 days before conception and during pregnancy. On days 13 and 19 of gestation, liver function and hyperlipidemia were seen in the fetuses. Altered liver function was evidenced by the increased activity of alcohol dehydrogenase, aldehyde dehydrogenase, glutamic oxaloacetic transaminase (GOT) (aspartate amino transferase), and glutamic pyruvic transaminase (GPT) (alanine amino transferase).

Myocardial infarction

Coconut oil, administered orally to rabbits with myocardial infarction induced by isoproterenol, produced a higher level of phospholipids in the heart and aorta. The concentrations of cholesterol and triglycerides were lower in the safflower oil fed group.

Nasal absorption

Sucrose ester of coconut fatty acid in aqueous ethanol solution (sucrose cocoate SL-40) administered intranasally to anesthetized male Sprague–Dawley rats at a dose of 0.5% sucrose cocoate with insulin, produced a rapid and significant increase in plasma insulin level with a concomitant decrease in blood glucose levels. Administration of a dose of 0.5% sucrose cocoate with calcitonin produced a rapid increase in plasma calcitonin levels and a concomitant decrease in plasma calcium levels.

Nephrotoxic activity

Fruit juice, administered by intravenous infusion to dogs at a dose of 3 mL/minute for 100 minutes, produced weak activity. Albuminuria was observed just before the end of infusion administration.

Neutrophil functions

Oil, administered to rats 21 days old at a dose of 15% final fat content of the diet for 6 weeks, produced a reduction in spontaneous and phorbol myristate acetate-stimulated H_2O_2 generation in glycogen-elicited peritoneal neutrophils relative to neutrophils from rats fed the control diet. The activity of superoxide desmutase, glutathione peroxidase, and catalase did not change in animals fed the fat-rich diets. The initial rate of O_2 generation in both resting neutrophils and phorbol myristate acetate-stimulated cells was significantly reduced when animals were fed coconut oil.

Ophthalmic absorption

Sucrose ester of coconut fatty acid in aqueous ethanol solution (sucrose cocoate SL-40), administered ophthalmically to anesthetized Sprague–Dawley male rats at a dose of 0.5% sucrose cocoate with insulin, produced an increase in plasma insulin level and a decrease in blood glucose levels.

Ornithine decarboxylase activity

Fixed oil (4.5%), *Clupeidae brevortia tyrannus* (4%), and *Zea mays* (1.5%); fixed oil (7.5%), *Clupeidae brevortia tyrannus* (1%), and *Zea mays* (1.5%); fixed oil (8.5%) and *Zea mays* (1.5%), administered orally to mice for 1 year, were active vs benzoyl peroxide-induced ornithine decarboxylase activity. Oil, administered to 30 ultraviolet (UV)-irradiated Sencar and SKH-1 mice at doses of 1/14% (A diet), 7.9/7.1% (B diet), and 15/0% (C diet) corn oil/coconut oil for 6 weeks, produced no increase in enzyme activity. The level of ornithine decarboxylase activity in the UV-irradiated mice fed diet A was significantly higher than in mice fed the B or C diet. In the SKH-1 mice, ornithine decarboxylase activity was increased by 3 weeks and was significantly higher in mice fed diet C than in mice fed diet A. There was no significant effect of dietary fat on UV-induced skin tumor incidence.

Oxidative DNA damages

Oil, administered to Fischer F344 rats at a dose of 19.8% coconut oil and 2% corn oil for 12–15 weeks, produced an excretion of 8-oxo-7,8dihydro-2'-deoxyguanosine (8-oxodG) in male group equal to 954 ± 367 pmol/kg/ 24 hours in the coconut oil fed group compared to 403 ± 150 pmol/kg/24 hours in the control. Calculated per whole animal, the excretion was 328 ± 128 pmol/24 hours in the coconut oil-fed rats and 137 ± 51 pmol/24 hours in the control.

Phospholipidemic effect

Oil, administered to phospholipids transfer protein knockout (PLTP0)-deficient mice, produced an increase of phospholipids and free cholesterol in the VLDL–LDL region of PLTP0 mice. Accumulation of phospholipids and free cholesterol was dramatically increased in PLTP0/HL0 mice compared to PLTP0 mice. Turnover studies indicated that coconut oil was associated with delayed catabolism of phospholipids and phospholipids/free cholesterol-rich particles. Incubation of these particles with hepatocytes of coconut-fed mice produced a reduced removal of phospholipids and free cholesterol by SRBI, even though SRBI protein expression levels were unchanged.

Plasma fatty acids

Oil, administered to chicken at doses of 10% and 20% of the diet, produced an increase in the percentages of lauric and myristic acids in free fatty acid and triacylglycerol fractions in chick plasma, whereas these changes were less pronounced in phospholipids and cholesterol esters. The percentage of arachidonic acid was higher in plasma phospholipids than in the other fractions and was drastically decreased by coconut oil feeding. Linoleic acid, the main fatty acid of cholesterol esters, was increased. Oil, administered to 37 children 1 year of age in the form of full vegetable-fat milk (3.5 g fat/dL, 100% vegetable fat from palm, coconut, and soybean oils), produced higher amounts of plasma linoleic acid and a plasma α-tocopherol concentrations than in the other milks tested. Oil, administered to male Wistar rats at a dose of 40% of diet for 2 months, increased apo A-I concentration in plasma and did not change apo A-I mRNA level.

Oil, administered to 38 healthy children in a form of full vegetable-fat milk (3.5 g fat/dL, 100% vegetable fat from palm, coconut, and soybean oils), produced a significantly lower percentage of SUFAs in plasma triglycerides than in children fed standard-fat milk. Plasma PUFA levels were significantly higher than in children fed standard-fat milk. Oil, administered to 41 healthy adults, produced a decrease of plasma lathosterol concentration, the ratio plasma lathosterol/cholesterol, LDL cholesterol, and apo B. Plasma total cholesterol, HDL cholesterol, and apo A levels were not significantly different between butter and coconut diets. Oil, administered to male golden Syrian hamsters at a dose of 15% w/w for 4 weeks, produced the highest triglyceride levels of the diets studied. Oil, administered to male golden Syrian hamsters at a dose of 4 g/kg of diet (12:0 and 14:0) for 7 weeks, produced the highest plasma cholesterol concentration compared to rapeseed and sunflower seed oil diets. Biliary lipids, lithogenic index, and bile acid profile of the gallbladder bile did not differ significantly among the six diets.

Platelets aggregation stimulation

Fruit juice, administered intravenously by infusion to dogs at a dose of 5 mL/min, was active. Total infusion was 300 mL. Oil, administered orally to six New Zealand white rabbits fed a commercial diet supplemented with 60 g/kg of coconut oil low in all PUFA for 60 days, produced a platelets aggregation induced by both thrombin and collagen significantly lower with either fish or linseed oil (*n*-3 PUFA), than with corn oil (*n*-6 PUFA) or the low PUFA coconut oil.

Prostaglandin outflow

Oil was administered to weanling male rats fed *ad libitum* a semisynthetic diet supplemented with 10% by weight of primrose oil, replaced partly or completely (25, 50, 75, or 100%) by hydrogenated coconut oil for 8 weeks. The release of prostanoids from the mesenteric vasculature was significantly reduced in the animals on the diet with the oil replaced by coconut oil.

Protective effect of vitamin A

Vitamin A, 240,000 IU, predissolved in 11.7 g of coconut oil and bolused directly into the rumen of mature wethers along with 4 g of chromic oxide or predissolved in 11.7, 23.4, or 35 g of coconut oil, produced significantly higher recoveries of vitamin A when dissolved in coconut oil (55.6%) compared to safflower oil (35 5%). Recoveries in abomasal digesta increased linearly with the amount of carrier coconut oil.

Semen cryopreservation

Coconut water extender, administered with glycerol to the semen of six adult dogs at concentrations of 4, 6, and 8%, produced satisfactory effect. There was no difference among groups in motility and vigor. A smaller percentage of total and secondary abnormalities were observed using 6% glycerol.

Sensitization (skin)

Fruit juice, administered subcutaneously to guinea pigs at a dose of 02 mg/animal, was active on skin. Edema occurred at the site of injection and recovered within 200 minutes. Fruit juice, administered subcutaneously to adults at a dose of 02 mg/person, was active on skin. Inflammation occurred at the site of injection and recovered within 90 minutes. Aqueous extract of the husk fiber, administered externally to rabbits, produced no significant dermic or ocular irritation.

Sickness behavior

Hydrogenated oil, administered to Swiss Webster mice at a dose of 17% w/w for 6 weeks, produced the bioactivity of plasma TNF-α equal to 32.6 $\pm$ 3.6 ng/mL in mice fed coconut oil diet compared to mice fed fish oil (98.2 $\pm$ 5.1 ng/mL).

Spasmogenic activity

Ethanol (95%) extract of the fresh leaf and stem, administered to guinea pigs at a dose of 0.5 mL/L, was active on ileum. Water extract of the fresh leaf and stem, administered intraperitoneally to guinea pigs at a dose of 0.5 mL/L, was active on ileum.

Subcellular membrane-bound enzymes activity

Oil, administered to male CFY weanling rats at a dose of 20% for 16 weeks, produced an increase of synaptosomal acetylcholinesterase activity in the coconut oil-fed group. The Mg^{2+}-adenosine triphosphate (ATPase) activity was similar among all groups in all the brain regions.

Toxicity assessment

Ethanol extract of the leaf, administered intraperitoneally to mice, was active, LD_{50} 0.75 g/kg. Ethanol extract of the fresh leaf and stem, administered intraperitoneally to mice at the minimum toxic dose of 1 mL/animal, was active. Water extract of the fresh leaf and stem, administered intraperitoneally to mice at the minimum toxic dose of 1 mL/animal, was active. Aqueous extract of the husk fiber, administered orally to mice, was active, LD_{50} 2.30 g/kg.

Tricarboxylate carrier influence

Oil, administered to rats at a dose of 15% of the diet for 3 weeks, produced a differential mitochondrial fatty acid composition and no appreciable change in phospholipids composition and cholesterol level. Compared with coconut oil-fed rats, the mitochondrial tricarboxylate carrier activity was markedly decreased in liver mitochondria from fish oil-fed rats. No difference in the Arrhenius plot between the two groups was observed.

Tumor prevention

The effect of kernel fiber on metabolic activity of intestinal and fecal β-glucuronidase activity during 1,2-dimethylhydrazine (DMH)-induced colon carcinogenesis was studied. Inclusion of fiber supported lower specific activity and less fecal output of β-glucuronidase than did the fiber-free diet. Kernel, administered to animals treated with DMH, resulted in higher average weight. A decrease of cholesterol and increase of phospholipids and cholesterol/phospholipids ratio in most of tissues was found. HMG-CoA reductase activity was decreased in most of the tissues of the kernel and DMH, kernel and chili, and kernel, chili, and DMH groups. Histopathological studies showed that kernel-fed animals had fewer papillae, less infiltration into the submucosa, and fewer changes in the cytoplasm with decreased mitotic figures. Oil, administered to rats for 4–8 weeks, produced a small but statistically insignificant reduction in TNF production. After 8 weeks, coconut oil suppressed production of the cytokine. Coconut oil produced no modulatory effect on the interleukin production. Oil, administered orally at a dose of 20% of diet to virgin female Balb/c mice treated with 7,12-dimethylbenz[a]anthracene (DMBA), produced no effect on body weight, feed intake, or survival to 44 weeks of age and 36 weeks after the six DMBA doses. Mammary tumor incidence was the same in the coconut oil or menhaden oil but significantly higher in the corn oil group. Oil was administered to female Wistar rats before mating and throughout pregnancy and gestation, and the male offspring were supplemented from weaning until 90 days of age. They were inoculated subcutaneously with Walker 256 tumor cells. Supplementation of the diet with coconut oil did not change cancer cachexia, except for a small decrease in serum triacylglycerol concentration.

Tumor-promoting effect

Fixed oil (4.5%) with 4% *Clupeidae brevortia tyrannus* and 1.5% *Zea mays*; 7.5% fixed oil, 1% *Clupeidae brevortia tyrannus*, and 1.5% *Zea mays*; and 8.5% fixed and 1.5% *Zea mays*, administered to mice in the diet for 52 weeks, were active. Tumors were initiated with dimethylbenzanthracene and promoted with benzoyl peroxide for 52 weeks. Oil, administered orally to female Sencar mice at doses of 5, 10, 15, and 20%, with addition of 5% corn oil for 1 week after initiation with 7,12-dimethylbenzanthracene and 3 weeks before the start of promotion with 12-0-tetradecanoylphorbol- 13-acetate, produced no significant difference in latency or incidence of papillomas or carcinomas between the saturated fat diet groups. Oil, administered to 30 Sencar and SKH-1 mice at doses of 1:14% (A diet); 7.9%:7.1% (B diet) and 15:0% (C diet) corn oil/coconut oil for 3 weeks before UV irradiation, produced tumor incidence that reached a maximum of 60, 60, and 53% for diets A, B, and C, respectively, with an average one to two tumors per Sencar mouse. For the SKH-1 mice, the diet groups reached 100% incidence by 29 weeks, with approx 12 tumors per mouse. No significant effect of dietary fat was found for tumor latency, incidence, or yield in either strain. Oil (17%) with 3% of sunflower seed oil, administered to DMBA-treated female Sprague–Dawley rats in the diet, produced twice as many tumors as those fed 3% sunflower seed oil or 20% of either saturated fat alone. Tumor yields in the rats fed these mixed-fat diets were comparable to rats fed a 20% lard diet, which provided about the same amount of linoleic acid.

Uncoupling protein expression

Oil, administered to female Wistar rats fed *ad libitum* a high-fat diet with coconut oil for 7 weeks, promoted an increase in body fat content, body weight, and uncoupling protein levels. At the completion of experiment I, oil was administered to high-fat diet rats for 3 weeks. Adipose depots were strongly reduced in the rats fed the high fat diet enriched with coconut oil. Specific uncoupling protein was 3.4 times higher than in controls.

Vascular permeability increased

Fixed oil (4.5%), 4% *Clupeidae brevortia tyrannus*, and 1.5% *Zea mays*; 7.5% of fixed oil, 1% *Clupeidae brevortia tyrannus*, and 1.5% *Zea mays*; 8.5% fixed oil and 1.5% *Zea mays*, administered to mice in the diet for 52 weeks, was active vs vascular permeability induced by benzoyl peroxide.

Cannabis sativa

Cannabis sativa is an annual herb of the *Moraceae* family that grows to 5 m tall. It is usually erect; stems variable, with resinous pubescence, angular, sometimes hollow, especially above the first pairs of true leaves; basal leaves opposite, the upper leaves alternate, stipulate, long petiolate, palmate, with 3–11, rarely single, lanceolate, serrate, acuminate leaflets up to 10 cm long, 1.5 cm broad. Flowers are monoecious or dioecious, the male in axillary and terminal panicles, apetalous, with five yellowish petals and five poricidal stamens; the female flowers germinate in the axils and terminally, with one single-ovulate ovary. Fruit is brown, shining achene, variously marked or plain, tightly embraces the seed with its fleshy endosperm and curved embryo; late summer to early fall; year-round in tropics. Drug-producing selections grow better and produce more drugs in the tropics; oil- and fiber-producing plants thrive better in the temperate and subtropical areas. The form of the plant and the yield of fiber from it vary according to climate and particular variety. Varieties cultivated for their fibers have long stalks, branch very little, and yield only small quantities of seed. Oil seed varieties are small, mature early, and produce large quantities of seed. Varieties grown for the drugs are small, much branched with smaller dark-green leaves. Between these three main types of plants are numerous varieties that differ from the main one in height, extent of branching, and other characteristics.

Fig. 17.12. Cannabis sativa. A–Male shoot; B and C–Flowers.

Origin and distribution

Native to Central Asia and long cultivated in Asia, Europe, and China. Now a widespread tropical, temperate, and subarctic cultivar. *Cannabis sativa* has been cultivated for more than 4500 years for different purposes, such as fiber, oil, or narcotics. The oldest use of hemp is for fiber, and later the seeds were used for culinary purposes. Plants yielding the drug were discovered in India, cultivated for medicinal purposes as early as 900 BC. In medieval times, it was brought to North Africa, where currently it is cultivated exclusively for hashish or kif.

Traditional uses

Afghanistan. Hot water extract of the resin is taken orally to induce abortion.

China. Hot water extract of the inflorescence is taken orally for wasting diseases, to clear the blood, to cool the temperature, to relieve fluxes, for rheumatism, to discharge pus, and to stupefy and produce hallucinations. The seed is taken orally as an emmenagogue. Decoction of the seed is taken orally as an anodyne, an emmenagogue, a febrifuge, for migraine, and for cancer. It is taken orally as a hallucinogen and externally for rheumatism.

Guatemala. The leaves are used externally to relieve muscular pains.

India. Hot water extract of the dried entire plant is taken orally as a narcotic and to relieve pain of dysmenorrhea. Hot water extract of the dried flower and leaf is taken orally for dyspepsia and gonorrhea and as a nerve stimulant. Hot water extract of the inflorescence of female plants is taken orally as an abortifacient. Hot water extract of the leaf is taken orally to relieve menstrual pain. For cuts, boils, and blisters, leaf paste is applied topically for 4 days. Hot water extract of the bark is taken orally for hydrocele and other inflammation. Extract of the leaves is used as an insect repellant. Hot water extract of the seed is taken orally as an emmenagogue. The powdered seed is taken orally as an aid in conception. One gram of seeds is powdered, then mixed with water, and given to women in the morning before breakfast for 7 days after menstruation.

The use of pepper and cane sugar is avoided. Paste of dried leaves is applied over the anus in the morning and evening for piles. The dried leaf juice is used externally on cuts and piles and taken orally as an anthelmintic. To eliminate cough, bronchitis, and other respiratory ailments, a half tablespoonful of powdered dried leaves is mixed with an equal amount of honey and taken orally three times daily. Seed oil is used externally for burns. The oil is extracted by roasting the seeds. Seeds are taken orally for diabetes, hysteria, and sleeplessness. The aerial parts are smoked to decrease nausea and vomiting induced by anticancer drugs. Hot water extract of the aerial parts is taken orally by males as an aphrodisiac. The dried aerial parts are smoked by women to increase their amorous prowess. The fresh leaves are taken orally for hemorrhoids. Hot water extract of the dried leaf and seed is taken orally for stomach troubles and indigestion. Fresh leaf juice is administered intraural to treat earache. The fruit is used externally for skin diseases. The unripe fruit is taken orally to induce sleep.

Iran. Fluidextract of the dried flowering top or the dried fruit is taken orally for abdominal pain associated with indigestion, for pain associated with cancer, for rheumatoid arthritis, for gastric cramps or neuralgia, for coughing, and as a hypnotic. Fluidextract of the dried fruit is taken orally for whooping cough, as a hypnotic, and a tranquilizer. The dried seed is taken orally as a diuretic. An infusion is taken orally as an analgesic in rheumatism or rheumatoid arthritis, a sedative, a diaphoretic, and for hysteric conditions, gout, epilepsy, and cholera. The seed oil is administered *per rectum* to reduce cramps associated with lead poisoning associated with constipation and vomiting. To reduce breast engorgement or reduce milk secretion, the seed oil is applied topically. In some cases, it would completely stop milk secretion. One to 2 g of seed oil is taken orally several times a day for urinary incontinency.

Jamaica. Hot water extract of the flower, leaf, and twig is taken orally as an antispasmodic and anodyne. Hot water extract of the resin is taken orally for diabetes.

Mexico. The aerial parts are smoked as a hallucinogen.

Morocco. The aerial parts are taken orally as a narcotic.

Nepal. Decoction of the leaf is taken orally by adults as an anthelmintic. The powdered leaf is mixed with cattle feed as a treatment for diarrhea. For headache, the dried leaves are ground with *Datura stramonium* leaves and *Picrorhiza schrophulariflora* stem and water then applied externally. The leaf juice is used externally as an antiseptic, as a hemostat on cuts and wounds, and to treat swelling of sprained joints. The seeds are crushed, mixed with curd, and taken orally for dysentery. Decoction of the seed is taken orally as an anthelmintic. To aid in parturition, 2 teaspoonfuls of powdered seeds are made into a paste with sesame oil (*Sesamum indicum* L.) and applied intravaginally during labor.

Pakistan. Hot water extract of the entire plant is taken orally as a parturifacient. Infusion of the leaf is taken orally for general weakness.

Saudi Arabia. The aerial parts, mixed with honey, sugar, and nutmeg, are taken orally as a psychotropic.

Senegal. The seed is taken orally as an emmenagogue.

South Africa. Hot water extract of the entire plant is taken orally for asthma[CS107]. Hot water extracts of the root and seed are taken orally to induce abortion, labor, and menstruation.

United States. Fluidextract of the inflorescence is taken orally as a narcotic, antispasmodic, analgesic, and aphrodisiac. Hot water extract of the flowering top is taken orally as a potent antispasmodic, anodyne, and narcotic. One teaspoon of plant material is steeped in 2 cups of boiling water, and 1 tablespoonful is taken two to four times a day. The dried aerial parts are smoked by both sexes as an aphrodisiac.

Vietnam. The seeds are taken orally as an emmenagogue.

West Indies. Hot water extract of the entire plant is taken orally as an antispasmodic.

Yugoslavia. Hot water extract of the seed is taken orally for diabetes.

Zimbabwe. Hot water extract of the aerial parts is taken orally as a treatment for malaria.

Medicinal values

Abortifacient activity

Alcohol extract of the dried leaf, administered intragastrically to pregnant rats at a dose of 125 mg/kg, produced teratogenic effects. Water extract of the dried leaf, administered intragastrically to pregnant rats at variable dosage levels on days 6–15 of pregnancy was active.

Acute cardiovascular fatalities

Six cases of possible acute cardiovascular death in young adults were reported where very recent cannabis ingestion was documented by the presence of Δ-9-tetrahydrocannabinol (Δ-9-THC) in postmortem blood samples. A broad toxicological blood analysis could not reveal other drugs.

Acute panic reaction (Koro)

Koro, an acute panic reaction related to the perception of penile retraction, was once considered limited to specific cultures. Over 70 American men responded by telephone to report negative reactions to cannabis. Three of them (Caucasians aged 22–26 years with considerable experience with cannabis) spontaneously mentioned experiencing symptoms of Koro after smoking cannabis. All three cases occurred after the participants had heard about cannabis-induced Koro and used the drug in a novel setting or atypical way. Two of the men had body dysmorphia, which may have contributed to symptoms. All three decreased their cannabis consumption after the Koro experience. Several factors may have interacted to create the symptoms. These include previous knowledge of cannabis-induced Koro, the use of cannabis in a way that might heighten a panic reaction, and poor body image.

Adverse effects

A causal role of acute cannabis intoxication in motor vehicle and other accidents has been shown by the presence of measurable levels of Δ-9-THC in the blood of drivers in the absence of alcohol or other drugs, by surveys of driving under the influence of cannabis, and by significantly higher accident culpability risk of drivers using cannabis. Evidence demonstrated that cannabis dependence, both behavioral and physical, occurred in about 7–10% of regular users, and that early onset of use—especially of weekly or daily use—is a strong predictor of future dependence. Cognitive impairments of various types are readily demonstrable during acute cannabis intoxication, but there is no suitable evidence yet available to permit a decision as to whether long-lasting or permanent functional losses can result from chronic heavy use in adults.

The gender effects on progression to treatment entry and on the frequency, severity, and related complications of the *Diagnostic and Statistical Manual of Mental Disorders*, 3rd edition revised drug and alcohol dependence among 271 substance-dependent patients (mean age: 32.6 years; 156 women) was studied. There was no gender difference among patients in the age at onset of regular use of any substance. Women experienced fewer years of regular use of opioids and cannabis and fewer years of regular alcohol drinking before entering treatment. Although the severity of drug and alcohol dependence did not differ by gender, women reported more severe psychiatric, medical, and employment complications. In a 3-day, double-blind, randomized, counterbalanced study, the behavioral, cognitive, and endocrine effects of 2.5 and 5 mg intravenous Δ-9-THC were characterized in 22 healthy individuals, who had been exposed to cannabis but had never been diagnosed with a cannabis abuse disorder. Prospective safety data at 1,3, and 6 months post-study was also analyzed. Δ-9-THC produced schizophrenia-like positive and negative symptoms, altered perception, increased anxiety and plasma cortisol, euphoria, disrupted immediate and delayed word recall, sparing recognition recall, impaired performance on tests of distractibility, verbal fluency, and working memory, but did not impair orientation.

This study examined the behavioral and neurochemical (cannabinoid *CB1* receptor gene expression) changes induced by spontaneous cannabinoid withdrawal in mice. Cessation of CP-55,940 treatment in tolerant mice induced a spontaneous time-dependent behavioral withdrawal syndrome consisting of marked increases (140%) in motor activity, number of rearings (170%), decreases in grooming (57%), wet-dog shakes (73%), and rubbing behaviors (74%) on day 1, progressively reaching values similar to vehicle-treated mice on day 3. This spontaneous cannabinoid withdrawal resulted in *CB1* gene expression up-regulation (20–30%) in caudate-putamen, ventromedial hypothalamic nucleus, central amygdaloid nucleus, and CA1, whereas in the CA3 field of hippocampus, a significant decrease (15–20%) was detected.

Alcohol interaction

The complementary DNA and genomic sequences encoding G protein-coupled cannabinoid receptors (CB1 and CB2) from several species were cloned. This has facilitated discoveries of endogenous ligands (endocannabinoids). Two fatty acid derivatives characterized to be arachidonylethanolamide and 2-arachidonylglycerol isolated from both nervous and peripheral tissues mimicked the pharmacological and behavioral effects of Δ-9-THC. The down-regulation of CB1 receptor function and its signal transduction by chronic alcohol was demonstrated. The observed down-regulation of CB1 receptor-binding and its signal transduction resulted from the persistent stimulation of receptors by the endogenous CB1 receptor agonists arachidonylethanolamide and 2-arachidonylglycerol, whose synthesis is increased by chronic alcohol treatment. The deletion of CB1 receptor has been shown to block voluntary alcohol intake in mice.

Allergenic effect

An "All India Coordinated Project on Aeroallergens and Human Health" was undertaken to discover the quantitative and qualitative prevalence of aerosols at 18 different centers in the country. Predominant airborne pollens were *Holoptelea*, *Poaceae*, *Asteraceae*, *Eucalyptus*, *Casuarina*, *Putanjiva*, *Cassia*, *Quercus*, *Cocos*, *Pinus*, *Cedrus*, *Ailanthus*, *Cheno/Amaranth*, *Cyperus*, *Argemone*, *Xanthium*, *Parthenium*, and others. Clinical and immunological evaluations revealed some allergenically important taxa. Allergenically important pollens were *Prosopis juliflora*, *Ricinus communis*, *Morus*, *Mallotus*, *Alnus*, *Querecus*, *Cedrus*, *Argemone*, *Amaranthus*, *Chenopodium*, *Holoptelea*, *Brassica*, *Cocos*, *Cannabis*, *Parthenium*, *Cassia*, and grasses. In the multitest routine skin-test battery, 78 of 127 patients tested (61%) were cannabis-test positive. Thirty of the 78 patients were randomly selected to determine if they had allergic rhinitis and/or asthma symptoms during the cannabis pollination period. By history, 22 (73%) claimed respiratory symptoms in July through September. All 22 of these subjects were also skin test-positive to weeds pollinating during the same period as cannabis (ragweed, pigweed, cocklebur, Russian thistle, marsh elder, or kochia).

Amnesic syndrome

A 26-year-old woman suffered disseminated intravascular coagulation (DIC) and a brief respiratory arrest following recreational use of 3,4-methylenedioxymethamphetamine (MDMA, or "ecstasy") together with amyl nitrate, lysergic acid (LSD), cannabis, and alcohol. She was left with residual cognitive and physical deficits, particularly severe anterograde memory disorder, mental slowness, severe ataxia, and dysarthria. Follow-up investigations have shown that these have persisted, although there has been some improvement in verbal recognition memory and in social functioning. Magnetic resonance imaging and quantified positron emission tomography investigations revealed severe cerebellar atrophy and hypometabolism accounting for the ataxia and dysarthria; thalamic, retrosplenial, and left medial temporal hypometabolism to which the anterograde amnesia can be attributed. There was some degree of frontotemporal–parietal hypometabolism, possibly accounting for the cognitive slowness. The putative relationship of these abnormalities to the direct and indirect effects of MDMA toxicity, hypoxia, and ischemia was considered.

Amyotrophic lateral sclerosis

One hundred thirty one respondents with amyotrophic lateral sclerosis—13 of whom reported using cannabis in the last 12 months—were examined. The results indicated that cannabis might be moderately effective at reducing symptoms of appetite loss, depression, pain, spasticity, and drooling. Cannabis was reported ineffective in reducing difficulties with speech and swallowing, and sexual dysfunction. The longest relief was reported for depression (approx 2–3 hours).

Analgesic activity

Ethanol (50%) extract of the entire plant, administered intra-peritoneally to mice at a dose of 250 mg/kg, was active vs tail pressure method. Flavonoid fraction of the leaf, administered intraperitoneally to mice, was active. The inflorescence, administered orally to male rats, produced weak activity vs paw pressure test, effective dose $(ED)_{50}$ 35.5 mg/kg and hot plate method, ED_{50} 53 mg/kg. Petroleum ether and ethanol (95%) extracts of the dried aerial parts, administered intragastrically to mice, was active vs phenylbenzoquinone-induced writhing, inhibitory concentration $(IC)_{50}$ 0.013 mg/kg and 0.045 mg/kg, respectively.

Analgesic effect

Ajulemic acid (AJA, CT-3, or IP-751), administered to healthy human adults and patients with chronic neuropathic pain, demonstrated a complete absence of psychotropic actions. It proved to be more effective than placebo in reducing this type of pain as measured by the visual analog scale. Signs of dependency were not observed after withdrawal at the end of the 1-week treatment period. Forty women undergoing elective abdominal hysterectomy were investigated in a randomized, double-blind, placebo-controlled, single-dose trial. Randomization took place when postoperative patient-controlled analgesia was discontinued on the second postoperative day. When patients requested further analgesia, they received a single, identical capsule of either 5 mg of oral Δ-9-THC ($n = 20$) or placebo ($n = 20$) in a double-blind fashion.

The primary outcome measure was summed pain intensity difference (SPID) at 6 hours after administration of the study medication derived from visual analog pain scores on movement and at rest. Secondary outcome measures were time-to-rescue medication and adverse effects of study medication. Mean (standard deviation [SD]) visual analog scale pain scores before medication in the placebo and Δ-9-THC groups were 6.3(2.6) and 6.4(1.3) cm on movement, and 3.2(1.9) and 3.3(0.9) at rest, respectively. There were no significant differences in mean (95% confidence interval [CI] of the difference) SPID at 6 hours between the groups (placebo 7.9, Δ-9-THC 4.3[–1.8 to 9] cm per hour on movement; placebo 8.8, Δ-9-THC 4.9[–0.2 to 8.1] cm per hour at rest) and time to rescue analgesia (placebo 217, Δ-9-THC 163[−22 to 130] minutes). Increased awareness of surroundings was reported more frequently in patients receiving Δ-9-THC (40 vs 5%, $p = 0.04$). There were no other significant differences with respect to adverse events.

THC, morphine, and a THC–morphine combination were administered to 12 healthy subjects using experimental pain models (heat, cold, pressure, and single and repeated transcutaneous electrical stimulation). THC (20 mg), morphine (30 mg), THC–morphine (20 mg THC + 30 mg morphine), or placebo were given orally as single dose. Reaction time, side effects (visual analog scales), and vital functions were monitored. For the pharmacokinetic profiling, blood samples were collected. THC did not significantly reduce pain. In the cold and heat tests, it even produced hyperalgesia, which was completely neutralized by THC–morphine. A slight additive analgesic effect was observed for THC–morphine in the electrical stimulation test. No analgesic effect resulted in the pressure and heat test, with neither THC nor THC–morphine.

Psychotropic and somatic side effects (sleepiness, euphoria, anxiety, confusion, nausea, dizziness, etc.) were common, but usually mild. Three cannabis-based extracts Δ-9-THC, cannabidiol [CBD], and a 1:1 mixture of them both) were given over a 12-week period in a randomized, double-blind, placebo-controlled, crossover trial. Extracts, which contained THC, proved most effective in symptom control. Regimens for the use of the sublingual spray emerged and a wide range of dosing requirements was observed. Side effects were common, reflecting a learning curve for both patient and study team. These were generally acceptable and little different to those seen when other psychoactive agents are used for chronic pain. Over a 6-week period 209 chronic noncancer pain patients were studied. Seventy-two (35%) subjects reported ever having used cannabis. Thirty-two (15%) subjects reported having used cannabis for pain relief (pain users), and 20 (10%) subjects were currently using cannabis for pain relief. Thirty-eight subjects denied using cannabis for pain relief (recreational users). Compared with nonusers, pain users were significantly younger ($p = 0.001$) and were more likely to be tobacco users ($p = 0.0001$). The largest group of patients using cannabis had pain caused by trauma and/or surgery (51%), and the site of pain was predominantly neck/upper body and myofascial (68 and 65%, respectively). The median duration of pain was similar in both pain users and recreational users (8 vs 7 years; $p = 0.7$). There was a wide range of amounts and frequency of cannabis use. Of the 32 subjects who used cannabis for pain, 17 (53%) used four puffs or less at each dosing interval, eight (25%) smoked a whole cannabis cigarette (joint), and four (12%) smoked more than one joint. Seven (22%) of these subjects used cannabis more than once daily, five (16%) used it daily, eight (25%) used it weekly, and nine (28%) used it rarely. Pain, sleep, and mood were most frequently reported as improving with cannabis use, and "high" and dry mouths were the most commonly reported side effects. Patients with chronic pain completed a questionnaire about the type of cannabis used, the mode of administration, the amount used and the frequency of use, and their perception of the effectiveness of cannabis on a set of pain-associated symptoms and side effects. Fifteen patients (10 males) were interviewed (median age, 49.5 years; range, 24–68 years). All patients smoked herbal cannabis for therapeutic reasons (median duration of use, 6 years; range, 2 weeks–37 years). Seven patients only smoked at night (median dose eight puffs, range two to eight puffs), and eight patients used cannabis mainly during the day (median dose of three puffs; range, two to eight puffs); the median frequency of use was four times per day (range, 1 to 16 times/day). Twelve patients reported improvement in pain and mood, whereas 11 reported improvement in sleep. Eight patients reported a "high;" six denied a "high." Tolerance to cannabis was not reported. THC was administered to six patients with chronic pain at doses 5–20 mg/day. A sufficient pain relief had been achieved in three patients. The other three suffered from intolerable side effects, such as nausea, dizziness, and sedation without a reduction of pain intensity. In these cases, the treatment was continued with other analgesics.

Ankylosing spondylitis

Ankylosing spondylitis is a systemic disorder occurring in genetically predisposed individuals. The disease course appears to be characterized by bouts of partial remission and flares. There were 214 patients questioned (169 men, 45 women; average disease duration, 25 years; age of disease onset, 22 years). The main symptoms of flare were pain (all groups), immobility (90%), fatigue (80%), and emotional symptoms, such as depression, withdrawal, and anger, (75%). All of patients experienced between one and five localized flares per year. Fifty-five percent of the groups contained patients ($n = 85$) who experienced a generalized flare. The main perceived triggers of flare were stress (80%) and "overdoing it" (50%). Patients reported that a flare might last anywhere from a few days to a few weeks and relief from flare were by analgesic injections (including opiates), relaxation, sleep, and cannabis. Three quarters of the groups agreed that there was no long-term effect on the ankylosing sponylitis following a flare.

Anti-arthritic effect

Oral administration of AJA, a cannabinoid acid devoid of psychoactivity, reduced joint tissue damage in rats with adjuvant arthritis. Peripheral blood monocytes (PBM) and synovial fluid monocytes (SFM) were isolated from healthy subjects and patients with inflammatory arthritis, respectively, treated with AJA (0–30 m*M*) in vitro, and then stimulated with lipopolysaccharide. Cells were harvested for messenger RNA (mRNA), and supernatants were collected for cytokine assay. Addition of AJA to PBM and SFM in vitro reduced both steady-state levels of interleukin-1γ (IL-1γ) mRNA and secretion of IL-1γ in a concentration-dependent manner. Suppression was maximal (50.4%) at 10 m*M* AJA ($p < 0.05$ vs untreated controls, $n = 7$). AJA did not influence tumor necrosis factor-α (TNF-α) gene expression in or secretion from PBM.

Anticonvulsant activity

Ethanol (95%) extract of the entire plant, administered subcutaneously to male mice and rats at a dose of 2–4 mL/kg, was active vs metrazole and electroshock, respectively. A dose of 4 mL/kg was inactive vs strychnine convulsions in mice. The entire plant, smoked by 29 patients with epilepsy under the age of 30 years, was active. It must be noted that in some species, cannabinoids can precipitate epileptic seizures. Tincture of the resin, administered intraperitoneally to mice at a dose of 25 mg/kg, produced 80% protection vs pentyle-netetrazole convulsions.

Anti-emetic activity

In a qualitative study of self-care in pregnancy, birth, and lactation within a nonrandom sample of 27 women in British Columbia, Canada, 20 women (74%) experienced pregnancy- induced nausea. Ten of these women used antiemetic herbal remedies, which included ginger, peppermint, and cannabis. Only ginger has been subjected to clinical trials among pregnant women, although the three herbs were clinically effective against nausea and vomiting in other contexts, such as chemotherapy-induced nausea and post-operative nausea. CBD, a major non- psychoactive cannabinoid administered by oral infusion to rats with nausea elicited by lithium chloride, and with conditioned nausea elicited by a flavor paired with lithium chloride, was active[CS382]. Oral nabilone, oral dronabinol (THC), and intramuscular levonantradol were administered to 1366 patients. Cannabinoids were more effective antiemetics than prochlorperazine, metoclopramide, chlorpromazine, thiethylperazine, haloperidol, domperidone, or alizapride. Relative risk was 1.38 (95% CI 1.18–1.62), number-needed-to-treat (NNT) was 6 for complete control of nausea; relative risk was 1.28 (CI 1.08–1.51), NNT 8 for complete control of vomiting. Cannabinoids were not more effective in patients receiving very low or very high emetogenic chemotherapy. In crossover trials, patients preferred cannabinoids for future chemotherapy cycles: relative risk 2.39 (2.05–2.78), NNT 3. Some potentially beneficial side effects occurred more often with cannabinoids: "high" 10.6 (6.86–16.5), NNT 3; sedation or drowsiness 1.66 (1.46–1.89),

NNT 5; euphoria 12.5 (3–52.1), NNT 7. Harmful side effects also occurred more often with cannabinoids: dizziness 2.97 (2.31– 3.83), NNT 3; dysphoria or depression 8.06 (3.38–19.2), NNT 8; hallucinations 6.10 (2.41–15.4), NNT 17; paranoia 8.58 (6.38–11.5), NNT 20; and arterial hypotension 2.23 (1.75–2.83), NNT 7. Patients given cannabinoids were more likely to withdraw because of side effects (relative risk 4.67 [3.07–7.09]; NNT 11).

Antifungal activity

Ethanol (50%) extract of the dried leaf was active on *Rhizoctonia solani*, mycelial inhibition was 65.99%. Water extract of the fresh leaf on agar plate at a concentration of 1:1 was active on *Fusarium oxysporum*. The water extract also produced strong activity on *Ustilago maydis* and *Ustilago nuda*. Water extract of the fresh shoot on agar plate was inactive on *Helminthosporium turcicum*.

Antiglaucomic activity

Water extract of the dried entire plant, administered intravenously to Rhesus monkeys and rabbits at a dose of 0.01 μg/animal, was active. The intraocular pressure rose for 24 hours postinjection, then fell for 3 days. A dose of 25 μg/animal, administered intravenously to rabbits, was also active. The effect was not influenced by atropine, scopolamine, methysergide, haloperidol, chlorpromazine, spironolactone, yohimbine or dexamethasone. Partial inhibition was seen when galactose, glucose or mannose were administered intravenously, concurrently. Water extract of the dried leaf and stem, applied opthalmically to rabbits was active.

Antigonadotropin effect

Ethanol (80%) extract of the dried aerial parts, administered intra-gastrically to male langurs at a dose of 14 mg/kg daily for 90 days produced equivocal effect.

Anti-inflammatory activity

Petroleum ether and ethanol (95%) extracts of the dried aerial parts, applied externally on mice at a dose of 100 μg/ear, was active vs tissue plasminogen activator-induced erythema of the ear. CBD was administered orally to rats at doses of 5–40 mg/kg daily for 3 days after the onset of acute inflammation induced by intraplantar injection of 0.1 mL carrageenan (1% w/v in saline). CBD had a time- and dose-dependent antihyperalgesic effect after a single injection. Edema following carrageenan peaked at 3 hours and lasted 72 hours. A single dose of CBD reduced edema in a dose-dependent fashion and subsequent daily doses produced further time- and dose-related reductions. There were decreases in prostaglandin E2 (PGE2) plasma levels, tissue cyclo-oxygenase activity, production of oxygen-derived free radicals, and nitric oxide ([NO], nitrite/nitrate content) after three doses of CBD. The effect on NO seemed to depend on a lower expression of the endothelial isoform of NO synthase.

Antispermatogenic effect

Sixteen healthy chronic marijuana smokers were associated with a decline in sperm concentration and total sperm count during the fifth and sixth weeks after 4 weeks of high-dose smoking (8–20 cigarettes/day). The dried aerial part, taken by inhalation daily, decreases the quantity as well as quality of spermatozoa. Ethanol (80%) extract of the dried aerial parts, administered intragastrically to langurs at a dose of 14 mg/kg daily for 90 days, was equivocal. Ethanol (95%) extract of the dried aerial parts, administered intraperitoneally to mice at a dose of 2 mg/animal daily for 45 days, produced a complete arrest of spermatogenesis. The effect was reversible.

Antistress activity

The leaf smoke, in combination with hashish smoke, administered to rats housed in a wire cage inside a larger cage with a cat, was equivocal. The rats' brains were dissected and measured for protein and catecholamine levels.

Anti-tumor activity

Arachidonyl ethanolamide, in three cervical carcinoma (CxCa) cell lines at increasing doses with or without antagonists to receptors to arachidonyl ethanolamide, induced apoptosis of CxCa cell lines via aberrantly expressed vanilloid receptor-1. Arachidonyl ethanolamide-binding to the classical CB1 and CB2 cannabinoid receptors mediated a protective effect. A strong expression of the three forms of arachidonyl ethanolamide receptors was observed in ex vivo CxCa biopsies. Three cannabis constituents, CBD, Δ-8-THC, and cannabinol displayed anti-proliferative activity in several human cancer cell lines in vitro. They were oxidized to their respective paraquinones 2,4, and 6. Quinone 2 significantly reduced cancer growth of HT-29 cancer in nude mice. Δ-9-THC binds and activates membrane receptors of the 7-transmembrane domain, G protein-coupled superfamily. Several putative endocannabinoids have been identified, including anandamide (AEA), 2-arachidonyl glycerol, and noladin ether. Synthesis of numerous cannabinomimetics has expanded the repertoire of cannabinoid receptor ligands with the pharmacodynamic properties of agonists, antagonists, and inverse agonists. These ligands have proven to be powerful tools both for the molecular characterization of cannabinoid receptors and the delineation of their intrinsic signaling pathways. Much of the understanding of the signaling mechanisms activated by cannabinoids has been derived from studies of receptors expressed by tumor cells. Cannabinoids and their derivatives exerted palliative effects in cancer patients by preventing nausea, vomiting, and pain and by stimulating appetite. These compounds have been shown to inhibit the growth of tumor cells in culture and animal models by modulating key cell-signaling pathways. Cannabinoids are usually well tolerated, and do not produce the generalized toxic effects of conventional chemotherapies.

Anxiolytic activity

AEA, a primary endogenous ligand of the brain cannabinoid receptors, is released in selected regions of the brain and is deactivated through a two-step process consisting of transport into cells followed by intracellular hydrolysis. Pharmacological blockade of the enzyme fatty acid amide hydrolase (FAAH), which is responsible for intracellular AEA degradation, produced anxiolytic-like effects in rats without causing the wide spectrum of behavioral responses typical of direct-acting cannabinoid agonists. These findings suggest that AEA contributes to the regulation of emotion and anxiety, and that FAAH might be the target for a novel class of anxiolytic drugs.

Attention deficit hyperactivity disorder

Attention defict hyperactivity disorder has been considered a mental and behavioral disorder of childhood and adolescence. It is being increasingly recognized in adults, who may have psychiatric comorbidity with secondary depression, or a tendency to drug and alcohol abuse. A 32-year-old woman known for years as suffering from borderline personality disorder and drug dependence (including cannabis, LSD, and ecstasy) and alcohol abuse that did not respond to treatment was reported. Only when correctly diagnosed as attention defict hyperactivity disorder and appropriately treated with the psychotropic stimulant methylphenidate (Ritalin®), was there significant improvement. She succeeded academically, which had not been possible previously, her craving for drugs diminished, and a drug-free state was reached.

Auditory function

Eight male subjects (aged 22–30 years) who had previously used cannabis were investigated. They performed air conduction pure tone audiometry in both ears over 0.5–8 kHz. A simple test of frequency selectivity by detecting a 4-kHz tone under two masking noise conditions was also carried out in one ear. Three test sessions at weekly intervals were carried out, at the start of which they ingested a capsule containing either placebo, 7.5, or 15 mg of THC. These were administered in a randomized cross-over, double-blind manner. Auditory testing was carried out 2 hours after ingestion. Blood samples were also obtained at this time point and assayed for Δ-9-THC and 11-hydroxy-THC levels. No significant changes in threshold or frequency resolution were seen with the dosages employed in this study.

Behavioral effect

A four-page, self-completed questionnaire was designed to determine the drugs used (licit, illicit, and doping substances) along with beliefs about doping and the psychosociological factors associated with their consumption. The questionnaire was distributed to high school students enrolled in a school sports association in eastern France. The completed forms were received from 1459 athletes: 4% stated that they had used doping agents at least once in their life (their main source of supply being peers and health professionals). Thirty-four percent of the sample smoked some tobacco, 66% used alcohol, 19% used cannabis, 4% took ecstasy, 10% took tranquillizers, 9% used hypnotics, 4% used creatine, and 41% used vitamins against fatigue. Beliefs about doping did not differ among doping agent users and nonusers, except for the associated health risks, which were minimized by users. Users of doping agents stated that the quality of the relations that they maintained with their parents was sharply degraded, and they reported that they were susceptible to influence and difficult to live with. More often than nondoping-agent users, these adolescents were neither happy, nor healthy, although paradoxically, they seemed less anxious and were more self-confident. Maternal exposure to Δ-9- THC in rats resulted in alteration in the pattern of ontogeny of spontaneous locomotor and exploratory behavior in the offspring. Adult animals exposed during gestational and lactational periods exhibited persistent alterations in the behavioral response to novelty, social interactions, sexual orientation, and sexual behavior. They also showed a lack of habituation and reactivity to different illumination conditions. Adult offspring of both sexes also displayed a characteristic increase in spontaneous and water-induced grooming behavior. Some of the effects were dependent on the sex of the animals being studied, and the dose of cannabinoid administered to the mother during gestational and lactational periods. Maternal exposure to low doses of THC sensitized the adult offspring of both sexes to the reinforcing effects of morphine, as measured in a conditioned place preference paradigm.

β-Endorphin interaction

Δ-9-THC administered to rats produced large increases in extracellular levels of β-endorphin in the ventral tegmental area and lesser increases in the shell of the nucleus accumbens (Nac). In rats that had learned to discriminate injections of THC from injections of vehicle, the opioid agonist morphine did not produce THC-like discriminative effects, but markedly increased discrimination of THC. The opioid antagonist naloxone reduced the discriminative effects of THC. Bilateral microinjections of β-endorphin directly into the ventral tegmental area, but not into the shell of the Nac, markedly increased the discriminative effects of ineffective threshold doses of THC, but had no effect when given alone. The increase was blocked by naloxone.

Binocular depth inversion reduction

A study to assess whether the binocular depth inversion illusion (BDII) could detect subtle cognitive impairment owing to regular cannabis use was conducted. Ten regular cannabis users and 10 healthy controls from the same community sources, matched for age, sex, and premorbid intelligence quotient (IQ) were evaluated. The subjects were also compared on measures of executive functioning, memory, and personality. Regular cannabis users were found to have significantly higher BDII scores for inverted images. This was not to the result of a problem in the primary processing of visual information, as there was no significant difference between the groups for depth perception of normal images. There was no relationship between BDII scores for inverted images and time since the last dose, suggesting that the measured impairment of BDII more closely reflected chronic than acute effects of regular cannabis use. There were no significant differences between the groups for other neuropsychological measures of memory or executive function. A positive relationship was found between psychoticism as defined by the revised Eysenck Personality Questionnaire and cannabis, tobacco, and alcohol use. Cannabis users also used significantly larger amounts of alcohol. No relationship was found between BDII scores and drug use other than cannabis or psychoticism. Nabilone, a psychoactive synthetic 9-*trans*-ketocannabinoid, CBD, and a combined oral application of both substances on binocular depth inversion and behavioral states were investigated in nine healthy male volunteers. A significant impairment of binocular depth perception was found when nabilone was administered, but combined application with CBD revealed reduced effects on binocular depth inversion.

Birth-weight effect

A total of 32,483 cannabis-using women giving birth to live-born infants were investigated. The largest reduction in mean birth-weight for any cannabis use during pregnancy was 48 g (95% CI, 83–14 g), with considerable heterogeneity among the five studies. Mean birth-weight was increased by 62 g (95% CI, 8-g reduction – 132-g increase; *p* heterogeneity, 0.59) among infrequent users (≤ weekly), whereas cannabis use at least four times per week had a 131-g reduction in mean birth- weight (95% CI 52–209-g reduction; *p* heterogeneity, 0.25). From the five studies of low birth-weight, the pooled odds ratio for any use was 1.09 (95% CI 0.94–1.27; *p* heterogeneity, 0.19). In a cohort study consisted of a multiethnic population of 7470 pregnant women. Information on the use of drugs was obtained from personal interviews at entry to the study and assays of serum obtained during pregnancy. Pregnancy outcome data (low birth-weight [<2500 g], pre-term birth [<37 weeks gestation], and abruptio placentae) were obtained with a standardized study protocol. A total of 2.3% of the women used cocaine and 11% used cannabis during pregnancy. Cannabis use was not associated with low birth-weight (1.1, 0.9–1.5), pre-term delivery (adjusted odds ratio [OR] 1.1, CI 0.8–1.3), or abruptio placentae (1.3, 0.6–2.8).

Bladder dysfunction

Two whole-plant extracts of *Cannabis sativa* were administered to patients with advanced multiple sclerosis (MS) and refractory troublesome lower urinary tract symptoms. The patients took the extracts containing Δ-9-THC and CBD (2.5 mg of each per spray) for 8 weeks followed by THC-only (2.5 mg THC per spray) for a further eight weeks, and then into a long-term extension. Assessments included urinary frequency and volume charts, incontinence pad weights, cystometry, and visual analog scales for secondary troublesome symptoms. Twenty-one patients were recruited and data from 15 were evaluated. Urinary urgency, the number and volume of incontinence episodes, frequency, and nocturia all decreased significantly following treatment ($p < 0.05$, Wilcoxon's signed rank test). Daily total voided, catheterized and urinary incontinence pad weights also decreased significantly for both extracts. Patient self-assessment of pain, spasticity, and quality of sleep improved significantly ($p < 0.05$, Wilcoxon's signed rank test) with pain improvement continuing up to a median of 35 weeks. There were few troublesome side effects, suggesting that cannabis-based medicinal extracts are a safe and effective treatment for urinary and other problems in patients with advanced MS.

Blood pressure stress reactivity effect

Data from an ascorbic acid (AA) trial (Cetebe 3 g/day for 14 days, $n = 108$) were compared by substance use level regarding systolic blood pressure (SBP) stress reactivity to the anticipation and actual experience phases of a standardized psychological stressor (10 minutes of public speaking and arithmetic). Self-reported never users of cannabis, persons not currently smoking tobacco, and persons consuming three or more caffeine beverages daily all exhibited AA SBP stress reactivity protection to the actual stressor, but not during the anticipation phase. Self-reported ever cannabis users, current tobacco smokers, and persons consuming less than three caffeine beverages daily exhibited the AA SBP protection during the anticipation phase, but only the lower caffeine consumption group exhibited AA protection during both phases. Covariates (neuroticism, extraversion and depression scores, age, sex, body mass index) were not significant.

Blood-borne sexually transmitted infections

Substance use, including alcohol and illicit drugs, increases the risk for the acquisition and transmission of sexually transmitted infection (STI). The prevalence of blood-borne STI including human immuno-deficiency virus (HIV), human T-cell lymphotrophic virus type 1, hepatitis B virus, and syphilis in residents of a detoxification and rehabilitation unit in Jamaica were investigated. The demographic characteristics and the results of laboratory investigations for STI in 301 substance abusers presented during a 5-year period were reviewed. The laboratory results were compared with those of 131 blood donors. The substances used by participants were alcohol, cannabis, and cocaine. None of the clients was an intravenous drug user. Female substance abusers were at higher risk for STI.

The prevalence of STI in substance abusers did not differ significantly from that in blood donors (12% vs 10%). The prevalence of syphilis in substance abusers was significantly higher than that in blood donors (6% vs 3%, $p < 0.05$). The prevalence of syphilis was dramatically increased in female substance abusers and female blood donors (30%, $p < 0.001$ and 13%, $p < 0.05$, respectively). An excess of human T-cell lymphotrophic virus type 1 was also observed in female compared with male substance abusers. Unemployment was identified also as a risk factor for sexually transmitted disease in substance abusers.

Brain aging effect

The impact of duration of education, cannabis addiction and smoking on cognition and brain aging was studied in 211 healthy Egyptian volunteers with mean age of 46.4 ± 3.6 years (range, 20–76 years). The subjects were classified into two groups: Gr I (n = 174; mean age, 49.9 ± 3.8 years; range, 20–76 years), nonaddicts, smokers, and nonsmokers, educated and noneducated, and Gr II cannabis addicts (n = 37; mean age, 43.6 ± 2.6 years; range, 20–72 years) all smokers, educated and noneducated. Outcome measures included the Paced Auditory Serial Addition test for testing attention and the Trailmaking test A and Trailmaking test B (TMb) for testing psychomotor performance. Age correlated positively with score of TMb in the nonaddict group and in the addict group (Trailmaking test A and TMb). Years of education correlated negatively with scores of TMb in the nonaddict group (Gr I) but not the addict group (Gr II). Cannabis addicts (Gr II) had significantly poorer attention than nonaddict normal volunteers (Gr I). It was determined that impairment of psychomotor performance is age related whether in normal nonaddicts or in cannabis addicts. A decline in attention was detected in cannabis addicts and has been considered a feature of pathological aging.

Brain cannabinoid receptor

In humans, psychoactive cannabinoids produce euphoria, enhancement of sensory perception, tachycardia, antinociception, difficulties in concentration, and impairment of memory. The cognitive deficiencies persist after withdrawal. The toxicity of cannabis has been underestimated for a long time, since recent findings revealed that Δ-9-THC-induced cell death with shrinkage of neurons and DNA fragmentation in the hippocampus. The acute effects of cannabinoids, as well as the development of tolerance, are mediated by G protein-coupled cannabinoid receptors. The CB1 receptor and its splice variant, CB1A, are found predominantly in the brain with highest densities in the hippocampus, cerebellum, and striatum. The CB2 receptor is found predominantly in the spleen and in hemopoietic cells and has only 44% overall nucleotide sequence identity with the CB1 receptor. The existence of this receptor provided the molecular basis for the immunosuppressive actions of cannabis. The CB1 receptor mediates inhibition of adenylate cyclase, inhibition of N- and P/Q-type calcium channels, stimulation of potassium channels, and activation of mitogen-activated protein kinase. The CB2 receptor mediates inhibition of adenylate cyclase and activation of mitogen-activated protein kinase. The discovery of endogenous cannabinoid receptor ligands, AEA (*N*-arachidonyl-ethanolamine), and 2-arachidonylglycerol made the notion of a central cannabinoid neuromodulatory system plausible. AEA is released from neurons on depolarization through a mechanism that requires calcium-dependent cleavage from a phospholipid precursor in neuronal membranes.

The release of AEA is followed by rapid uptake into the plasma and hydrolysis by fatty-acid amidohydrolase. The psychoactive cannabinoids increase the activity of dopaminergic neurons in the ventral tegmental area–mesolimbic pathway. Because these dopaminergic circuits are known to play a pivotal role in mediating the reinforcing (rewarding) effects of the most drugs of abuse, the enhanced dopaminergic drive elicited by the cannabinoids is thought to underlie the reinforcing and abuse properties of cannabis. Thus, cannabinoids share a final common neuronal action with other major drugs of abuse such as morphine, ethanol, and nicotine in producing facilitation of the mesolimbic dopamine system.

Hippocampal slices from humans, guinea pigs, rats, and mice, and cerebellar, cerebro-cortical, and hypothalamic slices from guinea pigs were incubated with [^{3}H] noradrenaline and then superfused. Tritium overflow was evoked either electrically (0.3 or 1 Hz) or by introduction of Ca^{2+} ions (1.3 *μM*) into Ca^{2+}-free, K^+-rich medium (25 *μM*) containing 1 *μM* of tetrodotoxin. The cyclic adenosone monophosphate (cAMP) accumulation stimulated by 10 *μM* of forskolin was determined in guinea pig hippocampal membranes. The following drugs were used: the cannabinoid receptor-agonists (-)-*cis*-3-[2-hydroxy-4-(1,1- dimethylheptyl) phenyl]-*trans*-4-(3-hydroxypropyl)cyclo-hexanol (CP-55,940) and R(+) -[2,3 -dihydro-5-methyl-3 - [(morpholinyl) methyl]pyrrolo[1,2, 3-de]-1 ,4-benzoxazinyl]-(1-naphthalenyl)methanone (WIN 55,212-2 [WIN]), the inactive *S*(-)-enantiomer of the latter (WIN 55,212-3) and the CB1 receptor antagonist *N*-piperidino-5-(4-chlorophenyl)-1-(2,4-dichlorophenyl) -4-methyl-3-pyrazole-carboxamide (SR 141716). The electrically evoked tritium overflow from guinea pig hippocampal slices was reduced by WIN (peak inhibitory concentration 30%, 6.5) but not affected by WIN 55,212-3 up to 10 m*M*.

The concentration–response curve of WIN was shifted to the right by SR 141716 (0.032-*μM*) (apparent pA2 8.2), which by itself did not affect the evoked overflow. WIN (1 *μM*) also inhibited the Ca^{2+}-evoked tritium overflow in guinea pig hippocampal slices and the electrically evoked overflow in guinea pig cerebellar, cerebro-cortical, and hypothalamic slices, as well as in human hippocampal slices, but not in rat and mouse hippocampal slices. SR 141716 (0.32 *μM*) markedly attenuated the WIN-induced inhibition in guinea pig and human brain slices. SR 141716 (0.32 *μM*) by itself increased the electrically evoked tritium overflow in guinea pig hippocampal slices, but failed to do so in slices from the other brain regions of the guinea pig and in human hippocampal slices, but failed to do so in slices from the other brain regions of the guinea pig and in human hippocampal slices. The cAMP accumulation stimulated by forskolin was reduced by CP-55,940 and WIN.

The concentration-response curve of CP-55,940 was shifted to the right by SR 141716 (0.1 *μM*; apparent pA2 8.3), that by itself did not affect cAMP accumulation. In conclusion, cannabinoid receptors of the CB1 subtype occur in the human hippocampus, where they may contribute to the psychotropic effects of cannabis, and in the guinea pig hippocampus, cerebellum, cerebral cortex, and hypothalamus. The CB1 receptor in the guinea pig hippocampus is located presynaptically, was activated by endogenous cannabinoids, and may be negatively coupled to adenylyl cyclase. The acute administration of AEA or THC in rats increased the maximum binding capacity (B_{max}) of cannabinoid receptors in the cerebellum and, particularly, in the hippocampus. This effect was also observed after 5 days of a daily exposure to AEA or THC. The increase in the B_{max} after the acute treatment seemed to be caused by changes in the receptor affinity (high K_d). The increase after the chronic exposure may be attributed to an increase in the density of receptors. The [^{3}H]CP-55,940 binding to cannabinoid receptors in the striatum, the limbic forebrain, the mesencephalon, and the medial basal hypothalamus was not altered after the acute exposure to AEA or THC. The chronic exposure to THC significantly decreased the B_{max} of these receptors in the striatum and nonsignificantly in the mesencephalon. This effect was not elicited after the chronic exposure to AEA and was not accompanied by changes in the K_d.

Cannabinoid hyperemesis

Nineteen patients were identified with chronic cannabis abuse and a cyclical vomiting illness. Follow-up was provided with serial urine drug screen analysis and regular clinical consultation to chart the clinical course. Of the 19 patients, five refused consent and were lost to follow-up, and five were excluded based on cofounders. In all cases, chronic cannabis abuse predated the onset of the cyclical vomiting illness. Cessation of cannabis abuse led to cessation of the cyclical vomiting illness in seven cases. Three cases did not abstain and continued to have recurrent episodes of vomiting. Three cases rechallenged themselves after a period of abstinence and suffered a return to illness. Two of these cases abstained again and became, and remain, well. The third case did not and remains ill. A novel finding was that 9 of the 10 patients, including the previously published case, displayed an abnormal washing behavior during episodes of active illness.

Cannabinoid-induced fos expression

Cannabinoid CB1 receptor agonist CP-55,940 in Lewis and Wistar rats was investigated. A moderate (50 μg/kg) and a high (250 μg/kg) dose level were used. The 250-μg/kg dose caused locomotor suppression, hypothermia, and catalepsy in both strains, but with a significantly greater effect in Wistar rats. The 50-μg/kg dose provoked moderate hypothermia and locomotor suppression but in Wistar rats only. CP-55,940 caused significant Fos immunoreactivity in 24 out of 33 brain regions examined. The most dense expression was seen in the paraventricular nucleus of the hypothalamus, the islands of Callej a, the lateral septum (ventral), the central nucleus of the amygdala, the bed nucleus of the stria terminalis (lateral division), and the ventrolateral periaqueductal gray. Despite having a similar distribution of CP-55,940- induced Fos expression,

Lewis rats showed less overall Fos expression than Wistar rats in nearly every brain region counted. This held equally true for anxiety-related brain structures (e.g., central nucleus of the amygdala, periaqueductal gray, and the paraventricular nucleus of the hypothalamus) and reward-related sites (Nac and pedunculopontine tegmental nucleus). In a further experiment, Wistar rats and Lewis rats did not differ in the amount of Fos immunoreactivity produced by cocaine (15 mg/kg). These results indicate that Lewis rats are less sensitive to the behavioral, physiological and neural effects of cannabinoids.

Cannabis withdrawal effect

A 35-year-old male was cognitively assessed prior to cessation of 18 years of daily cannabis use and monitored for several weeks post- cessation. Brain event-related potential measures of selective attention reflecting a difficulty in filtering out complex irrelevant information showed no indication of improvement over 6 weeks of abstinence. When tested in the acutely intoxicated state prior to cessation of use, a dramatic normalization of the event-related potential signature was observed. A treatment program based on supportive–expressive psychotherapy was administered and depression, anxiety, and general psychological health were monitored over the course of withdrawal from cannabis.

Cannabis–amphetamine interaction

Cannabinoid–amphetamine interactions were studied as follows:

1. 30 minutes after acute injection of (-)-Δ-9-THC (0.1 or 6.4 mg/kg, intraperitoneally).

2. 30 minutes after the last injection of 14-daily treatment with (-)-Δ-9-THC (0.1 or 6.4 mg/kg).
3. 24 hours after the last injection of 14- daily treatment with (-)-Δ-9-THC (6.4 mg/kg).

Acute cannabinoid exposure antagonized the amphetamine-induced dose-dependent increase in locomotion, exploration, and the decrease in inactivity. Chronic treatment with (-)-Δ-9-THC resulted in tolerance to this antagonistic effect on locomotion and inactivity but not on exploration, and potentiated amphetamine-induced stereotypes. Lastly, 24 hours of withdrawal after 14 days of cannabinoid treatment resulted in sensitization to the effects of D-amphetamine on locomotion, exploration, and stereotypes.

Cannabis-induced coma

Two cases of cannabis-induced coma were reported following accidental ingestion of cannabis cookies. The possibility of cannabis ingestion should be considered in cases of unexplained coma in a previously healthy young child if signs of conjunctival hyperemia, pupillary dilatation, and tachycardia were present and other causes, such as central nervous system infection or trauma were unlikely.

Cannabis-related arteritis

A 19-year-old man who presented with plantar claudication associated with necrosis in a toe underwent diagnostic arteriography and surgery for popliteal artery entrapment type III was studied. Surgical clearance resolved the popliteal artery entrapment but left the clinical symptoms unchanged. Closer questioning disclosed a history of cannabis consumption and intravenous vasodilatory therapy was started. After the 21-day course of vasodilator agents, the pain disappeared and the toe necrosis regressed. The patient stopped taking cannabis and had no signs of recurrence. A 24-year-old woman who was a heavy cannabis smoker with progressive Raynauld's phenomenon and digital necrosis, was investigated. Systemic sclerosis and other connective tissue disorders, as well as arteriosclerosis and arterial emboli were excluded with appropriate laboratory examinations. Arteriography revealed multiple forearm, palmar and digital occlusions with corkscrew-shaped vessels. Based on the characteristic arteriography and clinical findings, the diagnosis of cannabis arteritis was retained. With careful necrectomy, conservative wound dressings and secondary prostacyclin therapy a complete healing of digital necrosis was observed.

There was no recurrence during the 6-month follow-up. Young men were presented with distal arteriopathy of the lower limbs in three cases, and of the left upper limb in the remaining patient. Symptoms occurred progressively, distal pulses had disappeared, and distal necrosis was constant. Three patients suffered from Raynauld's phenomenon, none of them presented with venous thrombosis. Radiological evaluation revealed distal abnormalities in all cases, and proximal arterial thrombosis in one case. The four patients were cannabis smokers for at least four years. With cannabis interruption and symptomatic treatment, lesions improved for three patients. For one of them, recurrence of arteriopathy occurred when he resumed smoking cannabis. For the fourth who never stopped cannabis, an amputation was necessary. Ten male moderate tobacco smokers and regular cannabis users with a median age of 23.7 years, developed subacute distal ischemia of the lower or upper limbs, leading to necrosis in the toes and/or fingers and sometimes to distal limb gangrene. Two of the patients also presented with venous thrombosis and three patients were suffering from a recent Raynauld's phenomenon. Biological test results did not show evidence of the classical vascular risk factors for thrombosis.

Arteriographic evaluation in all of the cases revealed distal abnormalities in the arteries of feet, legs, forearms, and hands resembling those of Buerger's disease. A collateral circulation sometimes with opacification of the vasa nervorum was noted. In some cases, arterial proximal atherosclerotic lesions and venous thrombosis were observed. Despite treatment with ilomedine and heparin in all cases, five amputations were necessary in four patients. The vasoconstrictor effect of cannabis on the vascular system has been known for a long time. It has been shown that Δ-8-THC and Δ-9-THC may induce peripheral vasoconstrictor activity. Cannabis arteritis resembles Buerger's disease, but patients were moderate tobacco smokers and regular cannabis users.

Cannabis-related flashback

A young man who offended a friend without any objective reason was reported. The report of the forensic psychiatrist demonstrated that the offense was committed under the influence of a cannabis flashback. The last time the offender had consumed cannabis was 2 weeks before the acts. A plasmatic detection was realized and showed a level of 6 ng/mL, 30 minutes after the beginning of the flashback.

Capgras syndrome

A report describes an apparently greater incidence of Capgras syndrome among the Maori population compared with the European population. Five cases of Capgras syndrome were identified in the eastern catchment area where 19% of the population identified as Maori, 75% as European, and 6% as other or nonspecified. All of the cases occurred in Maori patients. No cases were identified in the western catchment area where 12% of the population identified as Maori, 87% as European, and 1% as other or nonspecified. Four of five cases were females. Two cases had a history of cannabis use. Three cases had exhibited dangerous behavior towards family members.

Carcinogenic activity

The dried leaf, administered intraperitoneally to rats of both sexes at a dose of 7 mg/kg/week, was active. The animals were irradiated with y radiation between 40 and 50 days of age and observed for 78 weeks. There was a greater incidence of tumors in animals given marijuana extract and γ radiation than either marihuana or γ radiation alone.

Cardiorespiratory effect

Fifty stable patients (25 males, 25 females) with methadone maintenance treatment (MMT) programs were investigated. Forty-six MMT patients were current tobacco smokers, 19 were current cannabis users, and none were currently using opioids other than prescribed methadone. Abnormalities of respiratory function were defined as those results outside the 95% confidence interval of reference values for normal subjects adjusted for age, weight, height, and sex. Thirty-one (62%) MMT patients had reduced carbon monoxide transfer factor; 17 (34%) had elevated single breath alveolar volume, and 43 (86%) had a reduced carbon monoxide transfer factor–alveolar volume ratio. Six patients (12%) had reduced forced expiratory volume in 1 second (FEV1); one (2%) had reduced forced vital capacity (FVC); and nine (18%) had an obstructive ventilatory defect. Ten (20%) patients had arterial CO_2 pressure higher than 45 mmHg and 14 (28%) had alveolar to arterial oxygen gradient higher than 15 mmHg.

Chest X-ray, echo-cardiography, and electrocardiogram showed no significant abnormalities. The potent cannabinoid receptor agonists WIN55,212-2 (0.05, 0.5, or 5 pmol/50 nL) and HU-210 (0.5 pmol/50 nL) or the CB1 receptor antagonist/inverse agonist AM28 1 (1 pmol/100 nL) were microinjected into the rostral ventrolateral medulla oblongata (RVLM) of urethane-anesthetized, immobilized and mechanically ventilated male Sprague–Dawley rats (n = 22). Changes in splanchnic nerve activity, phrenic nerve activity, mean arterial pressure, and heart rate in response to cannabinoid administration were recorded.

The CB1 receptor gene was expressed throughout the ventrolateral medulla oblongata. Unilateral microinjection of WIN 55,212-2 into the RVLM evoked short-latency, dose-dependent increases in splanchnic nerve activity (0.5 pmol; 175 ± 8%, $n = 5$) and mean arterial pressure (0.5 pmol; 26 ± 3%, $n = 8$), and abolished phrenic nerve activity (0.5 pmol; duration of apnea: 5.4 ± 0.4 seconds, $n = 8$), with little change in heart rate ($p < 0.005$). HU-210, structurally related to Δ-9-THC, evoked similar effects when microinj ected into the RVLM ($n = 4$). Prior micro-injection of AM281 produced agonist-like effects, and significantly attenuated the response to subsequent injection of WIN (0.5 pmol, $n = 4$).

Cardiovascular effects

The leaf, smoked by adults of both sexes, at a dose of 600 mg/ person (1–1.5% THC), produced no adverse effects on blood pressure, electrocardiogram, and the heart. Cannabis and Δ-9-THC increase heart rate, slightly increase supine blood pressure, and on occasion produced marked orthostatic hypotension. Cardiovascular effects in animals are different, with bradycardia and hypotension the most typical responses. Cardiac output increases, and peripheral vascular resistance and maximum exercise performance decrease. Tolerance to most of the initial cardiovascular effects appears rapidly. With repeated exposure, supine blood pressure decreases slightly, orthostatic hypotension disappears, blood volume increases, heart rate slows, and circulatory responses to exercise and Valsalva maneuver are diminished, consistent with centrally mediated, reduced sympathetic, and enhanced parasympathetic activity. Receptor-mediated and probably nonneuronal sites of action account for cannabinoid effects. The endocannabinoid system appears important in the modulation of many vascular functions. Cannabis' cardiovascular effects are not associated with serious health problems for most young, healthy users, although occasional myocardial infarction, stroke, and other adverse cardiovascular events are reported.

CB1 cannabinoid receptor in human placenta

CB1 (G protein-coupled) receptor and FAAH expression in human term placenta were investigated by immunohistochemistry. CB1 receptor was found in all layers of the membrane, with particularly strong expression in the amniotic epithelium and reticular cells and cells of the maternal decidua layer. Moderate expression was observed in the chorionic cytotrophoblasts. The expression of FAAH was highest in the amniotic epithelial cells, chorionic cytotrophoblast, and maternal decidua layer. The results suggest that the human placenta is a likely target for cannabinoid action and metabolism. This is consistent with a placental site of action of endocannabinoids and cannabis being responsible, at least in part, for the poor outcomes associated with cannabis consumption and pathology in the endocannabinoid system during pregnancy.

Central nervous system depressant activity

Fluidextract of the aerial parts, administered intraperitoneally to rats at a dose of 25 mg/kg, was active. The fluidextract, administered orally to dogs, produced ataxia. The leaf, smoked by human adults, produced a decrease in psychomotor performance.

Central nervous system effect

Δ-9-THC activates the two G protein-coupled receptors CB1 and CB2. The endogenous ligands of these receptors were identified as lipid metabolites of arachidonic acid, named endocannabinoids. The two most studied endocannabinoids are AEA and 2-arachidonyl-glycerol. The CB1 receptor is massively expressed throughout the central nervous system, whereas CB2 expression seems restricted to immune cells. Following endocannabinoid binding, CB1 receptors modulate second messenger cascades (inhibition of adenylate cyclase, activation of mitogen-activated protein kinases and of focal-adhesion kinases), as well as ionic conductances (inhibition of voltage-dependent calcium channels, activation of several potassium channels). Endocannabinoids transiently silenced synapses by decreasing neurotransmitter release. They play major roles in various forms of synaptic plasticity because of their ability to behave as retrograde messengers and activate noncannabinoid receptors (such as vanilloid receptor type-1). Mice strain with a disrupted *CB1* gene (CB1 knockout mice) appeared healthy and fertile, but they had a significantly increased mortality rate. They also displayed reduced locomotor activity, increased ring catalepsy, and hypoalgesia in hotplate and formalin tests. Δ-9-THC-induced ring catalepsy, hypomobility, and hypothermia were completely absent in CB1 mutant mice. In contrast, Δ-9-THC-induced analgesia in the tail-flick test and other behavioral (licking of the abdomen) and physiological (diarrhea) responses after Δ-9-THC administration were found. Results indicate that most, but not all, central nervous system effects of Δ-9-THC are mediated by the CB1 receptor.

Central nervous system stimulant Activity

The resin, ingested by a 4-year-old girl, showed signs of stupor alternating with brief intervals of excitation and foolish laughing with atactic movements. Her temperature, blood pressure, pulse, hemoglobin, leukocytes, serum electrolytes, and serum urea were normal. Respiratory rate was 12 beats per minute. Blood sugar elevated. Recovery was complete within 24 hours with no treatment.

Cerebellar clock-altering effect

Twelve volunteers who smoked cannabis recreationally about once weekly, and 12 volunteers who smoked daily for a number of years performed a self-paced counting task during positron emission tomography imaging, before and after smoking cannabis and placebo cigarettes. Smoking cannabis increased regional cerebral blood flow in the ventral forebrain and cerebellar cortex in both groups, but resulted in significantly less frontal lobe activation in chronic users. Counting rate increased after smoking cannabis in both groups, as did a behavioral measure of self-paced tapping, and both increases correlated with regional cerebral blood flow in the cerebellum. Results indicate that smoking cannabis appears to accelerate a cerebellar clock-altering self- paced behaviors.

Clinical endocannabinoid deficiency

Clinical endocannabinoid deficiency, and the prospect that it could underlie the pathophysiology of migraine, fibromyalgia, irritable bowel syndrome, and other functional conditions alleviated by clinical cannabis were studied. Migraine has numerous relationships to endocannabinoid function. AEA potentiated 5 -hydroxytrayptamine (HT1A) and inhibited 5-HT2A receptors supporting therapeutic efficacy in acute and preventive migraine treatment. Cannabinoids also demonstrated dopamine-blocking and anti-inflammatory effects. AEA is tonically active in the periaqueductal gray matter, a migraine generator. THC modulated glutamatergic neurotransmission via *N*-methyl-D-aspartic acid-receptors. Fibromyalgia is now conceived as a central sensitization state with secondary hyperalgesia. Cannabinoids have similarly demonstrated the ability to block spinal, peripheral and gastrointestinal mechanisms that promote pain in headache, fibromyalgia, irritable bowel syndrome and related disorders.

Cognitive functioning

Cognitive performance was examined in 145 adolescents aged 13–16 years for whom prenatal exposure to cannabis and cigarettes had been ascertained. The subjects were from a low-risk, predominantly middle-class sample participating in an ongoing, longitudinal study. The assessment battery included tests of general intelligence, achievement, memory, and aspects of executive functioning. Consistent with results obtained at earlier ages, the strongest relationship between prenatal maternal cigarette smoking and cognitive variables was seen with overall intelligence and aspects of auditory functioning, whereas prenatal exposure to marijuana was negatively associated with tasks that required visual memory, analysis, and integration. A multisite, retrospective, cross-sectional, neuropsychological study was conducted among 102 near-daily cannabis users (51 long-term users: mean, 23.9 years of use; 51 shorter-term users: mean, 10.2 years of use), compared with 33 nonuser controls. Measures from nine standard neuropsycho-logical tests that assessed attention, memory, and executive functioning were administered prior to entry into a treatment program following a median 17-hour abstinence. Long-term cannabis users performed significantly less well than shorter-term users and controls on tests of memory and attention. On the Rey Auditory Verbal Learning Test, long-term users recalled significantly fewer words than either shorter-term users ($p = 0.001$) or controls ($p = 0.005$). There was no difference between shorter-term users and controls. Long-term users showed impaired learning ($p = 0.007$), retention ($p = 0.003$), and retrieval ($p = 0.002$) compared with controls. Both user groups performed poorly on a time estimation task ($p < 0.001$ vs controls). Performance measures often correlated significantly with the duration of cannabis use, being worse with increasing years of use, but were unrelated to withdrawal symptoms and persisted after controlling for recent cannabis use and other drug use. A patient with a history of traumatic brain injury along with current mood disorder and cannabis use was reported. The impact of cannabis use appeared to have a detrimental effect on his mood. Treatment of the mood disorder resulted in larger cognitive gains. Sixty healthy volunteers (a negative urine drug-screening test was prerequisite) were investigated. On the first day, baseline data were obtained from a physical examination and a psychological test battery for the investigation of visual and verbal memory and cognitive perceptual performance. On the second day, subjects received a regular cigarette or one containing 290 mg/kg body weight of THC. Physical and psychological assessments were performed immediately (15 minutes) after subjects smoked their cigarettes. Twenty-four hours later, physical and psychological examinations were repeated. Results suggest that perceptual motor speed and accuracy, two very important parameters of driving ability, seem to be impaired immediately after cannabis consumption. The analyses included 1318 participants under age 65 years who completed the Mini-Mental State Examination (MMSE) during three study waves in 1981, 1982, and 1993 1996. Individual MMSE score differences between waves two and three were calculated for each study participant. After 12 years, study participants' scores declined a mean of 1.20 points on the MMSE (standard deviation, 1.90), with 66% having scores that declined by at least one point. Significant numbers of scores declined by three points or more (15% of participants in the 18–29-year-old age group). There were no significant differences in cognitive decline between heavy users, light users, and nonusers of cannabis. There were also no male–female differences in cognitive decline in relation to cannabis use. From 250 individuals consuming cannabis regularly, 99 healthy, free of any other past or present drug abuse, or history of neuropsychiatric disease cannabis users were selected. After an interview, physical examination, analysis of routine laboratory parameters, plasma/urine analyses for drugs, and Minnesota Multiphasic Personality Inventory testing, users and respective controls were subjected to a computer-assisted attention test battery comprising visual scanning, alertness, divided attention, flexibility, and working memory. Of the potential predictors of test performance within the user group, including present age, age of onset of cannabis use, degree of acute intoxication (THC + THC–OH plasma levels), and cumulative toxicity (estimated total life dose), an early age of onset turned out to be the only predictor, predicting impaired reaction times exclusively in visual scanning. Early-onset users (onset before age 16; $n = 48$) showed a significant impairment in reaction times in this function, whereas late-onset users (onset after age 16; $n = 51$) did not differ from controls ($n = 49$). Male volunteers ($n = 5$) with histories of moderate alcohol and cannabis use were administered three doses of alcohol (0.25, 0.5, or 1 g/kg), three doses of cannabis (4.8, or 16 puffs of 3.55% Δ-9-THC), and placebo in random order under double blind conditions in seven separate sessions. Blood alcohol concentration (10–90 mg/dL) and THC levels (63–188 ng/mL) indicated that active drug was delivered to subjects dose dependently. Alcohol and cannabis produced dose-related changes in subjective measures of drug effect. Ratings of perceived impairment were identical for the high doses of alcohol and cannabis. Both drugs produced comparable impairment in digit–symbol substitution and word recall tests, but had no effect in time perception and reaction time tests. Alcohol, but not cannabis, slightly impaired performance in a number recognition test.

Comorbid dysthymia and substance disorder

A total of 642 patients were assessed. Thirty-nine had substance-related disorder and dysthymia (SRD-dysthymia) and 308 had SRD only. Data on past use were collected by a research associate using a questionnaire. The patients with SRD-dysthymia and SRD did not differ with regard to use of alcohol, tobacco, and benzodiazepines. The patients with SRD-dysthymia started caffeine use at an earlier age, had shorter "use careers" of cocaine, amphetamines, and opiates, and had fewer days of cocaine and cannabis use in the last year. They also had a lower rate of cannabis abuse/dependence. The results indicated that patients with dysthymia and SRD have exposure to most substances of abuse that was comparable to patients with SRD only. They selectively use certain substances less often than patients with SRD only. A course and severity of SRD among 642 patients with comorbid major depressive disorder (MDD) was analyzed by means of both retrospective and concurrent data. Data on course included lifetime use, age at first use, years of use, use in the last year, periods of abstinence, and current diagnosis. Data on severity included two measures of SRD-associated problems, substance abuse vs dependence, self-help activities, and number of substances being abused. SRD-MDD patients tended to manifest lower levels of cannabis, opiate, and cocaine use, and more SRD-only patients were abusing three or more substances. Men with SRD-MDD demonstrated longer mean durations of abstinence compared with men with SRD-only, whereas SRD-MDD women demonstrated shorter mean durations of abstinence, compared with women with SRD-only. MDD-SRD patients showed slightly less substance abuse, but SRD severity was comparable with SRD-only patients.

Covariation among risk behaviors

A sample of 913 sexually active high school students completed a self-administered questionnaire that required mainly "yes" or "no" answers to questions involving participation in a range of risk behaviors. Contraceptive nonuse was not significantly associated with use of cigarettes, alcohol, or inhalants; perpetration or being a victim of violence; exposure to risk of physical injury; and suicidality. For males only, there was a significant inverse association between contraceptive nonuse and use of cannabis in the previous month. This was not the case for lifetime cannabis use for either gender.

Cytochrome P450 and 2C6 expression

Hashish (cannabis) and heroin effect on the expression of cytochrome P450 2E1 (CYP 2E1) and cytochrome P450 2C6 (CYP 2C6) was measured after single (24 hours) and repeated-dose treatments (four consecutive days). The expression of CYP 2E1 was slightly induced after single-dose treatments and markedly induced after repeated-dose treatments of mice with hashish (10 mg/kg body weight). It is believed that *N*-nitrosamines are activated principally by CYP 2E1 and the activity of *N*-nitrosodimethylamine was found to be increased after single- and repeated-dose treatments of mice with hashish by 23 and 41%, respectively. Hashish treatments of mice increased the total hepatic content of CYP by 112 and 206%, respectively; aryl hydrocarbon hydroxylase activity by 110 and 165%, respectively; nicotinamide adenine dinucleotide phosphate–cytochrome c reductase activity by 21 and 98%, respectively, and glutathione level by 81 and 173%, respectively. The level of free radicals was potentially decreased after single- or repeated-dose treatments with either hashish or heroin.

Cytotoxic effect

THC, in leukemic cell lines (CEM, HEL-92, and HL60) and in peripheral blood mononuclear cells, 6 hours after exposure induced apoptosis, even at one times the IC_{50}. THC did not appear to act synergistically with cytotoxic agents, such as cisplatin. THC-induced cell death was preceded by significant changes in the expression of genes involved in the mitogen-activated protein kinase signal transduction pathways. Both apoptosis and gene expression changes were altered independent of p53 and the cannabinoid 1 and 2 receptors (CB1-R and CB2-R).

Depressant activity

Heavy cannabis use and depression are associated and evidence from longitudinal studies suggests that heavy cannabis use may increase depressive symptoms among some users. Participants (n = 1920) were reassessed as part of a follow-up study. The analysis focused on two cohorts: those who reported no depressive symptoms at baseline (n = 849) and those with no diagnosis of cannabis abuse at baseline (n = 1,837). Symptoms of depression, cannabis abuse, and other psychiatric disorders were assessed with the Diagnostic Interview Schedule.

In participants with no baseline depressive symptoms, those with a diagnosis of cannabis abuse at baseline were four times more likely than those with no cannabis abuse diagnosis to have depressive symptoms at the follow-up assessment, after adjusting for age, gender, antisocial symptoms, and other baseline covariates. These participants were more likely to have experienced suicidal ideation and anhedonia during the follow-up period. Among the participants who had no diagnosis of cannabis abuse at baseline, depressive symptoms at baseline failed to significantly predict cannabis abuse at the follow up assessment.

The relationship between depressive symptoms and polydrug use (alcohol, cannabis, and cocaine) among blacks in a high-risk community was studied. A street sample (n = 570) from four high-risk communities was collected through personal interviews. Interviewers asked respondents about their drug use behavior during the past 30 days and their depressive symptoms during the past week. Odds ratios and logistic regressions, adjusted for age and sex, were used to assess the relationship between depressive symptoms and drug and polydrug use (drug use involving cocaine). Results showed that depressive symptoms are significantly associated with polydrug use. Depressive symptoms were not associated with alcohol use or with the combination of alcohol and cannabis use.

Diabetic ketoacidosis

One hundred fifty-eight young adults, aged 16–30 years, with type 1 diabetes, attending an urban diabetes clinic, were sent an anonymous confidential postal questionnaire to determine the prevalence of street drug use. Eighty-five completed responses were received. Twenty-nine percent of respondents admitted to using street drugs.

Of those, 68% habitually took street drugs more than once a month. Seventy-two percent of users were unaware of the adverse effects on diabetes. Results indicated that the street drug usage in young adults with type 1 diabetes is common and may contribute to poor glycemic control and serious complications of diabetes.

Digital necrosis

An 18-year-old woman, with a history of severe anorexia nervosa of 5 years' duration, who acknowledged regular use of tobacco and cannabis, was hospitalized for necrosis of the left index and thumb that had occurred shortly after left radial artery puncture for blood gas analysis. Acrocyanosis of the four limbs had been present since the onset of anorexia nervosa. Arteriography of the upper limbs showed major spasm of the left radial and cubital arteries and thromboses in the left inter-digital arteries of the left index and thumb. The distal portions of the arteries were then on the left and on the right. The necrotic lesions healed after intravenous administration of ilomedine and interruption of tobacco and cannabis. Acrocyanosis of the four limbs persisted.

Discriminative stimulus effect

Rhesus monkeys, trained to discriminate Δ-9-THC from vehicle in a two-lever drug discrimination procedure, were tested with a variety of psychoactive drugs, including cannabinoids or drugs from other classes. The results indicated that Δ-9-THC discrimination showed pharmacological specificity, in that none of the noncannabinoid drugs fully substituted for Δ-9-THC. The classical cannabinoids, Δ-9-THC and Δ-8-THC, and the novel cannabinoids, WIN and 1-butyl-2-methyl-3-(1-naphthoyl)indole, produced full dose-dependent substitution for Δ-9-THC in all monkeys. A heptyl indole derivative failed to substitute for Δ-9-THC, but it also did not displace [^{3}H] CP-55,940 from its binding site.

Dopamine metabolism

The effect of repeated administrations of THC or WIN, a synthetic cannabinoid receptor agonist, on dopamine turnover in the prefrontal cortex, striatum, and Nac in rats, was investigated. THC or WIN (twice daily for seven or 14 days) caused a persistent and selective reduction in medial prefrontal cortical dopamine turnover. No significant alterations of dopamine metabolism were observed in the Nac or striatum. These dopaminergic deficits in the prefrontal cortex were observed after a drug-free period of up to 14 days. The cognitive dysfunction produced by heavy, long-term cannabis use may be subserved, in part, by drug-induced alterations in frontal cortical dopamine turnover. Two weeks' administration of THC to rats, reduced dopamine transmission in the medial prefrontal cortex, whereas dopamine metabolism in striatal regions was unaffected.

Dopamine release

A 38-year-old drug-free schizophrenic patient took part in a single photon emission computerized tomographic study of the brain, and smoked cannabis secretively during a pause in the course of an imaging session. Cannabis had an immediate calming effect, followed by a worsening of psychotic symptoms a few hours later. A comparison of the two sets of images, obtained before and immediately after smoking cannabis, indicated a 20% decrease in the striatal dopamine D2 receptor-binding ratio, suggestive of increased synaptic dopaminergic activity.

Dopamine transmission modulation

The endogenous cannabinoid system is a new signaling system composed by the central (CB1) and the peripheral (CB2) receptors, and several lipid transmitters including AEA and 2-arachidonylglycerol. Cannabinoid CB1 receptors are present in dopamine projecting brain areas. In primates and certain rat strains it is also located in dopamine cells of the A8, A9, and A10 mesencephalic cell groups, as well as in hypothalamic dopaminergic neurons controlling prolactin secretion. CB1 receptors co-localize with dopamine D1/D2 receptors in dopamine projecting fields. Manipulation of dopaminergic transmission is able to alter the synthesis and release of AEA, as well as the expression of CB1 receptors. CB1 receptors can switch their transduction mechanism to oppose to the ongoing dopamine signaling. Acute blockade of CB1 receptor potentiates the facilitatory role of dopamine D2 receptor agonists on movement. CB1 stimulation results in sensitization to the motor effects of indirect dopaminergic agonists.

Dyskinetic activity

A 4-week dose escalation study was performed to assess the safety and tolerability of cannabis in six patients with Parkinson's disease (PD) with levodopa (L-DOPA)-induced dyskinesia. Then a randomized, placebo-controlled crossover study was performed, in which 19 patients with PD were randomized to receive oral cannabis extract followed by placebo or vice versa. Each treatment phase lasted for 4 weeks with an intervening 2-week washout phase. The primary outcome measure was a change in Unified Parkinson's Disease Rating Scale (UPDRS) (items 32 to 34) dyskinesia score. Secondary outcome measures included the Rush scale, Bain scale, tablet arm drawing task, and total UPDRS score following a levodopa challenge, as well as patient-completed measures of a dyskinesia activities of daily living scale, the PDQ-39, on–off diaries, and a range of category rating scales. Seventeen patients completed the study. Cannabis was well tolerated and had no pro- or antiparkinsonian action. There was no evidence for a treatment effect on L-DOPA-induced dyskinesia as assessed by the UPDRS, or any of the secondary outcome measures. An anonymous questionnaire sent to all patients attending the Prague Movement Disorder Center revealed that 25% of 339 respondents had taken cannabis and 45.9% of these described some form of benefit. 2,4,5-Trihydroxyphenethylamine (6-hydroxydopamine)-lesioned rats were treated with the enantiomers of the synthetic cannabinoid 7-hydroxy-Δ6-THC 1,1-dimethylheptyl. Treatment with its (-)- (3R, 4R) enantiomer (code name HU-210), a potent cannabinoid receptor type 1 agonist, reduced the rotations induced by L-DOPA/carbidopa or apomorphine by 34 and 44%, respectively. Treatment with the (+)-(3S, 4S) enantiomer (code name HU-211), an *N*-methyl-D-aspartate antagonist, and the psychotropically inactive cannabis constituent: CBD and its primary metabolite, 7-hydroxy-cannabinol, did not show any reduction of rotational behavior. The results indicate that activation of the CB1 stimulates the dopaminergic system ipsilaterally to the lesion, and may have implications in the treatment of PD.

Dystonic activity

The neural mechanisms underlying dystonia involve abnormalities within the basal ganglia—in particular, overactivity of the lateral globus pallidus. Cannabinoid receptors are located presynaptically on γ-aminobutyric acid receptor (GABA) terminals within the globus pallidus internus, where their activation reduces GABA reuptake. Cannabinoid receptor stimulation may thus reduce overactivity of the globus pallidus, and thereby reduce dystonia. A double-blind, randomized, placebo-controlled, crossover study using the synthetic cannabinoid receptor agonist nabilone in patients with generalized and segmental primary dystonia showed no significant reduction in dystonia following treatment with nabilone.

Endocrine effect

Animal models have demonstrated that cannabinoid administration acutely altered multiple hormonal systems, including the suppression of the gonadal steroids, growth hormone, prolactin, and thyroid hormone and the activation of the hypothalamic–pituitary–adrenal (HPA) axis. These effects were mediated by binding to the endogenous cannabinoid receptor in or near the hypothalamus. Despite these findings in animals, the effects in humans have been inconsistent, and discrepancies were likely owing in part to the development of tolerance. Intravenous administration of three cannabinoid agonists to nine castrated male calves under stress-free conditions provoked immediate increases of serum cortisol and respiration rate, and produced rapid hypoalgesia to cutaneous pain and thermal stimuli. AEA and methanandamide did not affect serum prolactin. Administration of WIN increased serum prolactin abruptly. None of the cannabinoid receptor agonists affected serum growth hormone.

Environmental stress and cannabinoids interaction

Anxiety and panic are the most common adverse effects of cannabis intoxication. Data suggest that cannabinoid CB1 receptor modulation of amygdalar activity contributes to these phenomena. Using Fos as a marker, it was tested the hypothesis that environmental stress and CB1 cannabinoid receptor activity interact in the regulation of amygdalar activation in male mice. Both 30 minutes of restraint and CB1 receptor agonist treatment (Δ-9-THC [2.5 mg/kg]) or CP-55,940 (0.3 mg/kg); by intraperitoneal injection) produced barely detectable increases in Fos expression within the central amygdala (CeA). The combination of restraint and CB1 agonist administration produced robust Fos induction within the CeA, indicating a synergistic interaction between environmental stress and CB1 receptor activation. An inhibitor of endocannabinoid transport, AM404 (10 mg/kg), produced an additive interaction with restraint within the CeA. In contrast, FAAH inhibitor-treated mice (URB597, 1 mg/kg) and FAAH (–/–) mice did not exhibit any differences in amygdalar activation in response to restraint compared with control mice. In the basolateral amygdala and medial amygdala, restraint stress produced a low level of Fos induction, which was unaffected by cannabinoid treatment. The CB1 receptor antagonist SR141 716 dose-dependently increased Fos expression in the BLA and CeA.

Epileptic effect

Δ-9-THC at a dose of 1 μM, significantly depressed evoked depolarizing postsynaptic potentials (PSPs) in rat olfactory cortex neurones. A standardized cannabis extract (SCE) and Δ-9-THC-free SCE significantly potentiated evoked PSPs (all results were fully reversed by the CB1 receptor antagonist

SR141716A, 1 μM). The potentiation by Δ-9-THC-free SCE was greater than that produced by SCE. On comparing the effects of Δ-9-THC-free SCE on evoked PSPs and artificial PSPs (aPSPs; evoked electrotonically following brief intracellular current injection), PSPs were enhanced, whereas aPSPs were unaffected, suggesting that the effect was not resulting from changes in background input resistance. Similar recordings made using CB1 receptor-deficient knockout mice and wild-type littermate controls revealed cannabinoid or extract-induced changes in membrane resistance, cell excitability and synaptic transmission in wild-type mice that were similar to those seen in rat neurones, but no effect on these properties were seen in CB1 receptor-deficient knockout mice cells.

Results indicated that the unknown extract constituent(s) effects over-rode the suppressive effects of Δ-9-THC on excitatory neurotransmitter release, which may explain some patients' preference for herbal cannabis rather than isolated Δ-9-THC (owing to attenuation of some of the central Δ-9-THC side effects) and possibly account for the rare incidence of seizures in some individuals taking cannabis recreationally. A SCE with pure Δ-9-THC, at matched concentrations of Δ-9-THC, and a Δ-9-THC-free extract (Δ-9-THC-free SCE) in in vitro rat brain slice model of epilepsy were examined. In the in vitro epilepsy model, in which sustained epileptiform seizures were induced by the muscarinic receptor agonist oxotremorine-M in immature rat piriform cortical brain slices, SCE was a more potent and again more rapidly-acting anticonvulsant than isolated Δ-9-THC. Δ-9-THC-free extract also exhibited anticonvulsant activity. CBD did not inhibit seizures, nor did it modulate the activity of Δ-9-THC in this model. These results demonstrated that not all of the therapeutic actions of cannabis herb might be a result of the Δ-9-THC content.

Estrogen receptors stimulating effect

THC, CBD, and desacetyllevonantradol, in estrogen-induced MCF-7 breast cancer cells at concentrations of no more than 10 μM, produced no effect. THC failed to antagonize the response to estradiol under conditions in which the antiestrogen LY156758 (keoxifene; raloxifene) was effective. The phytoestrogen formononetin behaved as an estrogen at high concentrations, and this response was antagonized by LY156758. THC, desacetyllevonantradol, or CBD did not stimulate transcription of an *EREtkCAT* reporter gene transiently transfected into MCF-7 cells.

Estrous cycle disruption effect

Ethanol (95%) extract of the dried aerial parts, administered intraperitoneally to gerbils at a dose of 2.5 mg/animal daily for 60 days, was active. Petroleum ether extract of the dried aerial, administered intraperitoneally to mice and rats at doses of 1 and 5 mg/animal, respectively, for 64 days, was active. Petroleum ether extract of the aerial parts, administered intraperitoneally to female rats, produced weak activity.

Petroleum ether extract of the entire plant, administered by gastric intubation to female mice at doses of 75 mg/kg and 150 mg/kg, was active. A dose of 3 mg/kg produced weak activity. Petroleum ether extract of the resin, administered intra-peritoneally to female rats at doses of 10 and 20 mg/kg, was active. Resin, administered orally to female rats at doses of 3, 15, and 75 mg/kg daily for 72 days, was active.

Familial mediterranean fever

A patient with familial Mediterranean fever was presented with chronic relapsing pain and inflammation of gastrointestinal origin. After determining a suitable analgesic dosage, a double-blind, placebo-controlled, crossover trial was conducted using 50 mg of Δ-9-THC daily in five doses in the active weeks and measuring effects on parameters of inflammation and pain. Although no anti-inflammatory effects of Δ-9-THC were detected during the trial, a highly significant reduction ($p < 0.001$) in additional analgesic requirements was achieved.

Food intake modulation

Cannabis sativa stimulates appetite, especially for sweet and palatable food. Cannabinoid action has proposed a central role of the cannabinoid system in obesity. Dronabinol, a commercially available form of a THC, has been used successfully for increasing appetite in patients with HIV wasting disease. Cannabinoid receptor antagonist may reduce obesity. To determine the prevalence of substance use in adolescents with eating disorders, the results of a data set of Ontario high school students were compared. One hundred and one female adolescents who met the *Diagnostic and Statistical Manual of Mental Disorders*, 4th edition's criteria for an eating disorder were followed up in a tertiary care pediatric treatment center. They were asked to participate in a cross-sectional study using a self-administered questionnaire assessing substance use and investigating reasons for use and nonuse; 95 agreed to participate and 77 completed the questionnaire (mean age, 15.2 years). The patients were divided into two groups: 63 with restrictive symptoms only, 17 with purging symptoms. The rates of drug use between subjects and their comparison groups were compared by *Z*-scores, with the level of significance set at 0.05. During the preceding year, restrictors used significantly less tobacco, alcohol, and cannabis than grade- and sex-matched comparison populations, and purgers used these substances at rates similar to those of comparison subjects. Other drugs seen frequently in the purgers included hallucinogens, tranquilizers, stimulants, LSD, phencyclidine, cocaine, and ecstasy. Both groups used caffeine and laxatives, but few used diet pills. Restrictors said they did not use substances because they were bad for their health, tasted unpleasant, were contrary to their beliefs, and were too expensive. Purgers generally used substances to relax, relieve anger, avoid eating, and "get away" from problems. Female adolescents with eating disorders who have restrictive symptoms use substances less frequently than the general adolescent population but do not abstain from their use. Those with purging symptoms use substances with a similar frequency to that found in the general adolescent population.

Gene expression effect

Cannabinoids can cross the placental barrier and be secreted in the maternal milk. Through this way, cannabinoids affect the ontogeny of various neurotransmitter systems leading to changes in different behavioral patterns. Dopamine and endogenous opioids are among the neurotransmitters that result more affected by perinatal cannabinoid exposure, which, when animals mature, produce changes in motor activity, drug-seeking behavior, nociception, and other processes. These disturbances are likely originated by the capability of cannabinoids to influence the expression of key genes for both neurotransmitters, in particular, the enzyme tyrosine hydroxylase and the opioid precursor proenkephalin. Cannabinoids seem to be able to influence the expression of genes encoding for neuroglia cell adhesion molecules, which supports a potential influence of cannabinoids on the processes of cell proliferation, neuronal migration or axonal elongation in which these proteins are involved. CB1 receptors, which represent the major targets for the action of cannabinoids, are abundantly expressed in certain brain regions, such as the subventricular areas, which have been involved in these processes during brain development. Cannabinoids might also be involved in the apoptotic death that occurs during brain development, possibly by influencing the expression of Bcl-2/Bax system. CB1 receptors are transiently expressed during brain development in different group of neurons which do not contain these receptors in the adult brain.

Glaucoma effect

Nine patients with glaucoma unresponsive to treatment were treated with orally administered Δ-9-THC capsules or inhaled cannabis in addition to their existing therapeutic regimen. An initial decrease in intraocular pressure was observed in all patients, and the investigator's therapeutic goal was met in four of the nine patients. The decreases in intraocular pressure were not sustained, and the patients elected to discontinue treatment within 1–9 months for various reasons.

Gliomatous effect

Gliomas, in particular glioblastoma multiform or grade IV astrocytoma, are the most frequent class of malignant primary brain tumors and one of the most aggressive forms of cancer. Cannabinoids and their derivatives slowed the growth of different types of tumors, including gliomas, in laboratory animals. Cannabinoids induced apoptosis of glioma cells in culture vía sustained ceramide accumulation, extracellular signal-regulated kinase activation and Akt inhibition. Cannabinoid treatment inhibited angiogenesis of gliomas in vivo. Cannabinoids killed glioma cells selectively and could protect nontransformed glial cells from death.

Gynecomastic effect

A retrospective analysis was carried out on 175 men over the age of 16 years who were presented with breast enlargement and/or "lumps" during a 7-year period to a single surgeon. The patients had complete biochemical assessment (liver function tests, γ-glutamyl transferase, prolactin, α-fetoprotein, and β-human chorionic gonadotropin), and mammography and/or ultrasound with fine-needle biopsy if indicated. Thirty-nine of the patients had bilateral true gynecomastia and 88 had unilateral gynecomastia (53% left). Carcinoma of the breast was diagnosed in eight, pseudo-gynecomastia in 18, 13 had physiological pubertal changes only, and 9 had other diagnoses. Adverse drug reactions were possibly implicated in the etiology of 47 patients, alcohol in seven patients, cannabis in one patient, testicular malignancy in four patients, and hepatocellular carcinoma in one patient. Five patients were found to have hyperprolactinemia. Twenty-four percent of patients were reassured without intervention; 18% failed to attend follow-up.

Hepatitis C risk factor

The study of a dually diagnosed population estimated the prevalence of hepatitis C virus (HCV) to be 29.7% or 16 times higher than that in the general population. A high correlation was found between the use of tobacco and HCV infection. This appears to be beyond the risk factor conveyed by intravenous drug use. Of the patients whose primary diagnoses were cocaine, opiate, amphetamine, or polysubstance dependence (drugs often used intravenously), 42% of the tobacco users were HCV-positive, whereas only 20% of the nontobacco using patients with similar primary diagnoses were HCV-positive. The association of tobacco use with HCV was found to be strong for females with alcohol, sedative/hypnotic, inhalant, or cannabis dependence, as none of the 17 nontobacco using female patients with these diagnoses were HCV-positive, whereas 14 of the 45 (31%) tobacco-using females with these diagnoses did test positive for HCV.

HIV involvement

The prevalence, predictors, and patterns of cannabis use—specifically medicinal cannabis use among patients with HIV—were examined. Any cannabis use in the year prior to interview and self- defined medicinal use were evaluated. A cross-sectional multicenter survey and retrospective chart review were conducted to evaluate overall drug utilization in HIV, including cannabis use. HIV-positive adults were identified through the HIV Ontario Observational Database; 104 consenting patients were interviewed. Forty-three percent of the patients reported cannabis use, whereas 29% reported medicinal use. Reasons for use were similar by gender although a significantly higher number of women used cannabis for pain management. The most commonly reported reason for medicinal cannabis use was appetite stimulation/weight gain. Male gender and history of intravenous drug use were predictive of any cannabis use. Age, gender, HIV clinical status, antiretroviral use, and history of intravenous drug use were not significant predictors of medicinal cannabis use. Despite the frequency of medicinal use, minimal changes in the pattern of cannabis use on HIV diagnosis were reported with 80% of current medicinal users also indicating recreational consumption. HIV patients (n = 252) were recruited via consecutive sampling in public health care clinics. Structured interviews assessed patterns of recent cannabis use, including its perceived benefit for symptom relief. Associations between cannabis use and demographic and clinical variables were examined using univariate and multivariate regression analyses. Overall prevalence of smoked cannabis in the previous month was 23%. Reported benefits included relief of anxiety and/or depression (57%), improved appetite (53%), increased pleasure (33%), and relief of pain (28%). Recent use of cannabis was positively associated with severe nausea (OR = 4, $p = 0.004$) and recent use of alcohol (OR = 7.5, $p < 0.001$) and negatively associated with being Latino (OR = 0.07, $p < 0.001$). No associations between cannabis use and pain symptoms were observed. No safety problems specific to HIV or protease inhibitors were found in a study in which volunteers stayed in a research hospital 24 hours a day and were randomly assigned to either smoke cannabis, take oral THC, or take an oral placebo. Cannabis and THC use was associated with weight gain.

Hyperglycemic activity

Ethanol (95%) extract of the leaf, administered intravenously to rats at a dose of 300 mg/kg, produced an increase of 40 mg percentage 2 hours postinjection and a corresponding decrease in liver glycogen. Ethanol (95%) extract of the dried leaf, administered by gastric intubation to rabbits, produced an increase followed by a gradual decrease in blood sugar levels. The dried leaves, smoked by human adults, produced elevated glucose levels in two out of four subjects and no impairment of insulin release or changes in growth hormone levels.

Hypoglycemic activity

Ethanol (95%) extract of the dried leaf, administered by gastric intubation to rabbits, produced an increase, followed by a gradual decrease, in blood sugar levels. Extract of the dried leaf, administered subcutaneously to rabbits at a dose of 0.5 mL/kg (approx 0.6 mg THC) for 9 weeks, further enhanced hypoglycemia induced by insulin. No hypoglycemic effect was seen in normal animals. Hot water extract of the resin, administered by gastric intubation to dogs at a dose of 20 g of air-dried resin/animal, produced weak activity. The dried leaf, smoked by adults at a dose of 2 g/person, was inactive.

Hypotensive activity

Ethanol (50%) extract of the entire plant, administered intravenously to dogs at a dose of 50 mg/kg, was active. Ethanol (95%) and water extracts of the dried aerial parts, administered intravenously to cats, were inactive. The ethanol extract stimulated respiration, and the water extract had no effect.

Ilicit drug in plasmapheresis donors

Seventy-five US plasma units from 10 different states in the United States and 75 German plasma units that had been analyzed principally for their protein composition were screened for drugs. Determinations were made, using automated immunoassays, of the presence of cannabis, cocaine, amphetamine, methamphetamine, MDMA, methyl enedioxy-ethylamphetamine (MDE), and opiates. Positive results were confirmed by gas chromatography–mass spectrometry. Eleven US plasma units were found to be positive for cocaine (14.6%), whereas all German samples were cocaine-negative ($p = 0.0007$). Fifteen US plasma units (20%) and one German unit (1,3%) were confirmed as positive for cannabis ($p = 0.0003$). Three out of 75 US plasma units were positive for both cannabis and cocaine. In none of the 150 samples were amphetamine, methamphetamine, MDMA, MDE, or opiates detected.

Immunomodulatory effect

The smoking of cannabis showed a significant local immuno-suppression of the bactericidal activity of human alveolar macrophages. In animal studies, cannabinoids were identified as potent modulators of cytokine production, causing a shift from T-helper-1 (Th1) to Th2 cytokines. In consequence, a compromised cellular immunity was observed in these animals, resulting in enhanced tumor growth and reduced immunity to viral infections. In vitro, immunosuppressive effects were shown in all immune cells, but only at high micro-molar cannabinoid concentrations not reached under normal clinical conditions. .

In conclusion, there was no evidence that cannabinoids induce a serious, relevant immunosuppression in humans, with the exception of cannabis smoking, which may affect local bronchoalveolar immunity. The immune function in 16 MS patients treated with oral cannabinoids was measured. A modest increase of tumor necrosis factor (TNF)-α in lipopolysaccharide-stimulated whole blood was found during cannabis plant-extract treatment ($p = 0.037$), with no change in other cytokines. In the subgroup of patients with high adverse event scores, an increase in plasma IL-12p40 was found ($p = 0.002$). The results indicate pro-inflammatory disease-modifying potential of cannabinoids in MS. THC and their metabolites inhibited production of IL-1 and γ-interferon, decreased a 33% of the lymphocytes activity and inhibited 66% of the lymphocytes adenylcyclase activity.

The consumption of cannabis decreased immunological competence of macrophages, and alternated their essential role of trophicity of the central nervous system. Inhibiting actions of cannabinoids on the cyclo-oxygenase, promoted production of arachidonic acid degradation products. This compound mimics the action of histamine, induced a raise of the vascular permeability and bronchospasm, and contributed at delayed reaction of anaphylaxia.

Infant mortality

For a period of 11 months, 2964 infants were enrolled and screened at birth for exposure to cocaine, opiate, or cannabinoid by meconium analysis. At birth, 44% of the infants tested positive for drugs, 30.5% positive for cocaine, 20.2% for opiate, and 11.4% for cannabinoids. Compared with the drug-negative group, a significantly higher percentage ($p < 0.05$) of the drug-positive infants had lower weight and smaller head circumference and length at birth and a higher percent of their mothers were single, multigravid, multiparous, and had little to no prenatal care. Within the first 2 years of life, 44 infants died: 26 were drug-negative (15.7 deaths per 1000 live births) and 18 were drug-positive (13.7 deaths per 1000 live births). The mortality rate among cocaine, opiate, or cannabinoid-positive infants were 17.7, 18.4, and 8.9 per 1000 live births, respectively. Among infants with birth-weight of 2500 g or less, infants who were positive for both cocaine and morphine had a higher mortality rate (OR = 5.9, CI = 1.4–24) than drug-negative infants. Eleven infants died from the sudden infant death syndrome (SIDS); 58% were positive for drugs, predominantly cocaine. The odds ratio for SIDS among drug-positive infants was 1.5 (CI = 0.46–5.01) and 1.9 (CI = 0.58–6.2) among cocaine-positive infants.

Infant neurobehavioral effect

The subjects and controls in this study were full- term infants of appropriate gestational age with no medical problems. At 1–2 days of age, 20 infants exposed to cocaine, alcohol, cannabis, and cigarettes, 17 infants exposed to alcohol and/or cannabis and cigarettes, and 20 drug-free infants were evaluated by using the Neonatal Intensive Care Unit Network Neurobehavioral Scale. Cocaine-exposed (CE) infants showed increased tone and motor activity, more jerky movements, startles, tremors, back arching, and signs of central nervous system and visual stress than unexposed infants. They also showed poorer visual and auditory following. There were no differences in how the examination was administered to CE and nonexposed infants. Reduced birth-weight and length were also observed in CE infants. Differences attributable to CE infants were related to muscle tone and motor performance, following during orientation, and signs of stress. CE infants were not more difficult to test, nor did they require an alteration in the examination. Both neurobehavioral patterns of excitability and lethargy were observed. The findings may have been a result of the synergistic effects of cocaine with alcohol and cannabis.

Inflammatory effect

A case of a 17-year-old male regular cannabis user who developed a large swollen uvula and partial upper airway obstruction after smoking cannabis was evaluated. Symptoms resolved with the administration of corticosteroids and antihistamines. A healthy 17-year-old man who inhaled cannabis prior to general anesthesia is described. In the recovery room, after an uneventful general anesthetic, acute uvular edema resulted in postoperative airway obstruction and admission to the hospital. The uvular edema was treated successfully with dexamethasone.

Information-processing effect

Information processes are thought to represent the basic building blocks of higher order cognitive processes. The inspection time task was used to investigate the effects of acute and subacute cannabis use on information processing in 22 heavy users compared with 22 nonusers. The findings indicated that users in the subacute state display significantly slowed information-processing speeds (longer inspection times) compared with controls. This deficit appeared to be normalized while users were in the acute state. These results may be explained as a withdrawal effect, but may also be owing to tolerance development because of long-term cannabis use.

Insecticidal activity

Leaf extract, administered to larvae of *Chironomus samoensis*, produced paralysis leading to death. The extract brought a drastic change in the morphology of sensilla trichoidea, the general body cuticle, and a significant reduction in the concentration of magnesium and iron, whereas manganese showed only slight average increase. Because the sensilla trichoidea has nerve connections, it was assumed that the toxic principle of the leaf extract has affected the central nervous system.

Intestinal motility activity

Rat intestinal epithelia mounted in an Ussing chamber attached with voltage/current clamp were used for measuring changes of the short-circuit current across the epithelia. The intestinal epithelia were activated with current raised by serosal administration of forskolin 5 μM. Ethanol extracts of cannabis augmented the current additively when each was added after forskolin. In subsequent experiments, ouabain, and bumetanide were added prior to ethanol extract of cannabis to determine their effect on Na^+ and Cl^- movement. The results suggested that the extract may affect the Cl^- movement more directly than Na^+ movement in the intestinal epithelial cells.

Intraocular pressure reduction

Polysaccharide fraction of the dried entire plant, administered intravenously to rabbits at a dose of 1 μg/animal was active. Water extract of the dried aerial parts, administered intravenously to rabbits at a dose of 250 μg/animal, was active. A dose of 5 μg/animal was inactive on Rhesus monkeys and active on rabbits. A dose of 10 mg/animal, administered *per rectum* to Rhesus monkeys and rabbits, was inactive.

IQ effect

Cannabis use for 70 individuals aged 17–20 years was determined through self-reporting and urinalysis. IQ scores were calculated by subtracting each person's IQ score at 9–12 years (before initiation of drug use) from his or her score at 17–20 years. The difference in IQ scores of current heavy users (at least five joints per week), current light users (less than five joints per week), former users (who had not smoked regularly for at least 3 months), and nonusers (who never smoked more than once per week and no smoking in the past 2 weeks) was compared. Current cannabis use was significantly correlated ($p < 0.05$) in a dose-related fashion with a decline in IQ over the ages studied. The comparison of the IQ difference scores showed an average decrease of 4.1 points in current heavy users ($p < 0.05$) compared with gains in IQ points for light current users (5.8), former users (3.5), and nonusers (2.6).

Lactate inhibition

The dried leaf, smoked by adults at a dose of 2 g/person, decreased blood lactic acid.

Leutinizing hormone-release inhibition

The dried aerial part, smoked by menopausal women at a dose of 1 g/person, was inactive. When administered to normal and castrated male rats, at a dose of 75 mg/kg, was active.

Lower limb occlusive arteriopathy

Seventy-three patients (60 males and 13 females less than 50 years of age) were divided into four groups: Buerger's disease (thromboangiitis obliterans [TAO]), atheromatous juvenile peripheral obstructive arterial diseases (POAD), autoimmune POAD, and arteriopathy of undetermined origin. The first symptoms occurred at 38 ± 8 years of age. Fourteen patients (20%) had TAO, 51 (70%) atheromatous POAD, 4 (5%) POAD with systemic or autoimmune disease, and 4 (5%) undetermined POAD. Age of onset was earlier in TAO (35 ± 8 vs 40 ± 8 years, $p = 0.046$), smoking was greater in the atheroma group (33 ± 16 vs 24 ± 14 pack/years, $p = 0.033$). Fifty-three patients with POAD had dyslipidemia and 26% had hypertension. Regular cannabis intake was more frequent in the TAO group (21% vs 8%). At the time of medical care,

Fontaine's stage was more frequently stage II in atheroma patients (57% vs 14%) and stage IV in TAO patients (86% vs 35%). TAO was diagnosed in 43% cannabis users and in 19% nonusers. Results indicated that the main etiology of juvenile POAD is atheroma, followed by TAO. Cannabis users accounted for at least 10% of these patients. They were characterized by lower tobacco intake, more distal lesions, more frequent involvement of the upper limbs. They presented more frequently as TAO. A case of a 30-year-old woman who smoked cannabis and developed intermittent claudication of the lower limbs was reported. Results indicated that cannabis could be involved not only in the pathogenesis of juvenile obstructive arteriopathy, but also in the development of atheromatous lesions.

Lung function

A group of over 900 young adults derived from a birth cohort of 1037 subjects were studied at age 18, 21, and 26 years. Cannabis and tobacco smoking were documented at each age using a standardized interview. Lung function, as measured by the FEV1–vital capacity (VC) ratio, was obtained by simple spirometry. A fixed effects regression model was used to analyze the data and to account for confounding factors. When the sample was stratified for cumulative use, there was evidence of a linear relationship between cannabis use and FEV1–VC ($p < 0.05$). In the absence of adjusting for other variables, increasing cannabis use over time was associated with a decline in FEV1–VC with time; the mean FEV1–VC among subjects using cannabis on 900 or more occasions was 7.2, 2.6 and 5% less than nonusers at ages 18, 21, and 26, respectively.

After controlling for potential confounding factors (age, tobacco smoking, and weight) the negative effect of cumulative cannabis use on mean FEV1–VC was only marginally significant ($p < 0.09$). Age ($p < 0.001$), cigarette smoking ($p < 0.05$), and weight ($p < 0.001$) were all significant predictors of FEV1–VC. Cannabis use and daily cigarette smoking acted additively to influence FEV1–VC. Results indicated that longitudinal observations over 8 years in young adults revealed a dose- dependent relationship between cumulative cannabis consumption and decline in FEV1–VC. When confounders were accounted for the effect was reduced and was only marginally significant, but given the limited time frame over which observations were made, the trend suggests that continued cannabis smoking has the potential to result in clinically important impairment of lung function.

Memory impairment

The effects of combined exposure to ethanol and Δ-9-THC in a memory task was investigated in rats. Ethanol, voluntarily ingested in alcohol-preferring rats, and THC, given by intraperitoneal injection, had a synergic action to impair object recognition when a 15-minute interval was adopted between the sample phase and the choice phase of the test. Ingestion of ethanol, or 2 or 5 mg/kg of THC were not able to modify object recognition in these experimental conditions. When voluntary ethanol ingestion was combined with administration of these doses of THC, object recognition was markedly impaired. THC impaired object recognition only at the dose of 10 mg/kg, when its administration was not combined with that of ethanol. The selective cannabinoid CB1 receptor antagonist SR 141716A (*N*-(piperidin-1-yl)-5-(4-chlorophenyl)-1(2,4-dichloro-phenyl)-4-methyl-1 H-pyrazole carboxamide HCl) at the dose of 1 mg/kg reversed the amnesic effect of 10 mg/kg of THC. This indicated that the effect is mediated by the receptor subtype. The synergism of ethanol and THC was not detected when an intertrial interval of 1 minute was adopted.

Memory improvement

Extract from fructus cannabis (EFC), administered intragastrically to mice with drug-induced dysmnesia at doses of 0.2, 0.4, and 0.8 g/kg, for 7 days, prolonged the latency and decreased the number of errors in the step-down test, and enhanced the spatial resolution of amnesic mice in water maze test. EFC at the dose of 0.2 g/kg overcame amnesia of three stages of memory process. EFC activated calcineurin activity at a concentration range of 0.01–100 g/L. The maximal value of EFC on calcineurin activity (35% ± 5 %) appeared at a concentration of 10 g/LCS320. EFC with activation of calcineurin, extracted from Chinese traditional medicine, was used to determine the effects on memory and immunity in mice. In the stepdown-type passive avoidance test, the plant extract (0.2 g/kg) significantly improved amnesia induced by drugs, and greatly enhanced the ability of cell-mediated type hypersensitivity and nonspecific immune responses in normal mice.

Mitochondrial Function Disruption

Δ-9-THC in the pulmonary transformed cell line A549 produced a rapid and extensive depletion of cellular energy stores. Adenosine 5'-triphosphatase levels declined dose dependently with an IC_{50} of 7.5 μg/mL of THC after 24 hours of exposure. Cell death was observed only at concentrations greater than 10 μg/mL. Studies using JC-1, a fluorescent probe for mitochondrial membrane potential, revealed diminished mitochondrial function at THC concentrations as low as 0.5 μg/mL. At concentrations of 2.5 and 10 μg/mL of THC, a decrease in mitochondrial membrane potential was observed 1 hour after THC exposure. Mitochondrial function remained diminished for at least 30 hours after THC exposure. Flow cytometry studies on cells exposed to particulate smoke extracts indicated that JC-1 red fluorescence was fivefold lower in cells exposed to cannabis smoke extract compared with tobacco smoke-exposed cells. Comparison with a variety of mitochondrial inhibitors demonstrated that THC produced effects similar to that of carbonyl cyanide *p*-trifluoromethoxyphenylhydrazone, suggesting uncoupling of electron transport. Loss of red JC-1 fluorescence by THC was suppressed by cyclosporin A, suggesting mediation by the mitochondrial permeability transition pore. This disruption of mitochondrial function was sustained for at least 24 hours after removal of THC by extensive washing.

Mitogenic effect

The resin was inactive on the human and rat white blood cells.

Molluscicidal activity

Ethanol (95%) and water extracts of the dried flowering tops, at a concentration of 1000 ppm, produced weak activity on *Biomphalaria straminea* and *Biomphalaria glabrata*. Water saturated with essential oil of the aerial parts, at a concentration of 1:2, produced weak activity on *Biomphalaria glabrata*.

Motor function

Nine cannabis smokers and 16 controls were studied to determine the attentional areas related to motor function, and primary and supplementary motor cortices. Echo planar images and high-resolution molecular resonance images were acquired. The challenge paradigm included left and right finger sequencing. Group differences in cerebral activation were examined for Brodmann areas (BA) 4, 6, 24, and 32 using region of interests analyses in statistical parametric mapping. Cannabis users, tested within 4–36 hours of discontinuation, exhibited significantly less activation than controls in BA 24 and 32 bilaterally during right- and left-sided sequencing and for BA 6 in all tasks except for left-sided sequencing in the left hemisphere. There were no statistically significant differences for BA 4. None of these regional activations correlated with urinary cannabis concentration and verbal IQ for smokers. The results suggested that recently abstinent chronic cannabis smokers produce reduced activation in motor cortical areas in response to finger sequencing compared with controls.

Multiple sclerosis

One hundred fifty-seven drug-naïve, first-episode schizophrenic patients were examined. A significantly elevated brain-derived neurotrophic factor (BDNF) serum concentrations in patients with chronic cannabis abuse ($n = 35$, $p < 0.001$) or multiple substance abuse ($n = 20$, $p < 0.001$) prior to disease onset were found. Drug-naive schizophrenic patients without cannabis consumption showed similar results to normal controls and cannabis controls without schizophrenia. Elevated BDNF serum levels were not related to schizophrenia and/or substance abuse itself but may reflect a cannabis-related idiosyncratic damage of the schizophrenic brain. Disease onset was 5.2 years earlier in the cannabis-consuming group ($p = 0.0111$). A cannabis-based medicinal extract (CBME) was administered to 160 patients with multiple sclerosis experiencing significant problems from at least one of the following: spasticity, spasms, bladder problems, tremor, or pain. The interventions were oromucosal sprays of matched placebo, or whole plant CBME containing equal amounts of Δ-9-THC and CBD at a dose of 2.5–120 mg of each daily, in divided doses. The primary outcome measure was a Visual Analogue Scale (VAS) score for each patient's most troublesome symptom. Additional measures included VAS scores of other symptoms, and measures of disability, cognition, mood, sleep and fatigue. Following CBME the primary symptom score reduced from mean 74.36 (11.1) to 48.89 (22.0) following CBME and from 74.31 (12.5) to 54.79 (26.3) following placebo. Spasticity VAS scores were significantly reduced by CBME (Sativex) in comparison with placebo ($p = 0.001$). There were no significant adverse effects on cognition or mood and intoxication was generally mild. A SCE with pure Δ-9-THC, at matched concentrations of Δ-9-THC, and a Δ-9-THC-free extract (Δ-9-THC-free SCE) in a mouse model of MS, were examined. Although SCE inhibited spasticity in the mouse model of MS to a comparable level, it caused a more rapid onset of muscle relaxation and a reduction in the time to maximum effect compared with Δ-9-THC alone. The Δ-9-THC-free extract or CBD caused no inhibition of spasticity. In an experimental allergic encephalomyelitis (EAE), an animal model of MS, it was demonstrated that the cannabinoid system is neuroprotective during EAE. Mice, deficient in the cannabinoid receptor CB1, tolerated inflammatory and excitotoxic insults poorly, and developed substantial neurodegeneration following immune attack in EAE. Exogenous CB1 agonists can provide significant neuroprotection from the consequences of inflammatory central nervous system disease in an experimental allergic uveitis model.

Mutagenic activity

Petroleum ether extract of the aerial parts, in the ration of *Drosophila* at concentrations of 0.5, 1, and 5% of the diet, was active. Petroleum ether extract of the dried leaf, administered by gastric intubation to male mice at a dose of 50 mg/kg, was active. Water and methanol extracts of the seed, on agar plate at a concentration of 100 mg/mL, were inactive on *Bacillus subtilis* H-17 (Rec+) and *Salmonella typhimurium* TA100 and TA98. Metabolic activation had no effect on the results.

Myocardial infarction

A young man who suffered a myocardial infarction after taking Viagra in combination with cannabis was investigated. Viagra is metabolized predominantly by the CYP450 3A4 hepatic microsomal isoenzyme. Cannabis is a known inhibitor of CYP450 3A4 isoenzyme. The effect of the Viagra was thus potentiated by the effect of cannabis.

Natural-killer cells effect

Leukemia susceptible BALB/c and resistant C57BL/6 mice were infected with Friend leukemia virus complex and its helper component Rowson-Parr virus. At different time points, their natural-killer cells were separated from spleens and treated with 0–10 μg/mL of THC, subsequently mixed with Yac-1 target cells for 4 and 18 hours. The natural-killer cell activity in both mouse strains infected by either virus complex or helper virus weakened on days 2–4 postinfection, normalized by day 8 and enhanced on days 11–14. Natural-killer cell activity on the effect of low concentration (1–2.5 μg/mL) of THC slightly increased in BALB/c, was unaffected in C57BL/6, especially in the 18 hour assays. In the combined effects of cannabis and retrovirus, damages by cannabis dominated over those of retroviruses. Inhibition or reactive enhancement of natural-killer cell activity on the effect of viruses were similar to those of infected but cannabis-free counterparts, but on the level of uninfected cells treated with cannabis. The effects of cannabis and retrovirus were additive resulting in anergy of natural-killer cells.

Neonatal abstinence syndrome

The relationship of maternal drug abuse to symptoms, the effectiveness of pharmacological agents in controlling symptoms, and the length of in-patient stay were investigated in infants with neonatal abstinence syndrome. Pharmacological treatment was oral morphine sulphate (0.2 mg four to six times hourly), phenobarbitone (3–7 mg/kg/day), or combination of the two were administered to infants with a serial Finnegan score greater than 8. The average maternal age was 24.6 years, (18–34 years). Drug use volunteered by the mothers was methadone alone in 6 cases, methadone and benzodiazepines in 14, methadone and heroin and benzodiazepines in 7, methadone and heroin in 10, heroin alone in 2, and other multiple drug use including oral morphine sulphate, dothiepin, and cannabis in 4. Average gestational age was 40.3 (35–42 weeks). The average birth-weight was 2.81 kg (1.89–3.91 kg). Time-to-onset of withdrawal symptoms was 2.8 (1–13) days. The duration of pharmacological treatment (oral morphine sulphate and/or phenobarbitone) was 21.8 (1–62) days. The total hospital stay for the 43 infants was 1011 days.

Neuroendocrine abnormalities

Prolactin response to D-fenfluramine was assessed in abstinent ecstasy (MDMA) users with concomitant use of cannabis only (13 males, 11 females) and in two control groups: healthy nonusers (13 females) and exclusive cannabis users. Prolactin response to D-fenfluramine was slightly blunted in female ecstasy users. Both male user samples exhibited a weak prolactin response to D-fenfluramine, but this was weaker in the group of cannabis users. Baseline prolactin and prolactin response to D-fenfluramine were associated with the extent of previous cannabis use. The results indicated that the endocrinological abnormalities of ecstasy users might be closely related to their coincident cannabis use.

Neurogenic symptoms alleviation

Whole-plant extracts of Δ-9-THC, CBD, 1:1 CBD: THC, or placebo were self-administered by sublingual spray to 24 patients with MS (n = 18), spinal cord injury (n = 4), brachial plexus damage

($n = 1$), and limb amputation owing to neurofibromatosis ($n = 1$), at doses determined by titration against symptom relief or unwanted effects within the range of 2.5–120 mg/24 hours for 2 weeks. The patients recorded symptoms, well-being, and intoxication scores on a daily basis using visual analog scales. At the end of each two-week period an observer rated severity and frequency of symptoms on numerical rating scales, administered standard measures of disability (Barthel Index), mood, cognition, and recorded adverse events. Pain relief associated with both THC and CBD was significantly superior to placebo. Impaired bladder control, muscle spasms, and spasticity were improved by cannabis medicinal extract (CME) in some patients with these symptoms. Three patients had transient hypotension and intoxication with rapid initial dosing of THC-containing CME. The results indicated that cannabis could improve neurogenic symptoms unresponsive to standard treatments. Unwanted effects were predictable and generally well tolerated.

Neuropathic pain relief

Forty-eight patients with at least one avulsed root and baseline pain score of four or more on an 11-point ordinate scale participated in a randomized, double-blind, placebo-controlled, three-period crossover study. The patients had intractable symptoms regardless of current analgesic therapy. They entered a baseline period of 2 weeks, followed by three, 2-week treatment periods; during each period they received one of three oromucosal spray preparations. These were placebo and two whole plant extracts of *C. sativa* L.: GW-1000-02 (Sativex), containing Δ-9-THC: CBD in an approx 1:1 ratio and GW-2000-02, containing primarily THC. The primary outcome measure was the mean pain severity score during the last 7 days of treatment. Secondary outcome measures included pain related quality of life assessments. The primary outcome measure failed to fall by the two points defined in our hypothesis. Both this measure and measures of sleep showed statistically significant improvements. The study medications were well tolerated with the majority of adverse events, including intoxication type mild to moderate in severity and resolving spontaneous reactions.

Neuroprotective effect

The effect of cannabidiol on β-amyloid peptide-induced toxicity in cultured rat pheocromocytoma PC12 cells was investigated. Following exposure of cells to β-amyloid peptide (1 μg/mL), a marked reduction in cell survival was observed. This effect was associated with increased reactive oxygen species production and lipid peroxidation, and caspase 3 (a key enzyme in the apoptosis cell-signalling cascade) appearance, DNA fragmentation, and increased intracellular calcium. Treatment of the cells with CBD (10^{-7}–10^{-4} mol) prior to β-amyloid peptide exposure, significantly elevated cell survival, whereas it decreased reactive oxygen species production, lipid peroxidation, caspase 3 levels, DNA fragmentation, and intracellular calcium. CBD and other cannabinoids were examined as neuroprotectants in rat cortical neuron cultures exposed to toxic levels of glutamate. The psychotropic cannabinoid receptor agonist Δ-9-THC and cannabidiol, reduced *N*-methyl-D-aspartate, α-amino-3 -hydroxy-5 -methyl-4-isoxazole propionic acid and kainate receptor mediated neurotoxicities. Neuroprotection was not affected by cannabinoid receptor antagonist, indicating a (cannabinoid) receptor-independent mechanism of action. CBD demonstrated a reduction in hydroperoxide toxicity in neurons. In this trial of the abilities of various antioxidants to prevent glutamate toxicity, cannabidiol was superior to both α-tocopherol and ascorbate in protective capacity.

Neuropsychological effect

Cerebral blood flow was measured in 12 long-term cannabis users shortly after cessation of cannabis use (mean 1.6 days). The findings showed significantly lower mean hemispheric blood flow values and significantly lower frontal values in the cannabis subjects compared with normal controls. The results indicated that the functional level of the frontal lobes was affected by long-term cannabis use.

Neurotransmission inhibition

The BLA or the medial prefrontal cortex (PFC) stimulation in urethane-anesthetized rats induced generation of action potentials in the Nac neurons. This excitatory effect was strongly inhibited by the synthetic cannabinoid agonists WIN (0.062–0.25 mg/kg, iv [intravenously]) and HU-210 (0.125–0.25 mg/kg, iv), or Δ-9-THC (1 mg/kg, iv). D1 or D2 dopamine receptor antagonists (SCH23390 0.5–1 mg/kg, sulpiride 5–10 mg/kg, iv) or the opioid antagonist naloxone (1 mg/kg, iv) were not able to reverse the action of cannabinoids. The selective CB1 receptor antagonist/reverse agonist SR141716A (0.5 mg/kg, iv) fully suppressed the action of cannabinoid agonists, whereas *per se* had no significant effect.

Nicotine and D-9-THC Interaction

Δ-9-THC administration to mice significantly decreased the incidence of several nicotine withdrawal signs precipitated by mecamylamine or naloxone, such as wet-dog-shakes, paw tremor, and scratches. In both experimental conditions, the global withdrawal score was significantly attenuated by Δ-9-THC administration. The effect of Δ-9-THC was not to the result possible adaptive changes induced by chronic nicotine on CB1 cannabinoid receptors. The density and functional activity of these receptors were not modified by chronic nicotine administration in the different brain structures investigated. The consequences of Δ-9-THC administration on *c-Fos* expression in several brain structures after chronic nicotine administration and withdrawal were examined. *c-Fos* was decreased in the caudate putamen and the dentate gyrus after mecamylamine precipitated nicotine withdrawal. Δ-9-THC administration did not modify *c-Fos* expression under these experimental conditions. Δ-9-THC also reversed conditioned place aversion associated to naloxone precipitated nicotine withdrawal. The results indicated that Δ-9-THC administration attenuated somatic signs of nicotine withdrawal and this effect was not associated with compensatory changes on CB1 cannabinoid receptors during chronic nicotine administration. Δ-9-THC also ameliorated the aversive motivational consequences of nicotine withdrawal.

Night vision improvement

In a double-blind study, graduated THC administration at doses of 0–20 mg (as Marinol) on measures of dark adaptometry and scotopic sensitivity was evaluated. Field studies of night vision were performed among Jamaican and Moroccan fishermen, and mountain dwellers with the LKC Technologies Scotopic Sensitivity Tester-1. Improvements in night vision measures were noted after THC or cannabis. The effect was dose-dependent and cannabinoid-mediated at the retinal level.

Nocturnal sleep effect

Eight healthy volunteers (four males, four females; aged 21–34 years) were taking placebo, 15 mg Δ-9-THC, 5 mg THC combined with 5 mg CBD, and 15 mg THC combined with 15 mg CBD. These were formulated in 50:50 ethanol to propylene glycol and administered using an oromucosal spray during a 30-minute period from 10 PM. Electroencephalogram was recorded during the sleep period (11 PM to 7 AM). Performance, sleep latency, and subjective assessments of sleepiness and mood were measured from 8:30 AM (10 hours after drug administration). There were no effects of 15 mg THC on nocturnal sleep. With the concomitant administration of the drugs (5 mg THC and 5 mg CBD to 15 mg THC and 15 mg CBD), there was a decrease in stage 3 sleep, and with the higher dose combination, wakefulness was increased. The next day, with a 15-mg THC dose, memory was impaired, sleep latency was reduced, and the subjects reported increased sleepiness and changes in mood. With the lower dose combination, reaction time was faster on the digit recall task, and with the higher dose combination, subjects reported increased sleepiness and changes in mood. Fifteen milligrams of THC appeared to be sedative, and 15 mg CBD appeared to have alerting properties as it increased waking activity during sleep and counteracted the residual sedative activity of the 15 mg THC.

Occipital stroke

A right occipital ischemic stroke occurred in a 37-year-old Albanese man with a previously uneventful medical history, 15 minutes after smoking a cigarette with approximately 250 mg of cannabis. Clinical manifestations of the stroke were left-sided hemiparesis, hemihypesthesia and blurred vision, which vanished spontaneously and almost completely after 3 days.

The patient has been smoking cannabis regularly from the age of 27, with a frequency of two to three cigarettes/cannabis per week during the 6 months that preceded his stroke. Except for cigarette smoking and slight dyslipidemia, classical risk factors for stroke/embolism were absent. The family history for cerebrovascular events, blood pressure, clotting tests, examinations for thrombophilia, vasculitis, extracranial and intracranial arteries, and cardiac investigations were normal or respectively negative; the stroke was attributed to the chronic cannabis consumption.

Oral cancer

A study of 116 patients aged 45 years and younger, diagnosed with squamous cell carcinoma of the mouth was conducted. Two hundred and seven controls who had never had cancer, matched for age, sex, and area of residence, were recruited. The self-completed questionnaire contained items about exposure to the following risk factors: tobacco products, cannabis, alcohol, and diet. Conditional logistic analyses were conducted adjusting for social class, ethnicity, tobacco, and alcohol habits. All tests for statistical significance were two-sided. The majority of oral cancer patients reported exposure to the major risk factors of tobacco and alcohol even at the younger age. The estimated risks associated with tobacco or alcohol were low among both males and females. Only smoking for 21 years or more produced significantly elevated odds ratios (OR = 2.1; 95% CI: 1.1–4). Exposure associated with other major risk factors did not produce significant risks in this sample. Long-term consumption of fresh fruits and vegetables in the diet appeared to be protective for both males and females.

Oral cytological effect

The effects of cannabis, methaqualone, or tobacco smoking on the epithelial cells in 16 patients were evaluated. The site samples included the buccal mucosa (left and right sides), the posterior dorsum of the tongue, and the anterior floor of the mouth. There was a significant prevalence of bacterial cells in the smears and a greater number of degenerate and atypical squamous cells in cannabis users compared with controls. Epithelial cells in smears taken from cannabis users and tobacco-smoking controls showed koilocytic changes.

Pancreatic effect

A 29-year-old man presented with acute pancreatitis after a period of heavy cannabis smoking. Other causes of the disease were ruled out. The pancreatitis resolved itself after the cannabis was stopped and this was confirmed by urinary cannabinoid metabolite monitoring in the community. There were no previous reports of acute pancreatitis associated with cannabis use in the general population. Drugs of all types are related to the etiology of pancreatitis in approximately 1.4–2% of cases.

Panic disorder

Sixty-six panic disorder patients were included in a study. All of whom met the DSM-IV diagnosis of panic disorder (n = 45) or panic disorder with agoraphobia ([PDA]; n = 21). Twenty-four patients experienced their first panic attack within 48 hours of cannabis use and then went on to develop panic disorder. All the patients were treated with paroxetine (gradually increased up to 40 mg/day). The two groups responded equally well to paroxetine treatment as measured at the 8 weeks and 12 months follow-up visits. There were no significant effects of age, sex, and duration of illness as covariates with response rates between the two groups. In addition, panic disorder or panic disorder with agoraphobia diagnosis did not affect the treatment response in either group. There were no significant differences in weight gain, sexual side effects, or relapse rates between patients according to gender or comorbid diagnosis.

Paroxysmal atrial fibrillation

A healthy young subject was observed for paroxysmal atrial fibrillation following cannabis intoxication. The abuse of this substance was the most possible and identifiable risk factor.

Place conditioning effect

THC was administered to female rats at doses of 1, 5, or 20 mg/kg) during gestation and lactation. Maternal exposure to low doses of THC (1 and 5 mg/kg), relevant for human consumption, produced an increased response to the reinforcing effects of a moderate dose of morphine (350 μg/kg), as measured in the place-preference conditioning paradigm (CPP) in the adult male offspring. These animals also displayed an enhanced exploratory behavior in the defensive withdrawal test. Only females born from mothers exposed to THC at a dose of 1 mg/kg exhibited a small increment in the place conditioning induced by morphine. The possible implication of the HPA was analyzed by monitoring plasma levels of adrenocorticotropic hormone (ACTH) and corticosterone in basal and moderate-stress conditions (after the end of the CPP test). Female offspring perinatally exposed to THC (1 or 5 mg/kg) displayed high basal levels of corticosterone and a blunted adrenal response to the HPA-activating effects of the CPP test. Male offspring born from mothers exposed to THC (1 or 5 mg/kg) displayed the opposite pattern: normal to low basal levels of corticosterone, and a sharp adrenal response to the CPP challenge. THC administration to rats at a low dose (1.5 mg/kg) resulted in failing to develop place conditioning, and developing a place aversion at a high dose (15 mg/kg). Administration of the cannabinoid antagonist SR141716A induced a CPP at both a low (0.5 mg/kg) and a high (5 mg/kg) dose.

Plant germination effect

Methyl chloride extract of the dried seed produced weak activity on *Amaranthus spinosus* (25.8%) inhibition. Methyl chloride extract of the dried leaves produced 17.5% inhibition of *Amaranthus spinosus*.

Plasma norepinephrine concentration

Forty-six newborn infants participated in a prospective study of the neonatal and long-term effects of prenatal cocaine exposure. Based on maternal self-report, maternal urine screening, and infant meconium analysis, 24 infants were classified as CE and 22 as unexposed. Between 24 and 72 hours postpartum, plasma samples for norepinephrine (NE), epinephrine, dopamine, and dihydroxyphenylalanine analysis were obtained. The Neonatal Behavioral Assessment Scale was administered at 1–3 days of age and at 2 weeks of age by examiners masked to the drug exposure status of the newborns. The CE newborns had increased plasma NE concentrations when compared with the unexposed infants (geometric mean, 923 pg/mL vs 667 pg/mL). There were no significant differences in plasma epinephrine, dopamine, or dihydroxyphenylalanine concentrations. Analysis for the effect of potential confounding variables revealed that maternal cannabis use was also associated with increased plasma NE, although birth-weight, gender, and maternal use of alcohol or cigarettes were not. Geometric mean plasma NE was 1164 pg/ mL in those infants with *in utero* exposure to both cocaine and cannabis compared to 812 pg/mL in those exposed to only cocaine and 667.0 pg/mL in those exposed to neither. Among the CE infants, plasma NE concentration correlated with an increased score for the depressed cluster ($r = 0.53$) and a decreased score for the orientation cluster ($r = -0.43$) of the Neonatal Behavioral Assessment Scale administered at 1–3 days of age. Adjusting for cannabis exposure had no effect on these relationships between plasma NE and the depressed and orientation clusters.

Pneumonic effect

A case–control study was conducted in 7001 individuals. Odds ratios were calculated by conditional logistic regression with substance use and social factors as cofounders. Pneumonia was not associated with kava use. Crude odds ratios = 1.26 (0.74–2.14, $p = 0.386$) increased after controlling for confounders (OR = 1.98, 0.63–6.23, $p = 0.237$) but was not significant. Adjusted odds ratios for pneumonia cases involving kava and alcohol users was 1.19 (0.39–3.62, $p = 0.756$). Crude odds ratios for associations between pneumonia and cannabis use (OR = 2.27, 1.18–4.37, $p = 0.014$) and alcohol use (OR = 1.95, 1.07–3.53, $p = 0.026$) were statistically significant and approached significance for petrol sniffing (OR = 1.98, 0.99–3.95, $p = 0.056$).

Postural syncope

Twenty-nine volunteers participated in a randomized, double-blind, placebo-controlled study. Cerebral blood velocity, pulse rate, blood pressure, skin perfusion on forehead and plasma Δ-9-THC levels were quantified during reclining and standing for 10 minutes before and after THC infusions and cannabis smoking. Both THC and cannabis induced postural dizziness, with 28% reporting severe symptoms. Intoxication and dizziness peaked immediately after drug. The severe dizziness group showed the most marked postural drop in cerebral blood velocity and blood pressure and showed a drop in pulse rate after an initial increase during standing. Postural dizziness was unrelated to plasma levels of THC and other indices.

Prenatal exposure

Data collected from the National Household Survey on Drug Abuse, a nationally representative sample survey of 22,303 non-institutionalized women aged 18–44 years, of whom 1249 were pregnant, were analyzed. During the 2-year study period, 6.4% of the non-pregnant women of childbearing age and 2.8% of the pregnant women reported that they used illicit drugs. Of the women who used drugs, the relative proportion of women who abstained from illicit drugs after recognition of pregnancy increased from 28% during the first trimester of pregnancy to 93% by the third trimester. However, because of postpregnancy relapse, the net pregnancy-related reduction in illicit drug use at postpartum was only 24%. Cannabis accounted for three-fourths of illicit drug use, and cocaine accounted for one-tenth of illicit drug use. Of those who used illicit drugs, over half of pregnant and two-thirds of non-pregnant women used cigarettes and alcohol.

Among the sociodemographic subgroups, pregnant and non-pregnant women who were young (18–30 years) or unmarried, and pregnant women with less than a high school education had the highest rates of illicit drug use. Over 12,000 women at 18–20 weeks of gestation were enrolled in an Avon Longitudinal Study of Pregnancy and Childhood. Five percent of the mothers reported smoking cannabis before and/or during pregnancy; they were younger, of lower parity, better educated, and more likely to use alcohol, cigarettes, coffee, tea, and hard drugs. Cannabis use during pregnancy was unrelated to risk of perinatal death or need for special care, but the babies of women who used cannabis at least once per week before and throughout pregnancy were 216 g lighter than those of nonusers, had significantly shorter birth lengths, and smaller head circumferences. After adjustment for confounding factors, the association between cannabis use and birth-weight failed to be statistically significant ($p = 0.056$) and was clearly nonlinear. The adjusted mean birth-weights for babies of women using cannabis at least once per week before and throughout pregnancy were 90 g lighter than the offspring of other women. No significant adjusted effects were seen for birth length and head circumference. In two hospitals, 12,885 pregnant women answered questionnaires regarding consumption of alcohol, tobacco, cannabis, and other drugs.

The prevalence of cannabis use was 0.8%. Women using cannabis, but no other illicit drugs were each retrospectively matched with four randomly chosen pregnant women in the same period and the same age group and with same parity. Eighty-four cannabis users were included. These women were socio-economically disadvantaged and had a higher prevalence of present and past use of alcohol, tobacco, and other drugs. No significant difference in pregnancy, delivery, or puerperal outcome was found. Children of women using cannabis were 150 g lighter, 1.2 cm shorter, and had 0.2 cm smaller head circumference than the control infants. A 27-year-old woman who smoked a joint (cannabis) and 20 cigarettes (tobacco) daily up to the time of a positive pregnancy test at 7 weeks and 4 days, was evaluated. On day 20 of pregnancy, she had a LSD minitrip. The patient had a spontaneous term delivery. The baby boy weight was between the 5th and the 50th percentile, length between the 50th and the 90th percentile, normal umbilical arterial and venous pH values, and Apgar scores of 7/9/10. There were no visible abnormalities, and behavior was normal. Eight hundred seven consecutive positive-pregnancy test urine samples were screened for a range of drugs, including cotinine as an indicator of maternal smoking habits. A positive test for cannabinoids was found in 117 (14.5%) of the samples. Smaller numbers of samples were positive for other drugs: opiates (11), benzodiazepines (4), cocaine (3), and one each for amphetamines and methadone.

Polydrug use was detected in nine individuals. Only two samples tested positive for ethanol. The proportion with a urine cotinine level indicative of active smoking was 34.3%. The outcome of the pregnancy was traced for 288 of the subjects. Cannabis use was associated with a lower gestational age at delivery ($p < 0.005$), an increased risk of prematurity ($p < 0.02$), and reduction in birth-weight ($p < 0.002$). Maternal smoking was associated with a reduction in infant birth-weight ($p < 0.05$). This was less pronounced than the effect of other substance misuse. A sample of low-income women attending a prenatal clinic was assessed. The majority of the women decreased their use of cannabis during pregnancy. The assessments of child behavior problems included the Child Behavior Checklist, Teacher's Report Form, and the Swanson, Noland, and Pelham checklist. Multiple and logistic regressions were employed to analyze the relations between cannabis use and behavior problems of the children at age 10, while controlling for the effects of other extraneous variables. Prenatal cannabis use was significantly related to increased hyperactivity, impulsivity, and inattention symptoms as measured by the Swanson, Noland, and Pelham, increased delinquency as measured by the Child Behavior Checklist, and increased delinquency and externalizing problems as measured by the Teacher's Report Form. The pathway between prenatal cannabis exposure and delinquency was mediated by the effects of cannabis exposure on inattention symptoms. Attention and impulsivity of prenatally substance-exposed 6-year-olds were assessed as part of a longitudinal study. Most of the women were light-to-moderate users of alcohol and cannabis who decreased their use after the first trimester of pregnancy. Tobacco was used by a majority of women and did not change during pregnancy.

The women, recruited from a prenatal clinic, were of low socioeconomic status. Attention and impulsivity were assessed using a Continuous Performance Task. Second and third trimester of tobacco exposure and first trimester of cocaine use predicted increased omission errors. Second trimester cannabis use predicted more commission errors and fewer omission errors. There were no significant effects of prenatal alcohol exposure. Lower Stanford-Binet Intelligence Scale composite scores, male gender, and an adult male in the household predicted more errors of commission. Lower Standford-Binet Intelligence Scale composite scores, younger child age, maternal work/school status, and higher maternal hostility scores predicted more omission errors. The neurophysiological effects of prenatal cannabis exposure on response inhibition were assessed in thirty-one participants aged 18–22. Ottawa Prenatal Prospective Study performed a blocked design Go/No-Go task while neural activity was imaged with functional magnetic resonance imaging.

The Ottawa Prenatal Prospective Study is a longitudinal study that provides a unique body of information collected from each participant over 20 years, including prenatal drug history, detailed cognitive/behavioral performance from infancy to young adulthood, and current and past drug usage. The functional magnetic resonance imaging results showed that with increased prenatal cannabis exposure, there was a significant increase in neural activity in bilateral PFC and right premotor cortex during response inhibition. There was also an attenuation of activity in left cerebellum with increased prenatal exposure to cannabis when challenging the response inhibition neural circuitry. Prenatally exposed offspring had significantly more commission errors than non-exposed participants, but all participants were able to perform the task with more than 85% accuracy. The findings were observed when controlling for present cannabis use and prenatal exposure to nicotine, alcohol, and caffeine, and suggest that prenatal cannabis exposure was related to changes in neural activity during response inhibition that last into young adulthood. The effects of prenatal cannabis and alcohol exposure on school achievement at 10 years of age were examined. Women were interviewed about their substance use at the end of each trimester of pregnancy, at 8 and 18 months, and at 3, 6, 10, 14, and 16 years.

The women were of lower socioeconomic status, high school-educated, and light-to-moderate users of cannabis and alcohol. At the 10-year follow-up, the effects of prenatal exposure to cannabis or alcohol on the academic performance of 606 children were assessed. Exposure to one or more cannabis joints per day during the first trimester predicted deficits in Wide Range Achievement Test- Revised reading and spelling scores and a lower rating on the teachers' evaluations of the children's performance. This relation was mediated by the effects of first-trimester cannabis exposure on the children's depression and anxiety symptoms. Second-trimester cannabis use was significantly associated with reading comprehension and under-achievement. Exposure to alcohol during the first and second trimesters of pregnancy predicted poorer teachers' ratings of overall school performance. Second-trimester binge drinking predicted lower reading scores.

There was no interaction between prenatal cannabis and alcohol exposure. Each was an independent predictor of academic performance. Pregnant rats were treated daily with Δ-9-THC from the fifth day of gestation up to the day before birth (GD21). Then rats were sacrificed and their pups removed for analysis of the neural adhesion molecule L1-mRNA levels in different brain structures. The levels of L1 transcripts were significantly increased in the fimbria, stria terminalis, stria medullaris, corpus callosum, and in gray-matter structures (septum nuclei and the habenula). It remained unchanged in most of the gray-matter structures analyzed (cerebral cortex, BAL nucleus, hippocampus, thalamic and hypothalamic nuclei, basal ganglia, and sub-ventricular zones) and also in a few white-matter structures (fornix and fasciculus retroflexus). The increase in L1-mRNA levels reached statistical significance only in Δ-9-THC-exposed males but not females, where only trends or no effects were detected.

The results supported evidence on a sexual dimorphism, with greater effects in male fetuses, for the action of cannabinoids in the developing brain. Fetal cannabis exposure has no consistent effect on outcome. Prenatal cocaine exposure has not been shown to have any detrimental effect on cognition, except as mediated through cocaine effects on head size. Although fetal cocaine exposure has been linked to numerous abnormalities in arousal, attention, and neurological and neurophysiological function, most such effects appear to be self-limited and restricted to early infancy and childhood. Opiate exposure elicits a well-described withdrawal syndrome affecting the central nervous, autonomic, and gastrointestinal systems, which is most severe among methadone-exposed infants. Executive functioning in cocaine/polydrug (cannabis, alcohol, and tobacco)-exposed infants was assessed in a single session, occurring between 9.5 and 12.5 months of age. In an A-not-B task, infants searched—after performance-adjusted delays—for an object hidden in a new location. The CE infants did not differ from non-CE controls recruited from the same at-risk population. Comparison of heavier-CE (n = 9) with the combined group of lighter-CE (n = 10) and non-CE (n = 32) infants revealed significant differences on A-not-B performance, as well as on global tests of mental and motor development.

Covariates investigated included socioeconomic status, marital status, race, maternal age, years of education, weeks of gestation, and birth-weight, as well as severity of prenatal cannabis, alcohol, and tobacco exposure. The relationship of heavier-CE status to motor development was mediated by length of gestation, and the relationship of heavier-CE status to mental development was confounded with maternal gestational use of cigarettes. The relationship of heavier-CE status to A-not-B performance remained significant after controlling for potentially confounded variables and mediators, but was not statistically significant after controlling for the variance associated with global mental development. Weight, height, and head circumference were examined in children from birth to early adolescence for whom prenatal exposure to cannabis and cigarettes had been ascertained. The subjects were from a low-risk, predominantly middle-class sample participating in an ongoing longitudinal study. The negative association between growth measures at birth and prenatal cigarette exposure was overcome, sooner in males than females, within the first few years, and by the age of 6 years, the children of heavy smokers were heavier than control subjects. Pre- and postnatal environmental tobacco smoke did not have a negative effect on the growth parameters; however, the choice of bottle-feeding or shorter duration of breastfeeding by women who smoked during pregnancy appeared to play an important positive role in the catch-up observed among the infants of smokers. Prenatal exposure to cannabis was not significantly related to any growth measures at birth, although a smaller head circumference observed at all ages reached statistical significance among the early adolescents born to heavy cannabis users.

Prolactin inhibition

The dried leaves, smoked by healthy female volunteers at a dose of 1 g/person, produced a decrease in plasma prolactin levels during the luteal phase of the menstrual cycle but not during the follicular phase. The results were significant at $p < 0.01$ level.

Propiospinal myoclonus

A 25-year-old woman with clusters of myoclonus induced by a single exposure to inhaled cannabis was evaluated. Investigations excluded a structural abnormality of the spine. Multichannel surface electromyogram with parallel frontal electroencephalogram recording confirmed the diagnosis of propriospinal myoclonus.

Psoriatic effect

Hot water extract of the dried seed, taken orally by 108 human adults with psoriasis at variable dosage level, was active. After 3–4 weeks of treatment, there was significant improvement. The extract was taken in combination with *Rehmannia glutinosa* (rhizome), *Salvia miltiorrhiza* (root), *Scrophularia ningpoensis* (root), *Isatis tinctoria* (branch and leaf), *Sophora subprostrata* (root), *Dictamnus dasycarpus* (rootbark), *Polygonum bistorta* (rhizome), and *Forsythia suspensa* (fruit).

Psychosocial morbidity association

Cannabis dependence is a prevalent comorbid substance use disorder among patients early in the course of a schizophrenia-spectrum disorder. Among 29 eligible patients, 18 participated in the study. First-episode patients with comorbid cannabis dependence ($n = 8$) reported significantly greater childhood physical and sexual abuse compared with those without comorbid cannabis dependence ($n = 10$). The result indicated the preliminary evidence of an association between childhood maltreatment and cannabis dependence among this especially vulnerable population. Childhood physical and sexual abuse may be a risk factor for the initiation of cannabis dependence and other substance use disorders in the early course of schizophrenia.

Psychotic effect

Thirty five hundred representatives 19 years of age were examined in a cohort study. The subjects completed a 40-item Community Assessment of Psychic Experiences, measuring subclinical positive (paranoia, hallucinations, grandiosity, first- rank symptoms) and negative psychosis dimensions, depression, and drug use. Use of cannabis was associated positively with both positive and negative dimensions of psychosis, independent of each other and of depression. An association between cannabis and depression disappeared after adjustment for the negative psychosis dimensions. First use of cannabis younger than age 16 years was associated with a much stronger effect than first use after age 15 years, independent of lifetime frequency of use. The association between cannabis and psychosis was not influenced by the distress associated with the experiences, indicating that self-medication may be an unlikely explanation for the entire association between cannabis and psychosis. Cross-sectional epidemiological studies indicated that individuals with psychosis use cannabis more often than other individuals in the general population. It has long been considered that this association was explained by the self-medication hypothesis, postulating that cannabis is used to self-medicate psychotic symptoms. This hypothesis has been recently challenged. Several prospective studies carried out in population-based samples, showed that cannabis exposure was associated with an increased risk of psychosis. A dose–response relationship was found between cannabis exposure and risk of psychosis, and this association was independent from potential confounding factors, such as exposure to other drugs and preexistence of psychotic symptoms. The brain mechanisms underlying the association have to be elucidated; they may implicate deregulation of cannabinoid and dopaminergic systems. Cannabis exposure may be a risk factor for psychotic disorders by interacting with a preexisting vulnerability for these disorders.

Refractory neuropathic pain

Seven patients (three women and four men) aged 60 ± 14 years suffering from chronic refractory neuropathic pain, received oral THC titrated to the maximum dose of 25 mg/day (mean dose: 15 ± 6 mg) during an average of 55.4 days (range: 13–128). Various components of pain (continuous, paroxysmal, and brush-induced allodynia) were assessed using visual analog scale scores. Health-related quality of life was evaluated using the Brief Pain Inventory, and the Hospital Anxiety and Depression scale was used to measure depression and anxiety. THC did not induce significant effect on the various pain, health-related quality of life and anxiety and depression scores. Numerous side effects (notably sedation and asthenia) were observed in five out of seven patients, requiring premature discontinuation of the drug in three patients.

Reproductive effect

Cannabis use during pregnancy in developed nations is estimated to be approx 10%. Recent evidence suggests that the endogenous cannabinoid system, now consisting of two receptors and multiple endocannabinoid ligands, may also play an important role in the maintenance and regulation of early pregnancy and fertility. Drugs of abuse, like alcohol, opiates, cocaine, and cannabis, are used by many young people for their presumed aphrodisiac properties. The opioids inhibit the hypothalamus–pituitary–gonads axis (HPG), and increase the prolactin levels, which interferes with the male and female sexual response. Cannabis, at high doses, could inhibit the HPG axis and reduce fertility. Cannabis initially increases libido and potency, but chronic use causes sexual inversion. Long-term use of cannabis has been found to cause physiological changes that can alter individual reproductive potential. The effects of cannabis depend on the dose and can include death from depression of the respiratory system. Cannabis is absorbed rapidly and eliminated very slowly. Δ-9-THC is highly liposoluble and fixes to the serum proteins, passing to the lungs and liver for metabolization and to the kidneys and liver for excretion. As with estrogens, there is an enterohepatic circuit for reabsorption and elimination. Ninety percent is eliminated in the feces, 65% within 48 hours. Because of the enterohepatic circuit and liposolubility, elimination requires 1 week for completion. The other important biotransformation of the active principle is hydroxylation. The hydroxylated derivatives are responsible for the psychoactivity of cannabis. Cannabis affects both neuroendocrine function and the germ cells. Studies on experimental animals have indicated that THC can cause a decline in the pituitary hormones, follicle stimulating hormone, luteinizing hormone, and prolactin, and in the steroids progesterone, estrogen, and androgens.

Human studies have shown that chronic users have decreased levels of serum testosterone. Because steroidogenesis can be restimulated with human chorionic gonadotropin, it appears that THC does not directly affect steroid production by the corpus luteum, but that its action is mediated by the hypothalamus. Because of its potent antigonadotropic action, THC is under study as an anovulatory agent. The same animal studies have shown that ovulation returns to normal 6 months after termination of use.

High rates of anovulation and luteal insufficiency have been observed in women smoking cannabis at least three times weekly. THC accumulates in the milk. Animal studies have shown that THC depresses the enzymes necessary for lactation and causes a diminution in the volume of the mammary glands. Significant amounts of the drug have been detected in both mothers' milk and the blood of newborns. Animal studies indicate that THC crosses the placenta, achieving concentrations in the fetus as high as those in the mother. Animal studies also demonstrated increasing frequency of abortions, intrauterine death, and declines in fetal weight. The effects were probably caused by an alteration in placental function. A human study likewise showed that cannabis use during pregnancy was significantly related to poor fetal development, low birth-weight, diminished size, and decreased cephalic circumference. Congenital malformations have been observed in experimental animals exposed to THC. Declines in sperm volume and count and abnormal sperm motility have been observed in chronic cannabis users. In vitro studies show that THC produces a marked degeneration of human sperm. Among sexually experienced girls, 39% (n = 123) reported using oral contraceptive pills(OCPs), 5.4% (n = 17) used Depo-Provera (medroxyprogesterone acetate) or Norplant (levonorgestrel), and 55.6% (n = 175) used no hormonal method. Logistic regression analysis revealed that the factors most significantly associated with the use of hormonal methods were older age (OR = 1.19; 95% CI, 1.07–1.33), not using a condom at last intercourse (OR = 0.55; CI, 0.34–0.90), and having had a well visit within 1 year (OR = 2.11; CI, 1.12–3.70). OCP users were less likely than Depo-Provera or Norplant users to have used alcohol (p = 0.041), cigarettes (p = 0.002), or cannabis (p = 0.018) in the past 30 days. OCP users were less likely than nonusers of hormonal methods to have smoked cigarettes (p = 0.034) or cannabis (p = 0.052). The school-based clinic had a greater proportion of subjects using long-acting progestins ($p < 0.001$).

Respiratory effect

Smoking a "joint" of cannabis resulted in exposure to significantly greater amounts of combusted material than with a tobacco cigarette. The histopathological effects of cannabis-smoke exposure included changes consistent with acute and chronic bronchitis. Cellular dysplasia has also been observed, suggesting that, like tobacco smoke, cannabis exposure has the potential to cause malignancy. Symptoms of cough and early morning sputum production are common (20–25%) even in young individuals who smoke cannabis alone. Almost all studies indicated that the effects of cannabis and tobacco smoking are addictive and independent. A small group of current male cannabis processors with a mean age of 43 years was studied. Questionnaire data, lung function, serial FEV1 and blood were collected from all workers. Seven workers (64%) complained of at least one respiratory symptom (one with byssinosis). The mean percentage predicted FEV1 was 91.5, FVC 97.7, peak expiratory flow 92.1, and forced expiratory flow between 25 and 75% of FVC 79.5. Serial FEV1 measurements in the two workers with work-related respiratory symptoms revealed a mean change in FEV1 on the first working day of –12.9%. This contrasted with +6.25% on the last working day. Respective values for the two workers without work-related symptoms were –1.4 and +3.2%. Nine hundred forty-three young adults from a birth cohort of 1037 were studied at age 21 years.

Standardized respiratory symptom questionnaires were administered. Spirometry and methacholine challenge tests were undertaken. Cannabis dependence was determined using DSM-III-R criteria.

Descriptive analyses and comparisons between cannabis-dependent, tobacco- smoking, and nonsmoking groups were undertaken. Adjusted odds ratios for respiratory symptoms, lung function, and airway hyperresponsiveness (PC20) were measured.

Ninety-one subjects (9.7%) were cannabis-dependent and 264 (28.1%) were current tobacco smokers. After controlling for tobacco use, respiratory symptoms associated with cannabis dependence included wheezing apart from colds, exercise-induced shortness of breath, nocturnal wakening with chest tightness, and early morning sputum production. These were increased by 61, 65, 72 (all $p < 0.05$), and 144% ($p < 0.01$) respectively, compared with non-tobacco smokers. The frequency of respiratory symptoms in cannabis-dependent subjects was similar to tobacco smokers of 1–10 cigarettes per day. The proportion of cannabis-dependent study members with an FEV1/FVC ratio of less than 80% was 36% compared with 20% for nonsmokers ($p = 0.04$). These outcomes occurred independently of coexisting bronchial asthma.

Reversal of cannabinoid addiction

Δ-9-THC was administered orally to mice at a dose of 10 mg/kg twice daily for 6 days to make them dependent on cannabinoids. Other groups of mice were administered orally with a Δ-9-THC and benzoflavone from *Passiflora incarnata* at doses of 10 or 20 mg/kg twice daily for 6 days. Mice receiving the Δ-9-THC and *Passiflora incarnata* extract developed significantly less dependence, worse locomotor activity, and less of typical withdrawal effects like paw tremors and headshakes, compared with mice receiving Δ-9-THC alone. Administration of SR-141716A, a selective cannabinoid-receptor antagonist (10 mg/kg, orally), to all groups on the seventh day resulted in an artificial withdrawal. Administration of 20 mg/kg of the *Passiflora incarnata* benzoflavone moiety to mice showing symptoms of withdrawal owing to administration of SR-141716A produced a marked attenuation of withdrawal effects.

Schizophrenic effect

The nerve growth factor (NGF) serum levels of 109 consecutive drug-naïve schizophrenic patients were measured and compared with those of healthy controls. The results were correlated with the long-term intake of cannabis and other drugs. Mean (± standard deviation) NGF serum levels of 61 control persons (33.1 ± 31 pg/mL) and 76 schizophrenics who did not consume illegal drugs (26.3 ± 19.5 pg/mL) did not differ significantly.

Schizophrenic patients with regular cannabis intake (> 0.5 g per day on average for at least 2 years) had significantly raised NGF serum levels of 412.9 ± 288.4 pg/mL ($n = 21$) compared with controls and schizophrenic patients not consuming cannabis ($p < 0.001$). In schizophrenic patients who abused not only cannabis, but also additional substances, NGF concentrations were as high as 2336.2 ± 1711.4 pg/mL ($n = 12$). On average, heavy cannabis consumers suffered their first episode of schizophrenia 3.5 years ($n = 21$) earlier than schizophrenic patients who abstained from cannabis. These results indicate that cannabis is a possible risk factor for the development of schizophrenia. This might be reflected in the raised NGF-serum concentrations when both schizophrenia and long-term cannabis abuse prevail.

Schizotypy correlation

Two hundred eleven healthy adults who used cannabis showed higher scores on schizotypy, borderline, and psychoticism scales than never- users. Multivariate analysis, covarying lie scale scores, age, and educational level indicated that high schizotypal traits best discriminated subjects who had used cannabis from never-users, whether or not they reported having used other recreational drugs. The results indicated that cannabis use was related to a personality dimension of psychosis-proneness in healthy people.

Sedative and stimulant effects

A double-blind, placebo-controlled study assessed subjective effects of smoking cannabis with either a long or short breath-holding duration. During eight test sessions, 55 male volunteers made repeated ratings of subjective "high," sedation, and stimulation, as well as rating their perceptions of motivation and performance on cognitive tests. The long, relative to the short, breath-holding duration increased "high" ratings after smoking cannabis, but not placebo. Cannabis smoking increased sedation and a perception of worsened test performance, and decreased motivation with respect to test performance. Paradoxical subjective effects were observed in those subjects reporting some stimulation, as well as sedation after smoking cannabis, particularly with the long breath-holding duration. Breath-holding duration did not produce any subjective effects that were independent of the drug treatment (i.e., occurred equally after smoking of cannabis and placebo).

Sexual receptivity

The effects of THC on sexual behavior in female rats and its influence on steroid hormone receptors and neurotransmitters in the facilitation of sexual receptivity was examined. Results revealed that the facilitatory effect of THC was inhibited by antagonists to both progesterone and dopamine D(1) receptors. To test further the idea that progesterone receptors (PR) and/or dopamine receptors (D[1]R) in the hypothalamus were required for THC-facilitated sexual behavior in rodents, antisense, and sense oligonucleotides to PR and D(1)R were administered intra-cerebroventricularly into the third cerebral ventricle of ovariectomized, estradiol benzoate-primed rats. Progesterone- and THC-facilitated sexual behavior was inhibited in animals treated with antisense oligonucleotides to PR or to D(1)R. Antagonists to cannabinoid receptor-1 subtype (CB1), but not to cannabinoid receptor-2 subtype (CB2) inhibited progesterone- and dopamine-facilitated sexual receptivity in female rats. Adult female and male rats that had been perinatally exposed to hashish extracts were investigated. Adult males perinatally exposed to hashish extracts exhibited marked changes in the behavioral patterns executed in the sociosexual approach behavior test; these changes did not exist in females.

Control males first visited the incentive male and took longer to visit the incentive female, whereas hashish-exposed males followed the opposite pattern. Hashish-exposed males spent more time in the vicinity of the incentive female, whereas they decreased their frequency of visits to, and the time spent in, the male incentive area. This behavior was observed during the first third of the test, but became normalized and even inverted during the last two-thirds. In the social interaction test, the normal reduction in the time spent in active social interaction following the exposure to a neophobic situation (high light levels) in controls did not occur in hashish-exposed males, although these exhibited a response in the dark-light emergence test similar to that of their corresponding controls. No changes were seen in spontaneous locomotor activity in both tests. These behavioral alterations observed in hashish-exposed males were paralleled by a significant decrease in L-3,4-dihydroxyphenylacetic acid contents in the limbic forebrain; this suggests a decreased activity of mesolimbic dopaminergic neurons. No effects were seen in females.

Smooth muscle relaxant activity

Ethanol (95%) and water extracts of the dried aerial parts, at a concentration of 1:1, produced weak activity on the rabbit duodenum. The ethanol extract was equivocal on the guinea pig ileum. Petroleum ether extract of the dried entire plant, administered intraperitoneally to rats at a dose of 0.89 mg/kg, was active vs corneo-palpebral reflex.

Spasticity treatment

Standardized plant extract was administered orally to 57 MS patients with poorly controlled spasticity, at a dose of 2.5 mg of THC and 0.9 mg of CBD. Patients in group A started with a drug escalation phase from 15 to a maximum of 30 mg of THC by 5 mg per day if well tolerated, being on active medication for 14 days before starting placebo. Patients in group B started with placebo for 7 days, crossed to the active period (14 days), and closed with a three-day placebo period (active drug-dose escalation and placebo sham escalation as in group A). Measures used included daily self-report of spasm frequency and symptoms, Ashworth Scale, Rivermead Mobility Index, 10-meter timed walk, nine-hole peg test, paced auditory serial addition test, and the digit span test.

There were no statistically significant differences associated with active treatment compared with placebo, but trends in favor of active treatment were seen for spasm frequency, mobility, and getting to sleep. In the 37 patients (per-protocol set) who received at least 90% of their prescribed dose, improvements in spasm frequency ($p = 0.013$) and mobility after excluding a patient who fell and stopped walking were seen ($p = 0.01$). Minor adverse events were slightly more frequent and severe during active treatment, and toxicity symptoms, which were generally mild, were more pronounced in the active phase. Six hundred thirty participants with stable MS and muscle spasticity were treated with oral cannabis extract ($n = 211$), Δ-9-THC ($n = 206$), or placebo ($n = 213$) for 15 weeks. Six hundred eleven of 630 patients were followed up for the primary end point.

No treatment effect of cannabinoids on the primary outcome ($p = 0.40$) was noted. The estimated difference in mean reduction in total Ashworth score for participants taking cannabis extract compared with placebo was 0.32 (95% CI, −1.04 to 1.67), and for those taking Δ-9-THC vs placebo it was 0.94 (−0.44 to 2.31). There was an evidence of a treatment effect on patient-reported spasticity and pain ($p = 0.003$), with improvement in spasticity reported in 61% ($n = 121$, 95% CI, 54.6–68.2), 60% ($n = 108$, 52.5–66.8), and 46% ($n = 91$, 39–52.9) of participants on cannabis extract, Δ-9-THC, and placebo, respectively.

Spatial working memory effect

Functional magnetic resonance imaging was used to examine brain activity in 12 long-term heavy cannabis users, 6–36 hours after last use, and in 10 control subjects while they performed a spatial working memory task. Regional brain activation was analyzed and compared using statistical parametric mapping techniques. Compared with controls, cannabis users exhibited increased activation of brain regions typically used for spatial working memory tasks (such as PFC and anterior cingulate). Users also recruited additional regions not typically used for spatial working memory (such as regions in the basal ganglia). The findings remained essentially unchanged when reanalyzed using the subjects ages as a covariate. Brain activation showed little or no significant correlation with subjects years of education, verbal IQ, lifetime episodes of cannabis use, or urinary cannabinoid levels at the time of scanning.

Spontaneous pneumomediastinum

Spontaneous pneumomediastinum is defined as pneumomediastinum in the absence of an underlying lung disease. It is the second most common cause of chest pain in young, healthy individuals (<30 years) necessitating hospital visits. Inhalational drug use (cocaine and cannabis) has been associated with a significant number of cases, although cases with no apparent etiological or incriminating factors are well-recognized. A case of an 18-year-old high school student with spontaneous pneumomediastinum was evaluated.

Sudden infant death syndrome.

In a nationwide case–control study of 369 cases and 1558 controls, two-thirds of SIDS deaths occurred at night (between 10 PM and 7:30 AM). The odds ratio (95% CI) for prone sleep position was 3.86 (2.67–5.59) for deaths occurring at night, and 7.25 (4.52–11.63) for deaths occurring during the day; the difference was significant. The odds ratio for maternal smoking and SIDS deaths occurring at night was 2.28 (1.52–3.42), and for the day, 1.27 (0.79–2.03). If the mother was single, the odds ratio was 2.69 (1.29–3.99) for a nighttime death, and 1.25 (0.76–2.04) for a daytime death. Both interactions were significant. The interactions between time of death and bed sharing, not sleeping in a cot or bassinet, ethnicity, late timing of prenatal care, binge drinking, cannabis use, and illness in the baby were also significant. All were more strongly associated with SIDS occurring at night. In a nationwide case–control study, 393 cases and 1592 controls were analyzed. Adjusting for ethnicity and maternal tobacco use, the SIDS odds ratio for weekly maternal cannabis use since the infant's birth was 2.23 (95% CI = 1.39, 3.57) compared with nonusers, and the multivariate odds ratio was 1.55 (95% CI = 0.87, 2.75).

Suicidal effect

Standardized interview assessments were conducted with 2311 youths aged 8–15 years who used drugs before age 16. Approximately 15 years after recruitment, 1695 persons (mean age = 21 years) were reassessed. One hundred fifty-five of them made suicide attempts (SA) and 218 had onset of depression-related suicide ideation (SI). The relative risk, from survival analysis and logistic regression models, to study early use of tobacco, alcohol, cannabis, and inhalants, with covariate adjustments for age, sex, race/ethnicity, and other pertinent covariates were examined. Early-onset of cannabis use and inhalant use for females, but not for males, signaled a modest excess risk of SA (cannabis-associated RR = 1.9; $p = 0.04$; inhalant-associated RR = 2.2; $p = 0.05$). Early-onset of cannabis use by females (but not for males) signaled excess risk for SI (RR = 2.9; $p = 0.006$). Early-onset alcohol and tobacco use were not associated with later risk of SA or SI. Two hundred seventy-seven same-sex twin pairs (median age: 30 years) discordant for cannabis dependence and 311 pairs discordant for early-onset cannabis use (before age 17 years) were examined.

Individuals who were cannabis-dependent had odds of SI and SA that were 2.5–2.9 times higher than those of their noncannabis-dependent co-twin. Cannabis dependence was associated with elevated risks of major depressive disorder (MDD) in dizygotic, but not in monozygotic twins. Twins who initiated cannabis use before age 17 years of age had elevated rates of subsequent SA (OR, 3.5, 95% CI, 1.4–8.6) but not of MDD or SI. Early MDD and SI were significantly associated with subsequent risks of cannabis dependence in discordant dizygotic pairs, but not in discordant monozygotic pairs. The results indicated that the comorbidity between cannabis dependence and MDD likely arises through shared genetic and environmental vulnerabilities predisposing to both outcomes. In contrast, associations between cannabis dependence and suicidal behaviors cannot be entirely explained by common predisposing genetic and/or shared environmental predisposition.

Synergic cytotoxicity

THC, in A549 lung tumor cells culture at concentrations of less than 5 μg/mL, produced no cytotoxic effect. At higher levels it induced cell necrosis, with a lethal concentration $(LC)_{50}$ of 16–18 μg/mL. Butylated hydroxyanisole ([BHA], a food additive)alone at concentrations of 10–200 μM, produced limited cell toxicity and significantly enhanced the necrotic death resulting from concurrent exposure to THC. In the presence of BHA at 200 μM, the LC_{50} for THC decreased to 10–12 μg/mL. Similar results were obtained with smoke extracts prepared from cannabis cigarettes, but not with extracts from tobacco or placebo cannabis cigarettes (containing no THC). Experiments were repeated in the presence of either diphenyleneiodonium or dicumarol as inhibitors of the redox cycling pathway. Neither of the compounds protected cells from the effects of combined THC and BHA, but rather enhanced necrotic cell death. Measurements of cellular ATP revealed that both THC and BHA reduced ATP levels in A549 cells, consistent with toxic effects on mitochondrial electron transport. The combination was synergistic in this respect, reducing ATP levels to less than 15% of the control. Exposure to cannabis smoke in conjunction with BHA may promote deleterious health effects in the lung.

Teratogenic activity

Resin, administered orally to pregnant rabbits at a dose of 1 mL/ kg, was active. Alcohol extract of the dried leaves, administered intragastrically to pregnant rats at a dose of 125 mg/kg from days 7 to 16 of gestation, was active. The fetuses showed several gross abnormalities, visceral anomalies, and skeletal malformations. Water extract of the dried leaf, administered intragastrically to pregnant rats at doses of 125, 200, 400, and 800 mg/kg, produced various types of malformations in the fetuses. Petroleum ether extract of the aerial parts, administered orally to rats and rabbits, was inactive.

Tourette syndrome

Tourette syndrome (TS) is a complex inherited disorder of unknown etiology, characterized by multiple motor and vocal tics. Involvement of the central cannabinoid (CB1) system was suggested because of therapeutic effects of cannabis consumption and Δ-9-THC-treatment in TS patients. The central cannabinoid receptor (CNR1) gene encoding the *CNR1* was considered as a candidate gene for TS and systematically screened by single-strand conformation polymorphism analysis and sequencing. Compared with the published *CNR1* sequence, three single-base substitutions were identified: 1326T→A, 1359G→A, 1419 + 1G→C. The change at position 1359 is a common polymor-phism (1359 G/A) without allelic association with TS. 1326T→A was present in only one TS patient and is a silent mutation, which does not change codon 442 (valine). 1419 + 1G→C affects the first nucleotide immediately following the coding sequence. It was first detected in three of 40 TS patients and none of 81 healthy controls. This statistically significant association with TS ($p = 0.034$) could not be confirmed in two subsequent cohorts of 56 TS patients (one heterozygous for 1419 + 1G→C) and 55 controls, and 64 patients and 66 controls (one heterozygous for 1419 + 1G→C), respectively. Transcript analysis of lymphocyte RNA from five 1419 + 1G→C carriers revealed no systematic influence on the expression level of the mutated allele.

In addition, segregation analysis of 1419 + 1G→C in affected families gave evidence that 1419 + 1G→C does not play a causal role in the etiology of TS. It was concluded that genetic variations of the *CNR1* gene are not a plausible explanation for the clinically observed relation between the cannabinoid system and TS. A single-dose, cross-over study in 12 patients, and a 6-week, randomized trial in 24 patients, demonstrated that Δ9-THC, the most psychoactive ingredient of cannabis, reduced tics in TS patients. No serious adverse effects occurred and no impairment on neuropsychological performance was observed. In the randomized, double-blind, placebo-controlled study, 24 patients with TS, according to DSM-III-R criteria, were treated over a 6-week period with up to 10 mg/day of THC. Tics were rated at six visits (visit 1, baseline; visits 2–4, during treatment period; visits 5–6, after withdrawal of medication) using the Tourette Syndrome Clinical Global Impressions scale (TS-CGI), the Shapiro Tourette-Syndrome Severity Scale (STSSS), the Yale Global Tic Severity Scale (YGTSS), the self-rated Tourette Syndrome Symptom List (TSSL), and a videotape-based rating scale. Seven patients dropped out of the study or had to be excluded, but only one because of side effects. Using the TS-CGI, STSSS, YGTSS, and video rating scale, there was a significant difference ($p < 0.05$) or a trend toward a significant difference ($p < 0.1$) between THC and placebo groups at visits 2, 3, and/or 4. Using the TSSL at 10 treatment days (between days 16 and 41) there was a significant difference ($p < 0.05$) between both groups.

Analysis of variance also demonstrated a significant difference ($p = 0.037$). No serious adverse effects occurred. In the randomized, double-blind, placebo-controlled study, the effect of a treatment with up to 10 mg Δ-9-THC over a 6-week period on neuropsychological performance in 24 patients suffering from TS was investigated. During medication and immediately, as well as 5–6 weeks after, withdrawal of Δ-9-THC treatment, no detrimental effect was seen on learning curve, interference, recall and recognition of word lists, immediate visual memory span, and divided attention. A trend towards a significant immediate verbal memory span improvement during and after treatment was found. A randomized double-blind placebo-controlled crossover single-dose trial of Δ-9-THC (5, 7.5, or 10 mg) in 12 adult TS patients was performed. Tic severity was assessed using the TSSL and examiner ratings (STSSS, YGTSS, TS-CGS).

Using the TSSL, patients also rated the severity of associated behavioral disorders. Clinical changes were correlated to maxi-mum plasma levels of THC and its metabolites 11-OH-THC and 11-nor-Δ-9-tetrahydrocannabinol-9-carboxylic acid. Using the TSSL, there was a significant improvement of tics ($p = 0.015$) and obsessive-compulsive behavior ($p = 0.041$) after treatment with Δ-9-THC compared with placebo. Examiner ratings demonstrated a significant difference for the subscore "complex motor tics" ($p = 0.015$) and a trend towards a significant improvement for the subscores "motor tics" ($p = 0.065$), "simple motor tics" ($p = 0.093$), and "vocal tics" ($p = 0.093$). No serious adverse reactions occurred. Five patients experienced mild, transient side effects. There was a significant correlation between tic improvement and maximum 11-OH-THC plasma concentration.

Toxic effect

Petroleum ether extract of the dried leaf, administered by gastric intubation to pregnant rats at a dose of 150 mg/kg, produced a reduction of food and water consumption and maternal weight gain. The weight of pups at birth was reduced by approx 10% of the litter size, and pup mortality at birth was not affected significantly. Water extract of the aerial parts, administered intravenously to male adults, was active. The resin, ingested by a 4-year-old girl, showed signs of stupor alternating with brief intervals of excitation and foolish laughing with atactic movements. Her temperature, blood pressure, pulse, hemoglobin, leukocytes, serum electrolytes, and serum urea were normal. Respiratory rate was 12 beats per minute. Blood sugar elevated. Recovery was complete within 24 hours with no treatment. Four patients suffered gastrointestinal disorders and psychological effects after eating salad prepared with hemp seed oil. The concentration of THC in the oil far exceeded the recommended tolerance dose. From January 1998 to January 2002, 213 incidences were recorded of dogs that developed clinical signs following oral exposure to cannabis, with 99% having neurological signs and 30% exhibiting gastrointestinal signs. The cannabis ingested ranged from 0.5 to 90 g. The lowest dose at which signs occurred was 84.7 mg/kg and the highest reported dose was 26.8 g/kg. Onset of signs ranged from 5 minutes to 96 hours, with most signs occurring within 1–3 hours after ingestion. The signs lasted from 30 minutes to 96 hours. Management consisted of decontamination, sedation (with diazepam as drug of choice), fluid therapy, thermo-regulation, and general supportive care. All followed animals made full recoveries. The suspension prepared from the benzene washing solution of cannabis seeds, administered intravenously to mice at a dose of 3 mg/kg, produced hypothermia, catalepsy, pentobarbital-induced sleep prolongation, and suppression of locomotor activity. These pharmacological activities of benzene washing solution of cannabis seeds were significantly higher than those of Δ-9-THC (3 mg/kg, iv).

Trauma injuries

An association between combat-related posttraumatic stress disorder (C-PTSD) and other mental disorders was studied in co-twin (male monozygotic twin pairs in the Vietnam Era Twin Registry). Logistic regression analyses demonstrated that combat exposure, adjusted for C-PTSD, was significantly associated with increased risk for alcohol and cannabis dependence and that C-PTSD mediated the association between combat exposure and both major depression and tobacco dependence. Sera from 111 patients with trauma injuries who presented during a 3 month period were screened for blood alcohol. Urine specimens were analyzed for metabolites of cannabis and cocaine. Sixty- two percent of patients were positive for at least one substance and 20% for two or more. Positivity rates were as follows: cannabis, 46%; alcohol, 32% (with 71% of these having blood alcohol levels >80 mg/ dL);

and cocaine (6%). Substance usage was most prevalent in the third decade of life. The patients who yielded a positive result were significantly younger than those negatives. There was no significant difference in age or substance usage between the victims of interpersonal violence or road traffic accidents. In the group designated "other accidents," patients were significantly older and had a lower incidence of substance usage than the other two groups. Cannabis was the most prevalent substance in all groups. Fifty and 55% of victims of road accidents and interpersonal violence, respectively, were positive for cannabis compared with 43 and 27% for alcohol, respectively. There was no significant difference in hospital stay or injury severity score between substance users and nonusers.

Trigeminovascular system effect

Arachidonylethanolamide is believed to be the endogenous ligand of the cannabinoid CB1 and CB2 receptors. Known behavioral effects of AEA are antinociception, catalepsy, hypothermia, and depression of motor activity, similar Δ-9-THC, the psychoactive constituent of cannabis. A role of the CB1 receptor in the trigeminovascular system, using intravital to study the effects of AEA against various vasodilator agents was examined. AEA inhibited dural blood vessel dilation brought about by electrical stimulation by 50%, calcitonin gene-related peptide (CGRP) by 30%, capsaicin by 45%, and NO by 40%. CGRP(8–37) attenuated NO-induced dilation by 50%. The AEA inhibition was reversed by the CB1 receptor antagonist AM251. AEA also reduced the blood pressure changes caused by CGRP injection, this effect was not reversed by AM251.

Tumor-promoting effect

A 28-year-old man who abused alcohol, nicotine, and cannabis for several years was investigated. He suffered simultaneously from a squamous cell carcinoma of the hypopharynx with bilateral cervical metastases, an adenocarcinoma of the transverse colon and a primary hepatocellular carcinoma. There were occurrences of three separate malignant tumors with different histologies in the aerodigestive tract, which could be related to a chronic abuse of cannabis.

Turning behavior

Cannabinoid agonists: WIN (1–100 ng/mouse), CP-55,940 (0.1– 50 ng/mouse), and AEA (0.5–50 ng/mouse), administered unilaterally into the mouse striatum, dose-dependently induced turning behavior. SR 141716A [*N*-(piperidin-1-yl)-5-(4-chlorophenyl)-1-(2,4-dichlorophenyl)-4-methyl-1H-pyrazole-3-carboxamide hydrochloride], the selective antagonist of CB1 receptor, antagonized the three cannabinoid receptor agonists-induced turning with similar effective $dose_{50}$ (0.13–0.15 mg/ kg, intraperitoneally). Spiroperidol (a D2 receptor blocker), (+)-SCH 23390 (a D1 receptor blocker), or prior 6-hydroxydopamine lesions of the striatum blocked WIN- and CP-55,940-induced turning, thus suggesting the involvement of DA transmission in cannabinoid-induced turning.

Uterine stimulant effect

Ethanol (50%) extract of the entire plant was inactive on the rat uterus. Ethanol (95%) and water extracts of the dried aerial parts, at a concentration of 1:1, produced strong activity on the non-pregnant rat uterus. Water extract of the flowering tops produced strong activity on the rat uterus.

Ventricular septal defect

A Birth Defect Case–Control Study was used to identify 122 isolated simple ventricular septal defect (VSD) cases and 3029 control infants. Exposure data on alcohol, cigarette, and illicit drug use were obtained through standardized interviews with mothers and fathers. Associations between lifestyle factors and VSD were calculated using maternal self-reports; associations were also calculated using paternal proxy reports of the mother's exposures. Maternal self-report of heavy alcohol consumption and paternal proxy report of the mother's moderate alcohol consumption were associated with isolated simple VSD. A twofold increase in risk of isolated simple VSD was identified for maternal self- and paternal proxy-reported cannabis use. Risk of isolated simple VSD increased with regular (≥3 days per week) cannabis use for both maternal self- and paternal proxy report, although the association was significant only for maternal self-report.

Visuospatial memory effect

Twenty-five college students who were heavy cannabis smokers (who had smoked a median of 29 of the last 30 days) were compared with 30 light smokers (1 day in the last 30 days). The subjects were tested after a supervised period of abstinence from cannabis and other drugs lasting at least 19 hours. Differences between the overall groups of heavy and light smokers did not reach statistical significance on the four subtests of attention administered. On examining data for the two sexes separately, marked and significant differences were found between heavy- and light-smoking women on the subtest examining visuospatial memory. On this test, subjects were required to examine a 6 × 6 "checkerboard" of squares in which certain squares were shaded. The shaded squares were then erased and the subject was required to indicate with the mouse which squares had formerly been shaded. Increasing numbers of shaded squares were presented at each trial. The heavy-smoking women remembered significantly fewer squares on this test, and they made significantly more errors than the light-smoking women. These differences persisted despite different methods of analysis and consideration for possible confounding variables.

Wilson's disease

A patient with generalized dystonia owing to Wilson's disease obtained mrked improvement in response to smoking cannabis.

Winiwarter-muerger disease

Two young men aged 18 and 20 years with juvenile endarteritis were evaluated. Both developed acute distal ischemia of the lower or upper limbs with arteriographic evidence suggestive of Winiwarter-Buerger disease. Both smoked regularly but not excessively, and both used cannabis regularly. In one case, the therapeutic response to with-drawal of cannabis was good. In the second, use of cannabis continued and arterial disease persisted. The main clinical and radiographical features in this condition are the same as in Winiwarter-Buerger disease.

INDEX